WAR DEPARTMENT TECHNICAL MANUAL

PERSHING HEAVY TANK T26E3

TECHNICAL MANUAL

RESTRICTED DISSEMINATION OF RESTRICTED MATTER. The information contained in restricted documents and the essential characteristics of restricted material may be given to any person known to be in the service of the United States and to persons of undoubted loyalty and discretion who are cooperating in Government work, but will not be communicated to the public or to the press except by authorized military public relations agencies. (See also par. 18b, AR 380–5, 28 Sep. 1942.)

by WAR DEPARTMENT • *15 JANUARY 1945*

©2013 Periscope Film LLC
All Rights Reserved
ISBN#978-1-937684-43-3
www.PeriscopeFilm.com

DISCLAIMER:

This manual is sold for historic research purposes only, as an entertainment. It contains obsolete information and is not intended to be used as part of an actual operation or maintenance training program. No book can substitute for proper training by an authorized instructor.

©2013 Periscope Film LLC
All Rights Reserved
ISBN#978-1-937684-43-3
www.PeriscopeFilm.com

WAR DEPARTMENT TECHNICAL MANUAL

TM 9-735

This manual supersedes the following publications insofar as they apply to this manual. These publications remain in force until such time as they are incorporated in other affected manuals or are specifically rescinded: TB ORD 20, dated 24 January 44; TB ORD 60, dated 13 March 44; TB ORD 88, dated 4 May 44; TB ORD 123, dated 17 July 44; TB ORD 126, dated 19 July 44, and TB ORD 217, dated 8 November 44.

HEAVY TANK
T26E3

WAR DEPARTMENT • 15 JANUARY 1945

RESTRICTED DISSEMINATION OF RESTRICTED MATTER. The information contained in restricted documents and the essential characteristics of restricted material may be given to any person known to be in the service of the United States and to persons of undoubted loyalty and discretion who are cooperating in Government work, but will not be communicated to the public or to the press except by authorized military public relations agencies. (See also par. 18b, AR 380-5, 28 Sep 1942.)

UNITED STATES GOVERNMENT PRINTING OFFICE
WASHINGTON : 1945

WAR DEPARTMENT
Washington 25, D. C., 15 January 1945

TM 9-735, Heavy tank T26E3, is published for the information and guidance of all concerned.

[A.G. 300.7 (11 Dec 43)
O.O.M. 061/Rar Ars (15 Jan 45)R]

BY ORDER OF THE SECRETARY OF WAR:

G. C. MARSHALL,
Chief of Staff.

OFFICIAL:
J. A. ULIO
*Major General,
The Adjutant General.*

DISTRIBUTION: AAF (10); AGF (10); ASF (2); AAF Comds (2); Arm & Sv Bd (2); Dept (10); S Div ASF (1); Tech Sv (2); Sv C (10); PC&S (1); PE "Mark for Ord O" (5); H&R (5); Dist O, 9 (5); Dist Br O, 9 (3); Reg O, 9 (3); Establishments, 9 (5); Decentralized Sub-O, 9 (3); Gen & Sp Sv Sch (10); USMA (2); A (10); CHQ (10); D (2); AF (2); R 17 (1); Bn 17 (1); T/O & E 9-7 (3); 9-9 (3); 9-37 (3); 9-57 (3); 9-65 (2); 9-67 (3); 9-76 (2); 9-127 (3); 9-197 (3); 9-317 (3); 9-325 (2); 9-327 (3); 9-328 (3); 9-377 (3).

(For explanation of symbols, see FM 21-6.)

CONTENTS

PART ONE — INTRODUCTION

		Paragraphs	Pages
SECTION	I. General	1– 2	1– 2
	II. Description and data	3– 4	3– 13
	III. Tools, parts, and accessories	5– 8	14– 29

PART TWO — OPERATING INSTRUCTIONS

		Paragraphs	Pages
SECTION	IV. General	9	30
	V. Service upon receipt of equipment	10– 13	30– 36
	VI. Controls and instruments	14– 15	36– 51
	VII. Operation under ordinary conditions	16– 22	52– 59
	VIII. Turret controls and operation	23– 27	59– 68
	IX. Operation of auxiliary equipment	28– 37	69– 80
	X. Operation under unusual conditions	38– 40	80– 87
	XI. Demolition to prevent enemy use	41– 42	87– 88

PART THREE — MAINTENANCE INSTRUCTIONS

		Paragraphs	Pages
SECTION	XII. General	43	89
	XIII. Special organizational tools and equipment	44	89– 91
	XIV. Lubrication	45– 46	92–110
	XV. Preventive maintenance services	47– 53	112–144
	XVI. Trouble shooting	54– 66	145–165
	XVII. Engine description and maintenance in vehicle	67– 70	166–175
	XVIII. Power unit removal and installation	71– 72	175–188
	XIX. Ignition system	73– 77	188–196
	XX. Starting system	78– 82	197–200

CONTENTS — Contd.

		Paragraphs	Pages
Section XXI.	Generating system	83–91	201–207
XXII.	Battery and lighting system	92–96	208–220
XXIII.	Instruments	97–100	220–234
XXIV.	Turret electrical system	101–109	234–243
XXV.	Radio interference suppression system	110–114	244–247
XXVI.	Fuel and air intake system, exhaust system, and cylinder head	115–133	247–283
XXVII.	Cooling system	134–144	283–302
XXVIII.	Lubrication systems	145–156	303–313
XXIX.	Angle drives and fan drive propeller shafts	157–158	314–315
XXX.	Transmission	159–161	315–320
XXXI.	Differential and final drives	162–168	320–347
XXXII.	Tracks and suspension	169–179	347–383
XXXIII.	Hull	180–192	383–398
XXXIV.	Turret	193–201	398–408
XXXV.	Auxiliary engine and controls	202–203	408–416
XXXVI.	Fire extinguisher system	204–205	416–420
XXXVII.	Accessory equipment	206	420–425

PART FOUR — AUXILIARY EQUIPMENT

Section XXXVIII.	General	207	426
XXXIX.	Armament description	208–209	426–429
XL.	Armament operating instructions	210–217	429–442
XLI.	Sighting and fire control equipment	218–224	442–451
XLII.	Ammunition	225–226	452–454
XLIII.	Communication systems	227–231	454–467

APPENDIX

Section XLIV.	Shipment and limited storage	232–234	468–472
XLV.	References	235–237	473–476
Index			477–488

RESTRICTED

TM 9-735
1-2

This manual supersedes the following publications insofar as they apply to this manual. These publications remain in force until such time as they are incorporated in other affected manuals or are specifically rescinded: TB ORD 20, dated 24 January 44; TB ORD 60, dated 13 March 44; TB ORD 88, dated 4 May 44; TB ORD 123, dated 17 July 44; TB ORD 126, dated 19 July 44; and TB ORD 217, dated 8 November 44.

PART ONE — INTRODUCTION

Section I

GENERAL

1. **SCOPE.**

   a. These instructions* are published for the information and guidance of the personnel to whom this equipment is assigned. They contain information on the operation and maintenance of the equipment, as well as descriptions of the major units, and their functions in relation to the other components of this vehicle. They apply only to Heavy Tank, T26E3 and are arranged in Four Parts: Part One, Introduction; Part Two, Operating Instructions; Part Three, Maintenance Instructions; Part Four, Auxiliary Equipment, and an Appendix.

   b. The appendix at the end of the manual contains instructions for shipment and limited storage, and a list of references including standard nomenclature lists, technical manuals, and other publications applicable to the vehicle.

   c. The stock and part numbers which appear throughout the manual are extracted from ORD 7, SNL G-226.

2. **RECORDS.**

   a. Forms and records applicable for use in performing prescribed operations are listed below with brief explanations of each:

   (1) W.D., A.G.O. FORM NO. 7360, ARMY MOTOR VEHICLE OPERATOR'S PERMIT. This form will be issued by commanding officers of posts, camps, stations, or organizations, to all operators of military vehicles who have passed the driver's examination (TM-21-300) and are qualified to drive the particular vehicles noted on the permit.

   (2) WAR DEPARTMENT LUBRICATION ORDER. War Department Lubrication Order No. LO 9-735, prescribes lubrication maintenance for this vehicle. A Lubrication Order is issued with each vehicle, and is to be carried with it at all times.

   (3) STANDARD FORM NO. 26, DRIVER'S REPORT, ACCIDENT, MOTOR TRANSPORTATION. One copy of this form will be kept with the vehicle at all times. In case of an accident resulting in injury or

---

*To provide operating instructions with the materiel, this technical manual has been published in advance of complete technical review. Any errors or omissions will be corrected by changes or, if extensive, by an early revision.

property damage, it will be filled out by the driver on the spot, or as promptly as practical thereafter.

(4) WAR DEPARTMENT FORM NO. 48, DRIVER'S TRIP TICKET AND PREVENTIVE MAINTENANCE SERVICE RECORD. This form, properly executed, will be furnished to the driver when his vehicle is dispatched on nontactical missions. The driver and the official user of the vehicle will complete in detail appropriate parts of this form. These forms need not be issued for vehicles in convoy or on tactical missions. The reverse side of this form contains the driver's daily and weekly preventive maintenance service reminder schedule.

(5) W.D., A.G.O. FORM NO. 478, MWO AND MAJOR UNIT ASSEMBLY REPLACEMENT RECORD. This form, carried with the vehicle, will be used by all personnel completing a modification or major unit assembly (engine, transmission, transfer case, and track replacement) to record clearly the description of work completed, date, vehicle hours and/or mileage, and MWO number or nomenclature of unit assembly. Personnel performing the operation will initial in the column provided. Minor repairs, parts, and accessory replacements will not be recorded.

(6) W.D., A.G.O. FORM NO. 6, DUTY ROSTER. This form, slightly modified, will be used for scheduling and maintaining a record of vehicle maintenance operations. It may be used for lubrication records.

(7) W.D., A.G.O. FORM NO. 462, PREVENTIVE MAINTENANCE SERVICE AND TECHNICAL INSPECTION WORK SHEET FOR FULL-TRACK AND TANK-LIKE WHEELED VEHICLES. This form will be used for all 50-hour (500-mile) or 100-hour (1,000 mile) services, and for technical inspections of these vehicles.

(8) W.D., A.G.O. FORM NO. 9-70, SPOT-CHECK INSPECTIONS REPORT FOR ALL MOTOR VEHICLES. This form may be used by all commanding officers or their staff representatives in making spot-check inspections on all vehicles.

(9) W.D., A.G.O. FORM NO. 468, UNSATISFACTORY EQUIPMENT REPORT. This form will be used for reporting manufacturing, design, or operational defects in materiel with a view of improving and correcting such defects, and for use in recommending modifications of materiel. This form will not be used for reporting failures, isolated materiel defects, or malfunctions of materiel resulting from fair wear and tear or accidental damage, nor for the replacement, repair, or the issue of parts and equipment. It does not replace currently authorized operational or performance records.

(10) W.D., A.G.O. FORM NO. 9-81, EXCHANGE PART OR UNIT IDENTIFICATION TAG. This tag, properly executed, may be used when exchanging unserviceable items for like serviceable assemblies, parts, vehicles and tools.

Section II
DESCRIPTION AND DATA

3. DESCRIPTION.

   a. General. Heavy Tank T26E3 is a heavily armored, full tracklaying, low silhouette combat vehicle with a 90-mm gun mounted in a fully enclosed power-operated turret which can be traversed 360 degrees. The vehicle is powered by an 8-cylinder, V-type, liquid-cooled, Ford, model GAF, gasoline tank engine.

   b. Distinguishing Features. The vehicle has a very low silhouette, a relatively flat rear deck, and the turret is mounted well forward of the center of the vehicle (fig. 5). A large projection at the rear of the rounded turret acts as a counterweight for the 90-mm gun. Six dual, independently sprung road wheels support the vehicle on each side. The upper portion of the track is covered by stowage boxes, fenders, and sand shields at the ends and sides. The tracks are driven by final drives at the rear of the vehicle, and run around idler wheels at the front of the vehicle. The sloping V-shaped front (figs. 1 and 4) adds to its low-slung appearance. The straight sides, without sponsons, make the vehicle seem to be suspended between the tracks and therefore considerably wider than other tanks when viewed from front or rear. One cal. .30 machine gun is mounted in the bow of the vehicle. A second cal. .30 machine gun is coaxially mounted on the left side of the 90-mm gun. A cal. .50 machine gun is mounted behind the loader's hatch. When not in use it is stowed in brackets on the rear of turret overhang or counterweight.

   c. Construction Features. The hull is an all-welded structure of heavy armor plate sides, with cast armor steel front and rear sections welded to an armor steel floor. The cast armor steel turret can be traversed either by hand or hydraulic power, and is provided with a gunner's seat and a small removable foot rest or platform (figs. 23 and 27). There is no turret basket. The vehicle is supported on its steel tracks by twelve dual road wheels with individual torsion bar springs, and four large cylindrical shock absorbers on each side. The power unit, consisting of the engine, transmission and differential, is removable as an assembly from the engine compartment at the rear of the vehicle (fig. 58). The cooling unit, consisting of two radiators, four fans, and four oil coolers, is removed from the engine compartment as an assembly (figs. 115 and 116). The drivers' and fighting compartments are provided with forced ventilation, and are heated. Separate heaters are provided for heating the drivers' and engine compartments (figs. 35 and 61). Dual driving controls are provided for the use of the driver or assistant driver (figs. 18 and 19). Four lifting eyes are provided on the hull for lifting the entire vehicle.

Part One—Introduction

Figure 1 — Heavy Tank T26E3, Front View — Gun in Firing Position

CAUTION: *Under no circumstances is a vehicle to be lifted by using the turret lifting eyes or any means other than the hull lifting eyes.*

4. TABULATED DATA.

   a. General.

Weight, fully equipped, including crew (estimated)	86,500 lb
Length (over-all—gun in firing position)	323 in.
Length (over-all—gun in traveling position)	268 in.
Width (over-all)	137 in.
Height (over-all)	109 in.
Ground clearance	$17 {11}/{16}$ in.

Description and Data

RA PD 344663

Figure 2 — Heavy Tank T26E3, Rear View — Gun in Traveling Position

Tread (center to center of tracks)	110 in.
Ground pressure (per sq in.) zero penetration	11.99
Weight of complete power unit (engine transmissions, differentials only)	4680 lb

b. **Engine.**

Type	60-degree V-type, 8 cylinder, 4-cycle gasoline
Model	GAF
Make	Ford
Maximum governed speed (no load)	2,800 rpm
Maximum governed speed (under load)	2,600 rpm
Maximum warm-up speed	1,200 rpm
Minimum idling speed	500 rpm

TM 9-735
4

Part One—Introduction

Figure 3 — Heavy Tank T26E3, Three-Quarter Right Rear View

TM 9-735
4

Description and Data

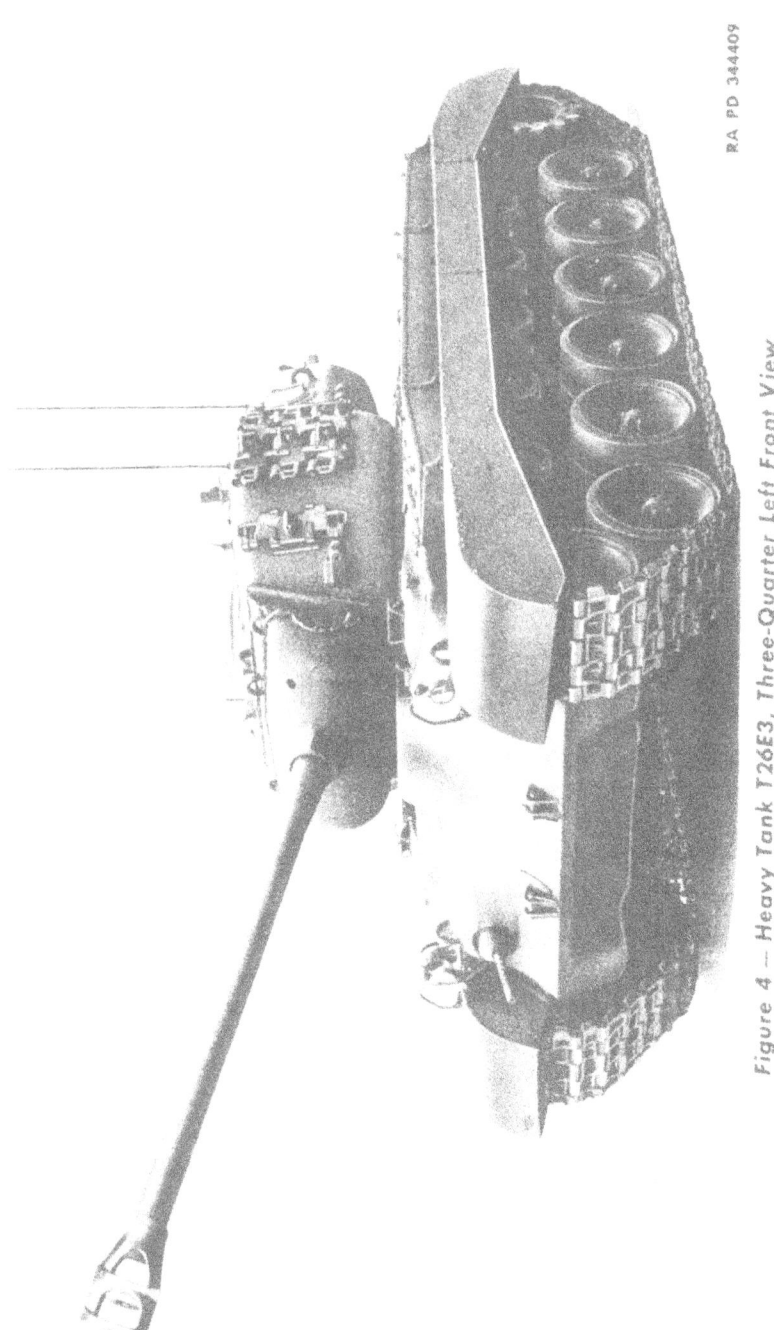

Figure 4 — Heavy Tank T26E3, Three-Quarter Left Front View

Part One—Introduction

c. **Fuel, Oil, and Water Capacities and Fuel Octane Rating.**

(1) OIL:

Engine	32 qt
Transmission (complete system)	54½ qt
Differential	72 qt
Auxiliary engine	3 qt
Final drive (one only)	7 qt
Air cleaners (each)	3½ qt

(2) FUEL:

Fuel tank (right—75½-gal capacity) fill to only	71½ gal
Fuel tank (left—116-gal capacity) fill to only	111½ gal
Octane rating of fuel	80 or higher
Fuel specification	U.S. Army Specifications 2-103 (grade C) latest issue

(3) COOLANT:

Water	22 gal

d. **Tracks (all steel).**

Track shoe width (tread)	24 in.
Number of links, or blocks, per track	82
Track pitch	6 in.
Ground contact	7,215 sq in.

e. **Performance.**

Maximum speed ranges (miles per hour):

Low speed range (first) (1)	0-9 mph
Intermediate speed range (second) (2)	6-19 mph
High speed range (third) (3)	12-30 mph
Reverse (R)	0-9 mph

f. **Limitations of Vehicles.**

Maximum recommended sustained vehicle speed (on hard road)	25 mph
Maximum speed (short periods—hard road)	30 mph

Number of miles without refueling:

Cross-country (approx)	75
Highway (approx)	110
Maximum grade-ascending ability	60 pct
Maximum grade-descending ability	60 pct
Maximum width of trench vehicle will cross	7 ft 11 in.
Maximum vertical obstacle (stone walls, trees, etc.)	46 in.
Maximum fording depth (at slowest speed forward)	48 in.

TM 9-735
4

Description and Data

Figure 5 — Heavy Tank T26E3, Right Side View

RA PD 335414

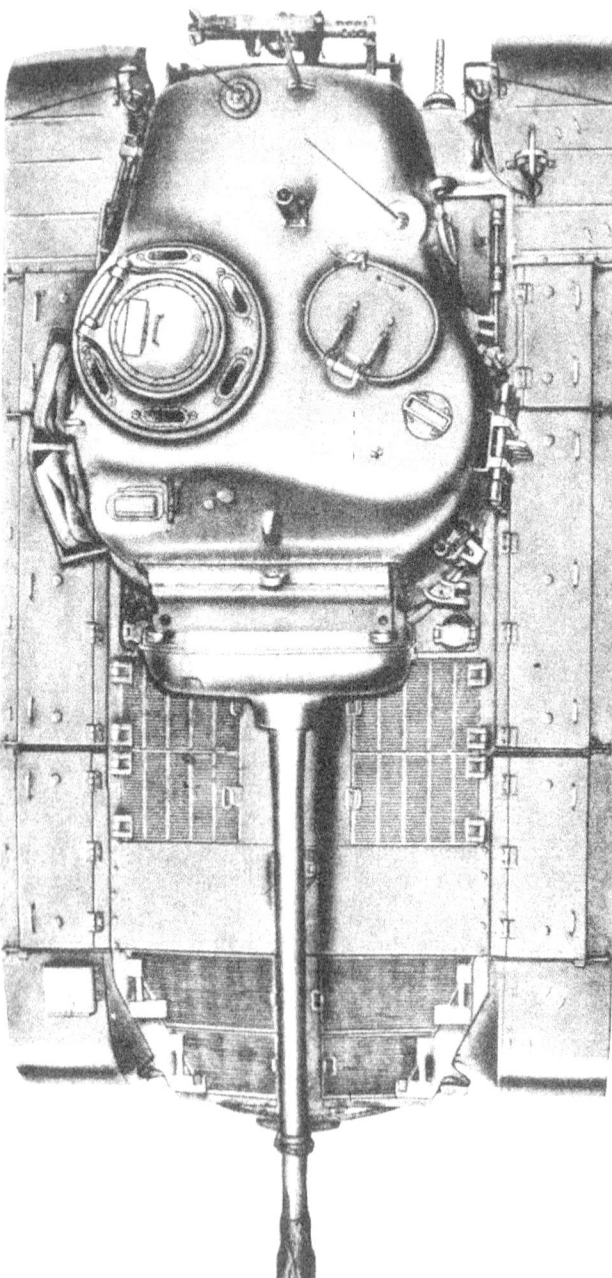

Figure 6 — Heavy Tank T26E3, Top View — Gun in Traveling Position

TM 9-735
4

Description and Data

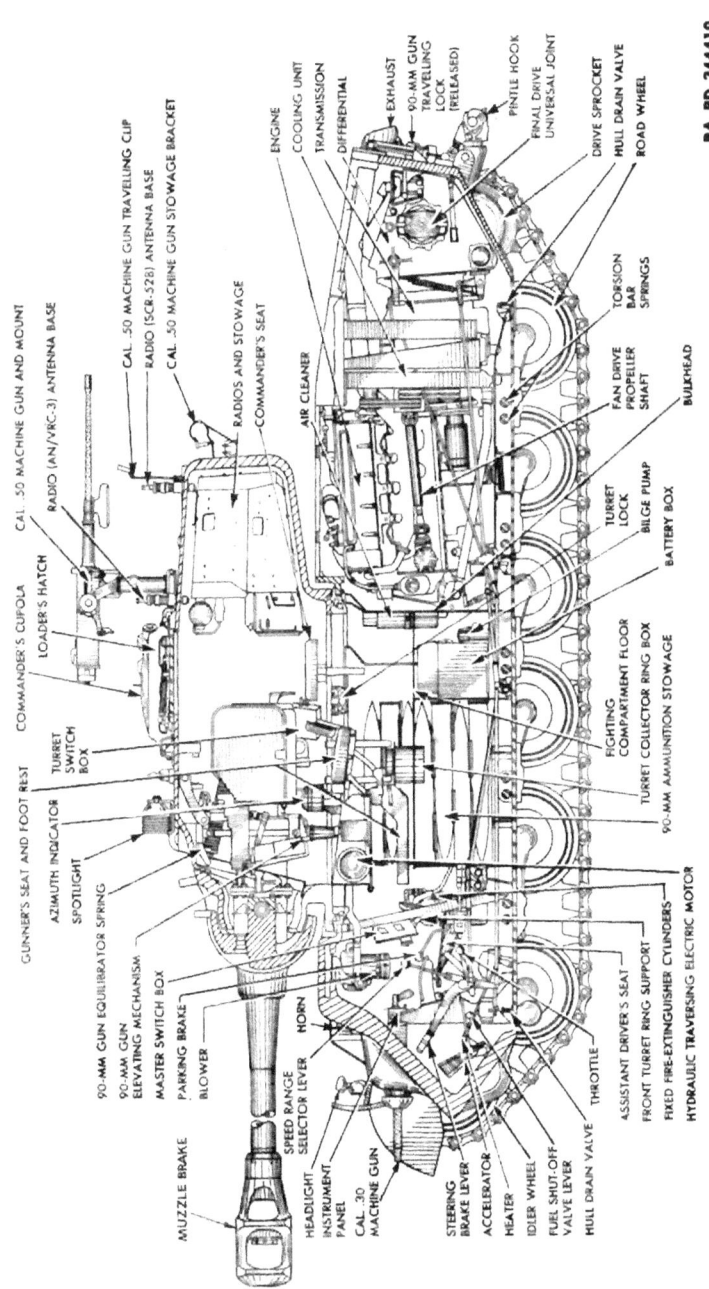

Figure 7 — Heavy Tank T26E3, Sectional View

TM 9-735
4
Part One—Introduction

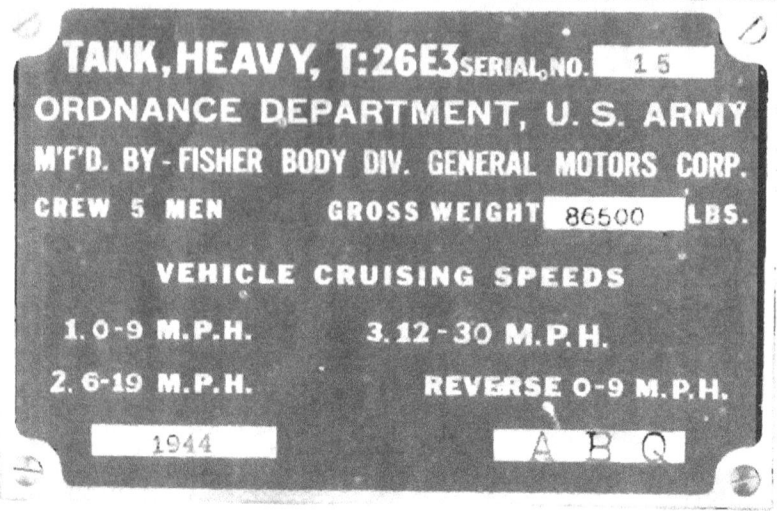

Figure 8 — Vehicle Serial Number Plate, Heavy Tank T26E3

   g. Lubrication (See Section XIV).

   h. Turret.

Hydraulic or manual operation	360 deg traverse

   i. Crew.

	5

   j. Communication.

Radios (sending and receiving) (2)	SCR-528 AN/VRC-3
Intravehicle	Interphone
Panel sets (AP-50A)	1 set
Flags, signal	1 set
Spotlight	1

   k. Seats. Adjustable padded seats with demountable backs are provided for the gunner, driver, and assistant driver. The latter two seats are equipped with safety belts. A padded height-adjustable seat of the snap-up type is provided for the vehicle commander and a padded demountable seat is furnished for the loader.

TM 9-735

Description and Data

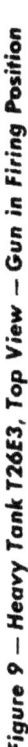

Figure 9 — Heavy Tank T26E3, Top View — Gun in Firing Position

TM 9-735
5—6

Part One—Introduction

RA PD 344465

Figure 10 — Heavy Tank T26E3, Front View — Gun in Traveling Position

Section III

TOOLS, PARTS, AND ACCESSORIES

5. PURPOSE.

a. The lists in this section are for information only, and must not be used as a basis for requisition.

6. ON-VEHICLE TOOLS.

a. Pioneer.

Quantity per Vehicle	Item Name and Stock No.	Stowage Location
1	Ax, chopping, 4 lb (41-A-1277)	R.H. front fender box
1	Crowbar, 5 ft long, pinch point (41-B-175)	L.H. front fender box
1	Handle, mattock (41-H-1286)	L.H. front fender box

14

TM 9-735
6

Tools, Parts, and Accessories

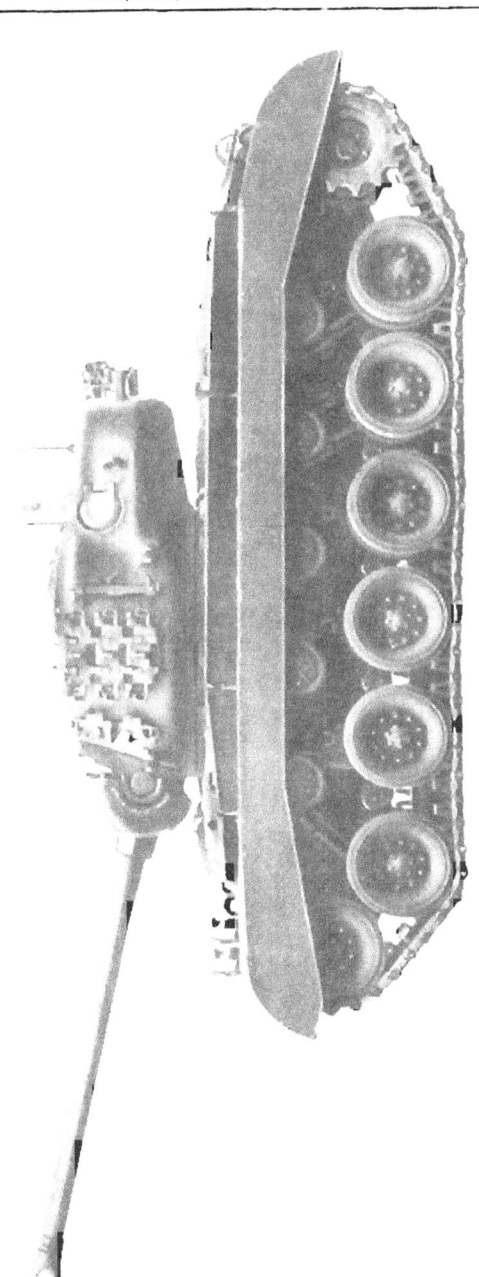

Figure 11 — Heavy Tank T26E3, Left Side View

15

TM 9-735

Part One—Introduction

Quantity per Vehicle	Item Name and Stock No.	Stowage Location
1	Mattock, pick, M1 (without handle) (41-M-722)	L.H. front fender box
1	Shovel, short-handled (41-S-3170)	L.H. front fender box
1	Sledge, Blacksmith's, dble face, 10 lb (41-S-3726)	R.H. front fender box

b. **Vehicle.**

1	Bar, cross, socket wrench 7/16 x 8 in. long (41-B-154)	Tool bag
1	Bar, socket wrench extension ½ sq dr 5 in. long (41-B-311-200)	Tool bag
1	Bar, socket wrench extension ½ sq dr 10 in. long (41-B-309)	Tool bag
1	Bar, extension ¾-in. sq dr 8½ in. long (41-B-309-20)	Tool bag
1	Bar, extension ¾-in. sq dr 16 in. long (41-B-309-25)	Tool bag
1	Chisel, cold, ¾ in. (41-C-1124)	Tool bag
1	Drift, track pin removing (41-D-1545-150)	R.H. front fender box
1	File, hand, smooth, 8 in. (41-F-1028)	Tool bag
1	File, 3-sq, smooth, 6 in. (41-F-1572)	Tool bag
2	Fixture, track connecting and link pulling (41-F-2995-155)	Side of turret
1	Hammer, machinist's ball peen, 32 oz. (41-H-527)	Tool bag
1	Handle, combination tee, ½ sq dr, 11 in. long (41-H-1509-55)	Tool bag
1	Handle, combination tee, ¾ sq dr, 17 in. long (41-H-1509-60)	Tool bag
1	Handle, flexible, ½ in. sq. 12 in. long (41-H-1502)	Tool bag
1	Handle, speeder, ½ in. sq dr, 17 in. long (41-H-1508)	Tool bag
1	Joint, universal, ½ in. sq dr (41-J-380)	Tool bag
1	Pliers, combination, slip joint, 8 in. (41-P-1652)	Tool bag
1	Pliers, side-cutting, 8 in. (41-P-1977)	Tool bag
1	Pilot, track pin inserting (41-P-402-400)	R.H. front fender box

Tools, Parts, and Accessories

Quantity per Vehicle	Item Name and Stock No.	Stowage Location
2	PULLER, track bushing (7043230)	Tool bag
1	RACHET, reversible, ½ in. sq dr, 9 in. (41-H-1505)	Tool bag
1	SCREWDRIVER, machinist's 5-blade (41-S-1385)	Tool bag
1	SCREWDRIVER, special purpose, 1½-in. blade (41-S-1062-90)	Tool bag
1	SCREWDRIVER, special purpose, 1¾-in. blade (41-S-1063)	Tool bag
1	TOOL, bucking assembly, track repair (41-T-3077)	R.H. center fender box
1	WRENCH, adjustable, single-end, 8 in. long (41-W-486)	Tool bag
1	WRENCH, adjustable, single-end, 12 in. long (41-W-488)	Tool bag
1	WRENCH, engineer's, dble hd, $5/16$ in. and ⅜ in. (41-W-990)	Tool bag
1	WRENCH, engineer's, dble hd, $7/16$ in. and ½ in. (41-W-1000)	Tool bag
1	WRENCH, engineer's, dble hd, $9/16$ in. and $11/16$ in. (41-W-1005-5)	Tool bag
1	WRENCH, engineer's, dble hd, ⅝ in. and ¾ in. (41-W-1008)	Tool bag
1	WRENCH, engineer's, dble hd, $13/16$ in. and ⅞ in. (41-W-1015)	Tool bag
1	WRENCH, engineer's, dble hd, $15/16$ in. and 1 in. (41-W-1021)	Tool bag
1	WRENCH, plug, $9/16$ in. hex, (for transmission) and oil drain plug (41-W-1960-175)	Tool bag
1	WRENCH, plug, ¾ in. hex, (for differential filler and drain plug) (41-W-1960-236)	Tool bag
1	WRENCH, safety screw, $3/32$ in. hex (41-W-2449)	Tool bag
1	WRENCH, safety screw, ⅛ in. hex (41-W-2450)	Tool bag
1	WRENCH, safety screw, $3/16$ in. hex (41-W-2452)	Tool bag

Quantity per Vehicle	Item Name and Stock No.	Stowage Location
1	WRENCH, safety screw, 1/4 in. hex (41-W-2454)	Tool bag
1	WRENCH, safety screw, 5/16 in. hex (41-W-2455)	Tool bag
1	WRENCH, safety screw, 3/8 in. hex (41-W-2458)	Tool bag
1	WRENCH, safety screw, 5/8 in. hex (41-W-2458-55)	Tool bag
1	WRENCH, socket, 1/2 in. sq dr, 3/8 in. sq (41-W-3001-200)	Tool bag
1	WRENCH, socket, 1/2 in. sq dr, 7/16 in. hex (41-W-3005)	Tool bag
1	WRENCH, socket, 1/2 in. sq dr, 1/2 in. hex (41-W-3007)	Tool bag
1	WRENCH, socket, 1/2 in. sq dr, 9/16 in. hex (41-W-3009)	Tool bag
1	WRENCH, socket, 1/2 in. sq dr, 5/8 in. hex (41-W-3013)	Tool bag
2	WRENCH, socket, 1/2 in. sq dr, 7/8 in. hex (41-W-3023)	Tool bag
2	WRENCH, socket, 1/2 in. sq dr, 15/16 in. hex (41-W-3025)	Tool bag
1	WRENCH, socket, 1/2 in. sq dr, 3/4 in. hex (41-W-3017)	Tool bag
1	WRENCH, socket, 1/2 in. sq dr, 1 in. hex (41-W-3027)	Tool bag
1	WRENCH, socket, 1/2 in. sq dr, 1 1/16 in. hex (41-W-3029)	Tool bag
1	WRENCH, socket, 1/2 in. sq dr, 1 1/8 in. hex (41-W-3031)	Tool bag
1	WRENCH, socket, 3/4 in. sq dr, 15/16 in. hex (41-W-3033-25)	Tool bag
1	WRENCH, socket, 3/4 in. sq dr, 1 5/16 in. hex (41-W-3040)	Tool bag
1	WRENCH, socket, 3/4 in. sq dr, 1 7/16 in. hex (41-W-3044)	Tool bag
1	WRENCH, socket, 3/4 in. sq dr, 1 1/8 in. hex (41-W-3036)	Tool bag
1	WRENCH, socket, 3/4 in. sq dr, 1 1/2 in. hex (41-W-3046)	Tool bag

Tools, Parts, and Accessories

Quantity per Vehicle	Item Name and Stock No.	Stowage Location
1	WRENCH, engineer's 15-deg, single hd, open-end, 42 in. long (41-W-1436-25)	L.H. center fender box
1	WRENCH, track-adjusting, 24 in. long (41-W-3295)	R.H. center fender box
c.	Gun.	
(1)	GUN, 90-MM, M3.	
1	BOLT, eye, breechblock removing (41-E-3150)	Gun spare parts box
1	TOOL, breechblock removing (41-T-3076-815)	Gun spare parts box
(2)	GUN, MACHINE, CAL. .30.	
2	SCREWDRIVER, common, 3-in. blade (41-S-1101)	Gun spare parts box
2	WRENCH, combination, M6 (41-W-867-32)	Gun spare parts box
1	WRENCH, socket, front barrel bearing plug (41-W-2574)	Gun spare parts box
(3)	GUN, MACHINE, CAL. .50, M2.	
1	WRENCH, combination, M2 (41-W-3249-850)	Gun spare parts box

7. **ON-VEHICLE EQUIPMENT.**

a. **Communications.**

2	ANTENNA, complete with cover, (SCD1241):— (Included with radio set) 1 Mast section, MS49 1 Mast section, MS50 1 Mast section, MS51 1 Mast section, MS52 1 Mast section, MS53	Turret antenna mtg.
1	FLAG SET, M238 Composed of: 1 Case, CS-90 3 Flagstaff, MC-270 1 Flag, MC-273 (red) 1 Flag, MC-274 (orange) 1 Flag, MC-275 (green)	Over radio

TM 9-735
7

Part One—Introduction

Quantity per Vehicle	Item Name and Stock No.	Stowage Location
12	FLARES	In flare box (7021296)
	3 White star, parachute flare, M17A1	
	3 White star, cluster flare, M18A1	
	3 Green star, parachute flare, M19A1	
	3 Green star, parachute flare, M21A1	
1	PANEL SET	R.H. center fender box
	Consisting of:	
	2 Case, CS-150	
	1 Panel, AL-140	
	1 Panel, AL-141	
1	RACK, engine oil (7021314)	On battery box
1	RADIO SET, either SCR-528, or British No. 19 w/5-position interphone complete as per signal corps parts	Turret bulge
1	RADIO SET, AN/VRC-3	Turret bulge
b.	**Extinguishers, Fire.**	
2	EXTINGUISHER, fire, 4 lb CO (58-E-200-65)	Driver's comp. Turret bulge
c.	**Rations.**	
10	TYPE "C" or "K" (5 men for 2 days; 60 cans "C" or 30 boxes "K")	Under turret platform
5	TYPE "D" (5 men for 1 day, 2 cans, 24 bars)	In field bags
d.	**Gun.**	
(1)	GUN, 90-MM, M3.	
1	BOOK, arty. Gun, O.O. form 5825	Manual box
1	BRUSH, bore, M19 w/staff and wiper ring (38-B-992-900)	R.H. center fender box
	Composed of:	
	1 Brush, bore, M19 (38-B-992-725)	
1	BRAKE, muzzle, M2, 90-mm.	
	1 Ring wiper (33-R-103)	
	3 staff section middle (41-S-4648-135)	
	1 Staff, section end (7042725)	
1	CASE, carrying, gunner's quadrant, M1	R.H. turret wall

20

Tools, Parts, and Accessories

Quantity per Vehicle	Item Name and Stock No.	Stowage Location
1	COVER, bore brush, M518	Bore brush
1	COVER, combination breech and empty shell case	Gun breech
1	COVER, muzzle (7046305)	Gun
1	GUN, lubricating oil (2 oz) (41-C-1362-500)	Gun spare parts box
1	HOSE, assembly (33-H-581-450)	Gun spare parts box
1	OIL, recoil, special (Spec. AXS-808) (In container 1 qt Type 1, Class D, Spec. 100-13)	Battery box rack
1	QUADRANT, elevation, M9	Gun
1	QUADRANT, gunner, M1	Quadrant case
1	RAMMER, cleaning and unloading	R.H. center fender box
1	SETTER, Fuze, M14	Turret bulge
1	SIGHT, bore (muzzle-RF11CD) (breech-RF11ZB)	Gun spare parts box
1	TARGET, testing (set of 4)	R.H. center fender box
1	WRENCH, Fuze, M7	Turret ammunition box
1	TABLE, firing FT 90-C-2	Manual box
(2)	GUN, MACHINE, CAL. .30, M1919A4 (FLEXIBLE).	
2	BAG, empty cartridge, cal. .30 (turret and bow gun)	Guns
20	BELT, ammunition, 250 rounds	Amm. boxes and guns
20	BOX, ammunition, cal. .30, M1	Turret, rack and guns
1	BRUSH, chamber cleaning, M6 (38-B-992-650)	Spare parts box
6	BRUSH, cleaning, cal. .30, M2 (38-B-992)	Spare parts box
1	CASE, cover group	Spare parts box
4	CASE, spare bolt, M2 (w/o contents)	Spare parts box
1	COVER, muzzle, cal. .30 (flex bow gun)	Gun
1	CHUTE, ammunition, bow gun (7021518)	
1	COVER, receiver, cal. .30 (turret)	Gun
1	COVER, receiver, cal. .30 (turret)	Gun
2	COVER, spare barrel, cal. .30	Spare barrels
1	HOOD, cal. .30 lt. M.G. tripod mount	Tripod mount

Part One—Introduction

Quantity per Vehicle	Item Name and Stock No.	Stowage Location
2	ENVELOPE, spare parts, M1 (w/o contents)	Gun spare parts box
2	EXTRACTOR, ruptured cartridge, MK. IV (41-E-557-200)	Oddment tray
1	MOUNT, tripod, machine gun, cal. .30, M2	L.H. front fender box
1	OILER, rectangular, 12 oz	Gun spare parts box
2	CASE, cleaning rod, M1 (24-C-534)	Spare parts box
1	REFLECTOR, barrel, cal. .30 (41-R-2330-975)	Gun spare parts box
2	ROD, cleaning, jointed, cal. .30, M1 (41-R-2567)	Gun spare parts box
2	ROLL, spare parts, M13 (w/o contents)	Gun spare parts box
2	ROLL, tool, M12 (w/o contents)	Gun spare parts box
(3)	GUN, MACHINE, CAL. .50, M2, H.B. FLEXIBLE (A.A.).	
1	BAG, empty cartridge (A.A. Gun)	Gun
5	BOX, ammunition (110 rd) cal. .50, M2	Turret bulge and gun
4	BRUSH, cleaning, cal. .50, M4	Gun spare parts box
1	CASE, cleaning, cal. .50, M15	Gun spare parts box
1	MOUNT, cal. .50 machine gun	Turret, rear
1	COVER, cal. .50 Machine gun	Gun
2	COVER, spare barrel, M13, 45 in.	Gun and spare barrel
2	ENVELOPE, spare parts, M1 (w/o contents)	Gun spare parts box
1	EXTRACTOR, ruptured cartridge	Turret oddment tray
1	OILER, filling, oil buffer	Gun spare parts box
1	ROD, jointed, cleaning, M7	Gun spare parts box
(4)	GUN, SUBMACHINE, CAL. .45, M3.	
5	CASE, cal. .45, S.M.G. clips	Turret bulge and front hull
(5)	CARBINE, CAL. .30, M1 (COMMAND TANK ONLY).	
1	COVER, carbine, (C153357)	On carbine
(6)	LAUNCHER, GRENADE, M8 (COMMAND TANK ONLY).	

e. **Ammunition.**

5000	ROUNDS, cal. .30 (in box, M1, D44070, and belt C3951)	Turret, hull, gun

Tools, Parts, and Accessories

Quantity per Vehicle	Item Name and Stock No.	Stowage Location
550	ROUNDS, cal. .50 (in box, M2, D73913)	Turret bulge and gun
900	ROUNDS, cal. .45 (in 30-round clips, D355506)	Turret bulge and front hull
70	ROUNDS, 90-mm	Center hull
12	GRENADES, hand	
	4 Fragmentation, MK 11	Turret bulge
	4 Smoke, WP, M-15	Turret bulge
	4 Smoke, Colored, M-16 or M-18	Turret bulge

f. Sighting.

1	BINOCULAR, M13, complete Composed of: 1 Binocular, M-13 1 Case, carrying, M-17	R.H. turret wall
15	HEAD (for periscope, M6) (spare)	In 5 periscope boxes
1	PERISCOPE M10F, if available, or	Gunner's periscope
1	PERISCOPE M4A1 (w/telescope, M77F)	Gunner's periscope
1	PERISCOPE M4A1 (w/telescope, M77F) (spare)	Under gun mount
6	PERISCOPE, M6 (in use)	Various periscope mounts
4	PERISCOPE, M6 (spare)	In periscope boxes
1	TELESCOPE, M71C or M70P, substandard	On telescope mount
1	TELESCOPE MOUNT T-90	On gun mount
1	HEADREST	On gunner's periscope mt.
1	MOUNT, periscope, T113 (with adapter for periscope M4A1) (7689765)	On gunner's periscope
10	BULB, lamp (for quadrant elevation and azimuth indicator)	In box (C101369)
4	LAMP, electric (for telescope reticle lights)	In box (C101369)
1	LIGHT, instrument, M30 (for elevation quadrant and periscope)	On elevation quadrant
1	LIGHT, instrument, M33 (for telescope)	On telescope mount

TM 9-735
7

Part One—Introduction

Quantity per Vehicle	Item Name and Stock No.	Stowage Location
1	HOLDER, periscope	Commander's periscope box
2	BLOCK, direct-vision (spare)	Turret bulge
1	COVER, telescope	On telescope
1	POST, aiming, M1, complete Consisting of: 1 Cover, aiming post, M401 2 Post, post aiming, M1	R.H. center fender box
1	SET, lights, Aiming Post, M14 Consisting of: 8 Battery, flashlight 1 Chest, M14 2 Lamp, electric, 3-volt 2 Lights, aiming post, M14	R.H. center fender box

g. Miscellaneous.

1	APPARATUS, decontaminating, 1½-qt, M2 (Spec. 197-54-113)	L.H. center fender box
1	BAG (for arm, spotlight, C100212)	Spotlight arm
5	BAG, canvas, field, O.D., M1936	Center fender boxes
1	BAG, tool, w/o contents (41-B-9-250)	L.H. fender box center
28	BATTERY (8 to be put in flashlight, 8 in. instruments, 12 spares)	In box (C101039)
1	BUCKET, canvas, folding, 18-qt	L.H. rear fender box
1	Box (battery) and bulb stowage	Inst. panel tray
1	Box, spare parts, gun (7021327)	Under turret platform
1	Box, flare (7021296)	Under turret platform
1	CABLE, steel towing 1⅛ in. x 20 ft (8-C-482-100)	On rear of hull
10	CANS, 4½-oz flash-burn prevention cream	Under turret platform
5	CANTEEN, M1941, with cup and cover, M1910	Near crew position
2	CONTAINER, water, 5-gal, (QMC standard)	Center fender boxes
1	CORRECTOR BARS (comp. compass) roll, w/bars (7069762)	Gun spare parts box
1	KIT, first-aid (24-unit), (Spec. 1553)	Driver's com. or fender box
1	COVER, azimuth indicator (7069999)	On indicator

TM 9-735

Tools, Parts, and Accessories

Quantity per Vehicle	Item Name and Stock No.	Stowage Location
2	Cover, hood assembly (D90240)	R.H. turret wall
2	Cover, light (7043573)	On headlight
1	Catalog, service parts SNL G197 and G198	Manual box
1	Cutter, wire, (QMC) M1938	R.H. fender box
1	Extension, lubr. gun, hose type, hydraulic (41-E-485-12)	R.H. front fender box
4	Flashlight, (Spec. 17-197)	Driver's comp., turret
1	Form No. 478, War Dept., A.G.O. in container, Form No. 478-1	Manual box
1	Gun, lubr. pressure (hand-operated) (41-G-1344-40)	R.H. front fender box
1	Hood, hatch, driver's right (7014433)	R.H. turret wall
1	Hood, hatch, driver's left (7014434)	R.H. turret wall
3	Lamp (spare for flashlight)	In box (C101039)
1	Lamp Bulb, inspection (A213670)	In box (C101039)
1	Lamp, inspection (B209297A)	L.H. front fender box
1	Lubrication Order, War Dept.	Engine bulkhead
1	Manual, field, for cal. .30 M.G., M1919A4	Manual box
1	Manual, field, for cal. .50 M.G., M2	Manual box
1	Manual, instruction (for Sherrill compass) (7069763)	Manual box
1	Manual, technical (for Waukesha engine)	Manual box
1	Manual, technical (for 90-mm gun)	Manual box
1	Manual, technical, (for hydraulic traverse oil gear), TM 9-1727K	Manual box
1	Manual, technical, for heavy tank, T26E1	Manual box
2	Mittens, asbestos (pr)	Turret oddment box
1	Net, camouflage, 45 ft x 45 ft, (Spec. T-1669)	R.H. fender
8 qt	Oil, engine, (Spec. 2-104) (in container, 1 qt) (type 1, class D, Spec. 100-13)	Battery box cover
1	Oil, hydraulic (Spec. 2-79) (in container, 1 qt (Type 1, class D, Spec. 100-13)	Battery box rack

Part One—Introduction

Quantity per Vehicle	Item Name and Stock No.	Stowage Location
1	Oiler, trigger-type, 1 pt	Battery box rack
7	Padlock, 1½, 2 keys (SLAX1A)	Fender boxes and loader's hatch
1	Paulin, 12 ft x 12 ft (24-P-29)	Left fender
5	Roll, blanket	R. and L. fender boxes
1	Stove, cooking, gasoline, 1 burner, M1942	R.H. center fender box
1	Tape, adhesive, olive drab, 4 in. wide x 15 yd long (Spec. AXS-871) (27-T-180-250)	R.H. front fender box
1	Tape, friction, ¾ in. wide, 30-ft roll (17-T-805)	Tool bag
2	Tube, flexible nozzle	Center fender boxes
1	Wire, soft iron, 14 ga, 10 ft (BFWX1C)	R.H. fender box

8. ON-VEHICLE SPARE PARTS.

a. Gun.

(1) Gun, 90-mm, M3.

1	Chain	Gun spare parts box
2	Gasket, recoil cylinder filling plug	Gun spare parts box
1	Mechanism, percussion, assembly Composed of:	Top of battery box
	1 Guide (firing pin)	
	1 Pin, firing	
	1 Pin, straight $\frac{3}{32}$ x ⅝ firing Pin guide (BFDX1BK)	
	1 Spring (firing pin retracting)	
	1 Stop (firing spring)	
3	Pin, cotter, ⅛ in. x ¼ in. (BFAX1DL)	Gun spare parts box
1	Pin, firing	Gun spare parts box
1	Pin, link	Gun spare parts box
2	Plug, filling, recoil cylinder	Gun spare parts box
1	Plunger, cocking lever	Gun spare parts box
1	Retainer, sear	Gun spare parts box
1	Spring, cocking lever plunger	Gun spare parts box
1	Spring, extractor plunger	Gun spare parts box
1	Spring, firing pin retracting	Gun spare parts box
2	Spring, firing	Gun spare parts box
1	Spring, sear	Gun spare parts box

Tools, Parts, and Accessories

Quantity per Vehicle	Item Name and Stock No.	Stowage Location
(2)	GUN, MACHINE, CAL. .30, M1919A4 (FLEXIBLE).	
2	BAND, lock, front barrel bearing	Gun spare parts box
1	BEARING, barrel, front	Gun spare parts box
2	BARREL	R.H. center fender box
2	BOLT, group	Gun spare parts box
	Consisting of:	
	1 Bolt, assembly	
	1 Extractor assembly	
	1 Lever, cocking	
	1 Pin, cocking lever	
	1 Pin, firing, assembly	
	1 Rod, driving spring, assembly	
	1 Sear	
	1 Spring, driving	
	1 Spring, sear, assembly	
1	COVER, group	Gun spare parts box
	Consisting of:	
	1 Cover, assembly	
	1 Lever, feed belt	
	1 Pawl, feed belt	
	1 Pin, belt feed pawl, assembly	
	1 Pivot, belt feed lever, group assembly	
	1 Slide, feed belt, assembly	
	1 Spring, cover extractor	
	1 Spring, feed belt pawl	
1	EXTENSION, barrel, group,	Gun spare parts box
	Consisting of:	
	1 Extension, barrel, assembly	
	1 Lock, breech	
	1 Pin, breech lock, assembly	
	1 Spring, locking barrel	
1	FRAME, lock, group,	Gun spare parts box
	Consisting of:	
	1 Accelerator	
	1 Frame, lock, assembly	
	1 Pin, accelerator, assembly	
	1 Pin, trigger	
	1 Plunger, barrel, assembly	
	1 Spring, barrel plunger	
	1 Spring, trigger pin	
	1 Trigger	

Part One—Introduction

Quantity per Vehicle	Item Name and Stock No.	Stowage Location
1	LEVER, cocking	Gun spare parts box
1	LEVER, feed belt	Gun spare parts box
1	PAWL, feed belt	Gun spare parts box
1	PAWL, holding belt	Gun spare parts box
1	PIN, accelerator, assembly	Gun spare parts box
1	PIN, belt holding pawl, split	Gun spare parts box
1	PIN, cocking lever	Gun spare parts box
1	PIN, firing, assembly	Gun spare parts box
1	PIN, trigger	Gun spare parts box
1	PLUG, front barrel bearing (B147157A)	Gun spare parts box
1	SCREW, belt feed lever pivot	Gun spare parts box
1	SPRING, barrel plunger	Gun spare parts box
1	SPRING, belt feed pawl	Gun spare parts box
2	SPRING, belt holding pawl	Gun spare parts box
1	SPRING, cover extractor	Gun spare parts box
1	SPRING, locking barrel	Gun spare parts box
2	SPRING, sear, assembly	Gun spare parts box
1	SPRING, trigger pin	Gun spare parts box
1	TRIGGER	Gun spare parts box
1	WASHER, lock, "tooth type" S., reg., No. 6 (BEAX1D)	Gun spare parts box
(3)	GUN, MACHINE, CAL. .50, M2, H.B. FLEXIBLE (A.A.).	
1	BARREL, assembly	R.H. center fender box
1	DISK, buffer	Gun spare parts box
1	EXTENSION, firing pin assembly	Gun spare parts box
1	EXTRACTOR, assembly	Gun spare parts box
1	LEVER, cocking	Gun spare parts box
1	PIN, cotter, belt feed lever pivot stud (BFAX1CE)	Gun spare parts box
1	PIN, cotter, cover pin (BFAX1DD)	Gun spare parts box
2	PIN, cotter, switch pivot (BFAX1BE)	Gun spare parts box
1	PIN, firing (7310080)	Gun spare parts box
1	PLUNGER, belt feed lever	Gun spare parts box
1	ROD, driving spring, w/spring, assembly	Gun spare parts box

Tools, Parts, and Accessories

Quantity per Vehicle	Item Name and Stock No.	Stowage Location

1	SLIDE, belt feed group
	Consisting of:
	  1 Arm, belt feed pawl
	  1 Pawl, feed belt, assembly
	  1 Pin, belt feed pawl, assembly
	  1 Slide, belt feed pawl
	  1 Spring, belt feed, pawl
		Gun spare parts box

1	SLIDE, sear	Gun spare parts box
1	SPRING, belt feed lever plunger	Gun spare parts box
1	SPRING, belt holding pawl	Gun spare parts box
1	SPRING, cover extractor	Gun spare parts box
1	SPRING, locking barrel	Gun spare parts box
1	SPRING, sear	Gun spare parts box
1	STUD, bolt	Gun spare parts box

PART TWO — OPERATING INSTRUCTIONS

Section IV
GENERAL

9. SCOPE.

a. Part Two contains information for the guidance of the personnel responsible for the operation of the equipment. It contains information on the operation of the equipment, with a description and location of controls and instruments.

Section V
SERVICE UPON RECEIPT OF EQUIPMENT

10. PURPOSE.

a. When a new or reconditioned vehicle is first received by the using organization, it is necessary for second echelon personnel to determine whether or not the vehicle has been properly prepared for service by the supplying organization, and to be sure it is in condition to perform any mission to which it may be assigned when placed in service. For this purpose inspect all assemblies, subassemblies, and accessories to be sure they are properly assembled, secure, clean, and correctly adjusted and/or lubricated. Check all tools and equipment against section III (tools and equipment stowage on vehicles), to be sure every item is present, in good condition, clean, and properly mounted or stowed.

b. In addition the using organization will perform a run-in test on all vehicles of at least 50 miles as directed in AR 850-15, according to procedures in paragraph 13 which follows.

c. Whenever practicable, the first echelon personnel crew will assist in the performance of these services.

11. CORRECTION OF DEFICIENCIES.

a. Deficiencies disclosed during the course of these services will be treated as follows:

(1) Correct any deficiencies within the scope of the maintenance echelons of the using organization before the vehicle is placed in service.

(2) Refer deficiencies beyond the scope of the maintenance echelons of the using organization to a higher echelon for correction.

Service Upon Receipt of Equipment

(3) Bring deficiencies of a serious nature to the attention of the supplying organization through proper channels.

12. **SPECIFIC PROCEDURES.**

a. **Preliminary Service.** Before the vehicle is moved to make the actual run-in test, certain inspections and services will be performed as follows:

(1) FIRE EXTINGUISHERS. See that portable and fixed system cylinders are securely stowed, and inspect the operating valves. If arrow on control head shaft of fixed cylinders (fig. 176) is not vertical, or red safety seals on valve bodies (fig. 176) are not intact, the cylinders have been discharged, and must be replaced with charged cylinders. Examine line, nozzles and control cables of fixed system to see if they are in good condition and that nozzles are not clogged. If seal on portable extinguisher (fig. 34) is broken, replace with charged extinguisher.

(2) FUEL, OIL AND WATER. Fill fuel tanks. Check radiator coolant supply and add as required. Check main engine and auxiliary engine oil supply. Add oil as directed on Lubrication Order (par. 46). During freezing weather, test antifreeze, and add as necessary to protect cooling system against freezing (par. 38). CAUTION: *If there is a tag on oil filler cap, follow instructions on tag before starting engine.*

(3) AUXILIARY ENGINE FUEL FILTER. Remove sediment bowl and screen (fig. 171). Wash parts in dry-cleaning solvent, dry and install. Inspect for leaks when engine is started in Item (6).

(4) BATTERIES. Make hydrometer and voltage test of batteries (par. 92) and, if necessary, add clean water to bring electrolyte to $\frac{3}{8}$-inch above cell plates.

(5) AIR CLEANERS AND BREATHERS CAPS. Examine main engine and auxiliary engine air cleaners and crankcase breathers, to see if they are in good condition, and secure. Remove elements and wash in dry-cleaning solvent, fill reservoirs to correct level with fresh oil, and reinstall securely, as directed on Lubrication Order (par. 46). Make sure all gaskets are in good condition, and that tubes and air connections are tight.

(6) ACCESSORIES AND BELTS. See that all units such as carburetors, auxiliary engine and generator, starter, generator, regulators, fans and shrouds, radiator and cooler cores, heater and ventilator blower are in good condition, clean and secure; and that all drive belts are not loose or damaged. Auxiliary engine drive belts should have $\frac{3}{8}$-inch, and fan drive belts $\frac{1}{4}$-inch finger-pressure deflection. Inspect angle drives and propeller shafts for looseness and damage. Start auxiliary engine with starter and by hand to test generator and engine for proper operation.

(7) ELECTRICAL WIRING. Inspect all exposed wiring conduits and connections to see that they are clean and secure, and well supported.

(8) TRACKS. Clean all foreign material from tracks and suspension units. Examine track links and lock nuts for looseness or damage. See that tracks are correctly assembled and mounted. Check track tension with 1-inch spacers between track and second and fourth support rollers to make sure there is ¼-inch clearance at third roller. Be sure spare track links are present, and properly mounted on turret.

(9) WHEEL AND HUB NUTS. See that sprocket, idler wheel, road wheel, support roller assembly mounting, hub nuts, and cap screws are all present and secure.

(10) FENDERS AND SAND SHIELDS. Inspect these units for looseness or damage. Include brush guards.

(11) TOWING CONNECTIONS. Inspect all tow shackles, pintle, and gun traveling lock assembly for looseness and damage. Be sure pintle hook and gun traveling lock operate properly and lock securely.

(12) HULL AND TARPAULIN. See that all hull attachments hardware, lift loops, doors, hatches, and their release and locking mechanisms, are in good condition, secure and operating properly. Examine all hull and turret attachments for looseness or damage. Be sure hull drain valves open and close freely. Inspect tarpaulin or camouflage net to see that they are in good condition, and mount or stow them securely. Make sure there are no bright spots to cause glare, and that all marking are legible.

(13) VISION DEVICES. Inspect periscopes and viewing prisms (mounted or spares) to see if they are in good condition and clean. See that mounted units are secure in holders, and that holders are properly mounted. Test each periscope to be sure it will elevate, traverse, and depress through full range. CAUTION: *Clean periscopes only with a soft cloth or brush.*

(14) LUBRICATE. Perform a complete lubrication of the vehicle, covering all intervals according to instructions on Lubrication Order (par. 46), except gear cases and units lubricated or serviced in items (1) to (13). Check all gear case oil levels, and add as necessary to bring to correct levels. Change only if condition of oil indicates the necessity, or if oil is not of proper grade for existing atmospheric temperature. NOTE: *Perform items (15) to (17) during lubrication.*

(15) SUSPENSIONS. Examine suspension arms, bumper springs and shock absorbers for looseness, damage or shock absorber fluid leaks. Inspect sprocket teeth for damage or excessive wear. Be sure road wheel and support roller tires are in good condition, and that

rubber is not separating from rims. While applying lubricant, examine seals for excessive leaks.

(16) STEERING LINKAGE. Inspect all shafts, arms, rods, connections, levers and grips to see if they are in good condition, correctly and securely assembled and mounted, and that they operate without excessive looseness or binding. Check brake linkage adjustment by pulling levers back evenly to make sure free travel is equal, and does not exceed 7 inches, and that brakes are fully applied before levers reach a vertical position (par. 164). Be sure parking brake mechanism locks the steering levers properly in applied position.

(17) FINAL DRIVE UNIVERSAL JOINTS. Examine joints for looseness and excessive grease leaks.

(18) VENTS AND BREATHERS. Examine breathers in differential, final drives, and transmission to be sure they are in good condition, secure, and not clogged.

(19) PRIMER. As engine is started in item (20), observe if primer action is satisfactory, and look for fuel leaks at primer pump or connections.

(20) ENGINE WARM-UP. Follow instructions and observe precautions in paragraph 17 and start engine. Note whether starter engages and disengages without excessive noise and develops adaquate cranking speed. Set hand throttle to run engine at 1,000 to 1,200 revolutions per minute during warm-up period.

(21) INSTRUMENTS.

(a) Oil Pressure Gage and Low Oil Pressure Warning Signal Light. Immediately after engine starts, observe if oil pressure is satisfactory (60 to 80 pounds at 2,800 revolutions per minute, and not less than 11 pounds at idle speed). Stop engine if pressure is not indicated in 30 seconds, or when warning signal light comes on at speeds above 1,000 revolutions per minute.

(b) Ammeter. Ammeter should show high positive (+) charge reading for a period after starting engine, until generators restore to batteries the current used in starting, then a slight charge or zero reading with all lights and accessories turned off and battery fully charged.

(c) Engine Temperature Gage. Gage should indicate increasing temperature until normal operating range of 160° F to 180° F is reached.

(d) Engine High Water Temperature Warning Signal. If warning signal (fig. 16) comes on, stop engine and investigate and correct cause of excessive temperature (par. 60).

(e) Transmission Low Oil Pressure and High Oil Temperature Warning Signal. If light fails to go out due to low oil pressure when engine is started, or comes on after engine is running, stop engine, and

investigate and correct cause of low oil pressure or high oil temperature (par. 61).

(*f*) *Fuel Gage.* With fuel gage control switch in each position, gage should register "FULL."

(*g*) *Tachometer.* Tachometer should indicate engine speed in hundreds of revolutions per minute, and record accumulating engine revolutions.

(22) ENGINE CONTROLS. Test operation of hand throttle and accelerator to make sure engine responds properly, and governor limits maximum speed to 2,800 revolutions per minute. Close throttle to make sure engine idles smoothly at 500 revolutions per minute. Operate fuel-cut off switch to test action of degassers, and again start engine.

(23) HORN AND WINDSHIELD WIPER. Be sure horn is in good condition and secure. If situation permits, test for proper operation and tone. Temporarily install driver's hood, and test operation of windshield wiper to make sure blade is in good condition and contacts glass evenly and firmly through its full stroke. Test defroster by feeling glass to make sure wires are heating properly. Also test operation of ventilator blower and heater motors.

(24) LAMPS (LIGHTS). Clean lenses and inspect all units for looseness and damage. If tactical situation permits, open and close all light switches to see if lamps respond properly.

(25) LEAKS, GENERAL. Look under vehicle, and within engine and fighting compartment for indication of fuel, oil, or coolant leaks. Trace any leaks found to source, and correct or report them.

(26) TOOLS AND EQUIPMENT. Check tools and On-vehicle Stowage Lists, section III, to be sure all items are present, and see that they are serviceable and properly mounted or stowed. CAUTION: *Before driving vehicle be sure traversing shift lever is in manual position (fig. 188), also that cradle traveling lock (fig. 182) gun traveling lock (fig. 182), and turret lock are in engaged position to prevent injury to personnel and damage to equipment.*

13. RUN-IN TEST.

a. **Procedures.** Perform the following procedures (1) to (9) inclusive during the road test of the vehicle. On vehicles which have been driven 50 miles or more before delivery to the using organization, reduce the length of the road test to the least mileage necessary to make observations listed below. CAUTION: *During road test, avoid continuous operation of the vehicle at speeds approaching maximums indicated on the caution plate.*

(1) INSTRUMENTS AND GAGES. When normal engine operating temperature of 160° F to 180° F has been reached, drive vehicle

TM 9-735
13

Service Upon Receipt of Equipment

observing precautions and instructions in paragraphs 16 through 21. Observe readings of all instruments and warning signals to be sure they indicate or record the proper functioning of the units to which they apply (par. 15). When vehicle is in motion, the speedometer should indicate miles per hour and record accumulating mileage. The differential low oil pressure warning signal (fig. 16) should go out. If signal comes on again stop vehicle and investigate and correct cause (par. 61).

(2) BRAKES; STEERING AND PARKING. Test steering brakes to see if they will stop vehicle effectively with levers even. Test steering action to make sure vehicle turns to right or left with equal braking effort.

(3) TRANSMISSION. Operate vehicle through each speed range of transmission. Observe if lever operates properly, and if there is any unusual noise or vibration in any range that might indicate damage, loose mountings or inadequate lubrication. Stop engine if low oil pressure or high oil temperature signal light comes on.

(4) ENGINE. Be on the alert for any unusual engine operating characteristics, or noise, such as lack of pulling power or acceleration, backfiring, misfiring, stalling, overheating, excessive exhaust smoke. Observe if engine responds properly to all controls.

(5) UNUSUAL NOISE. Be on the alert throughout road test for unusual noise or vibration from hull and attachments, suspensions, power train, or tracks, that might indicate looseness, damage, excessive wear or inadequate lubrication.

(6) HALT VEHICLE AT 10-MILE INTERVALS OR LESS FOR SERVICES (7) AND (8) FOLLOWING.

(7) TEMPERATURES. Cautiously hand-feel each road wheel, sprocket, idler, and support roller hub for abnormal temperatures. Examine transmission, and controlled differential and final drives, for indications of overheating or excessive leaks at seals, gaskets or vents.

(8) LEAKS. With engine running and fuel, engine oil and cooling systems under pressure, look within engine and fighting compartments and under vehicle for leaks.

(9) TRACK AND SUSPENSION—OPERATION. While vehicle is driven slowly forward, observe action of suspension units, and inspect track on each side of vehicle for broken or damaged parts, loose mountings or attachments, improper alinement, excessive looseness or binding, and shock absorber leaks.

(10) GUN: ELEVATING AND TRAVERSING MECHANISM. Disengage gun traveling lock and gun elevating lock. Place vehicle on a 10-degree lateral incline (tilted sideways). Disengage turret lock. Traverse turret 360 degrees in each direction to test action of hand

and hydraulic mechanism (par. 24) and to make sure turret does not bind. With gun forward or in traveling position elevate gun through its entire range to test for binding, excessive play, or erratic action. After tests, move traverse shift lever to manual position (fig. 188), secure the gun in the traveling lock, and engage the gun elevating and turret locks.

b. **Vehicle Publications and Reports.**

(1) PUBLICATIONS. See that vehicle Operator's Manuals, Lubrication Order, Standard Form No. 26 (Driver's Report—Accident, Motor Transportation) and W.D., A.G.O. Form No. 478 (MWO and Major Unit Assembly Replacement Record), are in the vehicle, legible, and properly stowed. NOTE: *U.S.A. registration number and vehicle nomenclature must be filled in on Form No. 478 for new vehicles.*

(2) REPORTS. Upon completion of the run-in test, correct or report any deficiencies noted. Report general condition of the vehicle to designated individual in authority.

Section VI
CONTROLS AND INSTRUMENTS

14. CONTROLS.

a. **Steering Levers.** The vehicle is equipped with dual controls for steering and stopping (figs. 18 and 19). Individual sets of steering levers are provided for the driver and assistant driver. When one set of steering levers is in use the other set is inoperative. To shift steering control the relief driver simply grasps his set of steering levers and begins steering. As soon as the relief driver's set of levers becomes operative the set formerly in use is returned to the forward, fully released position. Each set of levers is mounted on the frame of each driver's seat, one on each front corner.

b. **Service Brakes.** The same brakes used to steer the vehicle are used to stop it. To slow down or stop pull both steering levers back simultaneously and with equal force. When a stop is to be made remove foot from accelerator, leave speed range selector lever in gear and, as vehicle speed decreases, apply brakes as necessary to slow down or stop. When vehicle stops shift speed range selector lever into neutral. During braking some steering may be necessary to control the direction of the vehicle. When operating on hard pavement, or ice, use an on-and-off braking movement to reduce vehicle speed and thus reduce danger of skidding.

c. **Parking Brake.** The parking brake lever is located between the driver and assistant driver, near the floor in front of and attached to the turret ring support (fig. 134). To set parking brake move parking brake lever to center position. Pull back both steering levers simultaneously as far as they will go. Then release steering levers and return them to the full forward position. To release parking brake simply move parking lever from the center position all the way to either right or left.

d. **Speed Range Selector Lever.** The vehicle is equipped with dual speed range selector levers which are connected by linkage and a control rod to the speed range selector on the transmission (fig. 127). These levers permit either the driver or assistant driver, whoever is driving the vehicle, to select the different speed ranges required by changing terrain and operating conditions. The speed range selector levers are mounted on each end of a cross shaft between the instrument panel supports. Neutral, reverse, and three forward speeds are provided. These positions are marked on the dual speed range selector lever quadrant "3, 2, 1, N, R," reading from front to rear.

e. **Fuel Tank Shut-off Valve Levers.** There are two fuel tank shut-off valve levers. One lever is located at the base of each instrument panel support (fig. 15). Each lever is connected by a rod to its respective fuel tank shut-off valve installed in the fuel tank. The left lever controls the left fuel tank shut-off valve. The right lever controls the right fuel tank shut-off valve. To open either valve move lever sideways out of locking notch or detent and pull lever back as far as it will go. At end of pull, spring action will move lever sideways into OPEN locking notch. Moving lever sideways, and then forward as far as it will go, closes the valve, and spring action moves the lever sideways into CLOSED locking notch.

f. **Hand Throttle.** A dual-type hand throttle is provided, which can be operated by the driver or assistant driver (figs. 18, 19, and 127). Both throttles operate in unison. A friction mechanism in the linkage holds the throttle in any set position. To open driver's throttle pull up. When hand throttle is as far up as it will go, the carburetor throttle valves are wide open. To close hand throttle push down until positive stop is met. To open assistant driver's throttle push lever forward. Pull throttle back to vertical position to close. When starting engine advance throttle (open) so that engine will run at 1,000 to 1,200 revolutions per minute for the warm-up period. After engine has warmed up to 100° F to 110° F, close throttle. When throttle is fully closed engine will idle at its recommended idling speed of 500 revolutions per minute. CAUTION: *When stopping engine, leave throttle closed and allow engine to idle at 500 revolutions per minute for 5 minutes before stopping to uniformly cool engine parts.* The throttle is connected to the dual-type accelerator by linkage.

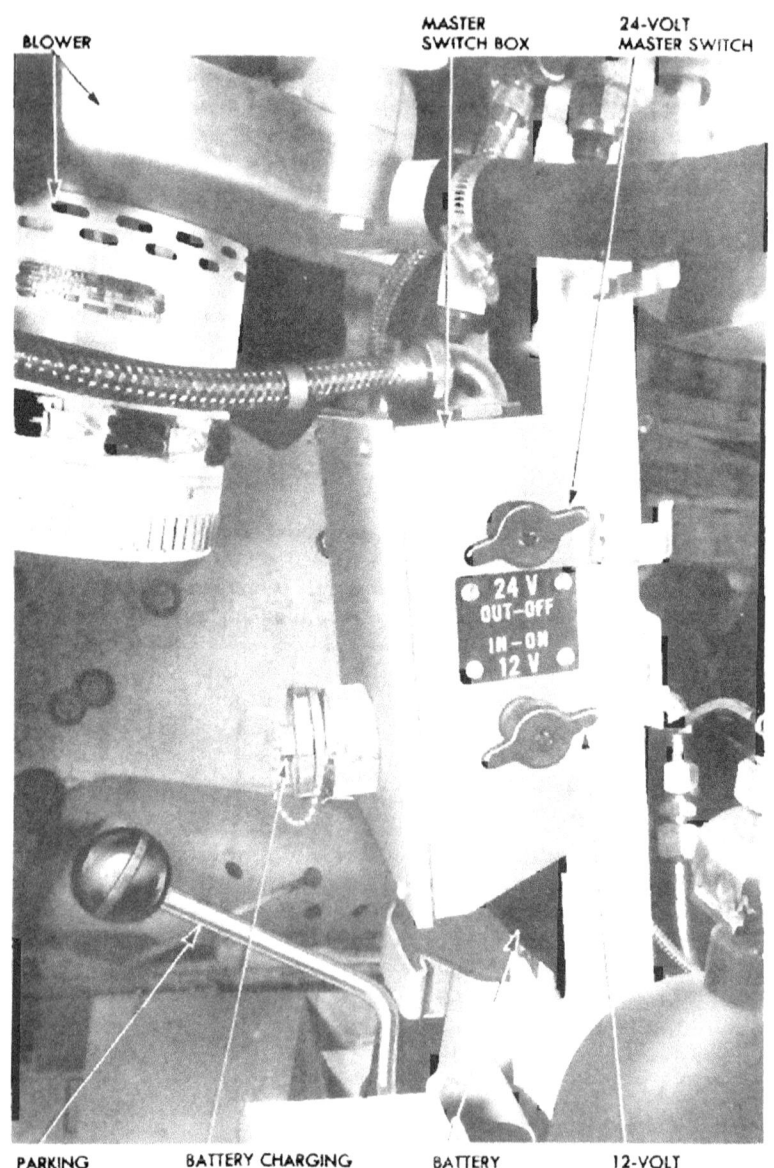

Figure 12 — Master Switch Box

Controls and Instruments

When the throttle is opened the accelerator moves down with the movement of the throttle. However, when the accelerator is depressed an over-running clevis permits the throttle to remain stationary. WARNING: *Do not operate vehicle with throttle.*

g. **Accelerator.** Step-type accelerator pedals are provided for driver and assistant driver (figs. 18, 19). Both pedals are mounted on a cross-shaft and operate in unison. The step is provided to facilitate pedal operation with the seat in either raised or lowered positions.

h. **Primer.** The engine primer, which takes the place of the conventional choke, is located in the fighting compartment on the right engine compartment bulkhead (fig. 29) and is operated from the fighting compartment by a member of the crew. A spring-loaded lever-type guard prevents the pump handle from being pulled out accidentally. Do not use primer at temperatures above freezing. For priming a cold engine use three to nine quick strokes, depending upon outside temperature. To operate primer, pull guard down and hold clear of pump handle. Move handle in and out with even action for desired number of strokes. When finished with primer, seat pump securely and release guard.

i. **24-volt Master Switch.** The 24-volt master switch controls the 24-volt current which is used to operate all the electrical units and equipment. Radios and intercommunication systems using 24-volts are controlled by the 24-volt master switch. It is the top switch located in the master switch box mounted on the front turret ring support (fig. 12). To turn switch on pull handle out, turn, and allow spring to pull switch inward into contact. To turn switch off pull handle outward and turn so handle stays in out position. CAUTION: *Do not turn 24-volt master switch off* while either main or auxiliary engine is running or radio set switch is on. To do so will damage the generator and/or radio set.

j. **12-volt Master Switch.** The 12-volt master switch controls the 12-volt current from the right-hand 12-volt battery to the radio (when a 12-volt British radio is installed). It is located directly below the 24-volt master switch in the master switch box (fig. 12). The 12-volt master switch is the same type and operates in the same manner as the 24-volt master switch (subpar. i preceding). CAUTION: *Do not turn 12-volt master switch off with radio set switch on.*

k. **Auxiliary Engine Fuel Shut-off Valve.** The fuel flow for the auxiliary engine is controlled by an electric shut-off valve which is mounted on a pipe nipple at the bottom of the right fuel tank (fig. 59). It is automatic in action, and is connected to the auxiliary engine ignition switch. When the switch is turned on the valve automatically opens. When the switch is turned off the valve automatically closes.

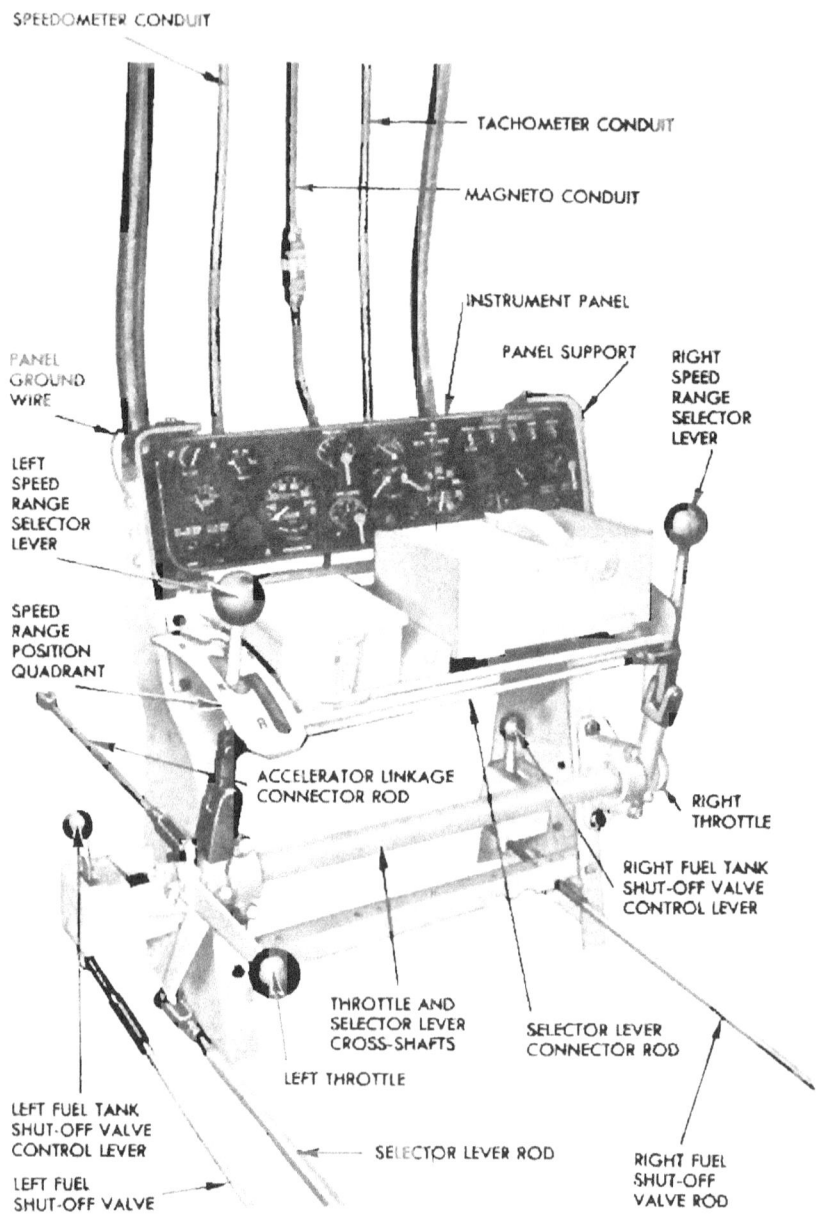

Figure 13 — Instrument Panel With Controls

Controls and Instruments

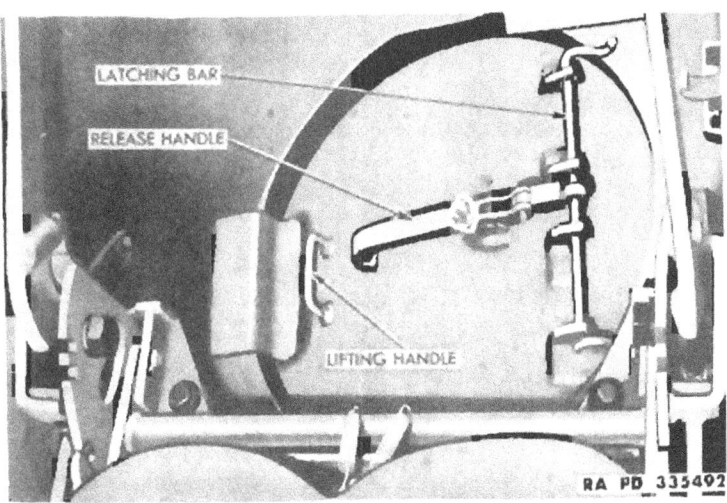

Figure 14 — Escape Door, Right Side

Valve can be blocked open when there is not sufficient current in the batteries to open it electrically (par. 29 d (1)).

l. **Service Headlights.** The vehicle is equipped with two service headlights (fig. 1) which are stowed inside the vehicle when not in use. The left headlight and the blackout driving light are stowed in brackets at the driver's feet. The right headlight is stowed in a bracket at the assistant driver's feet. Each headlight consists of a combination sealed beam driving light, and a blackout marker light. Sockets for lights are located on the front hull slope near each fender. Solid metal protector plugs (fig. 83) are provided to weather-seal the sockets, and are to be installed in sockets when service headlights are not in use. Plugs are chained to their stowage receptacles. To install service headlights unscrew locking handle at rear of each headlight bracket. Remove protector plug and install it in receptacle at front of bracket. Install headlight mount securely in socket. Screw in locking handle. At the same time move headlight as necessary to make sure point of locking handle pin seats securely in hole in headlight mount. Tighten locking hand-tight. If tactical situation permits operate driving light switch to test headlights and blackout marker lights. To remove headlight unscrew locking handle. Lift out headlight. Insert protector plug in socket and screw in locking handle until tight. Stow headlights in proper brackets inside vehicle.

TM 9-735
14

Part Two—Operating Instructions

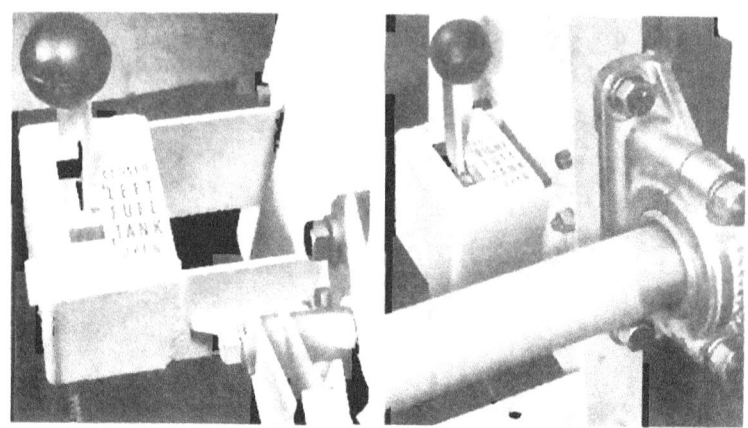

RA PD 344414

Figure 15 — Fuel Shut-off Valve Levers

m. **Blackout Driving Light.** The blackout driving light is a hooded, masked light (fig. 83) which is used to provide controlled illumination when service headlights are tactically inadvisable. It is installed only in the left headlight bracket. To install the blackout driving light follow procedure in subparagraph l preceding.

n. **Escape Doors.** An escape door is located in front of both the driver's and assistant driver's seats in the hull floor (fig. 14). To operate escape door pull up on quick-release handle, and allow door to drop to the ground. Do not pile equipment or objects on door. Keep door and operating mechanism well-oiled and free of mud, rust and dirt.

o. **Drivers' Doors.** Over each driver's seat is a heavy armor plate door (figs. 163 and 164) hinged so it swings up and outward. Doors are held in closed position by seating locking bolt securely in boss. They are held in open position by a spring-loaded catch. Pull out on catch handle to release catch. There is a periscope mount in the center of each door (fig. 163). See paragraph 183 a for further description. CAUTION: *Be sure doors are securely hooked if vehicle is to be operated with doors open.*

p. **Drivers' Seats.** A form-fitting, padded seat and back rest are provided for the driver and assistant driver. Each seat assembly is mounted on a stationary base and is adjustable in height. Control levers permit adjusting seats through four positions, low, intermediate low, intermediate high, and high. Pull lever back to raise or lower seat.

q. **Horn Button.** The horn button is mounted on the driver's right steering lever below the rubber grip (fig. 18). Sound horn by pressing button with right-hand little finger.

15. INSTRUMENT PANEL.

a. **Description.** The instrument panel (fig. 16) is located in the center of the driver's compartment close to the front slope. It accommodates the magneto switch which also includes the automatically controlled speedometer-tachometer switch, fuel cut-off switch, engine oil pressure gage, engine temperature gage, dual combination engine low oil pressure and high water temperature warning signals, speedometer, tachometer, dual combination transmission oil temperature and low transmission oil pressure warning signals, differential low oil pressure warning signal, heater switch, ammeter, main lights switch equipped with safety button, instrument panel light switch, two indirect instrument panel lights, two accessory outlets, and five circuit breaker buttons. The instrument panel is mounted on support brackets welded to the hull floor and is in vision range of both driver and assistant driver.

b. **Magneto Switch.** (MAGNETOS, fig. 16). The magneto switch is a four-position switch which controls the two engine magnetos. One magneto fires the right bank of four cylinders and the other magneto fires the left bank of four cylinders. When the switch lever is all the way to the left it is at the "OFF" position and both magnetos are grounded. When the lever is at "L" the left magneto only is on and the engine will run on the "left" four cylinders only. When the switch lever is at "R" the "right" magneto only is on, and the engine will run on the "right" four cylinders only. Moving the lever all the way to the right to "BOTH" turns on both magnetos, and the engine will run on both the "right" and "left" banks of cylinders. CAUTION: *Turn magneto switch off AFTER engine has stopped.* NOTE: *Operation of the magneto switch also automatically operates the speedometer-tachometer switch, turning it on and off.*

c. **Starter Switch.** (STARTER, fig. 16). The starter switch is controlled by a spring-loaded lever and has two positions, "ON" and "OFF." To operate, pull lever down and release when engine fires. When released, lever returns automatically to "OFF" position. WARNING: *Do not hold starter switch lever on longer than 30 seconds at a time, as this seriously depletes battery charge.*

d. **Booster Switch.** (BOOSTER, fig. 16). This switch is not used in the operation of Heavy Tank, T26E3.

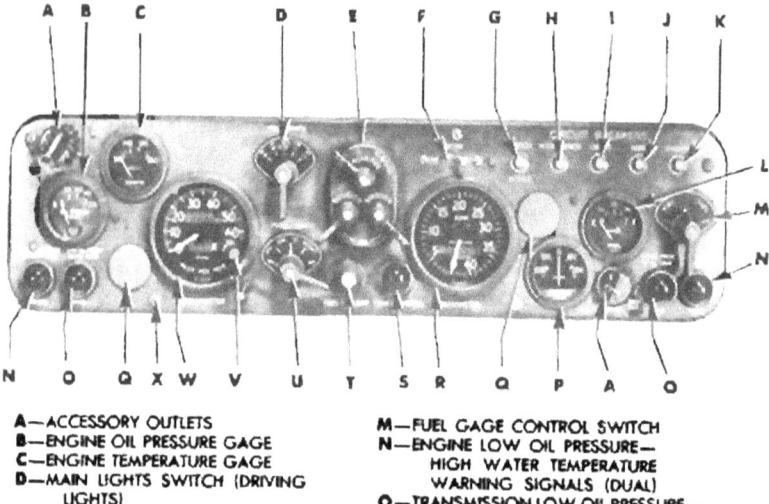

A—ACCESSORY OUTLETS
B—ENGINE OIL PRESSURE GAGE
C—ENGINE TEMPERATURE GAGE
D—MAIN LIGHTS SWITCH (DRIVING LIGHTS)
E—MAGNETO AND STARTER SWITCHES
F—DRIVERS' HEATER SWITCH
G—CIRCUIT BREAKER—SPEEDOMETER AND TACHOMETER ONLY
H—CIRCUIT BREAKER—ALL ACCESSORIES INCLUDING A, E AND T.
I—CIRCUIT BREAKER—HULL LIGHTS AND SWITCHES
J—CIRCUIT BREAKER—HORN ONLY
K—CIRCUIT BREAKER—ALL GAGES AND WARNING SIGNALS
L—FUEL GAGE
M—FUEL GAGE CONTROL SWITCH
N—ENGINE LOW OIL PRESSURE—HIGH WATER TEMPERATURE WARNING SIGNALS (DUAL)
O—TRANSMISSION LOW OIL PRESSURE—HIGH OIL TEMPERATURE WARNING SIGNALS (DUAL)
P—AMMETER
Q—INSTRUMENT PANEL LIGHTS
R—TACHOMETER
S—DIFFERENTIAL LOW OIL PRESSURE WARNING SIGNAL
T—FUEL CUT-OFF SWITCH
U—INSTRUMENT PANEL LIGHT SWITCH
V—RESET KNOB GUARD
W—SPEEDOMETER
X—MOUNTING SCREW   RA PD 344415

Figure 16 — Instrument Panel

e. **Fuel Gage and Fuel Gage Control Switch.** (FUEL, fig. 16). The fuel gage and the fuel gage control switch are mounted side by side. When the pointer of the switch lever is moved to "L" with the 24-volt master switch on, the level of fuel in the left fuel tank will register on the gage. When the pointer is moved to "R" the level of fuel in the right fuel tank will register on the gage. When the switch lever is in vertical position it indicates fuel level in the right tank. NOTE: *There is no "OFF" position on the switch.* The 24-volt master switch must be on before gage will register. Before operating vehicle or engine, check level of fuel in each fuel tank. When operating vehicle with full fuel tanks, run for 30 minutes with both fuel shut-off valves open to provide additional expansion space in each fuel tank. At the end of this period close one fuel valve and use fuel from one fuel tank at a time. Turn fuel gage control to position for fuel tank in use.

Controls and Instruments

f. **Fuel Cut-off Switch.** (FUEL CUT-OFF, fig. 16.) A bright metal button located in the lower center of the instrument panel operates the carburetor degasers which shut off the fuel to the idle fuel supply wells in the carburetors. To stop engine, hold button all the way in until engine stops. When engine stops, release button and turn off magnetos.

g. **Engine Oil Pressure Gage.** (OIL PRESS., fig. 16.) The engine oil pressure gage is calibrated from 0 to 120, and registers the pressure in the engine lubrication system. Normal oil pressure is indicated on gage between 60 to 110 pounds at 2,800 revolutions per minute, and approximately 20 pounds when idling at 500 revolutions per minute. If oil pressure of 11 pounds (or above) does not show on gage within 30 seconds after engine has started, stop engine immediately. If oil pressure falls below 11 pounds with engine running, the low oil pressure warning signals will light as described in subparagraph h following. Stop engine immediately.

h. **Dual Engine Low Oil Pressure Warning Signals.** (LOW OIL PRESSURE ENG., fig. 16.) In the lower right and left corners of the instrument panel there is a red jewel pressure-operated warning signal. These signals operate in unison, and light when engine oil pressure drops below 11 pounds, or when engine overheats as described in subparagraph j following. Observe both engine oil pressure and engine temperature gages to determine whether warning signals are indicating low oil pressure or overheated engine. The position of the pointer on the oil pressure gage will also act as a tell-tale to indicate drop in oil pressure. Signals light when 24-volt master switch is turned on, and remain lighted until engine oil pressure goes above 11 pounds, when they automatically go out. Under normal conditions signals will not light when engine is idling. After engine has been stopped and oil has drained back into engine oil pan, warning signals will light and will remain lighted until 24-volt master switch is turned off, or engine is restarted and normal oil pressure is again reached.

i. **Engine Temperature Gage.** (TEMP. °F, fig. 16.) The water temperature gage is calibrated from 60° F to 260° F. It registers the temperature of the coolant in the engine. The cooling system is sealed so that it operates above atmospheric pressure. Consequently the boiling point of the coolant is raised above the normal boiling point of water (212° F) so that overheating does not occur until 220° F is reached. In normal operation under maximum power on a level hard surface the engine temperature should not be greater than approximately 90° F higher than the normal atmospheric temperature. Warm up engine to from 110° F to 110° F before driving vehicle. Normal operating temperature is 160° F, to 180° F. The two thermo-

TM 9-735
15

Part Two—Operating Instructions

stats start to open at 150° F to 155° F and are fully open at 185° F.

j. **Dual Engine High Water Temperature Warning Signals.** (HI-WATER TEMP. ENG., fig. 16.) A temperature-controlled bi-metal switch also operates the low engine oil pressure warning signals (subpar. h preceding) to indicate engine high water temperature. If warning signals light observe both engine temperature and oil pressure gages to determine whether warning signals are indicating an overheated engine or low engine oil pressure. When the coolant in the cooling system reaches 220° F (danger point) signals will light and remain on until cause of overheating has been remedied. CAUTION: *Use great care when releasing expansion tank filler cap. Turn to first release position and vent steam before removing cap.*

k. **Speedometer-Tachometer Switch.** The speedometer-tachometer switch is automatically controlled by the operation of the magneto switch. When the magneto switch is turned on or off the speedometer and tachometer sending units become operative or inoperative.

l. **Speedometer.** (SPEEDOMETER, fig. 16.) The speedometer is calibrated from 0 to 60 miles per hour. It is equipped with an odometer which registers the total vehicle miles traveled, a trip mileage indicator and a trip mileage reset knob. The reset knob is mounted off-center in the face of the instrument and is operated by pushing in and turning. A reset knob guard protects the reset knob when panel is face down on work bench.

m. **Tachometer.** (TACHOMETER, fig. 16.) The tachometer is calibrated from 0 to 40. It registers the engine crankshaft speed in hundreds of revolutions per minute. Thus, if the engine is running at 2,800 revolutions per minute, it will register at 28 on the tachometer dial. A revolution counter, which records total crankshaft revolutions, is mounted on the magneto end of the "left" camshaft housing and is driven by the "left" cylinder head exhaust camshaft.

n. **Dual Transmission Low Oil Pressure Warning Signals.** (LOW OIL PRESSURE TRANS., fig. 16.) The red jewel warning signals located next to the engine warning signals, are provided to warn the driver when the transmission oil drops below correct operating pressure (60 to 65 pounds), or becomes overheated (high oil temperature warning signals (subpar. o following). Signals light when 24-volt master switch is turned on and remain lighted until the operation of the main engine has built up proper transmission oil pressure. WARNING: *Do not drive vehicle until transmission low oil pressure warning signals go out.* If signals light while transmission is in operation, stop engine immediately and check oil level with dipstick. Signals only give warning. No gage is provided.

o. **Dual Transmission High Oil Temperature Warning Signals** (HI-OIL TEMP. TRANS., fig. 16). The warning signals which indi-

TM 9-735

Controls and Instruments

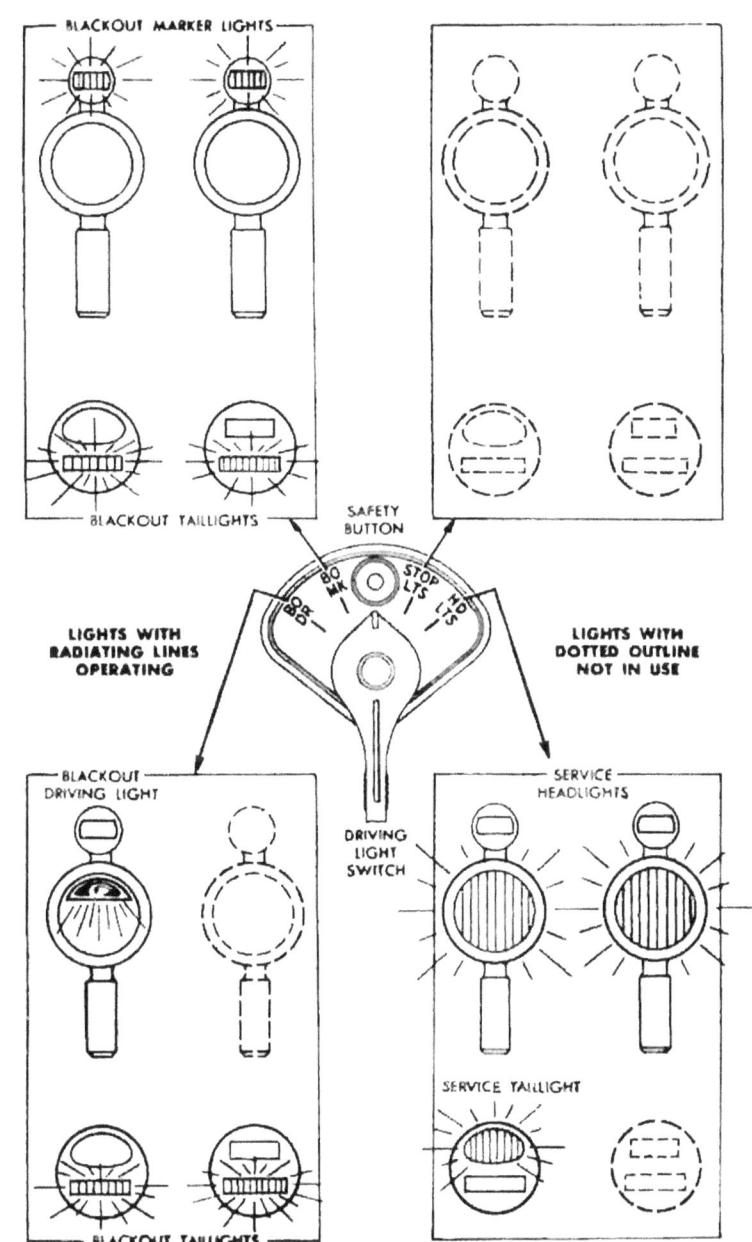

Figure 17 — Driving Light Chart

cate transmission low oil pressure (subpar. n preceding) also warn of transmission high oil temperature. If transmission oil temperature reaches 285° F, signals will light. Signals will automatically go out when cause of trouble has been removed and/or oil has cooled. If signals light with transmission in operation, stop engine immediately and hand-feel unit to see if it is abnormally hot. If it is, notify higher authority. Signals only give warning. No gage is provided.

 p. **Differential Low Oil Pressure Warning Signal (LOW OIL PRESS., fig. 16).** A red jewel warning signal located in the lower center of the instrument panel indicates low oil pressure in the differential. The signal lights when the 24-volt master switch is turned on, and goes out as soon as the engine turns over. Signal operates at 6 pounds pressure and, with differential in motion, lights if oil pressure drops below six pounds. If signal comes on when vehicle is in operation, stop immediately and investigate cause of trouble. Signal only gives warning. No gage is provided. Signal will light when vehicle is stopped and engine is idling.

 q. **Heater Switch (HEATER, fig. 16).** The heater switch is a toggle-type switch located directly above the tachometer and has two positions, "ON – OFF." It controls the heater fan motor. The heater is mounted in the angle of the front casting in the driver's compartment (fig. 35).

 r. **Ammeter (AMPERES, fig. 16).** The ammeter is calibrated with a range of 200-ampere charge (+) to 200-ampere discharge (−). However, full charge and full discharge are each approximately 150-amperes. The ammeter indicates the rate of current flow to and from the batteries. When the batteries are fully charged and all electrical units turned off, the ammeter will show a low rate of charge, with main or auxiliary engine running. Investigate a continually discharging battery.

 s. **Driving Light Switch (MAIN LIGHTS, fig. 16).** The driving light switch is to the left of the magneto switch and is a five-position, lever-operated type. When the lever is in vertical position the switch is "OFF." From this position the lever pointer can be moved only to the left to "B. O. MK" (blackout markers) without pushing in safety button. To move pointer to right to "STOP LT" (position inoperative), or to "HD LTS" (headlights), or to the left to "B. O. DR" (blackout driving light), the spring-loaded safety button above the "OFF" position must be pushed and held in before lever can be moved to these positions. The safety button prevents the service headlights from being accidentally turned on. Moving lever pointer to "OFF" position automatically locks the safety button, with the exception of movement to "B. O. MK." Refer to figure 17 for driving light chart. NOTE: *Although the stop light position is shown on*

TM 9-735
15
Controls and Instruments

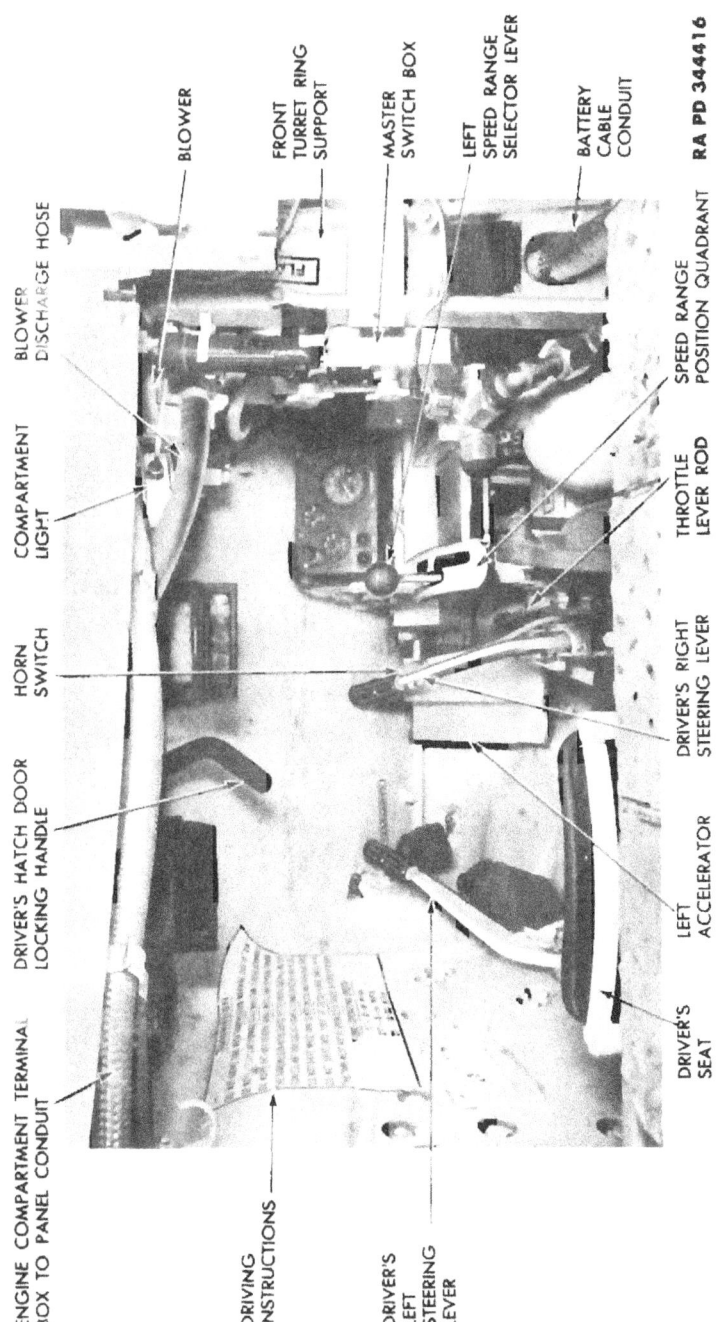

Figure 18 — Driver's Controls

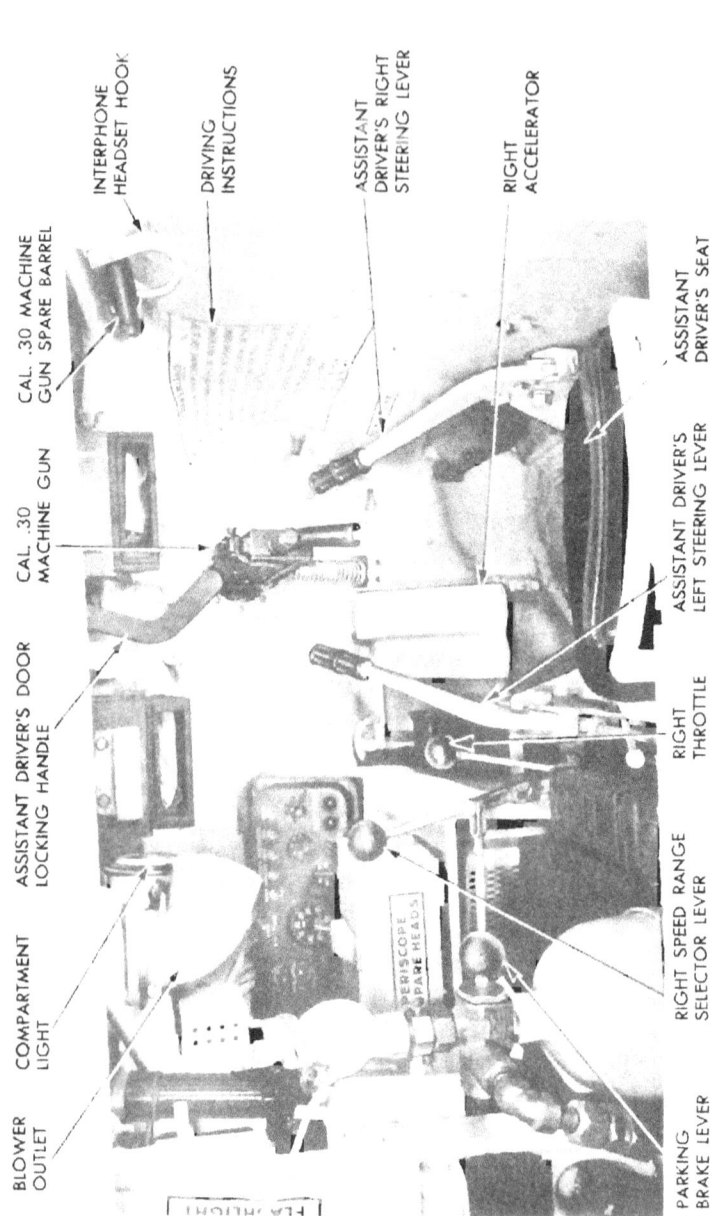

Figure 19 — Assistant Driver's Controls

Controls and Instruments

the main lights switch, there are no service or blackout stop lights on the vehicle.

t. **Instrument Panel Light Switch** (PANEL LIGHTS, fig. 16). The instrument panel light switch is located below the driving lights switch and is a lever-operated, rheostat-type. When the lever pointer is all the way to the left, the switch is "OFF." Moving lever pointer to the right to "DIM" turns panel lights on. Moving pointer from "DIM" to "BRIGHT" gradually increases their brilliance. When pointer is at "BRIGHT," instrument panel lights are fully on. Moving pointer from "BRIGHT" to "DIM" decreases light brilliance. Turn panel lights off by moving lever pointer from "DIM" to "OFF."

u. **Instrument Panel Lights** (PANEL LIGHTS, fig. 16). The gages and instruments on the panel are indirectly lighted by two recessed lights which are protected by removable snap-in covers (fig. 16).

v. **Accessory Outlets** (OUTLET, fig. 16). Two accessory outlets, equipped with chained protector plugs and gaskets, are provided. Outlets are for connecting inspection light, defrosters, windshield wipers, and other electrical accessories.

w. **Circuit Breakers** (SPEEDO. & TACH., ACCESSORIES, LIGHTS, HORN, GAGES, fig. 16). Five red circuit breaker reset buttons are grouped on the top right side of the panel. These buttons control the circuit breakers which take the place of conventional fuses. When a circuit becomes overloaded the circuit breaker opens and the red button snaps out. To reset a circuit breaker simply push button in. If button snaps out again it indicates trouble still exists. Investigate and correct or remove cause of trouble. CAUTION: *Do not push a circuit breaker button in and hold it. Serious damage or fire will result.*

(1) CIRCUIT BREAKER TABLE.

Button	Circuits Controlled
Speedo & tach	Speedometer and tachometer only.
Accessories	All accessories including accessory outlets, magneto switch, fuel cut-off, auxiliary engine control box feed, heater, compass, and speedometer-tachometer switch.
Lights	All hull lights (service, blackout, taillights, compartment lights and all light switches).
Horn	Horn only.
Gages	All gages and warning signal lights on instrument panel.

Part Two—Operating Instructions

Section VII

OPERATION UNDER ORDINARY CONDITIONS

16. USE OF INSTRUMENTS AND CONTROLS IN VEHICULAR OPERATION.

a. *Service Upon Receipt of Vehicle.* Before a new or reconditioned vehicle is placed in service, be sure that the services described in paragraphs 10 through 13 have been performed.

b. *Before-operation Service.* Perform the services in items 1 to 6 in paragraph 49 b before attempting to start the engine. Start and warm-up engine (par. 17 c (1) through (9) below) and complete the Before-operation Service.

17. STARTING THE ENGINE.

a. *General.* Before attempting to start the engine the driver must be familiar with all the instruments, gages and controls.

b. *Before-operation Service.* Complete the Before-operation Service (par. 16 b) before engine is started.

c. *Starting Engine at Above Freezing Temperature.* The procedure for starting the engine is given in the following steps:

(1) Set speed range selector lever in neutral (N).

(2) Set parking brake.

(3) Turn 24-volt master switch on.

(4) Determine quantity of fuel in both fuel tanks by operating fuel gage control switch. Leave switch on position of fuel tank to be used.

(5) Open fuel shut-off valve (or both valves when fuel tanks are full) on fuel tank to be used.

(6) Advance (open) throttle to run engine at 1,000 to 1,200 revolutions per minute during warm-up period.

(7) Turn magneto switch to "BOTH" position.

(8) Push starter switch lever down to "ON" position. When engine fires, release lever immediately. **WARNING:** *Do not hold lever at "ON" position longer than 30 seconds at a time.* Warm engine up to from 100° to 110° F at 1,000 to 1,200 revolutions per minute by setting throttle. When engine is warmed up close throttle. NOTE: *Do not idle engine at less than 500 revolutions per minute at any time.*

(9) Do not move vehicle until engine and transmission low oil pressure warning signals go out.

Operation Under Ordinary Conditions

d. **Starting Engine at Below Freezing Temperatures.**

(1) Follow starting procedure as outlined in subparagraphs b and c preceding.

(2) Before starter switch is operated, have member of crew in fighting compartment pump three to nine quick strokes on primer (depending on temperature conditions). After engine starts, an occasional stroke of primer may be required to keep engine running. CAUTION: *Do not overprime.*

(3) As engine fires, quickly depress accelerator two or three times to assist in keeping engine running.

e. **Engine Observation.** As soon as engine is running check oil pressure on engine oil pressure gage. Stop engine immediately if correct oil pressure (11 pounds or more) is not shown on the oil pressure gage (fig. 16) within 30 seconds. Investigate cause (par. 61 b). Run engine at 1,000 to 1,200 revolutions per minute to warm up to from 100° F to 110° F. When correct temperature has been reached, idle engine at 500 revolutions per minute. Carefully observe operation of gages and instruments. Test switches while engine is warming up. When engine has warmed up to from 100° F to 110° F, idle engine at 600 revolutions per minute and test magnetos. Move magneto switch to "L" and then "R," and compare both tachometer readings with tachometer reading when both magnetos are on. If when running on only one magneto a drop below 500 revolutions per minute is shown, investigate cause. Do not run engine for more than 1 minute on only one magneto. Observe oil pressure and engine temperature frequently during operation. Stop engine immediately if, during operation, engine, transmission, or differential warning signals go on, and investigate cause.

18. DRIVING THE VEHICLE.

a. **General.** Unless a driver is thoroughly competent to operate the vehicle, an instructor will occupy the assistant driver's seat to give advice until he is sufficiently proficient to handle the vehicle alone. Start engine as previously instructed (par. 17 c (1) through (9)) and release parking brake by moving lever from center position all the way to right or left. The vehicle is now ready to drive.

b. **Speed Range Data.** Before driving the vehicle the driver must be familiar with the speed ranges and operation of the transmission as given in the following paragraphs:

(1) SPEED RANGE TABLE.

Speed Range	Vehicle Speed
First (1)—Low	0 to 9 mph
Second (2)—Intermediate	6 to 19 mph
Third (3)—High	12 to 30 mph
Reverse (R)	0 to 9 mph

TM 9-735
18-19

Part Two—Operating Instructions

c. **Speed Range Selection.** There is no clutch or clutch pedal. The torque converter in the transmission eliminates the necessity of a clutch, as it transmits torque or power through the medium of oil. The correct speed ranges of the transmission must be preselected by the driver to correspond with the speed at which the vehicle is to be driven, and existing operating conditions. Running at top speed in a low speed range will overheat the transmission oil and waste fuel. (See Speed Range Table, subparagraph h (1) preceding). When decreasing the speed of the T26E3, or down-shifting, leave the speed range selector lever in the third (3) speed range position down to 14 miles per hour; in the second (2) speed range position down to 8 miles per hour. Shifting to the different speed ranges is accomplished by simply moving the speed range selector lever to the desired speed range. NOTE: *Any speed change (except reverse and neutral) to a higher or lower ratio can be made without changing the accelerator position.*

19. SPEED RANGES.

a. **Speed Ranges.**

(1) FIRST (1) OR LOW SPEED RANGE. First speed range provides an emergency power range for difficult pulling conditions. It is to be used when maximum pulling power is required in the T26E3 at speeds below 9 miles per hour. At speeds above 9 and up to 19 miles per hour operate in second (2) (intermediate) speed range.

(2) SECOND (2) OR INTERMEDIATE SPEED RANGE. Use second speed range for starting the vehicle on soft ground and on inclines. This speed range provides the greatest tractive effort for T26E3 operation at speeds between 6 and 19 miles per hour. Use intermediate (second) speed range whenever maximum performance is desired in this speed range. When the speed of the T26E3 exceeds 14 miles per hour shift speed range selector lever into third and operate in the high speed range. If, in heavy going, the T26E3 slows down to a speed of less than 9 miles per hour, shift from second (intermediate) into first (low) speed range.

(3) THIRD (3) OR HIGH SPEED RANGE. Use the third or high speed range for driving on good roads or on firm, level ground. Under favorable operating conditions the vehicle can be started in third or high speed range, and left in this range for all normal operating conditions. This range can also be used on grades up to approximately 10 percent incline. If the T26E3 speed is reduced below 14 miles per hour by steep hills, sand, mud, or snow, additional tractive effort can be obtained by operating in second (intermediate) speed range. WARNING: *Operate the vehicle in the various speed ranges within the allowable speeds (miles per hour) as listed in the speed range table*

Operation Under Ordinary Conditions

(par. 18 b (1)). *Operation in the wrong speed range for any length of time (particularly at full throttle) will cause excessive heating of the transmission oil, unnecessary fuel consumption and possible damage to the transmission.*

(4) REVERSE (R). Reverse is selected by moving the speed range selector lever into the position marked "R" on the quadrant. WARNING: *Bring vehicle to a complete stop before shifting into reverse. Stop vehicle completely before shifting from reverse to forward speed range. Do not shift into reverse speed range while vehicle is going forward to use transmission as a brake (except as described in subparagraph g following). Follow instructions stenciled on hull side plates in drivers' compartment.* When tactical situation permits, always station an observer in front of vehicle while backing.

h. **Driving Instructions.** When ready to start, move the speed range selector lever into speed range required for the anticipated driving conditions. Put vehicle in motion by slowly pushing down on accelerator as described in step (1) following. The vehicle is normally started in second (2) (intermediate) or third (3) (high) speed range depending on operating conditions and terrain. The T26E3 can be operated, at partial throttle under light load, in these ranges at speeds slightly lower than 6 and 14 miles per hour respectively. However, to obtain maximum performance and tractive effort, the vehicle must be operated in all speed ranges at the approved speeds listed in the speed range table (subpar. b (1) preceding) and stenciled on the hull side plates in the drivers' compartment.

(1) USE OF ACCELERATOR. Push slowly and firmly down on accelerator. As accelerator is depressed and engine speed increases the vehicle will begin to move forward. Hold accelerator in position to maintain speed. Maintain or increase engine speed as required when turning. CAUTION: *Do not pump accelerator at any time while driving.*

c. **Steering the Vehicle.** Steering the vehicle is accomplished by pulling back on either the right or left steering lever to brake or slow down the track on the side toward which the turn is to be made. The sharpness of the turn is regulated by how far back the lever is pulled, and the consequent braking action. To make a sharp turn pull the lever as far back as possible on the side toward which the turn is to be made, and hold in this position, until turn is completed. This will cause the braked track to slow down and the outer track to run faster. Pulling a steering lever back intermittently (on-off, on-off) will turn the vehicle in a series of short angles. Engine speed must be maintained while turning. CAUTION: *Do not ride steering levers. When not in use, steering levers must remain forward in fully released position.*

d. **Starting Vehicle on Upgrade.** If vehicle is headed up an incline or hill it will roll backward when the parking brake is released, unless increased engine power is applied to provide sufficient power through the transmission to move the vehicle forward. NOTE: *Do not attempt to hold the vehicle any length of time with this method as it tends to overheat the transmission.*

e. **Descending Steep Grades.** Only when descending particularly steep grades can the transmission be used as a brake to supplement the braking action of the steering brakes. Stop vehicle just over crest of hill or at top of grade. Hold vehicle with one steering brake so it will not roll forward, and shift speed range selector lever into reverse (R). Slowly release steering brake and at the same time increase engine speed as required to check vehicle forward speed to not more than 4 miles per hour. Continue to use steering brakes for braking, in addition to use of engine and transmission, until bottom of descent is reached. At bottom stop vehicle, shift transmission into desired forward speed range, and proceed. WARNING: *Do not make either shift while vehicle is in motion.* Follow instructions stenciled on hull side plates in drivers' compartment.

20. **DRIVING PRECAUTIONS.**

a. **General.** Until the driver (even one experienced on other types of track-laying vehicles) becomes thoroughly familiar with this type of vehicle, every precaution must be taken not to OVERDRIVE, or allow the vehicle to get out of control.

b. **Hard Pavement.** Great care in driving must be observed at all times when operating on hard surfaces or pavement. The vehicle, with its all-steel track, may normally be driven safely at recommended speeds, providing the driver has the proper skill and experience in operating this type of vehicle. Careless handling can result in loss of steering control, and consequent serious injury to personnel and damage to the vehicle.

c. **Ditches and Obstructions.** When crossing a ditch, shell hole, or trench, release the accelerator momentarily until the vehicle reaches the bottom, then bring the engine up to full power as the vehicle starts to climb. Shift down to a lower speed if necessary. When going over an obstruction release the accelerator and allow the vehicle to settle down over the obstruction before applying full power. In going over a high obstruction it may be necessary to use full power at the start, but upon reaching the crest of the obstruction release the accelerator until the vehicle is fully over the top of the obstacle. CAUTION: *Use care not to damage gun (if in-traveling position) and rear fenders when climbing out of ditches or shell holes or when pulling away from obstructions.*

Operation Under Ordinary Conditions

21. STOPPING THE VEHICLE.

a. Braking to a Stop. To stop the vehicle release the accelerator. Engine speed will return to normal idling speed and vehicle momentum will gradually decrease until vehicle is barely moving forward. Pull back both steering levers to stop the vehicle. Leave speed range selector in gear until vehicle stops, and then shift into neutral (N). Quicker stops can be made by releasing the accelerator and, at the same time, pulling both steering levers back together with equal force, or by using an on-off action. Always leave speed range selector in gear until vehicle stops, and then shift into neutral (N). During braking action some steering may be necessary to control direction of vehicle. Set parking brake after speed range selector lever has been shifted into neutral (N).

b. Setting Parking Brake. The parking brake is applied by moving parking brake lever to the center position and then pulling both steering levers back as far as they will go. Then release levers and return them to the full forward position. This sets parking brake.

c. Stopping the Engine. After completing an operation run engine at 500 revolutions per minute for 5 minutes to assure a gradual and uniform cooling of valves and other engine parts. When cooling period has elapsed, stop engine. Push and hold fuel cut-off button in until engine stops. Turn magneto switch off only after engine has stopped. NOTE: *If vehicle is to be stopped for any length of time turn 24-volt master switch off.*

22. TOWING THE VEHICLE.

a. General. The vehicle is equipped with four towing shackles and a steel wire rope towing cable. Two shackles are mounted on the front of the vehicle (fig. 1) and two on the rear (fig. 2). The towing cable is carried coiled in brackets on the rear of the vehicle (fig. 2).

b. Towing Vehicle To Start Engine. Under normal conditions and on level terrain an engine can be started by towing, providing the engine to be started is at normal operating temperature. Connect towing bar or cable. Place a driver in the towed vehicle and set speed range selector lever of towed vehicle in third (3) speed range. Turn magneto switch on to BOTH magnetos. Open throttle to run engine at 1,000 revolutions per minute. Start towing in a straight line and gradually increase vehicle speed up to 18 miles per hour. When this vehicle speed is reached shift into low (1) speed range. Engine should start immediately if in good running condition. NOTE: *A cold engine cannot be started by towing.*

c. Precautions. Maximum speed when towing other than to start engine must not exceed 12 miles per hour. A driver must be in con-

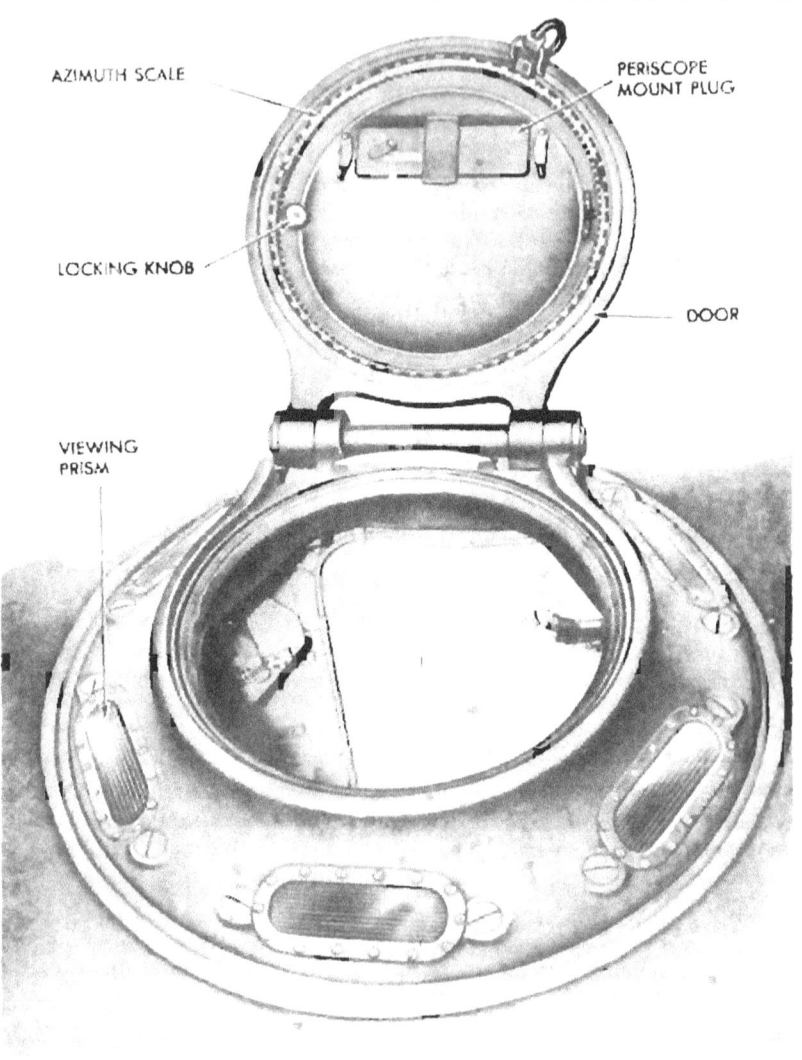

Figure 20 — Commander's Cupola

trol of the towed vehicle. Always place speed range selector lever of towed vehicle in neutral (N) before starting to tow. The towing cable must not be connected by any other means than by the cable eyes. Doubling cable causes sharp bends which break the wire strands, and leaves the cable weak and dangerous to handle. Changes in direction must be made by a series of short turns. Keep vehicles in line, especially before entering and when crossing wet terrain.

Section VIII
TURRET CONTROLS AND OPERATION

23. TURRET.

a. General. The turret is a one-piece casting of armor steel (fig. 3). The exterior has rounded surfaces and corners so that striking projectiles will be deflected with minimum damage. The rear portion of the turret projects beyond the turret race and acts as a counterweight to offset the weight of the 90-mm gun. The interior part of the counterweight is hollow, providing space for radio, submachine gun, ammunition stowage, grenades, spare periscopes and heads, and other items. Brackets for stowing the cal. .50 machine gun are welded on the rear of the turret overhang or projection. Turret illumination is provided by three compartment lights. There is one light over the gunner's station, one over the loader's station, and one behind the commander's station, in front of the radio. Each compartment light is provided with two lamps and a divided red and white lens. Light switch toggles are protected by safety latches to prevent accidental turning on of white lights during blackout or combat conditions. The same switch controls the current to both lamps. Two openings for the installation of radio mast bases are provided in the turret roof. A hooded spotlight mounting is also provided.

b. Turret Race. The turret is supported on the hull by means of a ball bearing race. The lower race ring is attached to the hull, and the upper race ring is attached to the turret. There are three lubrication fittings on the inside of the upper race ring. Lubricate turret race as outlined in paragraph 46.

c. Commander's Cupola. A dome-shaped casting of armor steel, known as the commander's cupola (fig. 20), is set into a circular opening in the top of the turret on the right. Six removable bulletproof glass prisms are equally spaced around the lower edge of the cupola. They permit the vehicle commander to scan any point of the horizon. In the center of the cupola there is a round hinged door, which is provided with an opening for installing a periscope.

TM 9-735
23

Part Two—Operating Instructions

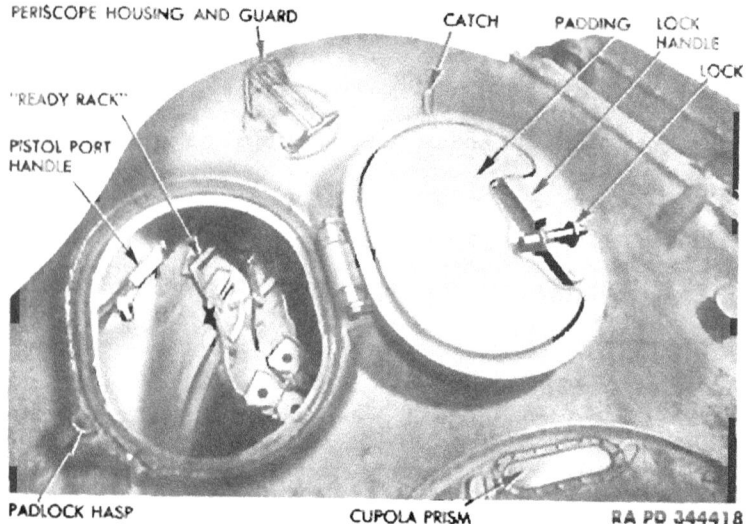

Figure 21 — Loader's Hatch, Open

When the periscope is not in use the opening is sealed with a metal plug. The door race plate rotates 360 degrees on ball bearings, and is equipped with an azimuth scale around the interior edge. A vane sight is bolted to the turret top ahead of the cupola and directly in line with the commander's periscope.

d. **Loader's Hatch.** The loader's hatch is an oval-shaped opening in the top of the turret on the left side, opposite the commander's cupola (figs. 21 and 22). A hinged, spring-controlled, armor steel door protects the opening. The door opens toward the front of the vehicle, and when in the fully opened position lies flat against the turret top. A spring-loaded catch holds the door in the open position. When closed the door is flush with the turret top and is locked from inside the turret. A hinged, cal. .50 machine gun mount is located behind the loader's hatch (fig. 186). When not in use the mount folds down against the turret top and is held down by a leather strap. A muzzle traveling clip, fold-down type, is located to the rear of the cal. .50 machine gun mount (fig. 186). A periscope for the loader's use is located in the turret top forward of the loader's hatch.

e. **Seats.** The gunner's seat, which is provided with a detachable back, can be raised and lowered or adjusted horizontally (fig. 23). The gunner's foot rest is removable. The commander's seat is the pad-type, adjustable in height, and can be folded up out of the way

TM 9-735
23—24

Turret Controls and Operation

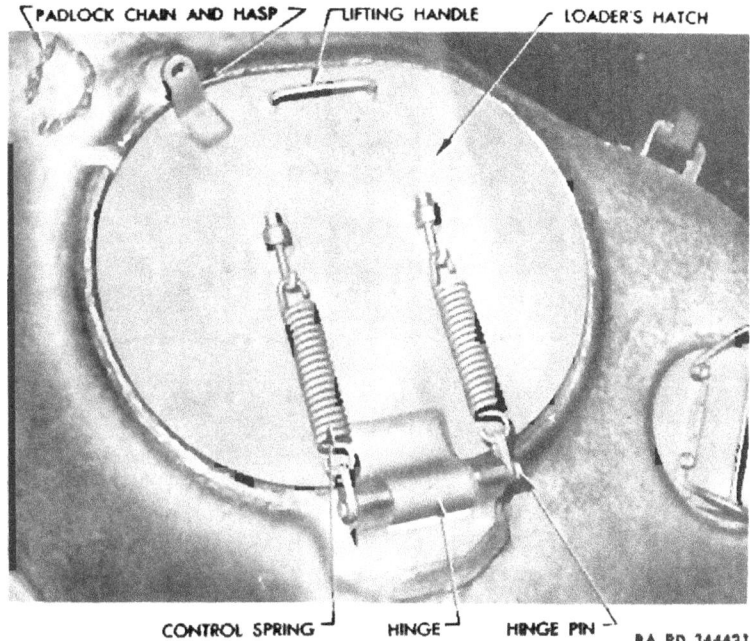

Figure 22 — Loader's Hatch, Closed

when not in use. The loader's seat is the same construction as the commander's seat but is without adjustment and is demountable. When not in use loader's seat is stowed behind the "ready-rounds" rack. All seats are suspended from the turret ring.

24. TURRET TRAVERSING MECHANISM.

a. **Before-operation Service.** Perform the Before-operation Service before traversing the turret (par. 49).

b. **General.** The turret can be traversed a full 360 degrees, either hydraulically or manually, by the gunner. The hydraulic traversing mechanism is also provided with a dual control for the use of the commander. The commander's turret control lever, equipped with safety latch to prevent accidental movement, is mounted on the turret roof ahead of the commander's station (fig. 24). The gunner's hydraulic control power traversing control handle is directly in front of the gunner's seat. It is equipped with electric switches for firing the 90-mm gun and the coaxially mounted cal. .30 machine gun. The turret is traversed manually by a manual traversing handle ahead of

TM 9-735

Part Two—Operating Instructions

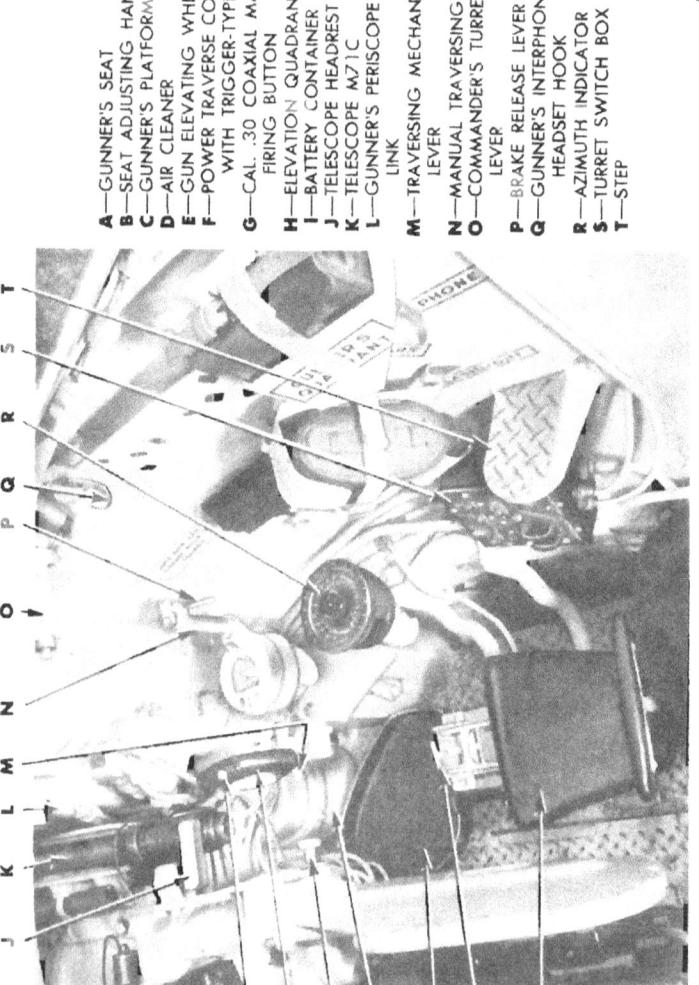

A—GUNNER'S SEAT
B—SEAT ADJUSTING HANDLE
C—GUNNER'S PLATFORM
D—AIR CLEANER
E—GUN ELEVATING WHEEL HANDLE
F—POWER TRAVERSE CONTROL HANDLE WITH TRIGGER-TYPE SWITCH
G—CAL. .30 COAXIAL MACHINE GUN FIRING BUTTON
H—ELEVATION QUADRANT M9
I—BATTERY CONTAINER
J—TELESCOPE HEADREST
K—TELESCOPE M71C
L—GUNNER'S PERISCOPE SYNCHRONIZING LINK
M—TRAVERSING MECHANISM GEARSHIFT LEVER
N—MANUAL TRAVERSING HANDLE
O—COMMANDER'S TURRET TRAVERSING LEVER
P—BRAKE RELEASE LEVER
Q—GUNNER'S INTERPHONE HEADSET HOOK
R—AZIMUTH INDICATOR
S—TURRET SWITCH BOX
T—STEP

Figure 23 — Gunner's Station

TM 9-735
24

Turret Controls and Operation

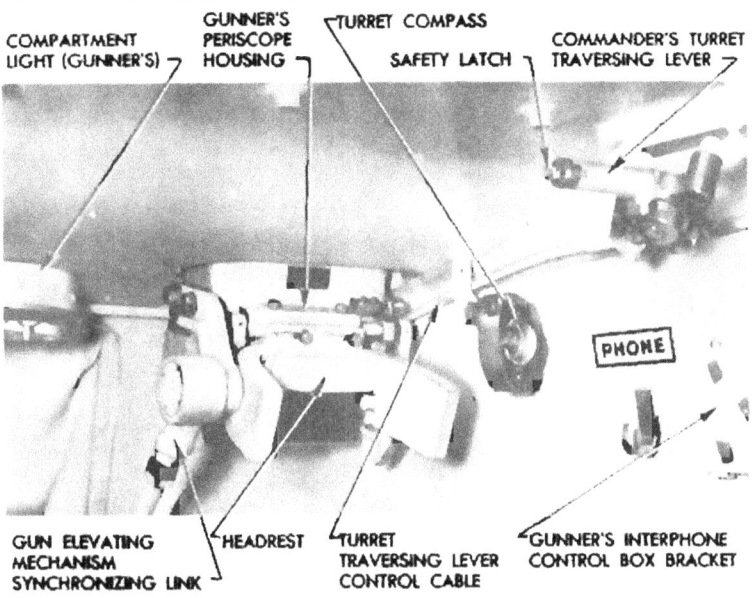

Figure 24 — Gunner's Periscope Housing and Commander's Turret Traversing Lever

and to the gunner's right. CAUTION: *The turret must never be traversed, either manually or mechanically, unless personnel in and on the vehicle are aware of operator's intention. Before traversing the turret close both doors. Also release the gun from the traveling lock and elevate it sufficiently to clear any obstructions.*

c. **Turret Traversing Gear Box and Manual Drive.** The gear box and manual drive assembly are mounted on the turret rim just ahead and to the right of the gunner's station (figs. 23 and 27). The assembly contains a train of gears which drive the traversing pinion which is in constant mesh with the turret ring gear. The turret can be traversed manually by releasing the brake and operating the manual traversing handle at the rear of the gear box. The brake, which is part of the manual traversing mechanism, holds the gears stationary and the turret in position when the shift lever is in manual drive position (step (2) following). Press brake lever in against manual traversing handle to release brake. A gearshift lever is provided at the bottom of the gear box housing for selecting either hydraulic or manual traversing (step (1) following).

d. **Engaging Traversing Gears.** Move gearshift lever (fig. 23) to "DOWN" position to engage the traversing gears for manual

TM 9-735

Part Two—Operating Instructions

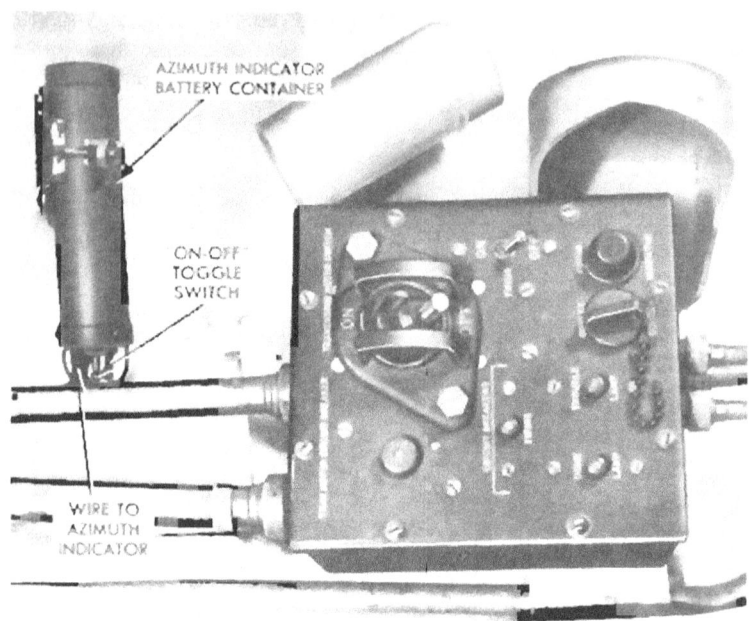

RA PD 344419

Figure 25 — Turret Master Switch Box

traversing. NOTE: *If gearshift lever does not move to full "DOWN" position, it means gears are not lined up for meshing. Turn manual traversing handle slightly while maintaining a downward pressure on gearshift lever. Always leave traversing mechanism in manual drive position, except when traversing turret hydraulically.* The "UP" position is for hydraulic traversing (step (3) following).

e. **Operating Turret Traversing Manual Control and Brake.** Traverse the turret manually by moving the gearshift lever "DOWN" and turning the manual traversing handle. Grasp turret brake release lever and manual traversing handle at the same time and squeeze them together (fig. 23). Brake release lever will move in against manual traversing handle and release brake. Hold lever in against traversing handle and turn handle to the right or left to traverse turret. The brake release lever is released by removing the hand. Spring action applies the turret brake.

f. **Operating Hydraulic Traversing Mechanism.** To operate the hydraulic traversing mechanism first make sure traverse pump

Turret Controls and Operation

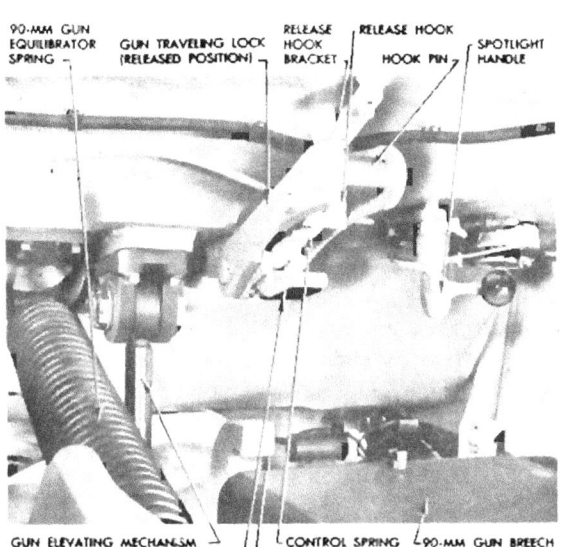

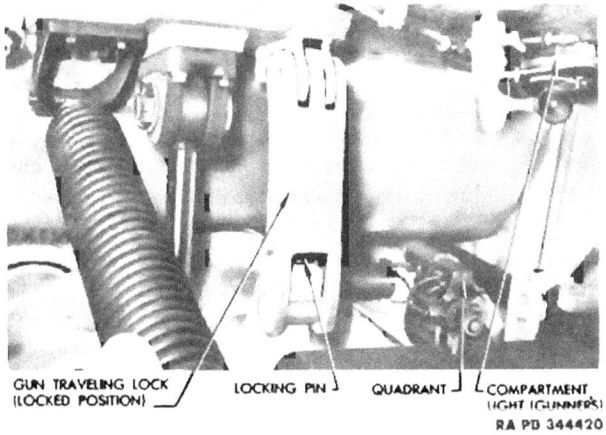

Figure 26 — Gun Elevating Lock and Spotlight Handle

control (pistol-grip) handle is in neutral, or vertical position (fig. 23). Elevate gun to clear drivers' doors and any other obstructions. Unlock turret. Turn 24-volt master switch on. Move gearshift lever on bottom of gear box to "UP" position. Turn on turret traversing motor switch located in turret control box on turret wall to gunner's right (figs. 23 and 25). Grasp turret power traversing control handle (pistol grip). Twist control handle to the right to traverse the turret to the right, or clockwise. Turn control handle to left to traverse the turret to left, or counterclockwise. Degree of turn of control handle regulates speed of traverse. NOTE: *If attempt is made to traverse turret hydraulically without unlocking turret lock, the electric motor, which operates the traversing pump, will stall and the circuit breaker on the control box will open.* If circuit breaker cuts out, turn off traversing motor switch and unlock turret lock. Reset circuit breaker and turn on switch. Operate hydraulic traversing mechanism as described. CAUTION: *Be sure drivers' doors are closed and gun is clear of traveling lock and any other obstructions before traversing turret.*

g. **After-operation.** Certain precautions must be taken before leaving the vehicle. CAUTION: *Turret lock must be in "LOCK" position.* Turret traversing gearshift lever must be in "DOWN" or manual position. Traversing motor switch must be off. Unless otherwise ordered, gun must be placed in traveling lock and also secured with gun elevating lock in turret (fig. 26). Turn off 24-volt and 12-volt master switches. WARNING: *Be sure gun breech is clear.*

25. TURRET LOCK.

a. **Description.** The turret lock is located to the right and slightly behind the gunner's seat (fig. 186). The pawl of the lock engages with the teeth of the turret ring gear, and holds the turret stationary when the turret lock is in locked position. CAUTION: *The turret lock must be in locked position whenever the vehicle is in motion unless the turret is to be traversed or the 90-mm gun is to be fired.* If it is desired to leave turret unlocked while changing position on uneven ground, place traversing shift lever in manual position to act as a brake or lock. Otherwise, turret will swing and cause serious damage to equipment and injury to personnel. WARNING: *Do not lock turret when firing gun.*

b. **Operating Turret Lock.** To unlock turret, grasp handle of turret lock, turn and pull straight out. Allow spring action to turn handle one-quarter turn to the left. Lock will automatically seat in "FREE" position. To lock turret, turn handle one-quarter turn to

TM 9-735

Turret Controls and Operation

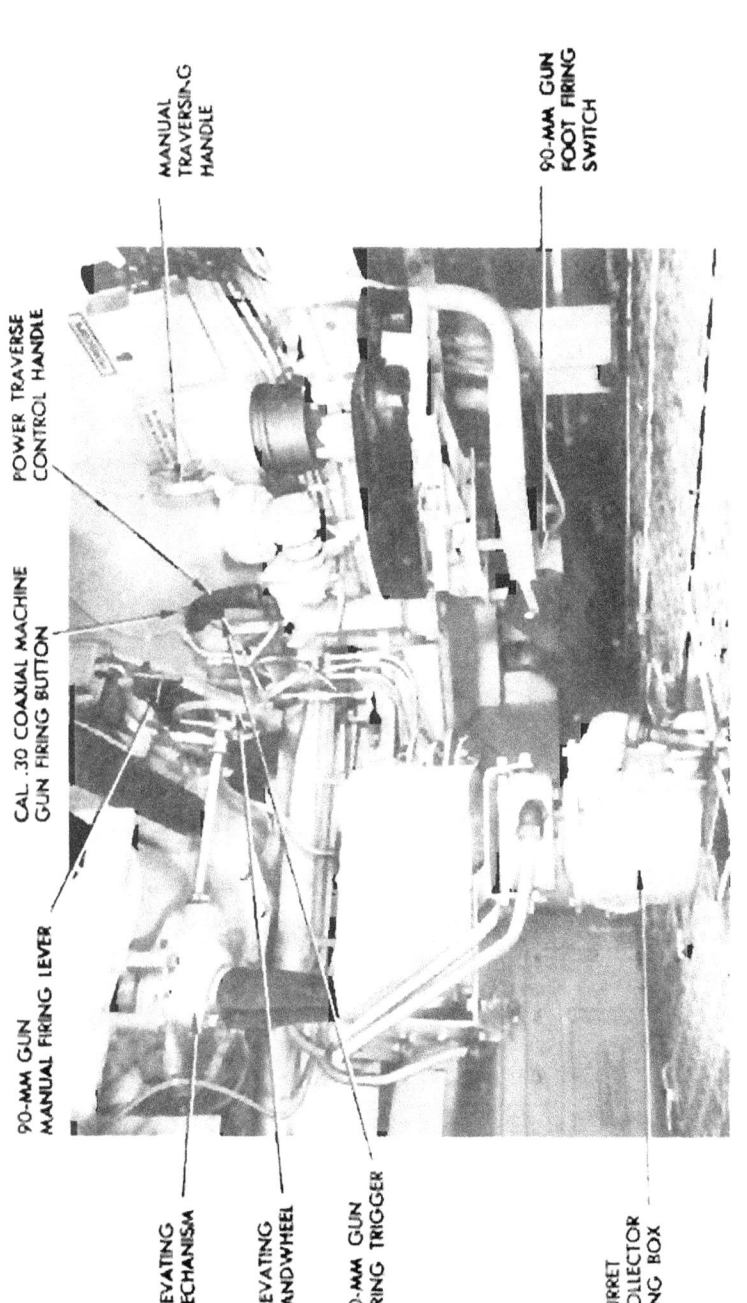

Figure 27 — Turret Controls, Firing, and Elevating Mechanism

TM 9-735
25-27

Part Two—Operating Instructions

Figure 28 — 90-MM Shell "Ready-Rack"

right and lock will snap into "LOCK" position. Seat lock by turning handle further to right, until positive stop is met. CAUTION: *If teeth of lock do not fully engage with teeth of turret ring gear, traverse turret slightly with the manual traversing handle until lock handle seats in lock housing indicating teeth are in mesh.*

26. ARMAMENT.

a. For complete details on armament see Part Four, Auxiliary Equipment.

27. PISTOL PORT.

a. Description. The vehicle is equipped with a round pistol port on the left side of the turret. Pull out on operating handle to release port door and swing door outward with handle. Lock handle to hold door in open position. Reverse procedure to close and lock pistol port door.

TM 9-735
28–29

Operation of Auxiliary Equipment

Section IX

OPERATION OF AUXILIARY EQUIPMENT

28. BEFORE-OPERATION SERVICE.

a. Before using or operating any of the auxiliary equipment units, perform the Before-operation Service where necessary or required (par. 49).

29. AUXILIARY ENGINE.

a. **General.** The auxiliary engine is located at the rear of the right fuel tank (fig. 56). The auxiliary engine control box is mounted on the engine compartment bulkhead above the bulkhead opening (fig. 173). The control box contains the auxiliary engine starter button, oil pressure gage, and ignition switch. Always check auxiliary engine oil level when checking main engine oil level.

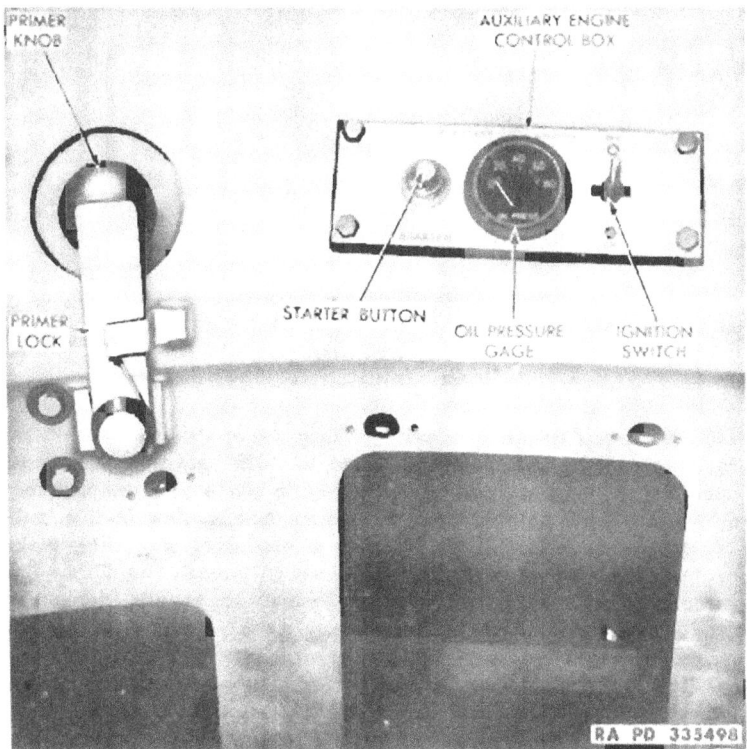

Figure 29 — Auxiliary Engine Control Box

TM 9-735
29

Part Two—Operating Instructions

RA PD 344474

Figure 30 — Hand Cranking Auxiliary Engine

b. **Starting Auxiliary Engine.** Turn 24-volt master switch on. Push toggle-type ignition switch down to "ON" position, and press starter button in to operate auxiliary engine starter. These switches are located in the auxiliary engine control box as described in subparagraph a preceding. After engine is operating smoothly, make sure the oil pressure gage indicates at least 15 pounds (fig. 29). The operating pressure range is between 15 and 35 pounds, depending upon temperature, length of time since last oil change, and engine wear. Be sure ammeter indicates generator is charging properly. CAUTION: *24-volt master switch must be on when auxiliary engine is operating, or it may cause damage to generator regulator unit.* Hand crank auxiliary engine (subpar. d following) if battery lacks power to operate cranking motor.

c. **Stopping Auxiliary Engine.** Push toggle-type ignition switch up to "OFF" position. If vehicle is not to be operated, or if there is

TM 9-735
29

Operation of Auxiliary Equipment

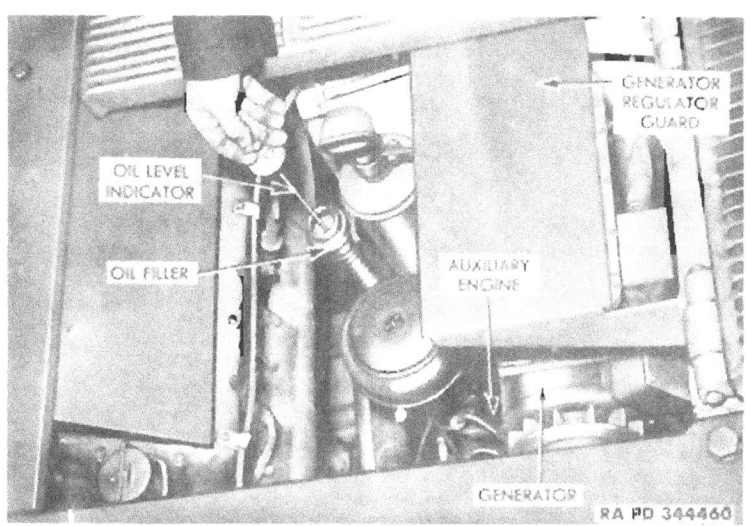

Figure 31 — Measuring Auxiliary Engine Oil Level

no immediate need of electrical current, turn 24-volt master switch off as soon as auxiliary engine stops.

d. **Hand Cranking Auxiliary Engine.** If necessary, the auxiliary engine may be started with the hand-cranking mechanism (fig. 30). Turn 24-volt master switch on. Turn ignition switch on (subpar. b preceding). Open right front intake door of engine compartment. Reach down and grasp hand cranking handle (fig. 30). Pull handle upward with quick movement and let go. Let spring action return handle to original position for another upward pull if engine fails to start on first attempt. Repeat pulls on handle as required to start engine, assuming engine to be in good operating condition. When engine starts, return handle to original position and close and secure intake door.

(1) AUXILIARY ENGINE EMERGENCY STARTING. If batteries are entirely discharged, the auxiliary engine fuel shut-off valve will not open electrically. This will be indicated if the auxiliary engine stops after being hand-cranked, after the gasoline in carburetor has been used. If this occurs, remove bulkhead opening cover and reach through opening to auxiliary engine fuel shut-off valve. Block valve handle up in fully raised position. Hold carburetor valve open, and hand-crank engine. Keep carburetor valve open until engine fires. Leave block under valve while engine is running. Remove block after

71

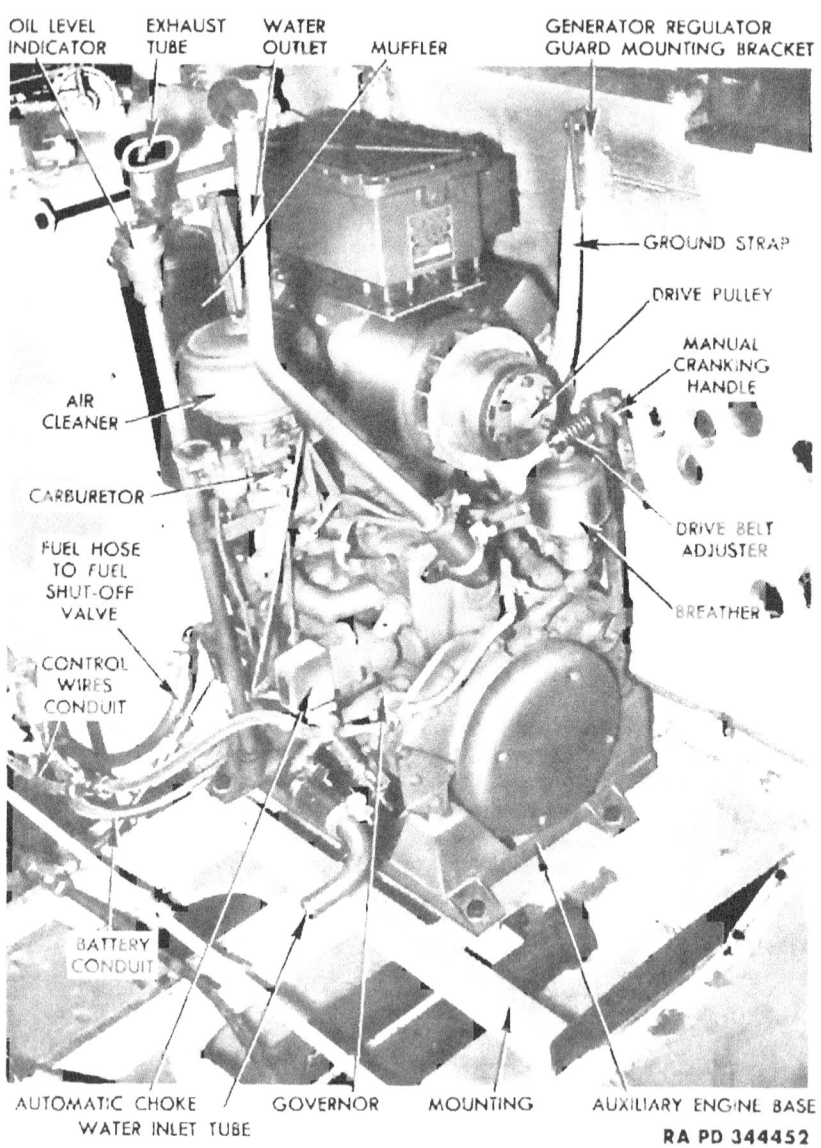

Figure 32 — Auxiliary Engine and Generator

TM 9-735
29-31

Operation of Auxiliary Equipment

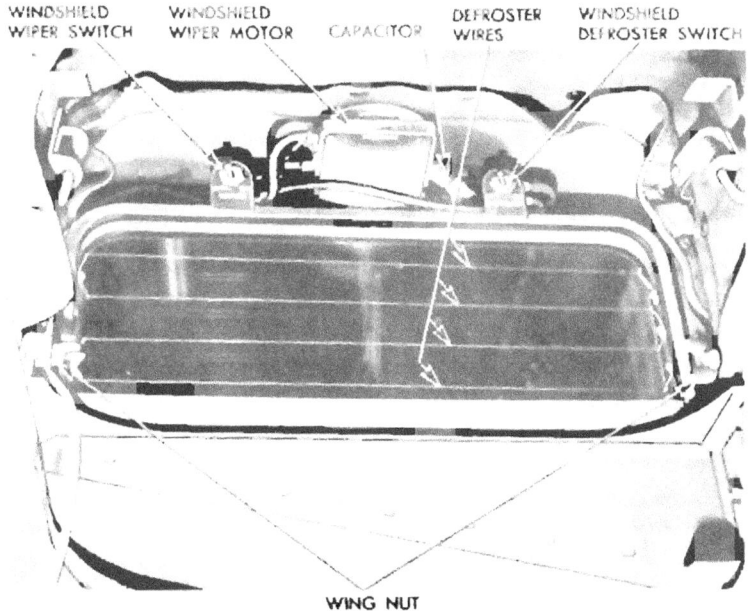

Figure 33 — Driver's Hatch Hood Control

auxiliary engine has been stopped. NOTE: *Use this method only in an emergency. In all other cases remove or recharge batteries.*

30. COMPASS.

a. General. There is one compass in the vehicle, located in the turret at the gunner's station (fig. 24). It is an automotive type and has no operating mechanism.

31. DRIVERS' HATCH HOODS.

a. General. Driving hatch hoods are provided for both the driver and assistant driver for use when operating the vehicle in rain or snow with the driver's hatches open. Hoods are equipped with an electrically operated windshield wiper and an electric defroster. Power to operate both the wiper and the defroster is obtained by plugging the extension wire into a utility outlet socket on the instrument panel. Separately mounted toggle switches are provided for the operation of the windshield wiper and defroster and are located just above the hood windshield (fig. 33).

b. Mounting Driver's Hatch Hood. Remove driver's hatch hood from bag carried in stowage rack on right side of turret (fig. 1). Open driver's door and hook securely. Place left-hand hood over driver's door opening. Insert adjustable side clip beneath torsion bar spring on driver's door. Hook toggle latch to projecting clip on the opposite side of door opening and fasten securely. Place wing bolt in slotted opening in front crash pad and tighten finger-tight (fig. 33). From inside vehicle pull hinged canvas hood top down to cover door opening snugly. Plug extension wire into an outlet socket on the instrument panel. Turn 24-volt master switch on. Test operation of windshield wiper and defroster by operating individual switches controlling them. Follow same procedure to mount right-hand assistant driver's hood.

c. *Operating Windshield Wiper.* See that the wiper blade is on the glass windshield, and push wiper switch to "ON" position (fig. 33).

d. *Operating Windshield Defroster.* Push defroster switch to "ON" position (fig. 33). The high resistance wires will heat the windshield glass and keep it free of snow and ice.

e. *Stowing Drivers' Hatch Hoods.* Remove hood wire from instrument panel. Fold canvas hood top up against windshield frame. Release side latch, loosen wing nut, and pull bolt from crash pad slot. Remove the hood(s), place in bag(s) and stow in rack on right side of turret.

32. FIRE EXTINGUISHERS.

a. *General.* The fixed fire-extinguisher system is for smothering fires in the engine compartment with carbon dioxide gas (figs. 38 and 39). The system consists of two 10-pound carbon-dioxide (CO_2) cylinders mounted side by side at the base of the front turret ring support, control head, exterior pull, copper tubing, and discharge nozzles in the engine compartment. Both cylinders are discharged simultaneously when the external remote control handle on the outside of the vehicle is pulled, or by operating the lever on the control head of the right-hand cylinder. For full description of fire-extinguisher system refer to paragraph 203 a. The maintenance of the system is described in paragraphs 203 and 204.

b. *Operation of Fixed Fire Extinguishers.* If a fire should occur in the engine compartment, pull out the remote control handle and let go, if outside the vehicle. If inside, operate the lever on the right-hand cylinder control head. Carbon dioxide will immediately be discharged from the discharge nozzles in the engine compartment (fig. 56) to smother the fire. CAUTION: *When replacing cylinders, handle with great care as they are as dangerous as a loaded shell.*

c. *Operation of Portable Fire Extinguishers.* The vehicle is equipped with two portable 4-pound fire extinguishers for fires which may occur in the drivers' or fighting compartments. One extinguisher

TM 9-735
32—33

Operation of Auxiliary Equipment

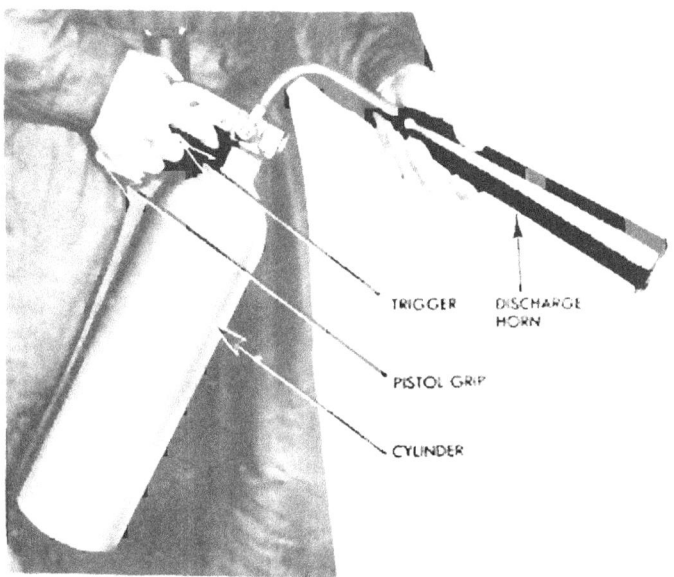

RA PD 36292

Figure 34.— Portable Fire Extinguisher

is mounted behind the commander's station, and one to the right of the driver. To operate, remove the portable extinguisher from its bracket, direct the discharge cone toward the base of the fire and pull the trigger (fig. 34). Move the cone slowly from side to side to cover burning area. Release the trigger to shut off the extinguisher discharge. CAUTION: *Handle extinguishers with great care. Do not jar, bump or drop.*

33. HEATER.

a. General. The hot water heater in the driver's compartment is controlled by the toggle-type heater switch on the instrument panel (fig. 16), and by the water shut-off valve in the tube on the outlet side of the heater (fig. 35). The heater fan motor circuit is protected by a circuit breaker (AUXILIARY) located on the instrument panel.

b. Operation of Heater. Turn 24-volt master switch on. Open heater shut-off valve. Push heater switch toggle to "ON" position.

TM 9-735
34-35

Part Two—Operating Instructions

Figure 35 — Heater

34. INSPECTION LIGHT.

a. **General.** The inspection light is for use in or around the vehicle where light is necessary. It is equipped with movable shutters to control the amount of illumination, and also with a 15-foot wire extension. The light can be plugged in the outlet sockets on the instrument panel, turret switch box, and the terminal box on top of the the left fuel tank in the engine compartment.

35. SPOTLIGHT.

a. **General.** The spotlight (fig. 7) can be controlled from either inside or outside of the vehicle. It is also used as a signal light. Normally, the spotlight is operated from inside the turret because in action sighting is done through the periscopes. When used in conjunction with the 90-mm gun the light beam is directed parallel with

TM 9-735
35–36

Operation of Auxiliary Equipment

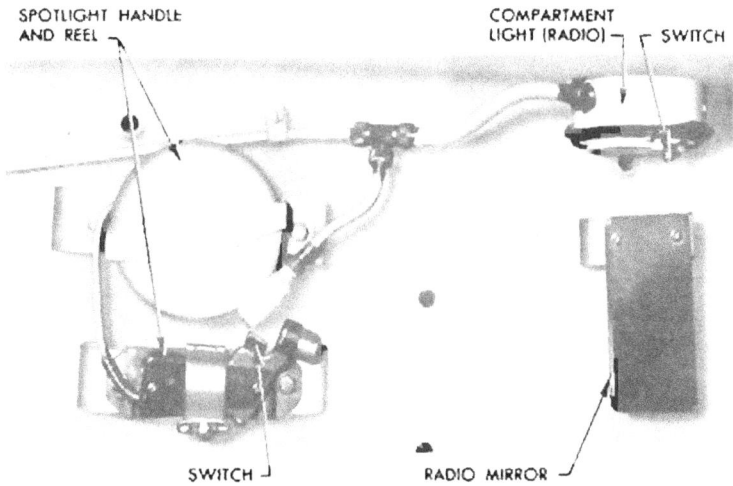

Figure 36 — Spotlight Signal Handle and Reel

the axis of the gun. When the light is on the target it is illuminated, as the gun is sighted. The spotlight may also be operated from inside or outside the turret as a signal light. When using the light as a signal light the spotlight is demounted and attached to the pistol-grip type handle (fig. 36) (subpar. c following).

b. **Operation as a Spotlight.** Rotate ball handle (fig. 26) to move spotlight head up or down. Swing ball handle from side to side to move spotlight head in a horizontal plane. To lock spotlight in line of fire, tighten wing bolt when light parallels axis of 90-mm gun. If necessary, tighten wing bolt to increase friction on rotation of the spotlight head in a horizontal plane. To operate spotlight from outside vehicle, grasp the knob on the back of the light head and aim as desired.

c. **Operation as a Signal Light.** Dismount spotlight head and attach pistol-grip type handle to spotlight base (par. 206 c (d) (2)). Aim light by using sights on top of shield. Press trigger switch (fig. 36) to operate light. When pressure on trigger switch is released light goes off. A flash simulates a dot and a longer period a dash for use in dot-dash code signals.

36. VENTILATOR BLOWER.

a. **General.** The ventilator blower (figs. 12, 18, and 19) is located in the drivers' compartment directly forward of the turret ring sup-

77

TM 9-735
36-37

Part Two—Operating Instructions

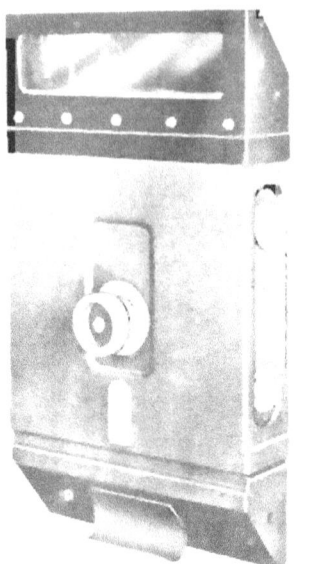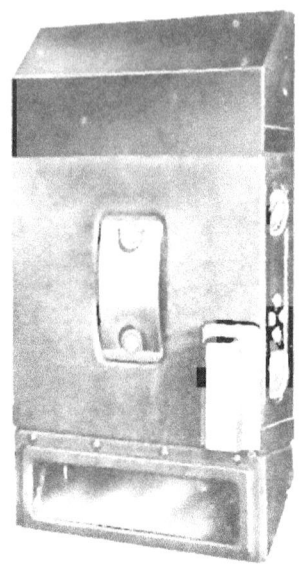

RA PD 302571

Figure 37 — Periscope

port. It is attached to the compartment roof and is controlled by a switch located on the battery master switch box (fig. 39). The blower operates at constant speed and the air output is controlled by a damper type valve mounted in the end of the discharge outlet. The blower motor circuit is protected by a circuit breaker located in the master switch box. For full description of ventilator, refer to paragraph 206 d (1).

b. **Operation of Blower.** Turn 24-volt master switch on. Push blower toggle switch on master switch box to "ON" position. Regulate and direct air output by setting outlet and damper in desired position. To stop blower push switch toggle to "OFF" position.

37. VISION EQUIPMENT.

a. **General.** For full description of the vision equipment, refer to paragraph 206 e (1). Periscope mounts are equipped with forehead rests. Place forehead against rest and turn periscope by moving head. Tightening lock screw secures periscope mount in a stationary

TM 9-735
37

Operation of Auxiliary Equipment

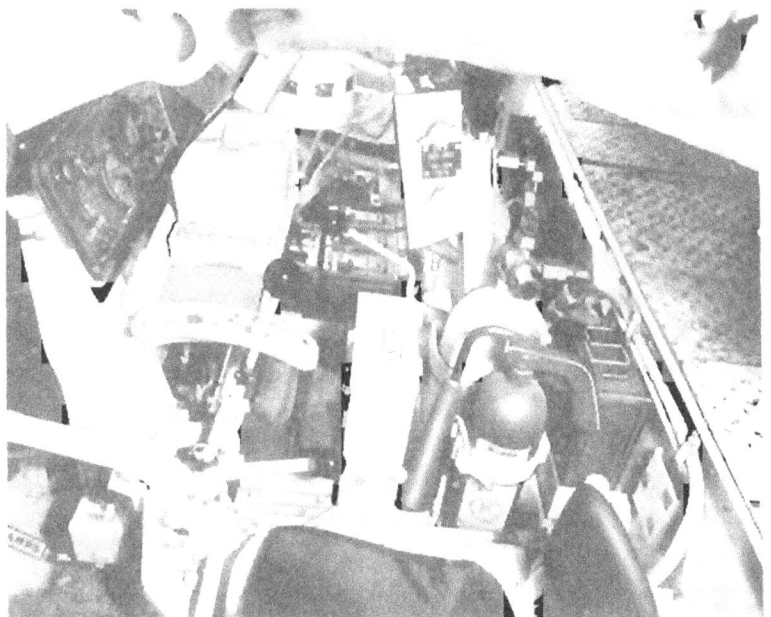

RA PD 344392

Figure 38 — Driver's Compartment Stowage, Viewed from Driver's Station

position. The gunner's telescopic periscope is linked to the 90-mm gun. The position of this periscope is always in constant relationship with center line of gun.

b. **Use of Periscopes.** To normally operate periscope (fig. 37), grasp sides of periscope mount and rotate, elevate, or depress as necessary to cover the desired field of vision. In extreme conditions, periscope can be moved by placing forehead against periscope protective pad and moving periscope as desired.

c. **Installing Periscope.** Back off knurled nut on back of periscope. Open locking latch. Insert periscope in mount and push up until knurled nut seats against mount. Close locking latch and securely tighten knurled nut.

d. **Periscope Wiper.** The rubber edge on the periscope mount can be used as a periscope lens wiper. Open locking latch, release knurled nut, and move periscope up and down in mount far enough so rubber edge will move fully across periscope glass and wipe it clean. When clean, reseat periscope, tighten knurled nut, and close locking latch.

TM 9-735
38

Part Two—Operating Instructions

Figure 39 — Driver's Compartment Stowage, Viewed from Assistant Driver's Station

Section X

OPERATION UNDER UNUSUAL CONDITIONS

38. OPERATION AT TEMPERATURES BELOW 0° F.

a. **General.** Low temperatures affect the flow and lubricating qualities of oils and greases, cause batteries to lose power, water to freeze, and metal to become brittle. Therefore, operation and maintenance of the vehicle at low temperatures must be modified. Extra precautions must be taken to minimize effects of low temperature which can cause damage to the vehicle or its parts, or failure of units to function properly. Refer to FM 31-15, Operation in Snow and Extreme Cold, for suggestions and instructions which can be applied to the operation and maintenance of the vehicle at temperatures from 0° F to -40° F.

b. **Fuel System Instructions.** Special precautions or instructions to maintain the efficiency of the fuel system when operating at low temperature are given in the following steps:

(1) GRADE OF FUEL. To reduce cold weather starting difficulties and assure maximum power, use winter grade gasoline, U.S. Army Specifications 2-103 (Grade C), latest issue.

(2) STORAGE OF FUEL. Condensation of moisture will cause

water to accumulate in tanks and drums. To reduce condensation keep tanks and drums as full as possible. Before filling containers with fuel make sure they are thoroughly clean and free from rust, or traces of other liquids. Covers, caps, and gaskets must be kept tight to prevent entrance of dirt, snow, or moisture.

(3) HANDLING OF FUEL. If possible, allow a container which has been filled, or moved, to stand long enough to permit any dirt in the fuel to settle before filling vehicle fuel tanks. Do not completely empty containers when presence of water or dirt in fuel is detected. Before pouring fuel from a container into the vehicle fuel tanks thoroughly wipe all dirt, snow, or ice from around filler opening and from dispensing can or hoze nozzle. CAUTION: *Static electricity builds up to greater potentials in cold weather. Therefore, maintain a positive metallic contact between container or hose nozzle and vehicle at all times when filling the fuel tanks, to ground the discharge. This will prevent a static spark which might cause a fire.* Strain fuel through filter paper, chamois skin, or other suitable material to prevent water in fuel entering fuel tanks. In cold weather pour 1 quart of grade 3 denatured alcohol into each fuel tank at the beginning of the cold weather, and 1 pint per month thereafter. This prevents ice forming in filler, tubes, or carburetors. NOTE: *Always keep vehicle fuel tanks full to reduce condensation.*

(4) PREPARATION OF FUEL SYSTEM. Keep fuel tank filters and auxiliary engine fuel pump sediment bowl clean to permit fuel to flow freely, and to eliminate ice formations. Make sure proper fuel pump vacuum and pressure is maintained to deliver adequate volume of fuel at low cranking speed. Keep carburetors adjusted to provide sufficient fuel for quick starting and smooth idling. Air cleaners must be kept clean and filled with proper grade of oil for prevailing temperature (par. 46). From 0° F to –40° F use hydraulic oil in air cleaners. WARNING: *Do not run engine with either or both air cleaner elements removed.*

c. **Engine Lubrication System Instructions.** If heated storage, winterization kit or heater is available or oil can be drained after operation, use SAE 30 engine oil undiluted. Otherwise, dilute as follows: Add SAE 30 engine oil to "FULL" mark. Then add 9½ quarts of gasoline. Run engine 5 to 10 minutes to thoroughly mix oil and gasoline. Stop engine and wait at least 3 minutes for oil to settle. Recheck oil level. Make scribe mark on dipstick to indicate oil level plus diluent. Check oil level frequently. Add SAE 30 engine oil as required to maintain level at "FULL" mark on bayonet gage. If vehicle is operated 4 hours or more, and then left standing unprotected for 5 hours or more, redilute oil. Add SAE 30 engine oil to "FULL" mark on dipstick, and then add gasoline to bring oil level up to dilution mark.

TM 9-735

Part Two—Operating Instructions

d. Transmission, Differential, and Final Drive Cold Weather Lubrication. Drain transmission every 1,000 miles and refill with SAE 10 engine oil for ALL temperatures. At temperatures below zero, after engine has warmed up, drive vehicle VERY SLOWLY for at least 100 yards to warm up transmission oil. Drain differential and final drives every 1,000 miles and refill with SAE 50 oil when operating at temperatures above 32° F. Use SAE 30 oil when operating at ALL temperatures below 32° F. Check all oil levels at regular intervals. WARNING: *Do not dilute oil in transmission, differential, or final drives.*

e. General Cold Weather Lubrication Instructions. Since various units require special attention when operating at temperatures from 0° F to −40° F, recommendations and suggestions for servicing these units are given in the following steps:

(1) TRACK SUSPENSION SYSTEM LUBRICATION. If general purpose grease No. 0 has been used for 1,000 miles in the road wheels, support rollers, and road wheel arm supports, no special precautions are necessary. If any of these parts are lubricated with general purpose grease No. 1, bearings must be removed, cleaned in dry-cleaning solvent, repacked with general purpose grease No. 0, and installed. In the case of track support rollers entire assembly must be removed, and an assembly lubricated with general purpose grease No. 0 installed.

(2) OTHER POINTS LUBRICATED WITH GENERAL PURPOSE GREASE. Lubricate all points requiring general purpose grease No. 0, for temperatures from 32° F to 0° F, with this same grease at temperatures below 0° F, except track suspension system points as noted in step (1) preceding.

(3) OILCAN POINTS. Lubricate oilcan points requiring engine oil at temperatures above 0° F with a light preservative lubricating oil at temperatures below 0° F.

(4) AUXILIARY ENGINE CRANKCASE. Dilute auxiliary engine lubricating oil with gasoline at temperatures below 0° F. Drain oil pan (par. 202 b (2)) and refill with 3 quarts of SAE 10 engine oil. Add 1 quart of gasoline and operate engine 5 minutes to mix diluent. Stop engine and allow at least 3 minutes for oil to settle. Check level and make scribe mark on bayonet gage at level of oil with diluent added. Recheck diluted oil level frequently when engine is operated. Add SAE 10 engine oil to maintain level at "FULL" mark on bayonet gage. If engine is operated 4 hours, it will be necessary to again dilute the oil. Add SAE 10 engine oil to raise level to "FULL" mark, and then add gasoline to raise level to diluted oil mark previously made on bayonet gage.

(5) HYDRAULIC TRAVERSING MECHANISM SYSTEM. When operating at temperatures consistently below 0° F, drain system (par.

Operation Under Unusual Conditions

196 b (2)) and refill with special recoil oil as per Lubrication Order (par. 46).

f. **Cooling System Cold Weather Instructions.** The proper preparation and care of the cooling system to prevent freezing at temperatures below 32° F, as given in the following steps, is of prime importance for trouble-free operation of the vehicle.

(1) PREPARATION OF COOLING SYSTEM FOR COLD WEATHER OPERATION. Thoroughly clean the cooling system of all rust, scale, or oil and eliminate all leaks before using antifreeze compound. Clean cooling system (par. 134 f) unless it has recently been cleaned and needs only to be flushed (par. 134 e). Carefully inspect all hoses, tubes, gasketed joints, and radiator cores for leaks. Remove all parts which are leaking and any hoses that have deteriorated and might leak, and install new parts or notify higher authority. Replace leaking radiators. All air or exhaust gas leaks into the cooling system must be eliminated or higher authority notified.

(2) INSTALLATION OF ANTIFREEZE COMPOUND. Use only ethylene-glycol type antifreeze compound, U.S. Army Specification AXS-864, in sufficient quantity to protect the cooling system against freezing at a temperature at least 10° F *below the lowest anticipated temperature.* Refer to the antifreeze compound chart following to determine the quantity of compound required.

ANTIFREEZE COMPOUND CHART
(For 22-gallon Capacity Cooling System)

To Prevent Freezing at Temperatures of:	To Make 1 Gallon of Solution	Quantity of Antifreeze Compound (ethylene-glycol type) Required — When Filling Cooling System
10° F	2 pt	22 qt
0° F	2½ pt	28 qt
–10° F	3 pt	33 qt
–20° F	3½ pt	39 qt
–30° F	4 pt	44 qt
–40° F	4½ pt	50 qt
–50° F	5 pt	55 qt

Under no circumstances use a solution consisting of more than 55 quarts of antifreeze and 33 quarts of water, as a more concentrated solution will merely raise the freezing point and decrease the protection. When the required quantity of antifreeze compound has been determined, mix thoroughly with water before pouring into the system. This will prevent concentrated quantities of compound being trapped in the cooling system. Fill system with solution until visible in bottom of surge tank. Start and run engine until normal operating temperature is reached to further thoroughly mix solution. Inspect cooling system for leaks while engine is operating. Stop engine and test degree of protection of solution (step (3) following).

(3) TEST ANTIFREEZE SOLUTION. When operating at temperatures below 32° F test and record the degree of protection of the antifreeze solution at regular intervals, or more frequently as required. Test only when solution is warm, using antifreeze solution hydrometer (18-H-940) in accordance with instructions furnished with the instrument. If the accuracy of the hydrometer is questioned, test it with a sample solution of one part antifreeze compound and two parts water; if correct, the hydrometer will show a protection reading of 0° F. If it is necessary to add a considerable amount of antifreeze compound, drain a somewhat greater amount of solution from the cooling system into clean containers. Add the required amount of antifreeze compound as determined in step (4) following, and fill system with drained solution until it is visible in bottom of surge tank. Start and warm up engine, and recheck solution.

(4) ADDING ANTIFREEZE COMPOUND. When some of the antifreeze solution has been lost the number of pints of antifreeze compound required to make 1 gallon of solution to protect against freezing at various temperatures can be determined by referring to the antifreeze compound chart in step (2) above. To increase the protection of the antifreeze solution refer to the desired antifreeze protection chart following.

DESIRED ANTIFREEZE PROTECTION CHART

Present Solution Protects Against Freezing At:	Quarts of Antifreeze Compound to Add To Protect Cooling System Against Freezing At:					
	+10 F	0 F	−10° F	−20° F	−30° F	−40° F
20° F	11	17	25	30	33	38
+10° F	—	11	17	22	28	33
0° F	—	—	8	14	19	25
−10° F	—	—	—	8	14	19
−20° F	—	—	—	—	8	11
−30° F	—	—	—	—	—	8

For example, if the hydrometer reading indicates the system is not protected down to 0° F (left hand column) and it is desired to increase the protection to −20° F (fourth column), 14 quarts of antifreeze compound must be added. These figures are approximate, and the solution must be checked with an accurate hydrometer after the indicated quantity has been added.

g. **Batteries and Electrical Systems Instructions.** Special precautions and instructions for preparing and servicing batteries, electri-

Operation Under Unusual Conditions

cal units, and systems for operation at low temperatures are given in the following steps:

(1) BATTERY CARE AT LOW TEMPERATURES. Low temperatures decrease battery power and efficiency and increase the danger of damage from freezing. Therefore, it is most important batteries be kept nearly fully charged with a corrected hydrometer reading from 1.300 to 1.275 to prevent damage, and to supply the additional power needed to crank the engine in cold weather. Do not attempt to crank engine if batteries have been chilled sufficiently to lower the temperature of the electrolyte to below −30° F. The batteries must first be warmed by operating the auxiliary engine, or assisted by connecting a warmed and fully charged "slave" battery to the battery charging receptacle on the master switch box (fig. 12). As indicated in the following chart, a battery which shows a fully charged hydrometer reading at 80° F is capable of delivering only 15 percent of its power at −30° F as shown in battery performance chart. This same percentage will apply if the battery is only partially charged. It is particularly important to adjust hydrometer readings (par. 92 c) during cold weather in accordance with the temperature correction chart shown in figure 77 to obtain a true indication of the relative charged condition of the battery and the temperature at which the electrolyte will freeze. A fully discharged battery will freeze and rupture at 5° F.

Temperature of Battery Electrolyte	Per Cent of Charge Output	Hydrometer Temperature Correction	Approximate Minimum Safe Corrected Reading to Prevent Freezing
0° F	43	−.032	1.165
−10° F	35	−.036	1.188
−20° F	25	−.040	1.205
−30° F	15	−.044	1.219
−40° F	10	−.048	1.229

The level of the electrolyte must be maintained ⅜ inch above the tops of the cell plates. At subzero temperatures add water only just before operating the vehicle to prevent the water from freezing before it is completely mixed with the electrolyte solution. Make sure that all battery cable and ground strap connections are clean and tight.

(2) GENERATOR AND STARTER INSTRUCTIONS. Since low temperature operation imposes greater loads on both the generator and starter, it is most important that brushes seat evenly and commutator be kept clean at all times. The generator regulator or generator should be replaced at the first indication that batteries are not receiving a normal or sufficient charge. Keep generator and auxiliary engine drive belts properly adjusted.

Part Two—Operating Instructions

h. Vehicle Cold Weather Operating Instructions. Special instructions and precautions to be taken when operating the vehicle and its units at low temperatures are given in the following steps:

(1) STARTING ENGINE IN COLD WEATHER. Refer to paragraph 17 d for instructions on starting engine in cold weather. Avoid racing or running engine at speeds above 1,000 revolutions per minute until temperature has reached 100° F. When a cold engine has been started it should not be stopped until the lubricating oil has been heated sufficiently to vaporize any water that might have collected in the oil.

(2) DRIVING VEHICLE IN COLD WEATHER. Do not drive vehicle until engine has warmed up to 100° F and transmission warning signals go out. Then operate transmission in first speed range for the first 5 minutes to further warm the oil. Avoid overloading engine, transmission, and differential or operating vehicle at high speed until all units are warmed up. If tracks are frozen to the ground break them loose with crowbar or mattock, rather than by breaking them loose with the vehicle power, thus throwing a shock load on power train and suspension system. At halts, free suspension system of ice.

(3) AFTER-OPERATION CARE OF VEHICLE IN COLD WEATHER. When possible store vehicle in a heated enclosure, or in a protected area, and cover it with the paulin. If engine oil has not been diluted, drain oil into clean containers and store in warm place. Tag battery master switch to indicate engine oil pan has been drained. If heated storage is not available, before pouring oil into engine heat oil to a temperature which will permit the bare hand to be submerged without burning.

39. OPERATION UNDER DUSTY CONDITIONS.

a. **General.** When operating under dusty or sandy conditions special precautions must be taken to prevent excessive wear and damage to the moving parts of the power unit and suspension system.

b. **Servicing Power Unit Under Dusty Conditions.** The instructions and precautions for servicing the power unit when operating under dusty conditions are given in the following steps:

(1) CARE OF AIR CLEANERS AND BREATHERS. Under extremely dusty conditions the air cleaner oil reservoirs must be cleaned every 2 to 4 hours, or more frequently as required; or whenever the trapped dirt exceeds ½ inch in depth. Air cleaner elements must be cleaned when inspection reveals that any appreciable quantity of dirt has accumulated to restrict the free flow of air, or whenever their capacity to trap the dust has been reached. Continued operation of the engine with dirty or saturated air cleaners will cause damage to the engine which will continue and increase long after the air cleaners have been

cleaned. The crankcase breather, auxiliary engine crankcase breather, transmission, differential, and final drive breathers must be cleaned more frequently as required when operating under dusty conditions, to prevent premature wear and damage to these units. Clean the engine oil filter more frequently as required to maintain its efficiency. Carefully examine all lubricating oil level indicators for evidence of discoloration, or gritty substance, that would indicate oil has become contaminated and must be changed.

(2) CARE OF COOLING SYSTEM. Inspect radiators and oil coolers frequently to make sure air passages are not restricted by an accumulation of dirt. Clean cores by flushing with water under pressure, or blow out with compressed air.

c. Care of Track Suspension System Under Dusty Conditions. Lubricate track suspension system more frequently to cleanse bearings of any sand or dirt that may have worked into hubs or housings. Inspect track suspension system units (including track links) for evidence of premature wear. Remove worn units promptly, and install new ones to prevent ultimate failure.

40. CARE OF VEHICLE AFTER FORDING OR LANDING.

a. General. After fording, landing, or other immersion in water stop vehicle at once, if tactical situation permits, and operate hull drain valves and bilge pumps to empty vehicle of any accumulated water. If the suspension system has been submerged for even a few minutes, lubricate all suspension system points to cleanse the bearings of water and grit.

Section XI

DEMOLITION TO PREVENT ENEMY USE

41. GENERAL.

a. Destruction of the vehicle when subject to capture or abandonment in the combat zone will be undertaken by the using arm only when, in the judgment of the military commander concerned, such action is necessary.

b. The instructions which follow are for information only. Certain of the methods of destruction outlined require TNT and incendiary grenades which may not be normal items of issue. The issue of these materials, and the conditions under which destruction will be effected *are command decisions in each case*, according to the tactical situation.

c. If destruction is resorted to, the vehicle must be so badly damaged that it cannot be restored to a usable condition in the combat zone either by repair or cannibalization. Adequate destruction requires that all parts essential to the operation of the vehicle be destroyed or damaged beyond repair. Equally important, the same essential parts must be destroyed on all like vehicles so that the enemy cannot construct one complete operating unit from several partially damaged ones.

42. **DETAILED INSTRUCTIONS.**

a. **Methods.** The following instructions apply to the Heavy Tank T26E3.

b. **Destruction of 90-mm Gun.**

(1) Insert four unfuzed incendiary grenades, M14 end to end halfway down the tube, with the tube at 0 degrees elevation. Ignite these grenades with a fifth grenade equipped with a 15-second safety fuze. Elapsed time: 2 to 3 minutes.

(2) The metal from the grenades will fuze with the tube and fill the grooves.

c. **Destruction of Vehicle.**

(1) At the time this manual was published, specific procedures for destroying the Heavy Tank T26E3 had not been determined. However, the basic principles of demolition apply to this vehicle. Destruction must be as complete as available time, equipment and personnel will permit.

(2) If thorough destruction of all parts cannot be completed, the most important features of the vehicle should be destroyed, and parts essential to operation which cannot be easily duplicated, should be ruined or removed.

(3) Smash, cut, burn, explode, or otherwise dispose of as many important components as time and personnel will permit. Be sure to leave the vehicle inoperative. Destruction of the same components in each vehicle is of prime importance to prevent the enemy from cannibalizing several partially destroyed vehicles and making up an operative vehicle.

PART THREE — MAINTENANCE INSTRUCTIONS

Section XII

GENERAL

43. SCOPE.

a. Part Three contains information for the guidance of the personnel of the using organizations responsible for the maintenance (first and second echelon) of this equipment. It contains information for the performance of the scheduled lubrication and preventive maintenance services, as well as description and maintenance of the major systems and units and their functions in relation to other components of the equipment.

Section XIII

SPECIAL ORGANIZATIONAL TOOLS AND EQUIPMENT

44. LIST OF SPECIAL TOOLS.

a. **Purpose.** This list of tools is for information only and is not to be used as a basis for requisition.

b. **Tools.**

Name	Fed. Stock Number	Mfgrs. Tool Number
ADAPTER, puller, main brg., stud and quill shaft upper and lower	41-A-18-226	
ADAPTER, pressure gage, tube to transmission	45-A-198-435	TEC-50-10
ADAPTER, socket wrench, ⅜-in. male sq. plug, with ½-in. female sq socket	41-A-20-200	SN-A4
ADAPTER, road wheel spindle arm removal		KM-J-4364
BAR, socket wrench, sliding, 22-in.	41-B-312-200	MTM-M3-16L
BOLT, eye, 1-in. 8NC-2 transmission lifting	41-B-1586-350	
BRAKE, operating shaft seal		
CABLE, extension rubber covered two conductor, stranded No. 1 gage, with plug on each end	17-C-568	
COMPRESSOR, valve spring	41-C-2559-40	KRW-T15
EYE, (plate) lifting final drive assembly	41-E-649-875	KM-J-4300

Part Three—Maintenance Instructions

Name	Fed. Stock Number	Mfgrs. Tool Number
Gage, transmission oil pressure	41-G-446	KM-J1467-M6
Handle, replacer oil seal	41-H-1296-175	TEC-2-303
Handle, ¾-in. for removers and replacers, length 6⅛-in.	41-H-1395-988	TEC-2-207
Head, sq 1-in. male	41-H-1779-50	MTM-M3-16E
Holder, magneto drive flange	41-H-2350	KRW-T31
Indicator, top dead center and timer	41-I-115	KRW-T77
Lifter, front road wheel		KM-J4366
Lifter, intermediate and rear road wheels		KM-J4367
Light, timing, Bosch magneto, complete with battery and neon bulbs	41-L-1439	TSE-5230
Puller, magneto drive flange	41-P-2941-750	KRW-T43
Puller, slide hammer type bogie gudgeon	41-P-2957-33	
Remover, track pin		KM-J4371
Remover and Replacer, brake band, link lock pin		KM-4359
Remover and Replacer, torsion bar	41-R-2378-950	KM-J4321
Remover and Replacer, controlled differential brake operating shaft, outer bearing		KM-4358
Replacer, oil seal, compensating arm spindle, track wheel	41-R-2392-65	KM-J4312
Replacer, oil seal, compensating arm spindle (use handle 41-H-1296-175)	41-R-2392-50	KM-J4315
Replacer, oil seal (inner) wheel arm spindle	41-R-2395-90	KM-J4318
Replacer, controlled differential		KM-J4357
Replacer, bearing cup (Timkin) inner wheel hub (use handle 41-H-1296-175)	41-R-2384-967	KM-J4308
Replacer, bearing cup (Timkin) outer wheel hub (use handle 41-H-1296-175)	41-R-2385-180	KM-J4307

Special Organizational Tools and Equipment

Name	Fed. Stock Number	Mfgrs. Tool Number
REPLACER, bearing (outer) fan hub, upper and lower (use handle 41-H-1395-988)	41-R-2381-366	KM-J4344
REPLACER, bearing (inner), fan hub, upper and lower (use handle 41-H-1395-988)	41-R-2381-365	KM-J4343
REPLACER, oil seal and arm spindle, track wheel	41-R-2395-465	KM-J4309
SLING, differential, track connection and bogie spring replacement	41-S-3830-30	KM-J3222
SLING, power train engine transmission and control differential		KMF-4349
STAND, engine traverse housing, radiator and cooler, overhaul	41-S-4942-300	KM-J-4347
TOOL, dressing breaker point, complete	41-T-3092-77	TSE-5229
WRENCH, socket, spark plug	41-W-3336-300	KRW-T5
WRENCH, tubular, pronged single end O. D., 2-in. length, 2½-in., 4 prongs	41-W-3734-200	KM-J3497
WRENCH, socket (detachable), 1-in. sq drive, 6 point opening, size 2½-in. (wheel spindle outer nut)	41-W-3058-470	SN-L803
WRENCH, sprocket retaining nut		KM-J4360
WRENCH, plug, straight bar, hex. size 2-in. x 2-in.	41-W-1961-125	KM-J4317
WRENCH, spanner, hook, circle dia. 5⅛-in. length, 14½-in. (wheel arm spindle nut)	41-W-3252-375	KM-J4319
WRENCH, cylinder head nut, long	41-W-866-200	KRW-T9
WRENCH, cylinder head nut, short	41-W-866-250	KRW-T29
WRENCH, socket, splined, turning camshaft	41-W-2964-300	KRW-T80
WRENCH, box, special offset, carburetor heat box nuts	41-W-639-850	KRW-T40
WRENCH, spanner, adjustable (offset) roller and track wheel brg. nut	41-W-3242-300	KM-J4350

Part Three—Maintenance Instructions

Section XIV

LUBRICATION

45. LUBRICATION ORDER.

a. Reproduction of War Department Lubrication Order LO 9-735 (figs. 40 and 41) prescribes first and second echelon lubrication maintenance.

b. A Lubrication Order is placed on or is issued with each item of materiel and is to be carried with it at all times. In the event the materiel is received without a copy, the using arm shall immediately requisition a replacement from the Adjutant General Depots. See lists in FM 21-6.

c. Lubrication instructions on the Lubrication Order are binding on all echelons of maintenance and there shall be no deviations from these instructions.

d. Service intervals specified on the Lubrication Order are for normal operating conditions and continuous use of the materiel with frequent firing. Reduce these intervals under extreme conditions such as excessively high or low temperatures, prolonged periods of high speed, traveling or firing, continued operation in sand or dust, immersion in water, or exposure to moisture, any one of which may quickly destroy the protective qualities of the lubricant and require servicing in order to prevent malfunctioning or damage to the materiel. Extend intervals when vehicle is not in use.

e. Lubricants are prescribed in the "Key" in accordance with three temperature ranges, "above +32°F," "+32°F to 0°F" and "below 0°F." The time to change grades of lubricants is determined by maintaining a close check on operation of the materiel during the approach to change-over periods. Sluggish starting is an indication that lubricants are thickening and the signal to change to grades prescribed for the next lower temperature range. Ordinarily it will be necessary to *change grades of lubricants only when expected air temperatures are consistently in the next higher or lower range.*

46. DETAILED LUBRICATION INSTRUCTIONS.

a. **Lubrication Equipment.** Each piece of materiel is supplied with lubrication equipment adequate to maintain the materiel. Be sure to clean this equipment both before and after use. Operate lubricating guns carefully and in such manner as to insure a proper distribution of the lubricant.

b. **Points of Application.**

(1) Lubrication fittings, grease cups, oilers and oil holes are

readily identifiable by reference to Lubrication Order (figs. 40 and 41). Wipe lubricators and the surrounding surface clean before applying lubricant.

(2) Where relief valves are provided, apply new lubricant until the old lubricant is forced from the vent. Exceptions are specified in notes on the Lubrication Order.

c. **Cleaning.** Unless otherwise specified, use dry-cleaning solvent, or Diesel fuel oil, to clean or wash all parts. Use of gasoline for this purpose is prohibited. After washing, dry all parts thoroughly before applying lubricant. See notes for cleaning gun and machine gun bores.

d. **Lubrication Notes on Individual Units and Parts.** The following instructions supplement those notes on the Lubrication Order which pertain to lubrication and service of individual units and parts.

(1) AIR CLEANERS AND BREATHERS.

(a) (Oil Bath Type) Daily, check level and refill engine air cleaner, auxiliary engine air cleaner and auxiliary engine crankcase breather oil reservoirs to bead level with used crankcase oil or engine oil. Every 500 miles, remove air cleaners and wash all parts.

(b) (Mesh Type) Crankcase, transmission, differential and final drive gear case breathers. Every 500 miles, daily for engine crankcase breather, remove element, wash, dry and reoil with used crankcase oil or engine oil.

(2) AUXILIARY ENGINE STARTER. When starter is removed, lubricate through oiler with engine oil above 0°F. Below 0°F use special preservative lubricating oil.

(3) BREECH AND FIRING MECHANISM. Daily and after firing, clean and oil all moving parts and exposed metal surfaces. Use special preservative lubricating oil for machine guns. CAUTION: *To insure easy breech operation and to avoid misfiring in cold weather, clean, dry and oil with special preservative lubricating oil.* To clean firing mechanism, remove and operate in dry-cleaning solvent.

(4) CRANKCASE (MAIN AND AUXILIARY ENGINES). Daily, check level and refill to "FULL" mark with engine oil. Every 500 miles, remove drain plug, (reach by removing cover plates in hull floor), and completely drain crankcase. Drain only when engine is hot. After thoroughly draining, replace drain plug and refill crankcase to "FULL" mark on gage with correct lubricant to meet temperature requirements. Run engine a few minutes, recheck level, and add oil if required. Be sure pressure gage indicates oil is circulating.

(5) CRANKCASE BELOW 0°F—(TANK ENGINE). If heated storage, winterization kit or heater are available or oil can be drained after operation, use SAE 30 engine oil undiluted. Otherwise, dilute

as follows: Add SAE 30 engine oil to "FULL" mark. Then add 9½ quarts of gasoline. Run engine 5 to 10 minutes. Mark new higher level on dipstick with file. During operation add SAE 30 engine oil to regular "FULL" mark as required. CAUTION: *Check level every 4 hours of operation.* After operation, check level. If level is below regular "FULL" mark, add SAE 30 engine oil to regular "FULL" mark. Then add gasoline to new higher level mark. If level is at or above regular "FULL" mark, add gasoline to new higher level mark. Run engine 5 to 10 minutes to mix oil and diluent. (Auxiliary Engine) If engine is to be idle 4 hours or more, drain crankcase. Before operation, refill with SAE 10 engine oil.

(6) FLYWHEEL PILOT BEARING. Whenever the engine and transmission are disconnected, remove, clean and repack bearing and fill cavity in crankshaft with prescribed lubricant.

(7) GEAR CASE. Daily check level of oil in transmission, and weekly check level of oil in differential and final drives. Add oil if necessary. Oil levels must be checked before operation with vehicle on level ground. Run engine a few minutes to fill torque converter before checking transmission oil level. Before operation and after draining, fill transmission to "COLD" mark on bayonet gage, differential to level mark on bayonet gage, and final drive to level of filler plug hole. Every 1,000 miles, drain and refill. Drain only after operation when oil is hot. After thoroughly draining, clean magnetic drain plugs in differential and final drives and reinstall. Refill gear cases with prescribed lubricant.

(8) GUN BORE.

(a) Immediately after firing, while the tube is still hot, swab with a solution consisting of ½ pound of soda ash to each gallon of warm water. Rinse with clean water, dry thoroughly and oil. Repeat this procedure on 3 consecutive days thereafter until there is no longer evidence of sweating.

(b) When gun is not being fired, daily wipe the bore clean and renew the oil film. Every 5 days, clean with dry-cleaning solvent, wipe thoroughly dry and oil.

(9) MACHINE GUN BORE. Immediately after firing and on 3 consecutive days thereafter, swab with rifle bore cleaner, dry thoroughly and oil. Every 5 days, when the gun is not being fired, wipe clean and renew the oil film. Whenever there is evidence of corrosion or gummy deposits resulting from congealed oil, swab with dry-cleaning solvent, wipe thoroughly dry and oil.

(10) OIL FILTER. Check operation of self-turning mechanism in oil filter. Remove cover plate on bulkhead. Remove manual turning nut, turn end for end and install. Run engine at speed required to develop approximately 30 pounds oil pressure. The manual turn-

Lubrication

ing nut will rotate slowly if filter is operating. If filter is not operating, remove it for cleaning and inspection. Install manual turning nut in original position and secure with locking wire. Weekly, remove filter element from housing, clean and inspect. Clean the filter element by washing in dry-cleaning solvent while rotating element with manual turning nut. CAUTION: *Do not use brush and do not blow element with compressed air. Install element when thoroughly dry.*

(11) SPEEDOMETER AND TACHOMETER CABLES. Remove core from cable conduit, wipe clean, coat lightly with general purpose grease and replace. Do not pack grease into cable conduit.

(12) SPEEDOMETER AND TACHOMETER SENDING UNITS. Remove two brass screws in each sending unit, apply one or two drops of engine oil and replace plugs. Below $0°$ F, use special preservative lubricating oil.

(13) TRANSMISSION AND DIFFERENTIAL OIL SCREENS. After draining and before refilling, remove screens. Use clean cloths to wipe out housings through screen openings. Wash screens, blow dry, and install, using new gaskets.

(14) UNIVERSAL JOINTS. Lubricate with general purpose grease. CAUTION: *Do not use excessive pressure on accessory universal joints as they are not provided with relief fittings.* Lubricate differential and final drive universal joints every 1,000 miles or when differential and final drives are removed, whichever occurs first. Remove two plugs in each universal joint, install fittings and lubricate. Remove fittings, replace plugs.

(15) OILCAN POINTS. Every 250 miles, oil door hinges and latches, bow gun ball mount, control rod clevises and pins, hinges and control levers, steering brake lever linkage, engine compartment grille hinges and latches, engine throttle control linkage. Weekly, oil gun traveling rest hinges and latch, elevating handwheel handle, escape hatch latches, hand traversing brake linkage, equilibrator mounting pins lock, loader's hatch and latches, etc.

(16) POINTS REQUIRING NO LUBRICATION. Parking brake mechanism, ventilator motor and blower, heater motor, main and auxiliary engine water pumps, track, and commanders cupola race ring do not require lubrication.

(17) POINTS TO BE DISASSEMBLED PERIODICALLY BY ORDNANCE MAINTENANCE PERSONNEL ONLY.

(a) *Accelerator, Throttle, Speed Range Selector Lever, and Steering Brake Lever Cross Shaft Bearings.* At time of overhaul, remove, clean and repack bearings with general purpose grease No. 1 above $32°$ F or No. 0 below $+32°$ F.

(b) *Fan, Differential Oil Pump, and Drive Pulley Hub Bearings.*

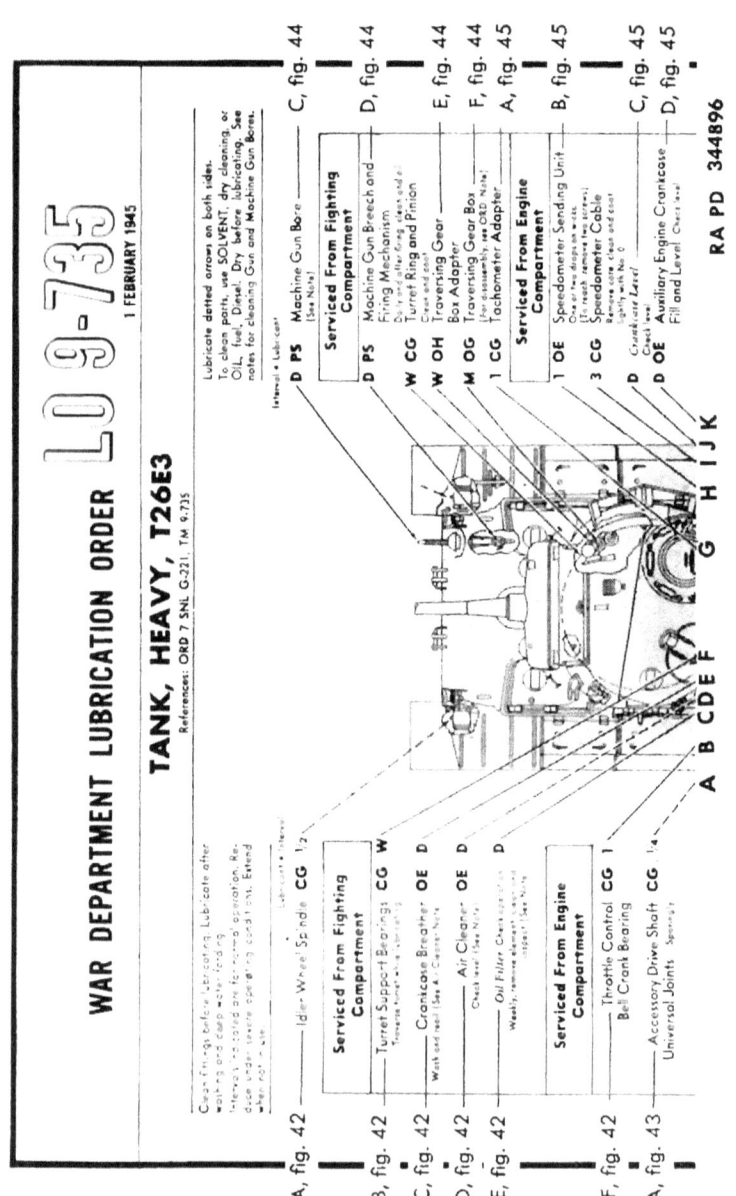

Figure 40

TM 9-735
46
Lubrication

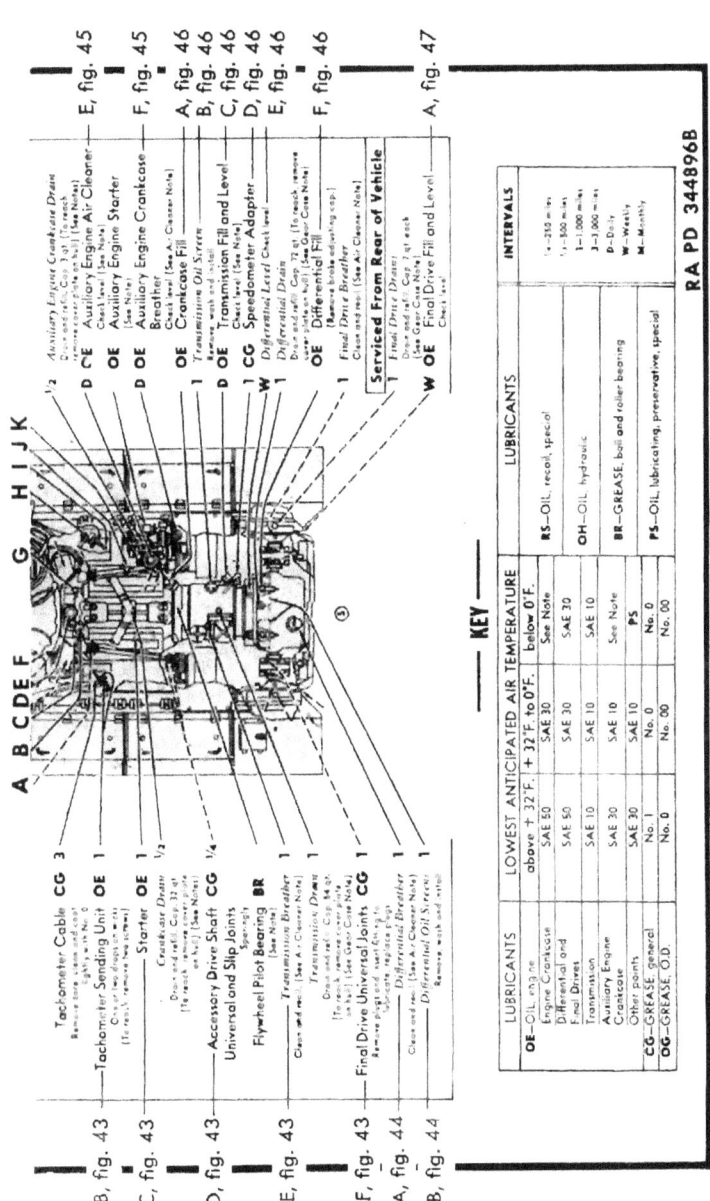

Lubrication Order

TM 9-735
46

Part Three—Maintenance Instructions

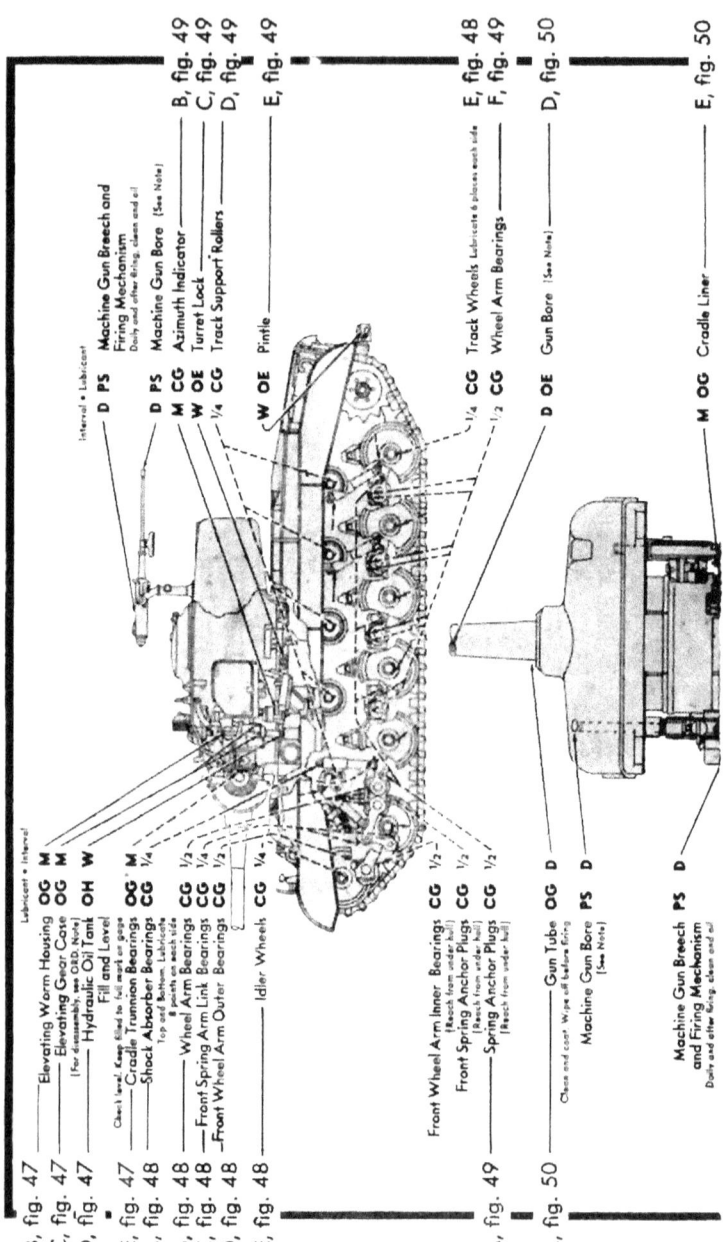

Figure 41

TM 9-735

Lubrication

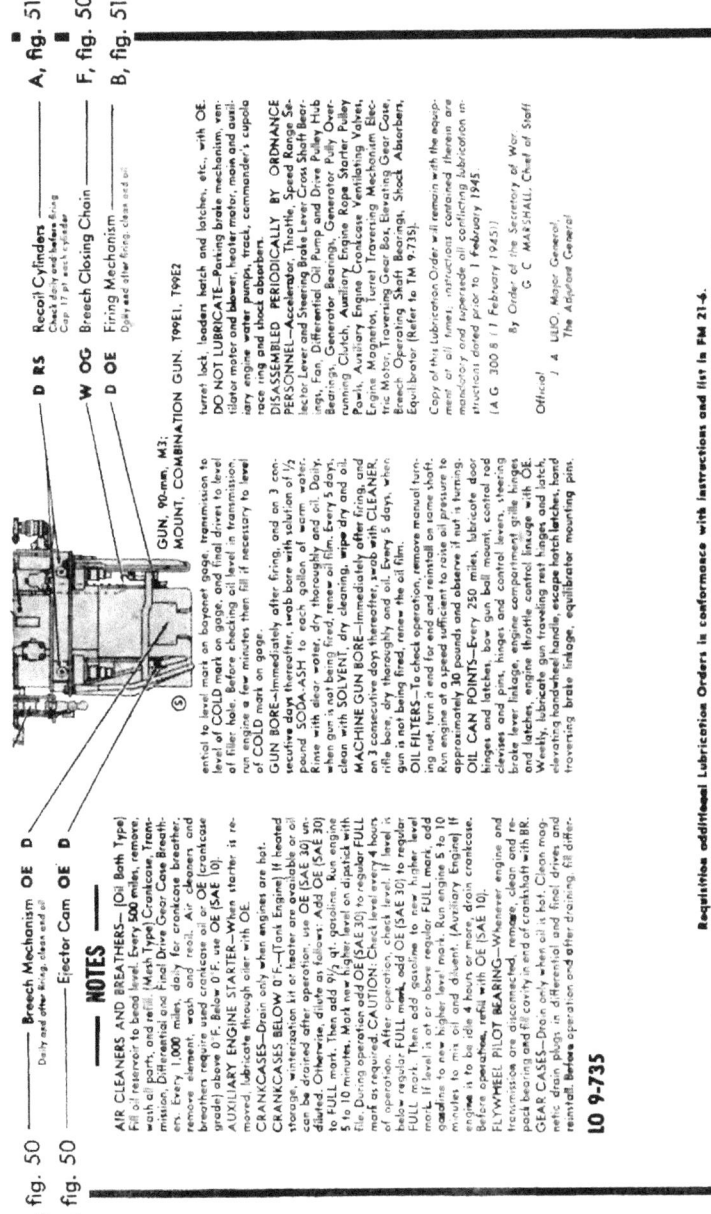

Lubrication Order

LO 9-735

C, fig. 50 — Breech Mechanism OE D
Daily and after firing, clean and oil

B, fig. 50 — Ejector Cam OE D

NOTES

AIR CLEANERS AND BREATHERS—(Oil Bath Type) Fill oil reservoir to bead level. Every 500 miles, remove, wash all parts, and refill. (Mesh Type) Crankcase, Transmission, Differential and Final Drive Gear Case Breathers. Every 1,000 miles, daily for crankcase breather, remove element, wash and reoil. Air cleaners and breathers require used crankcase oil or OE (crankcase grade) above 0°F, below 0°F, use OE (SAE 10).

AUXILIARY ENGINE STARTER—When starter is removed, lubricate through oiler with OE.

CRANKCASES BELOW 0°F.—(Tank Engine) If heated storage, winterization kit or heater are available or oil can be drained after operation, use OE (SAE 30) undiluted. Otherwise, dilute as follows: Add OE (SAE 30) to FULL mark. Then add 9½ qt. gasoline. Run engine 5 to 10 minutes. Mark new higher level on dipstick with file. During operation add OE (SAE 30) to regular FULL mark as required. CAUTION: Check level every 4 hours of operation. After operation, check level. If level is below regular FULL mark, add OE (SAE 30) to regular FULL mark. Then add gasoline to new higher level mark. If level is at or above regular FULL mark, add gasoline to new higher level mark. Running engine 5 to 10 minutes, may be idled without drain in crankcase. Before operation, refill with OE (SAE 10).

FLYWHEEL PILOT BEARING—Whenever engine and transmission are disconnected, remove, clean and repack bearing and fill cavity in end of crankshaft with BR.

GEAR CASES—Drain only when oil is hot. Clean magnetic drain plugs in differential and final drives and reinstall. Before operation add GO after draining fill orifice.

GUN, 90-mm, M3;
MOUNT, COMBINATION GUN, TYPE 1, TYPE 2

essential to level mark on bayonet gage, transmission to level of COLD mark on gage, and final drivers to level of filler hole. Before checking oil level in transmission, run engine a few minutes then fill if necessary to level of COLD mark on gage.

GUN BORE—Immediately after firing, and on 3 consecutive days thereafter, swab bore with solution of ½ pound SODA-ASH to each gallon of warm water. Rinse with clear water, dry thoroughly and oil. Daily, when gun is not being fired, renew oil film. Every 3 days, clean with SOLVENT, dry cleaning, wipe dry and oil.

MACHINE GUN BORE—Immediately after firing, and on 3 consecutive days thereafter, swab with CLEANER rifle bore, dry thoroughly and oil. Every 5 days, when gun is not being fired, renew the oil film.

OIL FILTERS—To check operation, remove manual turning nut, turn in end for end and reinstall on same shaft. Run engine at a speed sufficient to raise oil pressure to approximately 30 pounds and observe if oil is turning.

OIL CAN POINTS—Every 250 miles, lubricate door hinges and latches, bow gun ball mount, control rod clevises and pins, hinges and control levers, steering brake lever linkage, engine compartment grille hinges and latches, engine throttle control linkage with OE. Weekly, lubricate handwheel handle, escape hatch latches, hand traversing brake linkage, equilibrator mounting pins,

D RS — Recoil Cylinders A, fig. 51
Check daily and before firing
Cap 17 pt. each cylinder

W OG — Breech Closing Chain F, fig. 50

D OE — Firing Mechanism B, fig. 51
Daily and after firing, clean and oil

turret lock, loaders hatch and latches, etc., with OE. DO NOT LUBRICATE—Parking brake mechanism, ventilator motor and blower, heater motor, main and auxiliary engine water pumps, track, commander's cupola race ring and shock absorbers.

DISASSEMBLED PERIODICALLY BY ORDNANCE PERSONNEL—Accelerator, Throttle, Speed Range Selector Lever and Steering Brake Lever Cross Shaft Bearings, Fan, Differential Oil Pump and Drive Pulley Hub Bearings, Generator Bearings, Generator Pulley Overrunning Clutch, Auxiliary Engine Rope Starter Pulley Pawls, Auxiliary Engine Crankcase Ventilating Valves, Engine Magnetos, Turret Traversing Mechanism Electric Motor, Traversing Gear Box, Elevating Gear Case, Breech Operating Shaft Bearings, Shock Absorbers, Equilibrator (Refer to TM 9-735).

Copy of this Lubrication Order will remain with the equipment at all times. Instructions contained therein are mandatory and supersede all conflicting lubrication instructions dated prior to 1 February 1945.

(A.G. 300.8 (1 February 1945))

By Order of the Secretary of War:
G. C. MARSHALL, Chief of Staff

Official:
J. A. ULIO, Major General,
The Adjutant General

Requisition additional Lubrication Orders in conformance with instructions and list in FM 21-6.

RA PD 344897B

Whenever removed for any purpose, remove, clean and repack bearings with special high temperature grease.

(c) Generator Bearings. Once each year, remove generator, clean and repack bearings with ball and roller bearing grease.

(d) Generator Pulley Over-running Clutch. At time of disassembly for repair or overhaul, lubricate with SAE 10 engine oil by filling two-thirds full during reassembly of unit.

(e) Auxiliary Engine Rope Starter Pulley Pawls. At time of general overhaul, remove flywheel cover, clean and oil pawl pivots lightly with special preservative lubricating oil.

(f) Auxiliary Engine Crankcase Ventilating Valve. Every 6 months remove auxiliary engine crankcase ventilating valve, disassemble and clean thoroughly in dry-cleaning solvent. Dry, reassemble, and install.

(g) Engine Magnetos. At time of general overhaul, clean and repack bearings with ball and roller bearing grease. Lubricate cam wick on main engine magneto with one or two drops of SAE 10 engine oil.

(h) Shock Absorbers. These units are filled and sealed by manufacturer. They cannot be refilled without complete disassembly. If a leak develops replace faulty unit with a new one.

(i) Turret Traversing Gear Box. Every 6 months, disassemble, clean and repack gears with O. D. grease No. 0 above +32° F or No. 00 below +32° F. Also repack lower bearing on the pinion shaft.

(j) Turret Traversing Mechanism Electric Motor. When disassembled for repair or overhaul, clean and repack bearings with ball and roller bearing grease.

(k) Elevating Gear Case. Every 6 months, disassemble gear case, wash all parts and relubricate with O. D. grease No. 0 above +32° F or No. 00 below +32° F.

(l) Breech Operating Shaft Bearings. At initial assembly and every 6 months thereafter, remove shaft, clean shaft and bearings, coat with O. D. grease No. 0 above +32° F or No. 00 below +32° F.

(m) Equilibrator. When disassembled for repair, coat bushings and mating surfaces of tubes with O. D. grease No. 00.

e. **Reports and Records.**

(1) Report unsatisfactory performance of materiel to the Ordnance Officer responsible for maintenance in accordance with TM 38-250.

(2) A record of lubrication may be maintained in the Duty Roster (W. D., A. G. O. Form No. 6) and a record of changes in lubricants and recoil oils in the Artillery Gun Book for the materiel.

TM 9-735
46

Lubrication

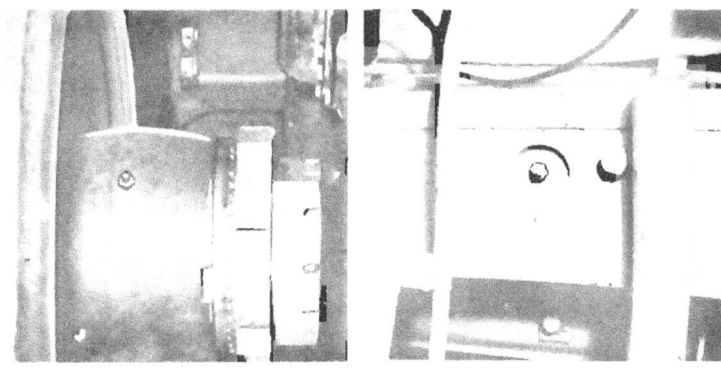
A—IDLER WHEEL SPINDLE

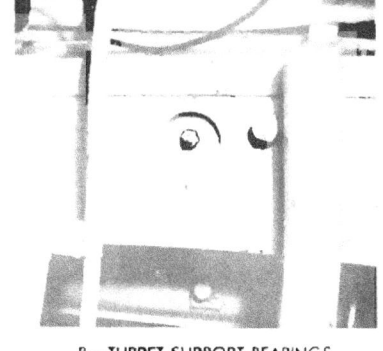
B—TURRET SUPPORT BEARINGS

C—CRANKCASE BREATHER

D—AIR CLEANER

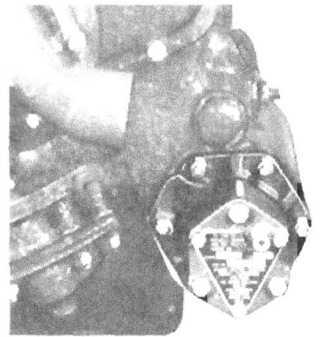
E—OIL FILTER

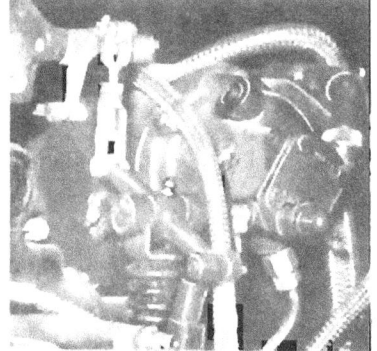
F—THROTTLE CONTROL
BELL CRANK BEARING

RA PD 344397

Figure 42 — *Localized Lubrication Views — A*

TM 9-735
46

Part Three—Maintenance Instructions

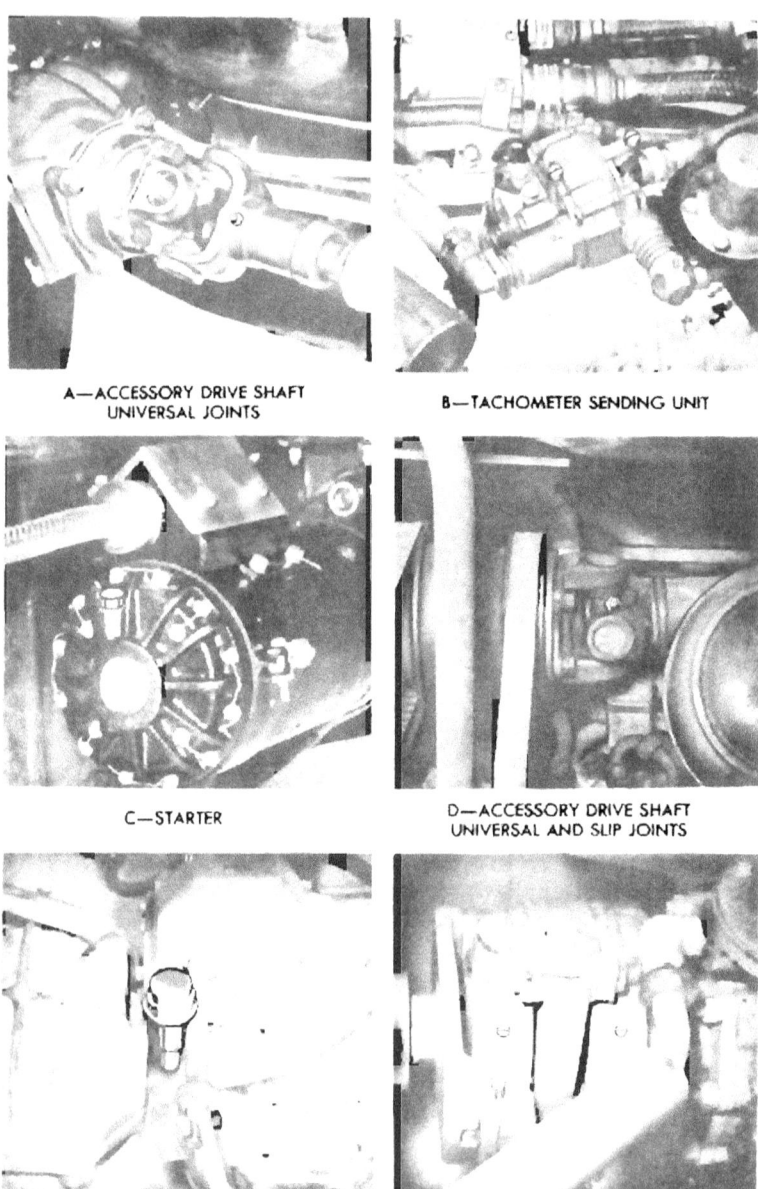

Figure 43 — Localized Lubrication Views — B

TM 9-735
46

Lubrication

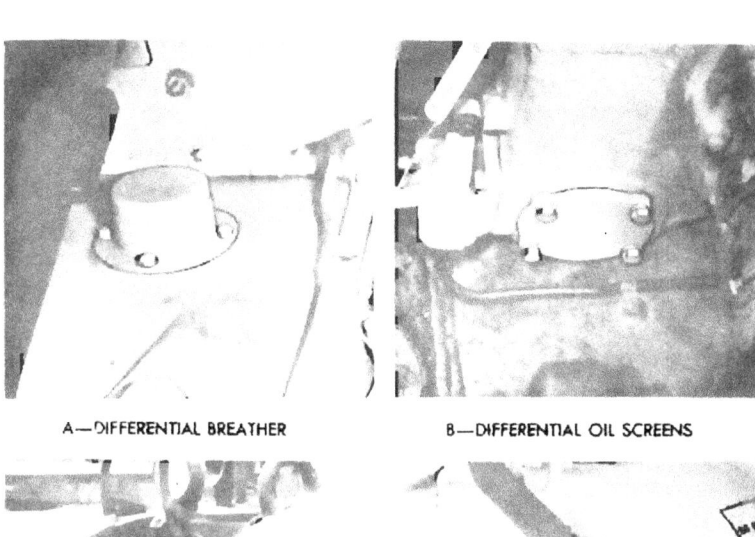

A—DIFFERENTIAL BREATHER   B—DIFFERENTIAL OIL SCREENS

C—MACHINE GUN BORE   D—MACHINE GUN BREECH
                           AND FIRING MECHANISM

 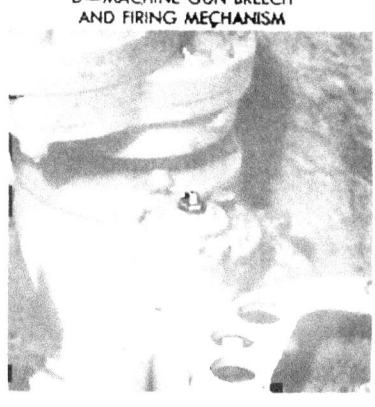

E—TRAVERSING GEAR   F—TRAVERSING GEAR BOX
BOX ADAPTER
                                              RA PD 344399

Figure 44 — Localized Lubrication Views — C

TM 9-735
46

Part Three—Maintenance Instructions

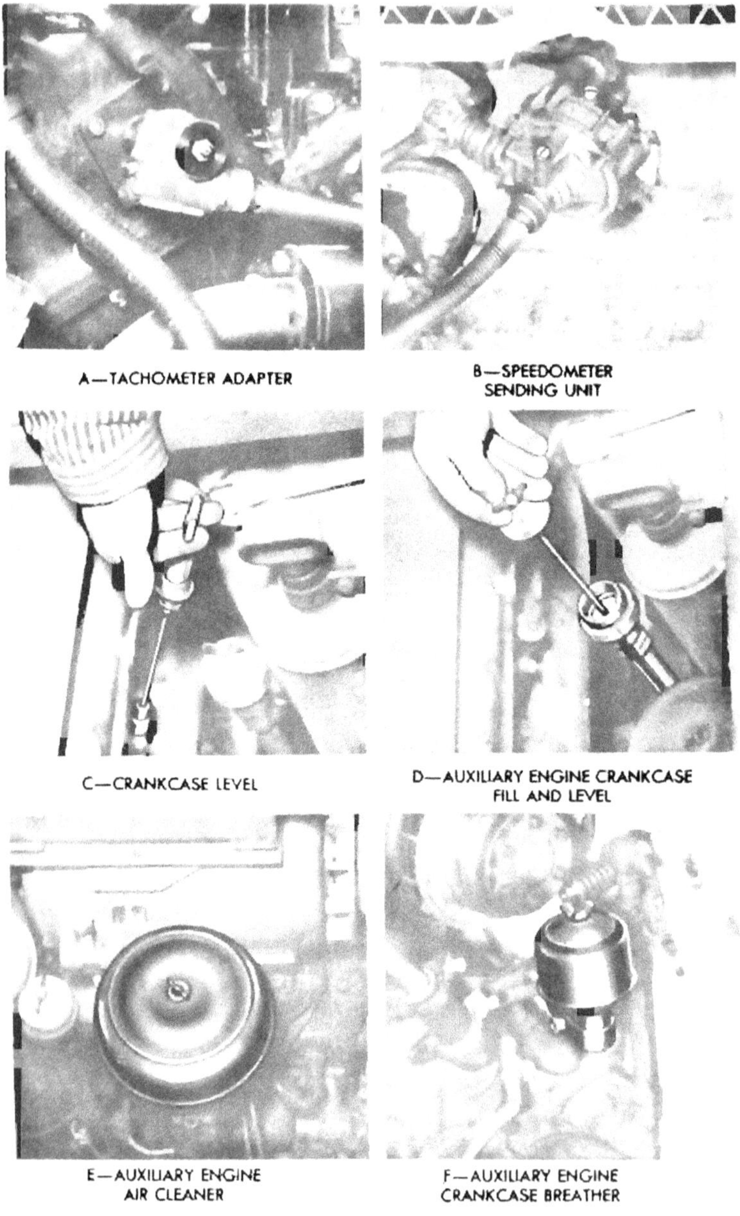

A—TACHOMETER ADAPTER

B—SPEEDOMETER SENDING UNIT

C—CRANKCASE LEVEL

D—AUXILIARY ENGINE CRANKCASE FILL AND LEVEL

E—AUXILIARY ENGINE AIR CLEANER

F—AUXILIARY ENGINE CRANKCASE BREATHER

RA PD 344400

Figure 45 — Localized Lubrication Views — D

TM 9-735
46

Lubrication

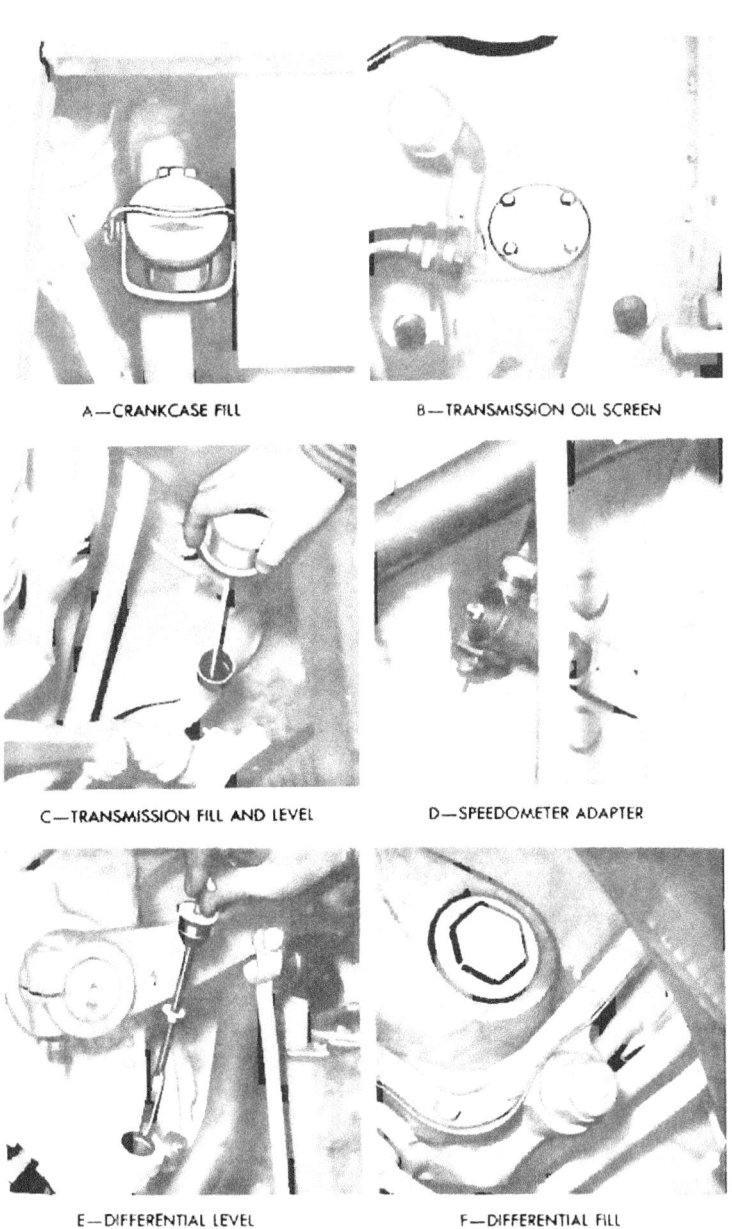

A—CRANKCASE FILL B—TRANSMISSION OIL SCREEN

C—TRANSMISSION FILL AND LEVEL D—SPEEDOMETER ADAPTER

E—DIFFERENTIAL LEVEL F—DIFFERENTIAL FILL

RA PD 344401

Figure 46 — Localized Lubrication Views — E

Part Three—Maintenance Instructions

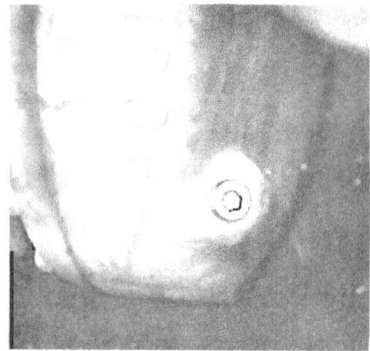

A—FINAL DRIVE LEVEL

B—ELEVATING WORM HOUSING

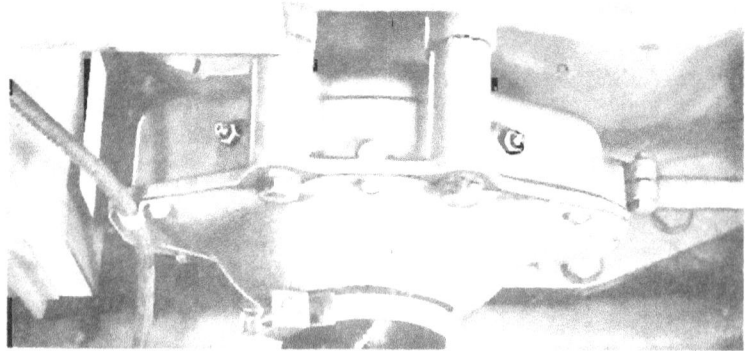

C—ELEVATING GEAR CASE

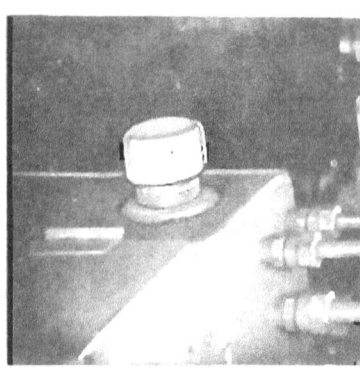
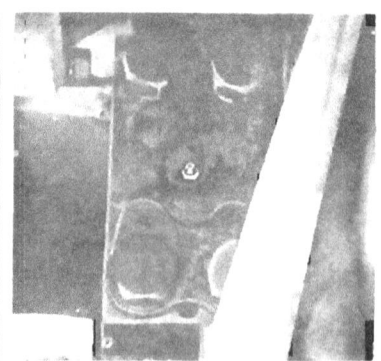

D—HYDRAULIC OIL TANK

E—CRADLE TRUNNION BEARINGS

RA PD 344402

Figure 47 — Localized Lubrication Views — F

A—SHOCK ABSORBER BEARINGS

B—WHEEL ARM BEARINGS

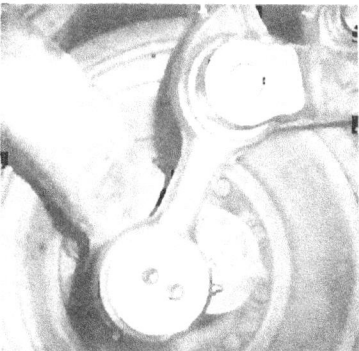
C—FRONT SPRING ARM LINK BEARINGS

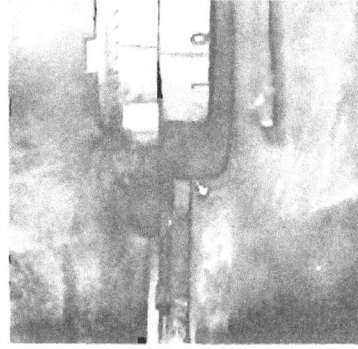
D—FRONT WHEEL ARM OUTER BEARINGS

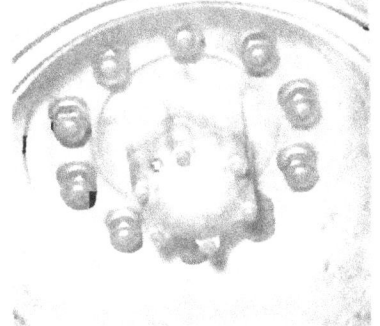
E—IDLER WHEELS AND TRACK WHEELS

Figure 48 — Localized Lubrication Views — G

TM 9-735

Part Three—Maintenance Instructions

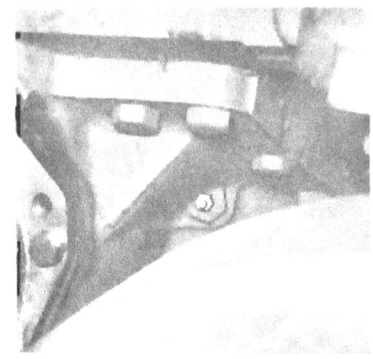
A—SPRING ANCHOR PLUGS

B—AZIMUTH INDICATOR

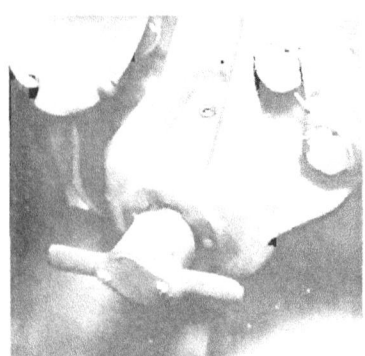
C—TURRET LOCK

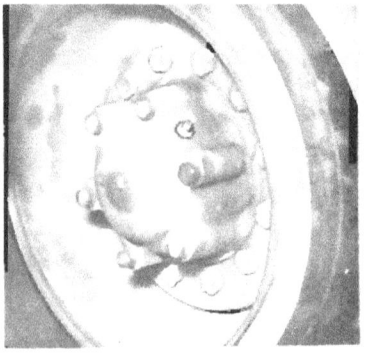
D—TRACK SUPPORT ROLLERS

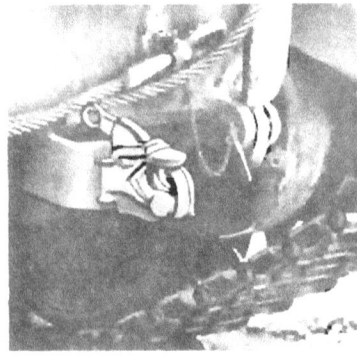
E—PINTLE

F—WHEEL ARM BEARINGS

RA PD 344404

Views — H

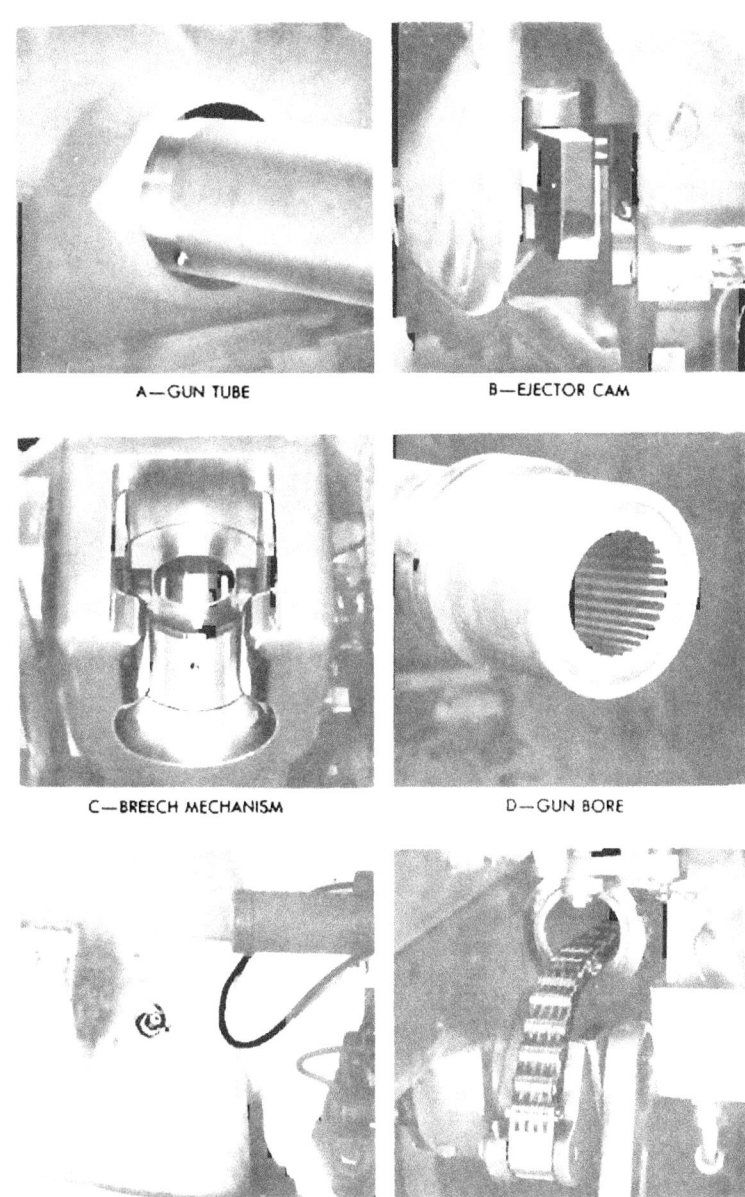

Figure 50 — Localized Lubrication Views — I

TM 9-735
46
Part Three—Maintenance Instructions

A—RECOIL CYLINDERS

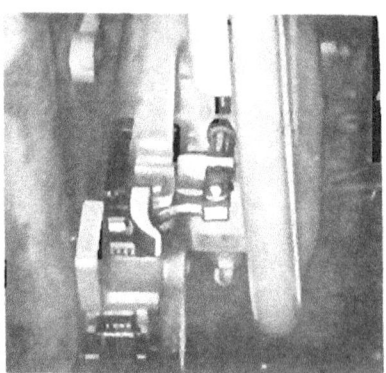
B—FIRING MECHANISM

RA PD 344406

Figure 51 — Localized Lubrication Views — J

TM 9-735
47-48

Part Three—Maintenance Instructions

Section XV

PREVENTIVE MAINTENANCE SERVICES

47. GENERAL INFORMATION.

a. **Responsibility and Interval.** Preventive maintenance services as prescribed by AR 850-15 are a function of using organization echelons of maintenance, and their performance is the responsibility of the commanders of such organizations. These services consist generally of Before-operation, During-operation, At-halt, After-operation, and Weekly Services performed by the operator, driver, or crew, and the scheduled services to be performed at designated intervals by organizational maintenance personnel.

b. **Definition of Terms.** The general inspection of each item applies also to any supporting member or connection, and is generally a check to see whether or not the item is in good condition, correctly assembled, secure, or excessively worn.

(1) The inspection for "good condition" is usually an external visual inspection to determine whether the unit is damaged beyond safe or serviceable limits. The term "good condition" is explained further by the following: not bent or twisted, not chafed or burned, not broken or cracked, not bare or frayed, not dented or collapsed, not torn or cut, not deteriorated.

(2) The inspection of a unit to see that it is "correctly assembled" is usually an external visual inspection to see whether it is in its normal assembled position in the vehicle.

(3) The inspection of a unit to determine if it is "secure" is usually an external visual examination; a wrench, hand-feel, or a prybar check for looseness. Such an inspection must include any brackets, lock washers, lock nuts, locking wires, or cotter pins used in assembly.

(4) "Excessively worn" will be understood to mean worn beyond serviceable limits, or to a point likely to result in failure if the unit is not replaced before the next scheduled inspection.

48. CREW MAINTENANCE (FIRST ECHELON).

a. **Purpose.** To ensure mechanical efficiency it is necessary that the vehicle be systematically inspected at intervals each day it is operated and weekly, so defects may be discovered and corrected before they result in serious damage or failure. Certain scheduled maintenance services will be performed at these designated intervals. Any defects or unsatisfactory operating characteristics beyond the scope of first echelon to correct must be reported at the earliest opportunity to the designated individual in authority. The services set forth in paragraphs 3, 4, 5 and 6, are those performed by the crew

Before-operation, During-operation, At-halt, and After-operation and Weekly.

b. **Use of W.D. Form No. 48.** Driver preventive maintenance services are listed on the back of "Driver's Trip Ticket and Preventive Maintenance Service Record" W.D., Form No. 48, to cover vehicles of all types and models. Items peculiar to this vehicle but not listed on W.D., Form No. 48, are covered in manual procedures under the items with which they are related. Certain items listed on the form that do not pertain to this vehicle are eliminated from the procedures as written into the manual. Every organization must thoroughly train each driver in performing the maintenance procedures set forth in this manual, whether or not they are listed specifically on W.D., Form No. 48. The items listed on W.D., Form No. 48 that apply to this vehicle are expanded in the manual, whether or not they are listed specifically on W.D., Form No. 48. The items listed on W.D., Form No. 48 that apply to this vehicle are expanded in this manual to provide specific procedures for accomplishment of the inspections and services. The services are arranged to facilitate inspection and conserve the time of the driver, and are not necessarily in the same numerical order as shown on W.D., Form No. 48. The item numbers, however, are identical with those shown on that form.

49. BEFORE-OPERATION SERVICE.

a. **Purpose.** This inspection schedule is designed primarily as a check to see that the vehicle has not been damaged, tampered with, or sabotaged since the "After-operation Service" was performed. Various combat conditions may have rendered the vehicle unsafe for operation and it is the duty of the crew to determine whether the vehicle is in condition to carry out any mission to which it is assigned. This operation will not be entirely omitted, even in extreme tactical situations.

b. **Procedures.** Before-operation Service consists of inspecting items listed below according to the procedure described, and correcting or reporting any deficiencies. Upon completion of the service, results will be reported promptly to the designated individual in authority.

(1) ITEM 1, TAMPERING AND DAMAGE. Open all top hatches and doors and examine entire vehicle, special equipment and armament for injury caused by tampering, sabotage, collision or shell fire since it was parked.

(2) ITEM 2, FIRE EXTINGUISHERS. See that fixed and portable extinguishers are in good condition and secure. If valves are damaged or appear to have been opened, report for exchange or refill. Be sure nozzles are not damaged or clogged.

TM 9-735

Part Three—Maintenance Instructions

(3) ITEM 3, FUEL, OIL AND WATER. Check fuel level in each tank. Gage should register full. Daily, check oil levels in main engine, auxiliary engine, transmission, and differential, and weekly, check oil level in final drives. Add oil as required according to instructions on Lubrication Order (par. 46). Check level of coolant and add as required. During freezing weather, if any great amount of coolant is added, make sure sufficient antifreeze compound is added to provide adequate protection. Investigate, correct, or report any unusual drop in levels since performing After-operation Service.

(4) ITEM 4, ACCESSORIES AND DRIVES. Examine units such as fuel pump, carburetors, generator, auxiliary engine, fan drive propeller shafts and U-joints, fans, shrouds, magnetos and air cleaners, ventilator blower and heater for looseness, damage or leaks. Start auxiliary engine. Oil pressure should be 15 to 35 pounds when warm. Be sure all drive belts have correct tension (par. 138 b).

(5) ITEM 16, STEERING LINKAGE. Inspect visible steering linkage, levers, and connections in drivers' and engine compartment for looseness or damage.

(6) ITEM 6, LEAKS—GENERAL. Look within engine and fighting compartments and under vehicle for indications of fuel, oil and coolant leaks. Trace any leaks found to source, and correct or report them.

(7) ITEM 8, PRIMER. When primer is used, observe if action is satisfactory and look for leaks at pump or connections.

(8) ITEM 7, ENGINE WARM-UP. Start engine and note starter action, particularly cranking speed. Run engine at 1000 to 1200 revolutions per minute until temperature reaches 100° F, then idle engine at 500 revolutions per minute during remainder of Before-operation Service. CAUTION: *If oil pressure is not indicated in 30 seconds, stop engine.*

(9) ITEM 9, INSTRUMENTS.

(a) Oil Pressure Gage and Low Oil Pressure Warning Signal Light. Gage should read between 60 and 80 pounds at 2800 revolutions per minute. Stop engine and investigate if low oil pressure warning signal light comes on at speeds above 1000 revolutions per minute.

(b) Ammeter. Ammeter may show high positive (+) charge for a period after starting, until generator restores to batteries, current used in starting, then only slight charge with lights and accessories turned off.

(c) Engine Temperature Gage. Gage should indicate increasing temperature until normal operating range of 160° F to 180° F is reached.

Preventive Maintenance Services

(d) Engine High Water Temperature Warning Signal. If warning signal (fig. 16) comes on, stop engine and investigate cause of excessive temperature (par. 60) and correct.

(e) Transmission Low Oil Pressure and High Oil Temperature Warning Signal. If light fails to go out due to low oil pressure when engine is started or comes on after engine is running, stop engine, investigate cause of low oil pressure or high oil temperature (par. 61) and correct.

(f) Fuel Gage. Gage should indicate correct level of fuel in each tank as selector switch is operated.

(g) Tachometer. Tachometer should indicate engine speed and record accumulating revolutions.

(h) Circuit Breaker Buttons. All circuit breaker buttons should remain at the "in" position. CAUTION: *Do not hold buttons "in."*

(10) ITEM 10, HORN, WINDSHIELD WIPERS, HEATER AND VENTILATOR. If tactical situation permits, test horn for proper operation and tone. If not in use, temporarily install hood and test operation of windshield wiper and defroster. Test operation of ventilator and heater motors.

(11) ITEM 11, GLASS. Clean, and inspect all glass in vision devices for looseness and damage.

(12) ITEM 12, LAMPS (LIGHTS). All lights must be secure, undamaged, and clean. If tactical situation permits, operate all light switches and observe if lights respond properly.

(13) ITEM 13, WHEEL AND HUB NUTS. See that all hub, sprocket, idler wheel and track wheel nuts and cap screws are present and secure.

(14) ITEM 14, TRACKS AND TIRES. Inspect all track wheel and support roller tires for unusual wear or damage. Look for loose track link pin nuts and evidence of loose track adjustment.

(15) ITEM 15, SUSPENSION. Inspect suspension arms, bumper springs, shock absorbers and track adjustment lock nut for good condition. Look for evidence of shock absorber fluid leakage.

(16) ITEM 17, FENDERS AND SAND SHIELDS. Inspect fenders and sand shields, horn and lamp guards, for looseness and damage.

(17) ITEM 18, TOWING CONNECTIONS. See that all tow shackles, gun traveling rest, and pintle hook are secure and in good condition, and that locking devices operate properly.

(18) ITEM 19, HULL, TARPAULIN AND CAMOUFLAGE NET. Inspect hatches and compartment doors for good condition. See that they open and close properly and latch securely. Test seat adjustment and locking devices. See that hull drain valves are closed. CAUTION: *Do not fully release bottom escape doors.* Be sure tarpaulin and camouflage are properly stowed or mounted.

(19) ITEM 20, DECONTAMINATOR. Examine for closed valve, full charge and secure mounting.

(20) ITEM 21, TOOLS AND EQUIPMENT. Be sure all items in tool and equipment stowage list (Section III) are present, serviceable, and properly mounted or stowed. Stowage box doors must close and latch securely.

(21) ITEM 22, ENGINE OPERATION. Engine should idle smoothly after warm-up period. Accelerate and decelerate several times and note any unusual noise or unsatisfacto. y operation.

(22) ITEM 23, DRIVER'S PERMIT AND FORM 26. Driver must have his operator's permit on his person. See that Form No. 26, vehicle manuals, Lubrication Order and W.D., A.G.O. Form No. 478 are present, legible and properly stowed.

50. DURING-OPERATION SERVICE.

a. **Observations.** While vehicle is in motion, listen for any sounds such as rattles, knocks, squeals, or hums that may indicate trouble. Look for indications of trouble in cooling system and smoke from any part of the vehicle. Be alert for odors indicating overheated components or units (such as generator or brakes) leaks in fuel system or exhaust system, or other trouble. When brakes are used, transmission shifted, or the vehicle turned, consider this a test and note any unsatisfactory or unusual performance. Watch the instruments constantly for unusual behavior indicating possible trouble in systems to which they apply.

b. **Procedures.** During-operation Services consist of observing items listed below according to the procedures following each item, and investigating any indications of serious trouble. Note minor deficiencies to be corrected or reported at earliest opportunity, usually the next scheduled halt.

(1) ITEM 26, STEERING BRAKES. When vehicle is put in motion, but before attaining any appreciable speed, test steering brakes to see if the vehicle can be stopped effectively; that levers meet resistance evenly and brakes are fully applied when levers are slightly ahead of vertical position and that there is a reserve of lever travel available. Operate levers independently to see if vehicle steers satisfactorily.

(2) ITEM 29, TRANSMISSION. Observe transmission operation with speed range selector lever in each position. In "first" (low speed), on level ground, vehicle should start to move when engine speed reaches 800 to 1000 revolutions per minute.

(3) ITEM 31, ENGINE AND CONTROLS. Be on the alert for lack of unusual engine power, misfiring, unusual noise or stalling, indication of overheating or unusual exhaust smoke. Notice whether accelerator

operates without binding. NOTE: *If radio interference is reported by operator, driver will cooperate in locating source of vehicular interference (par. 66).*

(4) ITEM 32, INSTRUMENTS. Look at all instruments frequently to see that they are indicating the correct operation of units to which they apply. Make sure speedometer indicates vehicle speed and records accumulating mileage.

(5) ITEM 34, RUNNING GEAR. Listen for unusual noises from tracks, wheels, rollers, and final drives.

(6) ITEM 36, GUNS: MOUNTS, ELEVATING, TRAVERSING AND FIRING CONTROLS. Before testing operation of gun, make sure gun is not loaded, turret lock, gun elevating, and gun traveling locks are released. After tests, place traverse shift lever in manual position, lock the turret and secure the gun, unless it is to be used. While vehicle is in motion, but before it is used in combat, test manual and hydraulic traversing controls and mechanism, gun elevating mechanism and both manual and electric gun firing controls to see that these units are operating properly.

51. AT-HALT SERVICE.

a. **Importance.** At-halt Services may be regarded as minimum maintenance procedures, and should be performed under all tactical conditions even though more extensive maintenance services must be slighted or omitted altogether.

b. **Procedures.** At-halt Services consists of investigating any deficiencies noted during operation, inspecting items listed below according to the procedures following the items, and correcting any deficiencies found. Deficiencies not corrected should be reported promptly to the designated individual in authority.

(1) ITEM 38, FUEL, OIL AND WATER. Check fuel and coolant supply and oil levels of main and auxiliary engines to see if adequate for operation to next scheduled stop. Replenish as needed.

(2) ITEM 39, TEMPERATURES. Cautiously hand-feel sprocket, wheel, idler, and roller hubs, and final drive housings for abnormal temperatures. Shock absorbers should feel warm when vehicle has been operated.

(3) ITEM 42, SUSPENSION. Look for damaged or loose suspension arms or shock absorbers. Remove trash lodged in suspension system.

(4) ITEM 43, STEERING LINKAGE. Observe accessible steering linkage to see that it is securely connected.

(5) ITEM 44, WHEEL AND HUB NUTS. See that all hub, sprocket, idler wheel, and road wheel nuts or screws are present and secure.

(6) ITEM 45, TRACKS AND TIRES. Inspect all road wheel and support roller tires for unusual wear or damage. Look for loose track link pin nuts and evidence of loose track adjustment.

(7) ITEM 46, LEAKS—GENERAL. Look in engine and fighting compartments and under and around vehicle for fuel, oil and coolant leaks.

(8) ITEM 47, ACCESSORIES AND BELTS. Examine units such as fuel pump, carburetors, generator, auxiliary engine, fan drive propeller shafts and U-joints, fans and shrouds for looseness, damage or leaks. Check all drive belts for satisfactory tension. If radio interference has been reported from operation of the engines, check radio noise suppression devices for looseness or damage.

(9) ITEM 48, AIR CLEANERS. When operating under extremely dusty or sandy conditions, inspect air cleaners and breather caps, on both main and auxiliary engines, at each halt to see that they are in condition to deliver clean air. Service as required, according to instructions in Lubrication Order (par. 46).

(10) ITEM 49, FENDERS AND SHIELDS. Inspect fenders and sand shields for looseness and damage.

(11) ITEM 50, TOWING CONNECTIONS. See that all tow shackles, gun traveling lock and pintle hook are secure and in good condition and that locking devices are operative.

(12) ITEM 51, HULL, TARPAULIN AND CAMOUFLAGE NET. Inspect for damage to hull or attachments. Clean engine compartment ventilator grilles. See that all attachments are secure. Be sure tarpaulin and camouflage net are securely mounted or stowed.

(13) ITEM 52, GLASS. Clean, and inspect all glass in vision devices for looseness or damage.

52. AFTER-OPERATION AND WEEKLY SERVICE.

a. Purpose. After-operation servicing is particularly important because at this time the crew inspects the vehicle to detect any deficiencies that may have developed, and to correct those they are permitted to handle. They should promptly report results of the inspection to the designated individual in authority. If this schedule is performed thoroughly, the vehicle should be ready to roll again on a moment's notice. The Before-operation Service, with a few exceptions, is then necessary only to ascertain whether the vehicle is in the same condition in which it was left upon completion of the After-operation Service. The After-operation Service should never be entirely omitted, even in extreme tactical situations, but may be reduced to the bare fundamental services outlined for the At-halt Service, if necessary.

b. Procedures. When performing the After-operation Service the crew must remember and consider any irregularities noticed in the Before-operation, During-operation, and At-halt Services. The After-operation Service consists of inspecting and servicing the following items. Those items of the After-operation Service that are marked by an asterisk (*) require additional Weekly Services, the procedures for which are indicated in step *(b)* of each applicable item.

(1) ITEM 56, INSTRUMENTS. Check pertinent instruments to see that they are operative and indicate normal function of units to which they apply.

(2) ITEM 55, ENGINE OPERATION. Accelerate the engine and note any tendency to miss or backfire, or any unusual noise or vibration that might indicate worn parts, loose mountings, incorrect fuel mixture or faulty ignition. Investigate and correct or report any deficiencies noted during operation. Idle the engine at 500 revolutions per minute for 5 minutes before stopping engine with fuel shut-off switch. Then turn ignition switch off and close both fuel shut-off valves.

(3) ITEM 54, FUEL, OIL AND WATER. Fill fuel tanks. If needed, add oil to main and auxiliary engines. Add coolant to bottom of filler neck screen. NOTE: *During freezing weather, add coolant made up of antifreeze and water in proper strength.*

(4) ITEM 57, HORN, WINDSHIELD WIPERS, HEATER AND VENTILATOR. Make sure horn is not loose or damaged. If tactical situation permits, test horn for proper operation and tone. When driver's hood is in use, test operation of windshield wiper and defroster. Test operation of ventilator blower and heater motors.

(5) ITEM 58, GLASS. Clean and inspect all glass in vision devices for looseness or damage.

(6) ITEM 59, LAMPS (LIGHTS). Clean all lenses. Examine for broken parts and security or mountings. Operate all light switches and see that lights respond.

(7) ITEM 60, FIRE EXTINGUISHERS. See if portable extinguisher is in good condition and secure. Inspect fixed system cylinders, valves, lines and nozzles for looseness or damage. Be sure nozzles are not clogged. If extinguishers have been used, or valve opened or damaged, report for exchange or refill.

(8) ITEM 61, DECONTAMINATOR. Inspect for damage and security of mountings.

(9) ITEM 62, *BATTERIES.

(a) Inspect batteries for leaks or damage, and for cleanliness and security of mountings and connections.

(b) Weekly. Clean batteries and box, and inspect for loose or corroded terminals. If terminals are corroded, remove, clean and apply a thin film of grease. Add clean water to bring level to $3/8$-inch

above plates. NOTE: *In freezing temperatures, do not add water until just before vehicle is to be operated.*

(10) ITEM 63, *ACCESSORIES AND BELTS.

(a) Examine all units such as carburetors, fuel pumps, generator, regulator, starter, air cleaners, fans, oil coolers and drive belts to see that they are in good condition and securely mounted.

(b) *Weekly.* Clean all trash from in and around oil cooler air passages. Be sure that fan and generator drive belts have ¼ inch and auxiliary engine drive belts has ⅜ inch finger pressure deflection, halfway between pulleys. Be sure accessible radio noise suppression units on accessories are securely connected and mounted.

(11) ITEM 65, AIR CLEANERS AND BREATHER CAPS. Inspect to see that they are in good condition, securely mounted and connected and not leaking. Examine for excessive dirt and proper oil level. If vehicle has been operated under extreme conditions of dust or sand, clean and service all air cleaners according to Lubrication Order (par. 46).

(12) ITEM 66, *FUEL AND OIL FILTER.

(a) Test operation of engine oil filter at manual turning nut (par. 147 b).

(b) *Weekly.* Remove and clean main engine oil filter element, also remove and clean auxiliary engine fuel pump sediment bowl and screen according to Lubrication Order (par. 46).

(13) ITEM 64, *ELECTRICAL WIRING.

(a) Inspect conduits for wear or abrasion, loose coupling nuts and loose or missing clamps or brackets.

(b) *Weekly.* Be sure all accessible radio noise suppression bond clips, straps, filters, and condensers or capacitors (par. 66) are securely connected or mounted. Tighten electrical conduit couplings nuts and wiring connections.

(14) ITEM 67, ENGINE CONTROLS. Look for worn or disconnected engine control linkage. Inspect tachometer and speedometer sending units for secure mountings and loose mechanical drive and electrical connections.

(15) ITEM 68, *TRACKS AND TIRES.

(a) Inspect all road wheel and support roller tires for unusual wear or damage. Look for loose track link pin nuts and evidence of loose track adjustment. Check track tension. With a 1-inch block between track and the second and fourth support rollers, third support roller should have ¼-inch clearance.

(b) *Weekly.* Tighten link pin nuts. If necessary adjust track tension according to paragraph 170 b.

(16) ITEM 69, *SUSPENSION.

(a) Inspect suspension arms, bumper springs and shock absorbers to see that they are in good condition and securely mounted. Be sure track adjustment lock nuts are secure. Look for evidence of shock absorber fluid leakage.

(b) Weekly. Tighten all mounting bolts and cap screws on suspension parts. Check for broken torsion bar using pry-bar method (par. 172 b).

(17) ITEM 70, STEERING BRAKE LINKAGE. Inspect levers, linkage and cross shafts for good condition, security, adequate lubrication and free operation. Investigate and correct or report any unsatisfactory condition noticed during operation. Steering levers must be parallel in released position. Pull both levers back slowly to make sure they have equal free travel of not more than 7 inches and that brakes are fully applied before levers reach a vertical position Test parking brake mechanism which must hold levers in fully applied position.

(18) ITEM 72, *TRANSMISSION AND DIFFERENTIAL BREATHERS.

(a) See that breathers are present, secure and not clogged.

(b) Weekly. Remove and clean breathers according to instructions in Lubrication Order (par. 46).

(19) ITEM 73, LEAKS—GENERAL. Inspect all lines, connections, plugs, and gasketed joints in fuel, oil and cooling systems for evidence of leakage. Correct or report any leaks found.

(20) ITEM 76, FENDERS AND SAND SHIELDS, HORN AND LAMP GUARDS. Inspect these items to see that they are in good condition and securely mounted.

(21) ITEM 77, *TOWING CONNECTIONS.

(a) See that all tow shackles, gun traveling lock, and pintle hook are secure and in good condition, and that locking devices are operative.

(b) Weekly. Examine tow cable for broken strands and see that supports hold cable securely. Inspect pintle draw bar and spring. Make sure pintle lock operates freely and locks securely.

(22) ITEM 78, HULL. Inspect entire hull for damage from shell fire or collision. Be sure escape doors are alined and that they are locked securely. Test bottom escape door release. CAUTION: *Do not fully release escape doors.* Tighten all exposed drain plugs and covers underneath hull. Operate hull drain valves to see that they open freely and close securely.

(23) ITEM 79, ARMOR. Look for fractures or damages that would render vehicle unsafe for combat use.

(24) ITEM 80, VISION DEVICES. Inspect periscopes and viewing prisms to see that they are in good condition, clean, secure in holders,

and that holders are not loose. Check spare prisms and their stowage boxes to see that they are in good condition, clean and secure. Make sure periscope holders operate freely and lock securely.

(25) ITEM 81; TURRET AND GUNS: MOUNTS, ELEVATING, TRAVERSING AND FIRING CONTROLS. Be sure all mounted guns are secure in their mounts, clean, lightly oiled (par. 46) and in condition for immediate use. Release gun from gun elevating and traveling lock and disengage turret lock until tests are completed. Test operation of both hand and hydraulic traversing mechanisms and controls by traversing turret through 360 degrees in each direction. Make sure turret does not bind and does not have excessive play. Observe action of azimuth indicator for proper operation. After operating hydraulic traversing mechanism, check level of hydraulic oil and add as required according to instructions on Lubrication Order (par. 46). Elevate gun through entire range to make sure it operates freely without excessive play. Report any deficiencies in operation of gun, sighting devices or fire control equipment noticed during combat. Make sure gun is not loaded, and test both manual and electrical gun firing controls. Be sure turret collector ring radio bond strap is in good condition and securely connected. NOTE: *Recoil tests can be made only under firing conditions and any deficiencies noted must be corrected or reported.* CAUTION: *Move traverse shift lever to manual position, secure the gun in the traveling lock and engage the elevating lock and turret lock.*

(26) ITEM 82, *TIGHTEN.

(a) Tighten any unit mounting or assembly nuts or cap screws when inspection has indicated a necessity.

(b) *Weekly.* Tighten universal joints, ammunition racks and clips, air cleaner and exhaust connections and any other points which inspection or experience indicate to be necessary on a weekly or mileage basis.

(27) ITEM 83, *LUBRICATE.

(a) Lubricate all points, except gear cases, requiring daily lubrication according to Lubrication Order (par. 46).

(b) *Weekly.* Lubricate all points of vehicle indicated on Lubrication Order paragraph 46, as necessary on a weekly or mileage basis.

(28) ITEM 84, *CLEAN ENGINE AND VEHICLE.

(a) Remove all empty shell casings and refuse from interior of vehicle. Wipe up oil or fuel drippings from driving compartment, turret, and engine compartment. See that engine compartment doors are clear of obstructions.

(b) *Weekly.* Wash exterior of vehicle and remove all dirt, mud and excess grease. If washing is impractical, wipe as clean as possible and watch for bright spots on finish or camouflage pattern that might cause glare. If compressed air is available, blow out all dust or insects from within compartment, and from in and around radiator and oil cooler core air passages.

(29) ITEM 85, *TOOLS AND EQUIPMENT.

(a) Check all On-Vehicle tools and equipment and spare parts against stowage lists (section III) to be sure all items are present and in serviceable condition and properly stowed or mounted.

(b) Weekly. Clean tools and equipment and mount or stow securely in designated locations. Tools with cutting edges, mounted on vehicle, must be sharp and bright surfaces treated to prevent glare.

53. ORGANIZATIONAL MAINTENANCE (SECOND ECHELON).

a. **Frequency.** The frequency of preventive maintenance services outlined herein is considered a minimum requirement for normal operation of vehicles. Under unusual operating conditions such as extreme temperatures, severe dust, sandy or extremely wet terrain, it may be necessary to perform certain maintenance services more frequently.

b. **First Echelon Participation.** The crews should accompany their vehicles and assist the mechanics while periodic second echelon preventive maintenance services are performed. Ordinarily the vehicle should be presented for a scheduled preventive maintenance service in a reasonably clean condition; that is, it should be dry, and not caked with mud or grease to such an extent that inspection and servicing will be seriously hampered. However, the vehicle should not be washed or wiped thoroughly clean, because certain types of defects, such as cracks, leaks, and loose or shifted parts or assemblies, are more evident if the surfaces are slightly soiled or dusty.

c. **Sources of Additional Information.** If instructions other than those contained in the general procedures in subparagraph d, or the specific procedures in subparagraph i, which follow, are required for proper performance of a preventive maintenance service or for correction of a deficiency, they may be secured from other sections of this manual, or from the designated individual in authority.

d. **General Procedures.** These general procedures are basic instructions which are to be followed when performing the services on the items listed in the specific procedures. NOTE: *The second echelon personnel must be thoroughly trained in these procedures so that they will apply them automatically.*

(1) When new or overhauled subassemblies are installed to correct deficiencies, care must be taken to see that they are clean, correctly installed, and properly lubricated and adjusted.

(2) When installing new lubricant retainer seals, a coating of the lubricant should be wiped over the sealing surface of the lip of the seal. When the new seal is a leather seal, it should be soaked in SAE 10 engine oil at least 30 minutes. The oil should be warm, if practicable. Then, the leather lip should be worked carefully by hand before installing the seal. The lip must not be scratched or marred.

e. **Definition of Terms.** Refer to paragraph 47 b.

f. **Special Services.** These are indicated by repeating the item numbers in the columns which show the interval at which the services are to be performed and show that the parts or assemblies are to receive certain mandatory services. For example, an item number in one or both columns opposite a TIGHTEN procedure, means that the actual tightening of the object must be performed. The special services include:

(1) ADJUST. Make all necessary adjustments in accordance with the pertinent section of this manual, special bulletins, or other current directives.

(2) CLEAN. Clean units of the vehicle with dry-cleaning solvent (SD) to remove excess lubricant, dirt, and other foreign material. After the parts are cleaned, rinse them in clean solvent and dry them thoroughly. Take care to keep the parts clean until reassembled and be certain to keep cleaning solvent away from rubber or other material which it will damage. Clean the protective grease coating from new parts, since this material is usually not a good lubricant.

(3) SPECIAL LUBRICATION. This applies both to lubrication operations that do not appear on the vehicle Lubrication Order, and to items that do appear on the Order, but which should be performed in connection with the maintenance operations if parts have to be disassembled for inspection or service.

(4) SERVE. This usually consists of performing special operations, such as replenishing battery water, draining and refilling units with oil, and changing or cleaning the oil filter, air cleaner, or cartridges.

(5) TIGHTEN. All tightening operations should be performed with sufficient wrench torque (force on the wrench handle) to tighten the unit according to good mechanical practice. Use a torque-indicating wrench where specified. Do not overtighten, as this may strip threads or cause distortion. Tightening will always be understood to include the correct installation of lock washers, lock nuts, lock wire, or cotter pins provided to secure the tightening.

g. **Special Conditions.** When conditions make it difficult to perform all preventive maintenance procedures at one time, they can sometimes be handled in sections, planning to complete all operations within the week if possible. All available time at halts and in bivouac

Preventive Maintenance Services

areas must be utilized, if necessary, to assure that maintenance operations are completed. When time is limited by the tactical situation, items with Special Services in the columns, should be given first consideration.

h. Work Sheet. The numbers of the preventive maintenance procedures that follow are identical with those outlined on W.D., A.G.O. Form No. 462, which is the "Preventive Maintenance Service Work Sheet for Full Track and Tank-like Wheeled Vehicles." Certain items on the work sheet that do not apply to this vehicle are not included in the procedures in this manual. In general, the numerical sequence of items on the work sheet is followed in the manual procedures, but in some instances there is deviation for conservation of the mechanic's time and effort.

i. Specific Procedures. The procedures for performing each item in the 50-hour (500-mile) and 100-hour (1000-mile) maintenance procedures, whichever shall occur first, are described in the following chart. Each page of the chart has two columns at its left edge corresponding to the 100-hour and the 50-hour maintenance respectively. Very often it will be found that a particular procedure does not apply to both scheduled maintenance. In order to determine which procedure to follow, look down the column corresponding to the maintenance due, and wherever an item number appears, perform the operations indicated opposite the number. NOTE: *Those procedures preceded by an asterisk (*) require additional services at each third 100-hour operation.*

MAINTENANCE		
100-Hour	50-Hour	
		ROAD TEST
		NOTE: *When the tactical situation does not permit a full road test, perform those items which require little or no movement of the vehicles. When a road test is possible, it should be for preferably 2 miles and not over 4 miles.*
1	1	**Before-operation Service.** Perform the Before-operation Service outlined in paragraph 49.
2	2	**Instruments and Gages.** Check as follows:
		OIL PRESSURE GAGES. Engine oil pressure must be indicated when engine is idling and 60 to 80 pounds at 2,800 revolutions per minute. Stop engine immediately when low oil pressure warning signal comes on.
		AMMETER. Ammeter may register a high positive (+) charge for a period after starting engine, until generators restore to batteries the current used in starting, then only slight charge or a zero reading

Part Three—Maintenance Instructions

MAINTENANCE	
100-Hour	50-Hour
3	3
5	5
7	7

with lights and accessories turned off, and batteries fully charged.

ENGINE TEMPERATURE GAGE. Engine temperature gage should indicate 160° F to 180° F. Engine high-water temperature warning signal comes on at maximum operating temperature, 235° F. CAUTION: *Do not move vehicle until engine temperature is above 100° F.*

TRANSMISSION OIL SIGNAL LIGHT. Transmission low oil pressure or high oil temperature warning signal should not burn during road test.

TACHOMETER. Tachometer should register engine speed in revolutions per minute and record accumulating revolutions.

FUEL GAGE. With selector switch in relative positions, gage should register level of fuel in each tank.

SPEEDOMETER. Speedometer should register vehicle speed, and record accumulating mileage.

Horn. If tactical situation permits, test siren for proper operation and tone. If not in use, temporarily install driver's hood and test operation of windshield wiper and defroster. Test operation of blower and heater motors.

Brakes (Steering). Accelerate vehicle to a moderate speed, release accelerator, apply both steering brakes observing whether or not brakes stop the vehicle effectively without leading to one side. Test steering action to be sure vehicle turns to right and left with equal braking effort. NOTE: *If vehicle tends to lead in either direction, it usually indicates either tight brake adjustment which will cause drag and excessive wear of brake lining, or unequal track tension.* With vehicle on a reasonable incline, parking brake mechanism must lock the levers in applied position and the brakes hold the vehicle effectively.

Transmission (Speed Range Selector Lever Action, Vibration and Noise). Operate vehicle through each speed range of transmission. Observe whether or not selector lever operates properly, and whether or not there are any unusual vibrations or noises in any speed range that might indicate damage, excessive wear, loose mountings, or improper

Preventive Maintenance Services

MAINTENANCE		
100-Hour	50-Hour	
		operation. NOTE: *If adjustment of linkage is necessary, adjust when performing Item 80.*
9	9	**Engine (Idle, Acceleration, Power, Noise, Smoke and Oil Consumption).** IDLE. With the vehicle stopped, observe if engine runs smoothly at normal idling speed (500 revolutions per minute). Throughout road test, observe if there is any tendency of engine to stall when accelerator is released and hand throttle closed. ACCELERATION, POWER, VIBRATION AND NOISE. Test engine for normal acceleration and pulling power in each speed. While testing in third speed range, accelerate from low speed with wide open throttle up to top speed. Listen for unusual engine noise, knock, whine, or vibration that might indicate loose, damaged, excessively worn, or inadequately lubricated engine parts or accessories or loose mountings or drive belts. Maximum governed engine speed should not exceed 2800 revolutions per minute. SMOKE. During operation, observe if there is excessive smoke from exhaust. OIL CONSUMPTION. Upon completion of road test, a check should be made to determine whether or not the engine has been consuming an excessive amount of oil.
10	10	**Unusual Noise (Final Drives, Sprockets, Road Wheels, Support Rollers, and Tracks).** During road test, listen for any unusual noise in these units that might indicate loose, damaged, inadequately lubricated or improperly adjusted parts.
11	11	**Temperatures (Final Drives, Hubs of Sprockets, Road Wheels and Rollers).** After operation, check by hand-feel for any abnormal temperatures of above units. Shock absorbers should feel warm after vehicle has been operated. NOTE: *If proper location is selected for this check, time will be saved in performing item 12.*
12	12	**Gun: Elevating and Traversing Mechanism.** Place vehicle in position where it is tilted sidewise about 10 degrees. Be sure all mounted guns are secure in their mounts, clean, lightly oiled (par. 46)

TM 9-735
53

Part Three—Maintenance Instructions

MAINTENANCE	
100-Hour	50-Hour
14	14
15	15
13	13
17	
18	18

and in condition for immediate use. Release gun from gun elevating and traveling lock and disengage turret lock until tests have been completed. Traverse turret through its full 360 degrees range by both hand and power controls. Check for binding, excessive lash, or erratic action. After operating, check hydraulic traversing mechanism for leaks and level of hydraulic oil and add as required on Lubrication Order (par. 46). Be sure turret lock operates properly. With gun pointed forward, elevate it through its full range and check for binding, excessive lash or erratic action.

Noise and Vibrations. (Engine, Mountings, Accessories, and Drives). While accelerating and decelerating the engine, listen for noise and notice any excessive vibrations in the above items that would indicate loose or damaged mountings, accessories or drives. CAUTION: *Before stopping engine, if it is at or beyond normal operating temperature, (180° F) allow engine to run 2 or 3 minutes at 800 revolutions per minute.* Stop engine.

MAINTENANCE OPERATIONS

Track Tension. Inspect tracks to see that the track tension adjustment is satisfactory (par. 170 b).

Leaks—Engine Oil, Water and Fuel. Inspect under vehicle, within engine and fighting compartments for evidence of oil, fuel, or water leaks.

Crankcase (Leaks and Levels). Remove engine compartment floor plates and inspect all accessible points of engine crankcase and oil pan for damage, leaking oil seals or gaskets.

SERVE. While engine is hot, examine sample of oil on indicator and if condition warrants, or oil change is due, drain engine oil pan and fill according to instructions on Lubrication Order (par. 46).

Side Armor (Fenders, Sand Shields, Guards, Paint and Markings, Shackles and Horn). Inspect these items to see that they are in good condition, and that armor, fenders, sand shields, guards, tow shackles and horn are secure, and that towing

Preventive Maintenance Services

MAINTENANCE		
100-Hour	50-Hour	
		shackles are not excessively worn. Observe condition of paint for rust or polished surfaces that may cause reflections, and check all vehicle markings to see that they are legible.
19	19	**Bottom** (Armor, Escape Doors, Inspection Plates, Drain Valves and Plugs). See that these items are in good condition and secure, that the bottom escape door latches operate properly, and are adequately lubricated and that bottom drain plugs and drain valves are tight. Tighten all bottom inspection plates securely. Apply few drops of oil to escape door latches. Test all drain valves in hull floor for proper operation.
20	20	**Differential and Final Drives.** Inspect housings for good condition and leakage; check lubricant level. See that all assembly and mounting bolts are secure.
20		SERVE. Drain differential, remove both oil screens and breather, clean in dry-cleaning solvent, install screens and breather and refill differential as directed on Lubrication Order (par. 46).
20		TIGHTEN. Tighten all external assembly and mounting bolts securely.
21	21	**Track Links.** Inspect tracks to see that links are in good condition, correctly assembled, and secure. Tighten track link pin nuts.
21		*NOTE: *Whenever the tracks are disconnected and removed from the sprockets, idler wheels and support rollers or at each third 100-hour maintenance service, the related items, 22 and 25 marked by an asterisk (*) on the Work Sheet, W.D., A.G.O. Form No. 462, should be inspected as described below in the asterisk-marked procedures. On the regular 50 to 100-hour maintenance services the tracks should not be removed unless repairs are needed.* CAUTION: *Whenever tracks are removed for repair or replacement, do not reinstall tracks until the services followed by the asterisk (*) in items 22 and 25 have been completed.*
22	22	**Idler Wheel Arm, Spindle, Adjustment and Lock Nuts.** Inspect these items to see that they are in good condition, correctly assembled, secure, and

Part Three—Maintenance Instructions

MAINTENANCE	
100-Hour	50-Hour
22		that grease is not leaking excessively from wheel bearing seals. Be sure track adjusting nuts are secure. Tighten all assembly and mounting bolts and nuts securely.

°NOTE: *In addition to the above, at each third 100-hour maintenance service or whenever the tracks are removed, check the idler wheel hub bearings for looseness or end play. Also spin the support rollers and listen for any unusual noise that might indicate damaged, excessively worn or inadequately lubricated bearings.* |
23	23	Suspension (Arms, Links, Torsion Springs, Wheel Arm Supports, Bumper Springs, Shock Absorbers, and Brackets). Inspect these items to see that they are in good condition, correctly assembled, and secure. Look for excessive leaks at seals and shock absorbers. Pry up on each road wheel to see that torsion springs are not broken.
23		TIGHTEN. Tighten all assembly and mounting nuts and cap screws securely.
24	24	Wheels (Tires, Support Rollers, and Skids). Inspect wheels and rollers for good condition, correct assembly and secure mounting. Pay particular attention to see that tire rubber has not separated from rim and that tires are not cut, torn, or excessively worn. Inspect for excessive lubricant leaks from bearings.
24		TIGHTEN. Raise road wheels (pars. 171 b (2) and d (2)) and examine bearings for looseness and end play. Spin wheels and listen for any unusual noise. Tighten assembly and mounting bolts securely.
24		°NOTE: *Whenever the tracks are removed, the above operation should be performed before the tracks are reinstalled.*
25	25	Sprockets (Hubs, Teeth, and Nuts). Inspect sprockets for good condition, correct assembly, and security of mounting bolts. Inspect sprocket teeth for excessive wear, and shaft flange gaskets, or oil seals for excessive leakage. If sprocket teeth are excessively worn, sprocket should be replaced or reversed (par. 168 b and c). Tighten assembly

Preventive Maintenance Services

MAINTENANCE		
100-Hour	50-Hour	
25		and mounting bolts securely. Make sure the hubs, and the nuts which attach hubs to final drive shafts, and bolts which join the hub sections, are tight and in good condition. *NOTE: In addition to the above, at each third 100-hour operation, or whenever the track is disconnected and removed from the sprocket, see that sprockets are well secured to the hubs, and that the hub to final drive shaft nuts are secure. Check the sprocket hub bearings for looseness and end play. After performing the above, reinstall the tracks and connect them securely.*
26	26	**Track Tension.** No. 3 roller should have ¼-inch clearance when 1-inch spacers are in place between track and No. 2 and No. 4 rollers. Adjust to this standard and lock securely (par. 170 h).
27	27	**Top Armor (Turret, Deck, Paint and Markings, Doors, Hatches and Latches and Antenna Mast).** Inspect these items to see that they are in good condition and secure. See that door and hatch hinges and latches operate properly, and that they are not excessively worn and are adequately lubricated. Make sure engine compartment door grilles are not obstructed. Examine paint for rust spots, or polished surfaces that may cause reflections, and see that vehicle markings are legible.
28	28	**Caps and Gaskets (Fuel and Radiator).** Inspect to see that fuel tank and expansion tank caps and gaskets are in good condition, secure, and not leaking. Be sure cap vents and pressure valve are not clogged.
30		**Engine Removal (When Required).** Remove power unit at 100-hour maintenance service only if inspections made in items 9, 13 and 17, and a check on oil consumption indicates a definite need. Clean exterior of power unit with dry-cleaning solvent and dry thoroughly, using necessary precautions against fire, and use care to keep solvent away from electrical wiring and equipment. When power unit is removed, remove and inspect fuel filter units in fuel tanks (par. 117 c). Clean in dry-cleaning solvent. Replace with new filter units if necessary.

Part Three—Maintenance Instructions

MAINTENANCE	
100-Hour	50-Hour
32	
33	33
34	34
34	
37	37
37	
38	38
42	42
42	42

32. **Spark Plugs.** Remove all spark plugs and examine them for cracked or broken insulators, excessive carbon deposits and burned electrodes. Unserviceable plugs and those having gaps in excess of .030 inch must be replaced. NOTE: *Perform item 33 before reinstalling plugs.*

33. **Compression Test.** Test engine compression in each cylinder. Normal compression is 155 to 180 pounds. Variation of 15 pounds between cylinders is not ordinarily detectable in performance. Record readings in space provided on War Department Form No. 462. Install spark plugs (par. 76 c).

34. **Starter.** Examine starter for loose mounting bolts or loose electrical connections. Be sure bond strap from engine to hull is secure.

CLEAN. Remove starter cover band and inspect for worn brushes, brush spring tension, dirty or scored commutator. Clean commutator if necessary. Tighten all mounting bolts and electrical connections.

37. **Magnetos (Points).** Inspect both magnetos to see that they are in good condition, and that mountings and connections are secure. Remove cap that covers distributor points and interrupter cam, and examine points for excessive burning or pitting, and uneven alinement of contacting surfaces, (par. 75 d) and replace magneto if necessary (par. 75 b and c).

ADJUST. Set magneto distributor point gaps to 0.016 inches (par. 75 d).

38. **Ignition Wiring and Conduits.** Inspect for good condition, cleanliness, correct assembly, tight connections, security of mountings and for chafing against other engine parts. Clean all ignition wiring shielding with a dry cloth and be sure all wiring harness coupling rings and nuts are secure. NOTE: *Do not disturb connections unless they are actually loose.*

42. **Breathers and Vents.** Inspect crankcase, transmission, differential, and final drive breathers to make sure they are in good condition.

SERVE. Remove crankcase breather element, clean in dry-cleaning solvent and reoil as instructed

Preventive Maintenance Services

MAINTENANCE		
100-Hour	50-Hour	
42		on Lubrication Order (par. 46). Install element making sure all connections are tight. SERVE. Remove transmission, differential, and final drive breathers, clean in dry-cleaning solvent and reoil as directed on Lubrication Order (par. 46). Install breathers making sure they are tight and gasket is not leaking.
43	43	**Air Cleaners (Carburetor).** Inspect both air cleaners to make sure they are in good condition, not leaking and that air intake valves operate freely and close securely. SERVE. Remove both air cleaners. Empty the oil reservoirs, clean reservoirs, screens and bodies with elements in dry-cleaning solvent, refill reservoirs to level marks as directed on Lubrication Order (par. 46). Assemble and install air cleaners making sure all joints, seals and connections are tight and not leaking.
44	44	**Carburetors (Governor, Primer, Degassers).** Inspect dual carburetors, degassers, and governor unit to see that they are in good condition, correctly assembled, securely mounted and connected, and that carburetors are not leaking. If engine operation in item 9 indicates need, check carburetor float level (par. 123 h (1)). See that primer and fuel tubes are in good condition, securely mounted and supported and not leaking and primer pump functions correctly.
45	45	**Manifolds.** Note whether manifolds and their gaskets are in good condition, correctly assembled, secure, and do not leak.
45		TIGHTEN. Tighten all loose assembly and mounting flange nuts evenly and securely.
46	46	**Cylinders (Heads and Gaskets).** Inspect externally to see that they are in good condition, secure, and whether there are indications of coolant or oil leakage or blow-by around studs or gaskets.
49	49	**Water Pump, Fans and Shrouds.** Inspect these units to see that they are in good condition, secure, that water pump is not leaking; and that fans are not obstructed and are alined so as not to interfere with shrouds.

Part Three—Maintenance Instructions

MAINTENANCE		
100-Hour	50-Hour	
50	50	**Accessory Drives (Belts, Pulleys, Shafts and Couplings).** Inspect all drive belts to see that they are in good condition. See that drive pulleys are secure and in alinement. Examine fan drive drive shafts, and universal joints to be sure they are in good condition, secure, and not leaking lubricant. Adjust drive belts so auxiliary engine belt has $\frac{3}{8}$ inch and all other belts have $\frac{1}{4}$ inch deflection under normal finger pressure (par. 83 c).
51	51	**Engine Compartment.** Inspect within engine compartment to see that floor, bulkheads, and all accessible mounting brackets or braces are in good condition and secure. Make sure that all control linkage is in good condition, securely connected and mounted. See that compartment is clean.
53	53	**Fuel (Tanks, Vents, Lines, and Pump).** Inspect fuel tanks, tubes and pump to see that they are in good condition, securely mounted and not leaking. Be sure fuel gage sending unit mountings and connections are secure and the vents are open. Check tank to hull bond straps to see that they are securely connected.
53		TIGHTEN. Tighten all accessible tank mountings, tubes, hoses, fittings and valves. Test operation of fuel valves.
53		CLEAN. If for any reason fuel tanks are removed, wash fuel filter elements in dry-cleaning solvent or replace with new elements if unserviceable.
16	16	**Fuel Pump Test.** Attach a fuel pump gage to the fuel line at carburetor. Run engine at idle speed to determine whether the fuel pressure is within the specified limits, $4\frac{1}{2}$ to 6 pounds. Disconnect gage, connect fuel hose, and run engine long enough to test connections for leaks.
54	54	**Engine Oil Filter.** Remove filter element from housing, and clean out all solids collected in filter sump. If element is serviceable, install; if not serviceable, replace with new element and test operation of filter (par. 147 b).
56	56	**Oil Coolers (Differential, Transmission).** Examine oil cooler cores and tubes to see that they are

Preventive Maintenance Services

MAINTENANCE		
100-Hour	50-Hour	
		in good condition, secure, and free from leakage. Clean obstructions from air passages.
57	57	**Exhaust Pipes and Outlet.** Inspect to see that they are in good condition, and securely assembled and mounted.
57		TIGHTEN. Tighten all mounting bolts and connections securely.
58	58	**Engine Mountings.** Inspect all mountings and bond straps to see that they are in good condition and secure.
58		TIGHTEN. Tighten all mountings and bond straps securely. NOTE: *When power unit is removed for repair or replacement, tighten the two rear differential mounting pads while the power unit is out of the vehicle. Also tighten the two front mounting pads to the engine support legs before installing the power unit.*
60	60	**Fire Extinguisher System (Tanks, Valves, Tubes, Nozzles, and Mountings).** Inspect tubes, connections and cylinders of fixed extinguishers to see that they are in good condition, securely mounted and fully charged. Full charge may be determined only by weighing. Examine control cables, while cylinders are disconnected, test to see that they operate properly. Be sure red safety seals are intact and that all nozzles are aimed correctly.
	60	TIGHTEN. Carefully tighten all assembly and mounting nuts or cap screws.
61		**Engine Power Unit (Install, Mountings, Lines and Fittings, Wiring, Oil and Water Supply).** If power unit was removed for repair or replacement, install at this time (par. 72). Tighten mountings securely. Connect all fuel, water and exhaust tubes or pipes and wiring connections and tighten securely. Be sure oil and coolant supply is adequate.
62	62	**Radiators (Cores, Mountings, Connections, Expansion Tanks, Antifreeze, Record).** Examine above listed units to see that they are in good condition, securely assembled and mounted, and not leaking. See that core external air passages are free of obstructions and inspect for badly bent cooling

Part Three—Maintenance Instructions

MAINTENANCE	
100-Hour	50-Hour
		fins. See that all bulkhead and shroud seals are in good condition to effectively seal against air leaks. Check coolant for excessive contamination. If antifreeze is in use, test its value and record in space provided on back of Work Sheet, W.D., A.G.O. Form No. 462. Add antifreeze as necessary to insure against freezing.
		SERVE. Clean out all insects and trash from in and around radiator core passages. If cooling system must be cleaned, clean according to instructions in paragraph 134 f. Use only specified cleaners and inhibitors, and during freezing weather, protect with ethylene glycol compound.
63	63	**Batteries (Cables, Hold-Downs, Carrier, Record Gravity and Voltage).** Inspect battery cases for cracks and leaks. Clean top of batteries. Inspect terminals, bolts, posts, straps and hold-downs to see that they are in good condition. Test specific gravity and voltage and record in space provided on back of Work Sheet W.D., A.G.O. Form No. 462. Gravity readings below 1.225 indicate batteries need recharge or replacement
		Perform high-rate discharge test according to instructions for "condition" test which accompanies test instrument. Cell variation should not be more than 30 per cent. NOTE: *Specific gravity must be above 1.225 to make test.*
63		CLEAN. Clean entire batteries and carrier, repaint carrier if corroded. Clean cable terminals, bolts and posts, and grease lightly; inspect bolts and nuts to see that they are in good condition; tighten terminals and hold-downs carefully to avoid damage to battery.
63	63	SERVE. Bring electrolyte up to ⅛ inch above plates with clean water.
64	64	**Accelerator and Linkage.** Examine accelerator pedals to see that they are in good condition and securely mounted. Inspect linkage to be sure it is securely connected; properly adjusted, paragraph 127 b, and not excessively worn.
65	65	**Starter (Primer and Instruments).** Start engine, observing all starting precautions in paragraph 17.

TM 9-735

Preventive Maintenance Services

MAINTENANCE		
100-Hour	50-Hour	
		In cold weather, test operation of primer pump and inspect for leaks. Note particularly if starter engages and disengages properly without unusual noise and has adequate cranking speed. Check all pertinent instruments to see if they indicate the proper functioning of the units to which they apply. Set throttle to run engine 1,000 to 1,200 revolutions per minute while warming up engine. CAUTION: *If oil pressure is not indicated in 30 seconds, stop engine and investigate cause.*
66	66	**Leaks (Engine Oil, Fuel and Water).** After oil, fuel and cooling systems are under normal operating pressure, check them thoroughly for leaks. Trace any leaks to their source, and correct them.
67		**Ignition Timing.** Check and set according to instructions (par. 75 c).
68	68	**Regulator Unit (Connections and Radio Suppression Units).** Inspect generator regulator to see that it is in good condition, securely connected and mounted. Be sure all radio noise suppression bond straps and capacitors are in good condition and securely connected.
68		SERVE. When regulator has reached normal operating temperature, connect the low voltage tester to see if generator output is properly controlled. Make test according to instructions in paragraph 62, and those which accompany the test instrument.
69	69	**Engine Idle.** Adjust engine idle speed and mixture for smooth operation (500 revolutions per minute) (par. 123 b).
70	70	**Throttle Synchronization.** Inspect throttle linkage connections and adjustment (par. 127 h).
71	71	**Fighting Compartment (Paint, Seats, Safety Straps, Crash Pads, Stowage Boxes, Ammunition Boxes, Clips and Racks).** Inspect to see that these items are in good condition, securely assembled and mounted; that fighting compartment is clean, and paint in satisfactory condition; that adjusting mechanisms of seats operate properly and are adequately lubricated. Pay particular attention whether dividers

TM 9-735
53

Part Three—Maintenance Instructions

MAINTENANCE	
100-Hour	50-Hour
72	72
73	73
75	75
75	
77	77
77	

and shell pads are all present and properly installed in ammunition boxes and racks, and that clips have sufficient tension to hold shells securely.

Turret (Seats, Platform, Lock). Examine these items to make sure they are in good condition, securely mounted, and operate properly.

Periscopes. Examine periscope and viewing prisms to see that they are in good condition, clean, secure in holders, and that holders are securely mounted; that their traversing, elevating and locking devices are free and not excessively worn. Examine spare periscopes and their stowage boxes to see that they are in good condition, clean and secure. CAUTION: *Prisms should be cleaned only with a soft cloth or brush.*

Brakes (Steering Levers, Linkage and Shafts). Examine and operate steering levers, linkage and shafts to see that they are in good condition, securely mounted or connected, operate properly and are correctly adjusted (par. 164 b). In released position, steering levers must be parallel and 13½ inches from front slope. Pull levers back evenly to make sure they have equal free travel of not more than 7 inches and that brakes are fully applied before levers reach a vertical position (par. 164 b). Correct any deficiencies noted during item 5 Road Test. Parking brake mechanism must be in good condition, securely mounted and securely lock the levers in applied position, as well as release the levers freely.

TIGHTEN. Tighten all assembly and mounting nuts and cap screws securely.

Differential and Breather. Inspect all differential and oil pump seals and gaskets for evidence of leaking. Examine all mounting bolts for looseness. Tighten as necessary.

CLEAN AND TIGHTEN. Remove breather, clean in dry-cleaning solvent and install as directed on Lubrication Order (par. 46). Tighten all external assembly and mounting bolts and cap screws securely.

Preventive Maintenance Services

MAINTENANCE		
100-Hour	50-Hour	
78	78	**Transmission and Breather.** Inspect all transmission seals and gaskets for evidence of leaking. Examine all attaching bolts for looseness. Tighten as necessary.
78		SERVE. Drain transmission, remove oil screen and breather, clean in dry-cleaning solvent, install screen and breather, and refill transmission as directed on Lubrication Order (par. 46).
80	80	**Transmission Controls and Linkage.** If during road test in item 7, the need is indicated, adjust transmission linkage according to instructions in paragraph 160 b. Lubricate linkage according to Lubrication Order (par. 46).
81	81	**Final Drives and U-Joints.** Inspect final drives to make sure they are in good condition, securely mounted, not leaking and that anchor bolts are tight. Inspect universal joints to see that they are in good condition, securely connected and are not leaking excessively.
81		SERVE. Drain final drives, remove breathers, clean in dry-cleaning solvent, install breathers, and refill final drives as instructed in Lubrication Order (par. 46).
81		TIGHTEN. Tighten the final drive anchor bolts and all U-joint assembly and yoke cap screws securely.
85	85	**Lamps (Lights) and Switches (Head, Tail, Stop, Blackout and Internal).** Test to see that switches and lamps (lights) operate properly. Inspect to see that all lights are in good condition and securely mounted.
85		ADJUST. Aim headlamp beams according to instructions (par. 96 a).
86	86	**Wiring (Junction and Terminal Blocks and Boxes, Circuit Breakers and Buttons).** Inspect accessible above listed items to see that they are in good condition, securely mounted and connected and clean. Examine all exposed wiring and conduits for damage, and see that they are well supported, securely connected and clean. Be sure all radio noise suppression bonding straps and all attaching bolts at terminals and wiring support clips are tight.

Part Three—Maintenance Instructions

MAINTENANCE	
100-Hour	50-Hour
87	87

87 — **Collector Ring (Brushes, Heads, Cylinder and Cover).** With battery and radio master switches off, remove collector box (slip ring) cover and examine to see that the above listed items are in good condition and clean. Be sure the brushes contact the cylinder evenly under normal spring tension. Install cover securely and see that bonding strap is in good condition and securely connected.

88 | 88 — **Radio Bonding (Suppressors, Capacitors, and Shielding).** Inspect all accessible radio suppression units not covered in the foregoing specific procedures to see that they are in good condition, securely mounted and connected. Be sure that all additional noise suppression bond straps and internal-external toothed washers listed in paragraph 66 are secure and their contact surfaces are clean.

GENERATOR AND AUXILIARY ENGINE

89 | 89 — **Auxiliary Engine (Mountings, Manifolds, Muffler, Oil Pan, Water Pump, Breather, Ventilator Valve).** Make sure mounting bolts are tight and tension adjuster is properly adjusted. Inspect water pump, hoses, connections and gaskets for good condition or leaks. Examine muffler and exhaust manifold to see that they are in good condition, securely mounted and that gaskets and joints are not leaking.

89 — SERVE. Remove crankcase breather and ventilator valve, clean in dry-cleaning solvent, fill breather oil reservoir to full mark as instructed on Lubrication Order (par. 46). Install breather and ventilator, making sure all joints and connections are tight and not leaking. Drain oil and refill oil pan as directed on Lubrication Order (par. 46). While drain plug cover is removed, inspect oil pan gaskets for leaks.

90 | 90 — **Spark Plugs.** Make sure ignition cable conduits are securely connected to spark plugs. Examine spark plugs for evidence of gasket leaks. Tighten plugs or replace gaskets. If engine operation in item 95 indicates faulty plugs, clean plugs or replace unserviceable plugs.

Preventive Maintenance Services

MAINTENANCE		
100-Hour	50-Hour	
90		SERVE. Remove, inspect and clean spark plugs. NOTE: *If engine has lacked power, a compression test may be performed while spark plugs are removed. At 300 to 400 revolutions per minute and at 70 degrees, compression should be approximately 95 pounds with not more than 10% to 20% variation between cylinders.*
91	91	**Magneto (Points and Wiring).** Examine magneto for secure mounting and oil leaks at gaskets. Make sure ground wire and ignition cable connections are secure and conduits are not damaged. If engine operation in item 95 indicates the need, inspect, adjust or replace magneto distributor points.
91		SERVE. Inspect magneto distributor points and if points are dirty, clean them. If points are worn, pitted or out of alinement, replace them. Adjust gap to 0.018 to 0.020 inch.
92	92	**Carburetor, Automatic Choke, and Air Cleaner.** Examine carburetor, intake manifold and automatic choke to make sure these items are in good condition, securely mounted and not leaking at gaskets or connections. If engine operation in item 95 indicates the need, adjust carburetor and automatic choke.
		SERVE. Remove air cleaner, clean in dry-cleaning solvent and refill reservoir to level marks as directed on Lubrication Order (par. 46). Install cleaner, making sure all connections and joints are tight and not leaking.
93	93	**Fuel (Pump, Tubes, Shut-Off Valve).** Inspect fuel pump, tubes, hoses, and valves to make sure they are in good condition, securely mounted, and not leaking at gaskets or joints.
93	93	CLEAN. Remove sediment bowl and screen, clean in dry-cleaning solvent and install, making sure gasket and connections are not leaking.
94	94	**Generator (Mount, Regulator, Commutator, Brushes).** Inspect generator to make sure it is in good condition and securely mounted and that all electrical conduits and connections are tight. Inspect generator regulator to make sure it is in good con-

TM 9-735
53

Part Three—Maintenance Instructions

MAINTENANCE	
100-Hour	50-Hour
94	
95	95
125	125
126	126

94 — dition and securely mounted, and that regulator and generator bond straps are in good condition and securely connected. If generator operation in item 95 indicates improper performance, refer to paragraph 62, or if excessive radio interference is noticed, refer to paragraph 66, and clean commutator or replace generator or regulator.

CLEAN. At each third 100-hour service, inspect brushes and brush holder and clean generator commutator.

95 — **Operation (Controls, Instruments, Engine, Generator).** Start engine with starter and observe if starter engages and disengages without excessive noise and develops adequate cranking speed. Make sure carburetor is properly adjusted for easy starting. Be sure oil pressure gage indicates 15 to 35 pounds pressure immediately after starting. Stop engine and use manual starter to see that it functions properly. Observe action of automatic choke to see that it opens and closes freely. Note action of governor to see that engine reaches sufficient speed to provide maximum generator output under full load, without reaching excessive speed. If radio interference is traced to the generator or auxiliary engine, refer to paragraph 66. NOTE: *Observe ammeter reading to see that generator functions properly and that the regulator correctly controls generator output. Inspect all fuel, oil and water connections and visible gaskets for looseness or leaks with engine operating. Listen for any unusual noises in engine, generator or drive pulley clutches which might indicate improper operation or lack of lubrication of these units. Stop engine.*

ARMAMENT

125 — **Bow and Turret, Gun Mounts** (Cal. .30 and Cal. .50). Inspect mounts for good condition and loose connections). Guns must move freely through full travel without binding.

126 — **Gun—90-mm** (Mount, Traversing and Elevating Mechanism and Firing Controls). Inspect to see that above units are in good condition, clean, well lubricated, correctly and securely assembled and not

MAINTENANCE	
100-Hour	50-Hour
129	129
130	130
131	131

excessively worn. Check to see that manual and electric firing controls are in good condition and secure, paying particular attention to wiring, switches, and connections. Operate both firing controls to see that they function properly. Operate hand elevating controls through entire range to see that they function properly. Traverse turret by hand to see if there is any binding, and that turret can be turned through its entire 360 degree range. See whether brake is effective when hand control brake lever is released. Inspect power traversing system, including motor, pump, reservoir, wiring and operating controls to see that they are in good condition, correctly assembled, secure, operate properly, and are not excessively worn. Examine hydraulic system for leaks and for proper level of oil. Add oil if required, according to instructions in Lubrication Order (par. 46). Make an operating check of traversing system by closing motor switch, placing shift lever in power operating position and turning operating control and moving commanders traversing control to right, left, and neutral positions to see if traversing mechanism responds properly and any overrun is properly controlled.

Spare Gun Barrels and Parts. Check stowage list, paragraph 8 a, to see that all items are present, and see that they are in good condition and properly stowed.

TOOLS AND EQUIPMENT

Tools (Vehicle Kit, and Pioneer). Check standard vehicle and pioneer tools against stowage lists (par. 6) to see that all items are present, in good condition and properly stowed or mounted. Any tools mounted on outside of vehicle, having bright or polished surfaces should be treated to prevent glare or reflections. Tools with cutting edges should be sharp and edges protected.

Equipment. Check special equipment items against vehicle stowage list (section III) to see if it is all present, in serviceable condition, and properly stowed or mounted.

TM 9-735
53

Part Three—Maintenance Instructions

MAINTENANCE		
100-Hour	50-Hour	
132	132	**Spare Track Links.** Inspect to see if they are all present, in good condition and securely mounted to turret.
133	133	**Spare Oil Supply (Recoil, Hydraulic, Engine).** Check to see that supply of listed spare oil is present and properly stowed. This supply should be maintained at all times.
134	134	**Decontaminator.** Examine to see that it is in good condition, and secure. Check for amount of charge by removing filler plug. NOTE: *Contents must be replaced each 90 days, as solution deteriorates.*
135	135	**Fire Extinguishers (Portable).** Inspect to see that extinguishers are fully charged, in good condition and securely mounted.
136	136	**Publications and Form No. 26.** All vehicle and equipment manuals, Lubrication Order, Form No. 26, W.D., A.G.O. Form 478 (MWO and Major Unit Assembly Replacement Record), must be in vehicle, legible, and properly stowed.
137	137	**Vehicle Lubrication.** If due, lubricate in accordance with Lubrication Order, paragraph 46, and current lubrication directives, using only clean lubricant and omitting items that have had lubrication during this service. Replace damaged or missing fittings, vents or plugs.
138	138	**Modifications (Modification Work Orders Completed).** Inspect vehicle to determine that all modification work orders have been properly completed. Enter any modifications or major unit assembly replacements during this service on Form No. 478.
139		**Final Road Test.** Make a final road test, rechecking items 2 to 15 inclusive. Recheck transmission and differential to see that lubricant is at correct level and that there are no leaks. Confine this road test to the minimum distance necessary to make satisfactory observations. While testing vehicle, operate it in a normal manner. NOTE: *Correct, or report any deficiencies found during final road test to designated authority.*

Section XVI
TROUBLE SHOOTING

54. GENERAL.

   a. Introduction. This section contains information for locating and correcting trouble which may develop in any part of the vehicle, except armament and communication. Information pertaining to the latter is handled in their respective sections.

   b. Instruction in Use of Trouble Shooting Section. To correct a given trouble it is first necessary to isolate it by its symptoms. Trouble symptoms are given under subparagraphs b, c, d, etc. immediately following the paragraph title. It is then necessary to read through each subparagraph and related steps to eliminate the possible causes and find the actual cause. Causes of trouble are given in the most likely order of their occurrence, under steps (1), (2), (3), etc. following each symptom. The numbered steps also give instructions for correction of the trouble, or refer to the particular part of the manual in which the subject is treated. Subparagraphs a contain information on related systems in which possible causes of trouble may be found.

55. ENGINE.

   a. General. Ignition, starting, fuel and air intake and exhaust systems, and engine cooling system are all regarded as a part of the engine, but are treated separately in trouble shooting to simplify procedure. Trouble in any of these systems will be reflected in faulty engine performance.

   b. Engine Fails to Turn When Starter Switch is Operated. Failure of engine to turn over when starter switch is turned on is due either to faults within the starting system, mechanical reasons preventing the turning over of the engine, or hydrostatic lock. As faults within the starting system are the most likely, refer to paragraph 58 before considering the following information:

   (1) JAMMED STARTER PINION OR BROKEN ENGINE PARTS. If starter is not heard to turn when starter switch is turned on, first determine if batteries are fully charged by turning on vehicle lights. If vehicle lights are bright, indicating batteries are fully charged, then the starter pinion may be jammed in contact with the flywheel ring-gear. Turn magneto switch off and place short bar in rear yoke of left hand fan drive propeller shaft and attempt to turn engine backward. CAUTION: *To avoid serious injury make sure magneto switch is off when hand-cranking engine.* To turn engine by hand maximum force on a 2-foot bar may be required. If turning engine backward frees starter pinion, attempt to again start engine. If starter jams again, replace starter. If flywheel ring gear is damaged, notify

higher authority. If engine cannot be turned backward manually the cause may also be due to broken engine parts. Notify higher authority.

(2) HYDROSTATIC LOCK. If engine does not turn and cause is not within the starting system, or mechanical failure of parts, it may be that the cylinders are full of water above the pistons, which is known as hydrostatic lock. This condition should be suspected if the vehicle has been submerged below the level of exhaust opening. Remove all spark plugs and hand-crank engine (step (1) preceding). If there is water in cylinders water will be forced from spark plug holes. If water comes out, crank engine with starter until all water has been forced out of all cylinders. Dry out cylinders and spark plugs with compressed air. When engine is started again moisture may cause it to misfire for a time. However, as the engine warms up the moisture will be dissipated by the heat. In a short time the engine will stop misfiring.

c. **Engine Turns But Fails to Start.** Failure of engine to start can be due to magneto switch off, lack of fuel, faulty carburetion, over- or underpriming. Ignition trouble can generally be disregarded. With a dual ignition system it is unlikely that both magnetos would fail at the same time. If faulty carburetion is suspected, refer to paragraph 59 before considering the following:

(1) OVER- OR UNDERPRIMING. If primer has been used it is possible engine is flooded. Turn magneto switch off and crank engine intermittently for several seconds. Fully open throttle to clear combustion chambers. If primer has not been used, it is possible that it should be. In excessively cold weather, more use of primer may be required. The primer should operate with a resistance of about 10 pounds. If pump handle goes in easily, it indicates a broken primer tube to the cylinders, defective pump, or clogged fuel supply to the pump. Investigate and correct cause of failure.

d. **Engine Runs But Lacks Power.** Lack of power can be caused by ignition faults, faulty carburetion, faulty transmission, worn, damaged or improperly adjusted engine parts, or steering brakes set too tight. If faulty ignition is suspected, refer to paragraph 57. Faulty carburetion will usually be indicated by backfiring into the carburetors, if the mixture is too lean, or by smoky exhaust if mixture is too rich. In either case refer to paragraph 59 d. Steering brakes that are too tight will cause apparent loss of power, reflected by sluggish operation of vehicle and lack of free travel of steering levers.

(1) WORN OR DAMAGED ENGINE PARTS. If this condition is suspected, make compression test as described in paragraph 70 c. A blown cylinder head gasket would cause loss of power and would be indicated by a noticeable hissing noise, particularly upon acceleration

from low speed or when engine is lugged down in a heavy pull in third (3) speed range. A visual examination will often locate gasket break, if it is blown externally. The compression test will determine which bank of cylinders is affected. Replace blown gasket as described in paragraph 133 b and c. Burned or sticking valves will also cause loss of power which a compression test will indicate. This condition is usually accompanied by engine running irregularly and misfiring. If valve trouble is suspected, notify higher authority.

e. **Engine Backfires Through Carburetors.** Backfiring of engine through carburetors is caused by faulty ignition timing or faulty carburetion, except that a cold engine will sometimes backfire when accelerated. As faulty carburetion is the most likely cause, refer to paragraph 123 b.

f. **Engine Runs Unevenly or is Noisy in Operation.** An engine running unevenly can be caused by faulty ignition, faulty carburetion, or mechanical failure. Refer to paragraph 57 for checking suspected ignition faults and to paragraph 59 b and c for checking suspected carburetion faults. Loose bearings and piston pins will not cause uneven running but they will cause noisy operation. Refer such condition to higher authority without further operation of vehicle. Noisy operation is also caused by improper engine lubrication indicated by low oil pressure on oil pressure gage. If this condition is suspected, refer to paragraph 145.

g. **Engine Stalls Frequently.** An engine which stalls frequently doubtless has carburetion trouble (par. 59 b and c).

h. **Engine Overheats.** Overheating of engine is usually caused by cooling system faults (par. 60) or one magneto may be timed late (par. 57 c (4)). A faulty temperature gage will indicate an apparent overheating of the engine (par. 60 b (5)).

i. **Smoky Exhaust.** A smoky exhaust can be caused by any one of the following:

(1) FAULTY CARBURETOR ADJUSTMENT. A smoky exhaust is usually caused by a carburetor mixture that is too rich. Adjust carburetor (par. 123 b).

(2) OIL LEVEL TOO HIGH. If engine oil level is excessively high it may cause a smoky exhaust. Check oil level and drain excess oil.

(3) CLOGGED OIL CONTROL RINGS OR WORN PISTONS OR CYLINDERS. Clogged oil control rings, pistons, or worn cylinder walls will cause excessive oil consumption resulting in smoky exhaust. Compare present consumption with previous rate and with another engine in service for a comparable period. If oil consumption is found excessive, notify higher authority.

TM 9-735
55–56

Part Three—Maintenance Instructions

j. **Fuel Cut-off Button Does Not Stop Engine.** If engine does not stop when fuel cut-off button is held in, refer to paragraph 59 h.
NOTE: *Engine can be "starved" by closing both fuel shut-off valves and allowing engine to run until fuel in idle fuel wells and fuel tubing is consumed when engine will stop from lack of fuel.*

56. TRANSMISSION.

a. This paragraph will include transmission assembly and all controls and linkage. Every instance of irregular operation which will involve disassembly should be referred to higher authority. Common trouble symptoms and remedies follow:

b. **Irregular Delivery of Power.** Recognized by a jerking action of the vehicle similar to operation of automotive vehicles with a bad ignition miss, or carburetion trouble. Condition is caused by air leakage at some external point in the transmission suction system.

(1) LOOSE SCREEN COVER OR POOR GASKET. Examine screen cover on the side of the transmission. See that cap screws are tight. If cover is tight trouble may be due to a poor screen cover gasket, or a loose external suction line.

(2) LOOSE OR DAMAGED FRONT SUCTION LINE OR GASKETS. Examine front suction tube located on right side of transmission toward the front (fig. 124). See if flange bolts are tight and if tube has been sprung out of shape. If bolts are tight and tube has not been sprung remove tube (par. 161 b (3)). Replace both gaskets and install tube securely.

c. **Power Loss or Slippage.** This symptom can be recognized by the engine turning at a speed faster than is normal for the existing driving condition. Trouble may be due to internal damage to the transmission or its components, in which case refer to higher authority.

(1) LOW OIL LEVEL. Check oil level and fill to required level.

(2) WATER IN OIL. Water in quantity in the transmission will form steam in the torque converter, the pressure of which will prevent the converter from filling properly with oil. Consequently there will be a power loss, or slippage in all transmission speeds. This condition will be indicated by steam issuing from the transmission breather. If water is indicated, completely drain transmission system (par. 151 c, and d) and refill to correct level with specified oil.

(3) LOW OIL PRESSURE. If oil pressure is low in either the inner clutch or outer clutch oil systems, it will cause sluggish clutch action and slow application of bands. Check oil level (par. 151 b (1) and (2)). If pressure is too low, refer to higher authority.

(4) DEFECTIVE INTERNAL PARTS. If this condition is suspected, notify higher authority. The only exception is slippage of clutch bands indicated by a rise in transmission oil temperature, slowness

of speed changes, and probably an odor of burning band linings. If any of these symptoms are noticed, refer to subparagraph d following.

d. **Slow Speed Changes (or Vehicle Suffers Power Loss in Second Speed, Third Speed or Reverse).** If bands are not tight enough, the shift from first to other speeds will be sluggish and soft, and the vehicle will have a tendency to slow down or stall.

(1) SLIPPING SECOND SPEED BAND. The second speed band is in use only in second speed range. A band not sufficiently tight will be indicated by a rapid rise in transmission temperature. Another indication will be the smell of burning brake linings. Adjust band (par. 159 b).

(2) SLIPPING THIRD AND REVERSE BAND. The third and reverse band is in action in these speed ranges only and if it is not sufficiently tight it will cause slippage. Refer to subparagraph d (1) preceding.

e. **Metal Particles and Band Debris Found in Oil Screens.** If particles of metal are found in the oil screen, it is an indication of internal damage. Notify higher authority immediately without further operation of vehicle. If pieces of band material are found it indicates one or both transmission bands are slipping badly. Adjust bands (par. 159 b). If proper adjustment cannot be made, notify higher authority.

57. IGNITION SYSTEM.

a. **General.** Because of the fact that the engine will start and run on only one bank of cylinders, ignition troubles are confined principally to erratic firing. If the engine cranks and fails to fire, the trouble is almost certain to be other than an ignition fault. It is unlikely that both magnetos will fail at the same time.

b. **Engine Starts, Runs Evenly, but Lacks Power.** If the engine starts but lacks power, the trouble is likely to be of an ignition nature. The causes will be either the failure of one magneto or the ground wire of one magneto shorted.

(1) SHORTED GROUND WIRE TO ONE MAGNETO. Turn magneto switch to "BOTH" position and start engine. If ground wire of one magneto is believed shorted, turn magneto switch to first "R" and then "L" positions. When dead bank of cylinders is located, remove ground wire from that magneto. If wire is shorted, engine will pick up speed as soon as wire is removed. Install new wire.

c. **Engine Misses or Backfires in One Bank of Cylinders.** Run the engine on first one bank of cylinders and then the other, to determine in which bank engine is missing and backfiring. Misfiring in only one bank is an indication of faulty spark plugs, faulty wiring, faulty magneto, blown cylinder head gasket, or bad valves in that bank of cylinders. If missing is accompanied by backfiring through the carburetors and is confined to one bank of cylinders, the magneto is probably out of time. Retime magneto (par. 75 c (4), (5) and (6)). Disregard improper valve timing except as a most remote possibility.

(1) FAULTY SPARK PLUGS. Test spark plugs (par. 76 c (1)) in bank of cylinders where misfiring is occurring. Test will indicate which cylinder is not firing. Trouble may be due to faulty plug, faulty spark plug wire, or to faulty terminal in the magneto. Remove old plug and install new spark plug.

(2) FAULTY SPARK PLUG WIRE. A wire from the magneto to a spark plug may be shorted against the conduit. Disconnect wire at spark plug and magneto. Temporarily attach a length of high-tension wire direct from magneto terminal to spark plug and run engine. If this eliminates the trouble, install new wiring harness (par. 77).

(3) FAULTY DISTRIBUTOR. If erratic firing exists in one bank of cylinders and is not due to causes in steps (1) and (2) preceding, the distributor points are probably faulty. Install new distributor points.

(4) FAULTY MAGNETO. Magneto trouble will usually result in misfiring not confined to any particular cylinder and occasional backfiring through the carburetors. If misfiring has been traced to the magneto and is not due to faulty breaker points install a new magneto (par. 75 c). If backfiring and missing continue in only one bank of cylinders and the trouble cannot be traced to ignition faults, refer to paragraph 57 c. In very rare instances terminals within the magneto may be faulty which will confine faulty ignition to one or more cylinders in one bank. If this condition exists install a new magneto (par. 75 c).

58. STARTING SYSTEM.

a. General. To determine and eliminate faults within starting system, batteries, wiring, starter switch, starter relay switch, and the flywheel ring gear with which the pinion meshes must be all considered and checked.

b. Starter Turns but Engine Does Not. This condition can usually be recognized by hearing the starter spin when the starter switch is turned on. It is usually caused by the starter speed being too slow to throw the pinion into engagement with flywheel. It can also be caused by dirty starter shaft and pinion, breakage within the starter, or by a section being stripped from the teeth of the flywheel ring gear.

(1) DEFECTIVE STARTER. If starter is heard to spin but engine does not crank, it will be due to either a dirty or defective starter, or to teeth being stripped from flywheel ring gear. The latter situation is recognized by a clashing sound when starter switch is turned

on and the starter pinion strikes the section of the flywheel ring gear from which the teeth are stripped. It can be definitely determined by turning the flywheel a few inches by the hand-cranking method. CAUTION: *To avoid serious injury, make sure magneto switch is off when hand-cranking.* If teeth are stripped from the flywheel ring gear the engine can be cranked by hand, but the starter will spin and a clashing sound will be heard when the defective spot in the flywheel ring gear is reached when the starter is used. If this condition exists, do not attempt to start engine. Notify higher authority. If starter turns but engine does not, and flywheel ring gear is not defective, the trouble is either a defective or dirty starter, in which case install a new starter (par. 81 c).

c. **Starter Engages but Engine Turns Too Slowly or Not At All.**

(1) If starter engages flywheel but does not turn engine, or turns it over slowly, the cause is due to discharged or defective batteries, high resistance electrical connections, defective starter, jammed starter pinion, or hydrostatic lock.

(2) DEFECTIVE BATTERIES OR LOOSE OR CORRODED WIRE TERMINALS. This condition can be recognized by turning on vehicle lights and operating starter switch lever. If brilliance of lights drops noticeably, usually to a dim glow, examine all connections of the large wires, remove corrosion and tighten connections. If this does not correct condition, batteries are probably defective. Install new battery or batteries (par. 92 e). If starter still does not turn properly it is probably defective.

(3) DEFECTIVE STARTER. If starter turns too slowly and none of the causes in steps (1) and (2) preceding apply, starter is defective. Install a new starter (par. 81 c).

d. **Starter Does Not Turn.** If starter does not turn when switch is turned on, it is due to lack of current to the system, a defective starter relay switch, or a defective starter. This condition can be recognized by lack of sound of the starter turning, and is due to one of the causes given steps (1), (2), (3) and (4) following.

(1) DISCHARGED BATTERIES OR BROKEN CONNECTIONS IN CABLES. Be sure 24-volt master switch is on. If starter does not turn when switch is turned on, turn on vehicle lights to determine if batteries are discharged; if there is a broken electrical connection in the heavy wire between the two batteries or a poor ground connection. Unless lights are very dim, the batteries are charged enough to turn starter, and the wires referred to are not defective. If there are no lights the batteries are either completely discharged (determine by hydrometer test, par. 92 c) or there is a broken wire which must be determined by examination of each wire.

(2) DEFECTIVE CIRCUITS OR STARTER SWITCH. If starter does not turn, there may be a loose wire running from starter relay switch to the starter switch, or the switch itself may be defective. Attach a jumper wire to the small terminal on the starter relay switch. Strike the other end of the jumper wire to metal, at the same time holding starter switch lever on. If this causes a spark, it shows that either the starter switch or the small wire from circuit breaker to relay switch is defective. Examine wire and terminals and install new wire or a new switch (par. 79 c) if necessary.

(3) DEFECTIVE STARTER RELAY SWITCH. If starter does not turn when starter switch is turned on, starter relay switch or the starter may be defective. Short-out the relay switch located in end of battery compartment by using a jumper of No. 2 wire. Place one end of wire against one of the large terminals on the starter relay switch and hold it firmly. Place other end of jumper wire firmly against second large terminal. This will cause starter to operate if relay switch is defective. Install a new starter relay switch (par. 80 c).

(4) DEFECTIVE STARTER. If starter does not operate when starter switch is turned on, and none of the causes in steps (1), (2), and (3) preceding apply, the fault lies within the starter. Install new starter (par. 81 c).

59. FUEL AND AIR INTAKE AND EXHAUST SYSTEMS.

a. General. This paragraph covers carburetion faults and trouble affecting engine compression such as blown gaskets in the exhaust manifold or faulty exhaust tubes. Failure of exhaust valves is the only other trouble that can develop in the exhaust system.

b. Engine Cranks but Fails to Start. If engine does not start when starter switch is turned on, it is almost certain to be carburetion trouble. The dual ignition system (magnetos) practically rules out the possibility of ignition faults causing starting failure. Over- or underpriming should also be considered. However, in this type of failure the engine will usually start, even though the fuel supply is restricted by lack of priming. Insufficient fuel, however, will cause erratic running or the engine will stall when accelerated.

(1) FUEL NOT TURNED ON OR EMPTY FUEL TANK. If engine fails to start and is properly primed, see that fuel tank shut-off valves are open and that there is a fuel in the fuel tanks.

(2) CLOGGED FUEL TUBES OR FILTER OR WATER IN FUEL. If it is suspected that carburetors are receiving insufficient fuel, disconnect fuel tube at carburetors and crank engine with starter. If fuel flow is restricted it may be due to clogged tubes or filter. Clean fuel tubes by blowing out with compressed air or, if necessary, clean

fuel filter (subpar. 117 c). If water or ice is found in the fuel system drain both fuel tanks (par. 116 c (5)). If fuel still does not flow the trouble is probably due to a faulty fuel pump.

(3) FAULTY FUEL PUMP. Disconnect fuel hose at carburetor and crank engine with starter. If fuel is not pumped when engine is being cranked, install a new fuel pump (par. 121 c).

(4) FAULTY CARBURETOR LINKAGE. Investigate possibility of faulty carburetor linkage. Rods may be jammed or bent preventing both carburetors from functioning. It is unlikely that both carburetors will be faulty at the same time so this condition as a failure to start can be eliminated. With carburetor linkage in proper condition and adjustment engine will start with only one carburetor functioning. Inspect and adjust linkage (par. 127 b (1)).

c. **Engine Runs Unevenly with Smoky Exhaust.** If the engine runs unevenly and has smoky exhaust after it has warmed up, it is due to one or both carburetor float levels being too high, to the fuel pump pressure being too high or to an excessively high engine oil level.

(1) CARBURETOR FLOAT LEVEL TOO HIGH. Check float level in both carburetors (par. 123 b (1)). If float level is too high in either carburetor, install a new carburetor (par. 123 d).

(2) EXCESSIVE FUEL PUMP PRESSURE. If carburetor float level is correct and engine has smoky exhaust and runs unevenly, the fuel pump may be delivering too much fuel. Check pressure of fuel pump (par. 70 e). Install a new fuel pump (par. 121 c) if pressure is incorrect.

(3) OIL LEVEL TOO HIGH. The engine oil level may be so high it is fouling the spark plugs. Check oil level and drain out excess engine oil.

d. **Engine Runs Unevenly and Backfires Through Carburetors.** If engine runs unevenly and backfires through carburetors after it has been warmed up, the trouble may be due to faulty carburetion or to faulty ignition timing.

(1) CARBURETOR FLOAT LEVEL TOO LOW. Check float level in both carburetors (par. 123 b (1)). If level is not correct, install another carburetor (par. 123 d).

(2) INSUFFICIENT FUEL PUMP PRESSURE. Check fuel pump pressure as described in paragraph 70 e and install another fuel pump (par. 121 c) if necessary.

(3) FAULTY CARBURETOR OR ADAPTER GASKET. Inspect carburetor gaskets and adapter gaskets. If faulty, or the connecting parts are loose, smoky discoloration at the gasket due to explosion blow-by when engine backfires through carburetors will be seen. If

gaskets are faulty, replace carburetor gasket or adapter gasket (par. 124 c).

e. **Engine Lacks Power.** Engine lack of power can be due to the failure of either carburetor or to one magneto not functioning at high engine speed. NOTE: *Misfiring of engine is usually due to the failure of one magneto.*

(1) FAULTY CARBURETORS. Adjust throttle linkage in accordance with instructions (par. 127 b). Check float level of carburetors (par. 123 b (1)).

(2) FAULTY MAGNETO. If it is suspected that one magneto is failing at high speeds, test magnetos by switching magneto switch to first "R" then "L" positions with engine running at a speed between 1,000 and 1,200 revolutions per minute. If one magneto is faulty the engine will have a tendency to stop when the faulty magneto is switched on. If magneto is faulty, inspect magneto points, or install magneto as outlined in paragraph 75 c.

f. **Engine Misses at All Speeds.** If engine misses at all speeds the cause is most likely to be faulty ignition (par. 57). If ignition is not at fault it may be due to improper carburetion (subpars. c and d above). Items under subparagraph c will cause smoky exhaust and items under subparagraph d will cause backfiring through carburetors in addition to misfiring. If cause of missing cannot be traced to items referred to, faulty intake or exhaust valves may be the cause.

(1) FAULTY INTAKE OR EXHAUST VALVES. If missing is believed caused by faulty intake or exhaust valves, check compression pressure (par. 70 c). If pressure of any cylinder is found incorrect, notify higher authority.

g. **Hissing Noise or Excessive Smoke in Engine Compartment.** If there is a hissing noise or unusually smoky condition in the engine compartment, with engine running, it may be due to either blown cylinder head gaskets or faulty gaskets in one of the exhaust manifolds. However, as blown cylinder head gasket would probably cause engine to misfire, the cause of the hissing noise or smoky condition is probably due to a loose manifold or exhaust tube, or to a blown gasket manifold. Inspect manifolds and install new gaskets if necessary (par. 130 c). Inspect exhaust tubes and install new gaskets or packing as required (par. 131 c).

h. **Engine Does Not Stop When Fuel Cut-off is Operated.** If engine does no stop when fuel cut-off button is pushed in, it may be due to defective circuits, or to one of the two degassers in either carburetor being defective. Be sure circuit breaker button on panel is in, and circuit is closed.

(1) DEFECTIVE FUEL CUT-OFF CIRCUIT. If circuit breaker button marked "ACCESSORIES" does not stay in, check circuit (fig. 91)

with test lamp to locate ground and repair or replace defective wire. If degassers still do not slow or stop engine, test for defective switch. Disconnect instrument panel to engine compartment terminal box conduit from back of instrument panel. Remove cover from master switch box. Connect jumper lead to No. 10 and No. 48 wire terminals (fig. 76). Turn 24-volt master switch on and start engine. Touch jumper lead to terminal "N" and to terminal "P" in receptacle at end of conduit removed. If engine slow or stops one or more of the degassers are operative and the switch must be replaced. If engine slows but does not stop proceed with test for defective degassers before connecting conduit and installing switch box cover.

(2) DEFECTIVE DEGASSERS. Turn magneto switch to "L" and run engine on "left" bank of cylinders. Touch the jumper lead to terminal "N" in end of conduit, previously removed. If engine slows down the "left" rear degasser is operative. Then touch jumper lead to terminal "P" to test "left" front degasser. Turn magneto switch to "R" to run engine on "right" bank of cylinders and repeat the tests. Replace defective degassers which do not slow down the engine as tests are made.

60. ENGINE COOLING SYSTEM.

a. **General.** The units comprising the cooling system that would be subject to trouble shooting, are the radiators, fans, fan belts, water pump, expansion tank, and the connecting tubes and hose.

b. **Engine Overheats.** Overheating of engine can be due to a number of causes, including atmospheric conditions or terrain, as well as mechanical faults not connected with the engine.

(1) INSUFFICIENT COOLANT OR IMPROPER CIRCULATION. If engine is overheating level of coolant may be low, due to improper filling, excessive engine use between fillings, or leaks in the cooling system and clogged passages. Refill cooling system to bottom of screen in expansion tank filler opening. Check for leaks. Make sure overheating is not caused by clogged water passages in the radiators or engine block. If this condition is suspected, drain and flush cooling system several times. Lack of circulation resulting from an inoperative water pump can also be a cause of overheating. However, eliminate all other causes of overheating before assuming water pump is at fault. If trouble is traced to water pump, install new pump (par. 143 c). NOTE: *Freezing of coolant in radiators and other passages will cause improper circulation.* Carefully thaw out cooling system and then refill with the correct percentage of coolant and antifreeze to provide sufficient protection. Thermostats occasionally become inoperative in the closed position which retards circulation. Remove both thermostats (par. 137 b). Heat a bucket of water to 165° F, using hydrometer (18-H-940). Immerse both thermostats.

If they do not fully open after 3 minutes, replace one or both thermostats as required (par. 137 c).

(2) INSUFFICIENT AIR CIRCULATION. Clogged passages in radiators and/or engine compartment doors will cause rise in engine temperature. Clean cores and openings. Loose fan belts are also a cause of overheating. Belt slippage prevents the fans being driven fast enough to force sufficient air through the radiators. Check and adjust fan belt tension (par. 138 b).

(3) LATE TIMING OF MAGNETO. Late ignition may possibly cause a rise in engine temperature, especially if atmospheric temperature is high. However, with the dual ignition system it is not likely that both magnetos will be out of time at the same time. Late timing of one magneto will not cause a noticeable rise in engine temperature without first having affected engine operation. If one magneto is believed timed late, check timing of both magnetos and retime the one at fault (par. 75 c).

(4) POWER TRAIN FAULTS. Binding of some part of the differential, final drives, or track suspension system can cause rise in engine temperature, if atmospheric temperature is high. However, such a condition would first cause a rise in transmission oil temperature and would also cause steering difficulties.

(5) FAULTY ENGINE TEMPERATURE GAGE. If engine temperature gage (or sending unit) is faulty, it will cause an apparent abnormal high engine temperature gage reading. If gage shows a high temperature, hand-feel the various parts of the engine, water jacket, and radiators. Normal temperatures will be just a little too hot to touch with comfort. If either engine temperature gage or sending unit is believed at fault, determine which one is faulty. To test gage disconnect wire from engine sending unit and temporarily attach a new engine temperature gage to sending unit wire. Attach a jumper wire to the other gage terminal and touch end of jumper wire to positive post on either battery. A lower reading on the new gage will show old gage is faulty. Install new gage (par. 136 b). A faulty sending unit will show the same indications on both old and temporary test gage. Install new sending unit (par. 136 e).

61. ENGINE LUBRICATION SYSTEM.

a. **General.** The engine lubrication system consists of those parts of the engine that supply engine lubrication, the oil pressure gage sending unit, and the dual low oil pressure signals and switch.

b. **Low Oil Pressure.** Low engine oil pressure will be indicated on the oil pressure gage and by the dual low oil pressure signals. It can be caused by a number of faults given in the following information. Noisy engine operation frequently results if engine oil pressure is low. NOTE: *Do not operate engine if the oil pressure is low. It will cause serious damage.*

(1) IMPROPER LUBRICATING OIL. Low engine oil pressure is usually caused by dilution of oil due to overpriming or engine oil not changed often enough. It can also be caused by using a lighter viscosity oil than grade specified (par. 46). Inspect engine oil while engine is hot. If it is thinner than operating experience indicates it should be, drain and refill with specified grade and quantity (par. 145 c and d).

(2) DEFECTIVE OIL PRESSURE GAGE OR LOW OIL PRESSURE WARNING SIGNAL SWITCH. A defective oil pressure gage or defective low oil pressure warning signal switch will show apparent low oil pressure either on the gage or by the warning signals lighting. Do not operate engine if there is any question regarding accuracy of either gage or signals. NOTE: *Accurate indication of oil pressure is of the utmost importance.* One unit may be defective and show low oil pressure, while the other does not. As a safety precaution it is best to assume that the oil pressure is low if one instrument so indicates. Test oil pressure gage, low oil pressure signals and switch. To test switch remove it (par. 150 e), and temporarily connect a new unit. If gage, or signals, still show incorrect indication, then the gage is at fault. Install a new gage (par. 150 b and c). If gage and switch are operating correctly and warning signals are not, replace low oil pressure warning signal lamp(s) (par. 150 h and i). If test proves switch is at fault, complete the installation of the new sending unit.

(3) DEFECTIVE OIL PUMP. If oil pump is believed defective, it can be checked only by removing pump and installing a new one.

(4) BROKEN OIL TUBE. A broken oil tube will cause low oil pressure and will, within a very short time, exhaust the supply in the engine. Locate a broken oil tube by examining hose and tubes at magneto end of the engine. If a broken oil tube or hose is found, install a new one.

(5) WORN OR BURNED-OUT BEARINGS. Worn or burned-out bearings will cause a low oil pressure reading on the oil pressure gage. This is particularly true if engine has been run for any length of time. Defective bearings can be determined by an excessive rattling sound. Run engine at 600 revolutions per minute *for a very short period* when trying to identify bearing noise. A burned-out bearing can be recognized by a more pronounced rattling sound than that produced by a worn bearing. Should either situation develop, stop engine immediately and notify higher authority.

c. **Excessive Oil Pressure.** Excessive oil pressure is unlikely to occur, but if it should, the correction of the trouble is important.

Damage to various parts of the engine can result from neglecting corrective measures.

(1) INCORRECT GRADE OF OIL. Most likely cause of excessive oil pressure is incorrect grade of oil (par. 46); a grade that is too heavy for the engine or temperature, particularly subzero temperatures. Examine oil and compare with proper seasonal grade. If incorrect oil is found in engine, drain and refill with prescribed grade and quantity (par. 145 c and d).

(2) OTHER CAUSES. The causes given in subparagraph b (2) preceding (defective gage) can also apply to excessive oil pressure, with the same corrective measures.

d. **Excessive Oil Consumption.** Do not ignore excessive oil consumption. It indicates serious trouble and may cause depletion of oil supply at a time when engine cannot be refilled.

(1) SERIOUS OIL LEAKS. If engine uses an excessive amount of oil, leaks can be suspected. Examine the various parts of the engine, including engine compartment floor. If any accumulation of oil is found on floor plates or parts, find the cause and correct before further use of engine.

(2) INCORRECT GRADE OF OIL. If oil is too light in body it will cause excessive oil consumption. Inspect oil in engine by comparison with known correct grade. Drain and refill with proper grade and quantity (par. 145 c and d).

(3) WORN OR SCORED PISTONS AND CYLINDERS. If pistons, rings, and cylinders are badly worn, it will cause excessive oil consumption. This condition is usually indicated by a smoky exhaust, fouled spark plugs, and loss of power. If this cause is suspected, make a compression test of each cylinder (par. 70 c). If pressure is low, notify higher authority.

e. **No Oil Pressure.** If no oil pressure is indicated immediately after engine is started, stop engine immediately. Check oil supply. If oil supply is adequate, start engine and run a little above idling speed; look for following causes:

(1) CLOGGED OIL TUBES. Oil tubes may be clogged or frozen. There may be an accumulation of water in the oil pan that has frozen. In any case do not run engine more than 1 minute. Shut it down and remove oil tube at fitting located just under the fan drive on left side of engine. Start engine; if oil comes out of tube with considerable quantity and force, it indicates oil supply is not shut off from pump, and that pump is not defective. If no oil comes out, pump is defective and must be replaced (notify higher authority), or oil is frozen (in cold weather). If latter cause is suspected contact higher authority for instructions for thawing out engine. It will also be necessary to determine cause of water in oil pan. This will prob-

ably be due to a cracked block or blown cylinder head gasket. Trace and locate trouble and notify higher authority.

62. GENERATING SYSTEM.

a. General. For trouble shooting purposes the generating system consists of the generator, generator regulator, circuit breaker, wiring, 24-volt master switch, and the ammeter (fig. 73).

b. Faulty Generator Regulator. If vehicle lamps burn out frequently the generator regulator probably is at fault, and must be replaced (par. 85 b and c).

c. Incorrect Indication of Ammeter. The ammeter indicates the amount of electrical current being generated, or flowing from the batteries, and will vary according to generation of current and the amount of current being used. All current consumption, except current used by the starter, is shown on the ammeter. If generator or generator regulator is faulty, or shorts or broken wires develop in the wiring system, such fault will be reflected by an ammeter discharge indication above normal. Faults that would probably cause abnormal ammeter indications follow:

(1) FAULTY GENERATOR. A faulty generator will cause a low or unsteady amperage indication on the ammeter. If the ammeter pointer drops below normal position (based on experience), with the auxiliary engine running, it shows that either the generator or regulator is at fault. NOTE: *If batteries are fully charged, ammeter indication will be correspondingly lower.* Examine generator brushes and adjust or install new brushes (par. 84 a and b). If incorrect indication on ammeter continues, it shows generator regulator unit is at fault. Install a new regulator unit. If this does not correct the trouble, the generator is faulty and must be replaced (par. 83).

(2) FAULTY GENERATOR REGULATOR. The generator regulator can also be faulty by reason of failure of the reverse current relay to open when engine is stopped. This will be shown by the ammeter pointer showing a heavy discharge with engine not running. Turn 24-volt master switch off at once. Hand-hold one of the generator pulleys and have the 24-volt master switch turned on. If pulley is felt to move, it indicates the generator regulator is faulty. Turn off 24-volt master switch. Replace the generator regulator (par. 85 b and c). After new generator regulator has been installed, position drive belt, and adjust (par. 86 c), start engine. Observe ammeter to make sure system is generating properly.

(3) FAULTY WIRING. Loose wiring in the generating system, including wires running to the ammeter, can cause erratic action of the ammeter. Examine all terminals and tighten as necessary. A short in any wire will be indicated by a released circuit breaker in the

faulty circuit. If the generator regulator unit is not faulty some wire within the wiring system may be shorted. Locate short by inspecting various wires in the faulty circuit. Test wires by disconnecting at terminals and pushing circuit breaker button in. Button will continue to snap out until wire containing short is disconnected. When this wire is disconnected and circuit breaker button is reset it will stay in. Replace shorted wire or damaged conduit.

(4) FAULTY AMMETER. If none of the causes in steps (1), (2), and (3), preceding apply, and the ammeter does not show a normal amperage indication for the condition at the time it is probably faulty. Attach two jumper wires to the posts of a new ammeter and temporarily connect the other ends to the two large posts of the ammeter shunt in the battery compartment. Start auxiliary engine. Observe indication on ammeter under various uses of current. Accelerate engine to show *charge*. Stop auxiliary engine. Turn on vehicle driving lights to show *discharge*. If this test shows the ammeter is faulty, replace it with a new ammeter (par. 89 b and c).

63. BATTERY AND LIGHTING SYSTEM.

a. General. The battery and lighting system consist of the two 12-volt batteries, the 24-volt master switch, all vehicle lights, compartment lights, and all light switches and necessary wiring (fig. 76). NOTE: *No reference is made within this paragraph to the batteries in connection with the starting system, as they are covered in paragraph 92.*

b. Improper Operation of Lighting System. If lamps do not light when various switches are turned on, or if the lamps are dim, it may be due to one of the following causes:

(1) DISCHARGED OR DEFECTIVE BATTERY. Make hydrometer test (par. 92 c). If necessary recharge or install fully charged battery or batteries. If lights dim to a low glow only when starter switch is operated, it indicates loose or corroded terminals or a defective battery. Examine, clean, and tighten terminals or replace battery with a new one (par. 92 d and e) if old one is believed defective.

(2) DEFECTIVE 24-VOLT MASTER SWITCH OR LIGHT SWITCH. If lamps do not light at all, the 24-volt master switch may be defective, the switch controlling a particular light may be faulty, or the lamp itself may be defective. If the 24-volt master switch is defective, the starter will not operate. Replace switch (par. 94 b and c). If a light switch is believed defective, first test condition of lamps (subpar. b (3) following) and then condition of wiring (subpar. b (4) following). If lamp and wiring are found in good condition light switch is defective. Install new switch (par. 96 e (1) and (2)).

(3) DEFECTIVE LAMP OR LAMP-UNIT. If lamps do not light, test the lamp by installing a new one (par. 96). If the lamp in a

sealed beam unit is suspected of being faulty the entire unit must be replaced (par. 96 a (2) and (3)).

(4) DEFECTIVE WIRING. If insulation has been broken in any part of the lighting system wiring, it will cause the circuit breaker which protects that particular wiring or circuit to open. Replace wiring and conduit as necessary (par. 95 c and d).

c. **Battery Uses Excessive Amount of Water.** Batteries will use considerably more water in summer than in winter, the amount used being in proportion to atmospheric temperature and use of batteries. If batteries use more water than is normal it is probably due to the fact that battery case is cracked. This will be indicated by wetness around the battery compartment (particularly on the floor). Remove batteries (par. 92 d) and wash out battery compartment with soda solution. Replace damaged battery (par. 92 e).

64. INSTRUMENTS.

a. **General.** Instruments covered in this paragraph are those on the instrument panel not covered in any of the paragraphs within this section.

b. **Defective Tachometer.** There are five parts of the tachometer assembly that can become inoperative: Tachometer head, tachometer adapter, tachometer drive cable (the flexible shaft connecting the tachometer adapter with the tachometer sending unit), sending unit mounted on top of the left fuel tank, the tachometer head in the instrument panel, the automatic speedometer-tachometer switch and connecting wires.

(1) BROKEN DRIVE SHAFT. If the tachometer does not indicate engine revolutions per minute, disconnect drive shaft housing from adapter by unscrewing knurled connector at the adapter. Try to pull drive shaft from housing. If it comes out in sections install a new shaft (par. 100 d and e).

(2) DEFECTIVE TACHOMETER ADAPTER. If drive shaft does not come out, start the engine and see if the slotted connection (drive quill) inside the adapter opening turns. If it does not turn, adapter is defective and must be replaced (par. 100 b and c).

(3) DEFECTIVE SENDING UNIT. If items in steps (1) and (2) preceding are not at fault, the sending unit on top of left fuel tank may be defective. Disconnect drive shaft housing and conduit from sending unit. Temporarily install a new sending unit without attaching it to fuel tank. If old sending unit proves defective, remove and install new unit (par. 100 f).

(4) DEFECTIVE AUTOMATIC SPEEDOMETER-TACHOMETER SWITCH. Disconnect switch and install a jumper wire. Start engine. If tachometer operates it indicates switch is faulty. Install a new switch.

If tachometer does not operate it indicates wire to tachometer head or the head itself is at fault.

(5) DEFECTIVE CONDUIT. If items in steps (1), (2), and (3) preceding are not at fault, the conduit from the sending unit to the tachometer head may be faulty. Unscrew conduit connector from tachometer at the back of instrument panel and at the sending units, and install a new conduit.

(6) DEFECTIVE TACHOMETER HEAD. If items in subparagraphs (1), (2), (3), (4), and (5) preceding are not at fault, remove tachometer head from instrument panel and install a new one (par. 98 f (1) and (2)).

c. Defective Speedometer. If the speedometer is defective, the same causes and remedies will apply as for the tachometer. The speedometer adapter is mounted on the right side of the transmission housing, and the sending unit is mounted on top of the right fuel tank.

d. Defective Transmission High Oil Temperature Signals. If either of the dual transmission high oil temperature signals on the instrument panel is defective, and does not light, it indicates a burned-out panel lamp. If both signals fail to light it may be due to a faulty switch. Install a new switch (par. 152 f and g).

e. Defective Fuel Gage, Fuel Gage Control Switch or Fuel Sending Unit. If the fuel gage does not indicate fuel level in either tank, it may be due to faulty fuel gage, control switch, or fuel gage sending unit.

(1) DEFECTIVE FUEL GAGE OR FUEL GAGE CONTROL SWITCH. To determine whether the fuel gage or the control switch is at fault, temporarily attach a new fuel gage in the fuel gage circuit after removal of the instrument panel. If the new gage does register correctly, it indicates that the control switch is at fault. Replace control switch (par. 119 d and e). If the fuel gage proves defective, replace it with a new one (par. 119 b and c).

(2) DEFECTIVE FUEL GAGE SENDING UNIT. If the right or left tank fuel gage sending unit registers incorrectly, with a known quantity of fuel in the tank, the fault lies in the float mechanism within the tank. Replace the fuel gage sending unit (par. 119 f and g).

f. Defective Circuit Breakers. If a circuit breaker releases it can be assumed that the trouble lies in the wiring of the circuit which that particular circuit breaker protects. Inspect the circuit wiring. If the wiring is not defective, then the circuit breaker is at fault. Replace the circuit breaker.

g. Defective Instrument Panel Lamps or Switch. If either instrument panel lamp does not light, replace the lamp, first making sure circuit breaker is operative. If neither of the lamps light, the trouble is probably due to a defective switch or wiring. Remove instrument panel (par. 97 h). Remove instrument panel light switch (par. 98 h (1)). Disconnect all wires from switch and temporarily attach wires to proper terminals of a new switch. If the lamps then light and their brilliancy can be regulated properly, it shows old switch is defective. Install new switch (par. 98 h (2)). If temporary installation of new switch does not remedy the trouble, it shows there is some defect in the circuit. Locate and correct.

65. TURRET.

a. General. The only faults that will be considered in the turret as far as trouble shooting is concerned are those in connection with the traversing mechanism. In vehicles equipped with hydraulic traversing mechanism, traverse turret manually before making further investigations. This is to make sure the turret will rotate smoothly in both directions and that the trouble is not caused by the turret binding at some point or points.

b. Traversing Pump Fails to Operate. If the traversing pump fails to operate, the fault lies within the items covered in the following subparagraphs:

(1) BATTERY WIRES SHORTED OR LOOSE CONNECTIONS. Examine wires and connections of all wires in the pump motor circuit. Particularly examine the connections and the switch in the switch box. Tighten connections, or replace defective wires.

(2) BURNED-OUT MOTOR. If no other cause can be found the motor may be burned out or otherwise defective. Notify higher authority.

(3) SHEARED COUPLINGS OR DEFECTS WITHIN PUMP, GEAR BOX OR CONTROLS. If the motor runs but turret does not traverse, or traverses in an erratic manner, refer to paragraph 196. If the turret drifts one way or the other with vehicle on level ground and control handle in neutral position, refer to paragraph 196 b.

66. RADIO INTERFERENCE SUPPRESSION.

a. Preliminary Instructions. When checking radio interference test the vehicle in a location free from high tension lines, other vehicles, machinery, and electrical equipment which could be a source of interference. Stop engine and turn off all vehicle electrical equipment. Turn on radio and check noise level. Listen carefully to type of noise present under these conditions. This registering of noise sounds is done so that when vehicle equipment is checked the presence of a new noise or interference can be immediately detected. Checking noise level is also done so that noises already present will not be attributed to lack of vehicle suppression. If noise level with all vehicle equipment shut off is too high due to atmospheric condi-

Part Three—Maintenance Instructions

tions or other outside causes, delay further checking, if tactical situation permits, until such time as moderate noise levels prevail. Disconnect the radio terminal box capacitor (terminal end). If radio interference does not increase, replace capacitor. If interference increases, as it should, leave terminal box capacitor disconnected for the remainder of the tests, or until any defective equipment has been located, replaced, and tested. Examine all shielded conduits and cables to make sure couplings are tight, and conduits and cables are clamped or bonded to hull at least every 2 feet.

b. **Noise Caused by Ignition System of Vehicle Engine.** Radio noise caused by the engine of the vehicle can be determined by running the engine on first one bank of cylinders and then on the other bank. It is very unlikely that noise will be present in both magneto systems at the same time. However, if such is the case, then the remedy will apply for both magneto systems as given in the following information:

(1) LOOSE OR DAMAGED CONDUITS OR WIRING. Examine conduit of ignition system and see that all coupling nuts and terminal connections are tight and that conduit is not broken or crushed. Replace any damaged conduit or wires.

(2) LOOSE BONDING CLIPS OR ENGINE GROUNDING STRAP. Tighten all bonding (holding) clips and tighten engine ground strap. NOTE: *If information given in subparagraphs b and b (1) preceding does not correct the trouble, and it is believed due to ignition faults, refer to higher authority.*

c. **Noise Caused by Generator or Auxiliary Engine Ignition System.** Radio noise can be caused by either the generator system or the ignition system of the auxiliary engine.

(1) FAULTY GENERATOR REGULATOR. If radio operator reports the radio interference as a clicking sound the trouble is probably due to a faulty generator regulator. Tighten bolts holding regulator unit to generator. Tighten bonding strap between generator and hull and between regulator unit and bracket. If noise continues, replace regulator (par. 85 b). NOTE: *If information given in subparagraph c (preceding) and c (2) and c (3) following does not correct the trouble, refer to higher authority.*

(2) IGNITION NOISE. To determine if ignition noise is the cause of the disturbance, push auxiliary engine STOP switch. If noise occurs while engine is running at charging speed, it shows that cause lies within generator system. Refer to step (3) following. If noise stops as soon as STOP switch is pushed, it shows cause of noise is in auxiliary engine ignition system. Examine condition of ignition wires and conduits and replace any damaged wiring harness.

(3) GENERATOR SYSTEM NOISE. If radio interference is caused by the generator system, tighten all bonding clips, coupling nuts and

the ground straps of the auxiliary engine. Examine conduit and if it is crushed, or cut, or in any way damaged, replace the conduit and/or wires (par. 91 b and c). Tighten conduit fittings in junction and control boxes. If noise continues, replace first one, and if necessary both, of the condensers which are found on the brush-holding plate of the generator. Replace them in the same position as the original ones were installed. If noise still continues, refer to higher authority.

d. **Noisy Turret Traversing Electrical System.** If radio noise is believed due to faults within the turret traversing electrical system it can be determined by traversing the turret at least one complete revolution with vehicle engine and auxiliary engine stopped. If radio interference is caused when turret is traversed, the fault lies within the following:

(1) FAULTY CONDUITS, WIRING, OR BONDING. Tighten all coupling nuts and bonding clips. Replace all defective conduit and/or wiring (par. 102). Tighten bonding strap and bonding cable under turret collector ring assembly. If either strap or cable is damaged, refer to higher authority. Tighten all conduit fittings in junction and control boxes and tighten switches and junction blocks in switch boxes. Replace the capacitor in switch box. If noise continues and trouble is not remedied refer to step (2) following.

(2) FAULTY TURRET MOTOR OR COLLECTOR RING. If cause or radio interference is not traceable to subparagraph d (1) preceding, it is probably due to a faulty motor or a faulty collector ring assembly. In either case refer to higher authority.

e. **Noise Caused by Miscellaneous Systems.** Noise may be caused by loose or damaged conduits or wires of the lighting systems; various instruments including their respective sending units such as fuel tank gages, engine temperature gage and signals, oil pressure gage and signals, and speedometer and tachometer. Various instruments on the instrument panel and in control boxes may become loose. Battery terminals and ground cable of the batteries may work loose or become corroded or otherwise damaged. The lights of the vehicle and their switches, as well as the capacitors in windshield wipers, are also sources of radio noise.

(1) LOCATION AND REMEDY OF MISCELLANEOUS RADIO NOISE. The same general rule for maintenance of various systems, instruments and devices must be followed as given previously concerning correction of radio interference. A process of elimination can be applied to location of trouble within any system.

f. **Radio Interference Not Traceable to Any Electrical System.** If radio interference persists in spite of remedies set forth in subparagraph b preceding, to and including subparagraph e (1) preceding, the trouble is probably due to defective radio apparatus, or to faults within the structure of the vehicle. Refer to higher authority.

Part Three—Maintenance Instructions

Section XVII
ENGINE DESCRIPTION AND MAINTENANCE IN VEHICLE

67. ENGINE.

a. **Description.** Heavy Tank T26E3 is powered by a Ford, 60-degree, V-type, 8-cylinder, 4-cycle, valve-in-head, liquid-cooled, model GAF gasoline engine (figs. 52, 53 and 54). The cylinder block and crankcase are cast in one piece of aluminum, with steel sleeves in the cylinder bores. The water jackets extend the full length of the cylinders. Each cylinder bank has two overhead camshafts, one intake and one exhaust, and each cylinder has two intake and two exhaust valves (fig. 108). The engine, transmission, and differential are mounted in the engine compartment at the rear of the vehicle (fig. 56) and are considered as one unit (power unit) when removing and installing. The engine is supported by two legs, one on each side at the middle of the oil pan and cast integral with the pan (fig. 54). Rubber shock mountings are used between the engine support legs and the engine mounting brackets on the hull floor (fig. 54). Two magnetos, one for each bank of cylinders, are mounted on the front or magneto end of the engine (fig. 52). They are located at each end of cross shafts which are driven by gears from the main crankshaft. The water pump (fig. 52) on the magneto end of the engine, is driven from the crankshaft. An accessory angle drive and a fan drive propeller shaft are located on each side of the engine (figs. 53 and 54). They are driven from the main crankshaft by accessory drive gears in the engine, accessory drive shafts and universal joints. Four pusher-type, 5-bladed fans, two on each side of the cooling unit, are driven through double V-type belts and pulleys by the fan drive propeller shafts (figs. 55 and 115). The generator is mounted on top of the auxiliary engine and is driven by either the auxiliary engine or the main engine through V-belts and two overrunning pulleys. The gear type oil pump (par. 149) is located at the bottom of the oil pan on the magneto end of the engine (fig. 52). The carburetors, air intake manifold, and the fuel pump are mounted on top of the engine between the cylinder banks (fig. 56).

b. **Engine Nomenclature.** Throughout this manual the front, or flywheel end of the engine, will be referred to as the "flywheel end," because the engine is mounted in the tank with the front or flywheel end toward the *rear* of the vehicle. The *rear* end of engine will be referred to as the "magneto end," because the rear of the engine faces the *front* of the vehicle (fig. 55). NOTE: *The terms "right" and "left" used with reference to the engine are opposite to the right and left of the vehicle.* Standing in the fighting compartment and facing the magneto end of the engine, the cylinder bank on the right

TM 9-735
67

Engine Description and Maintenance in Vehicle

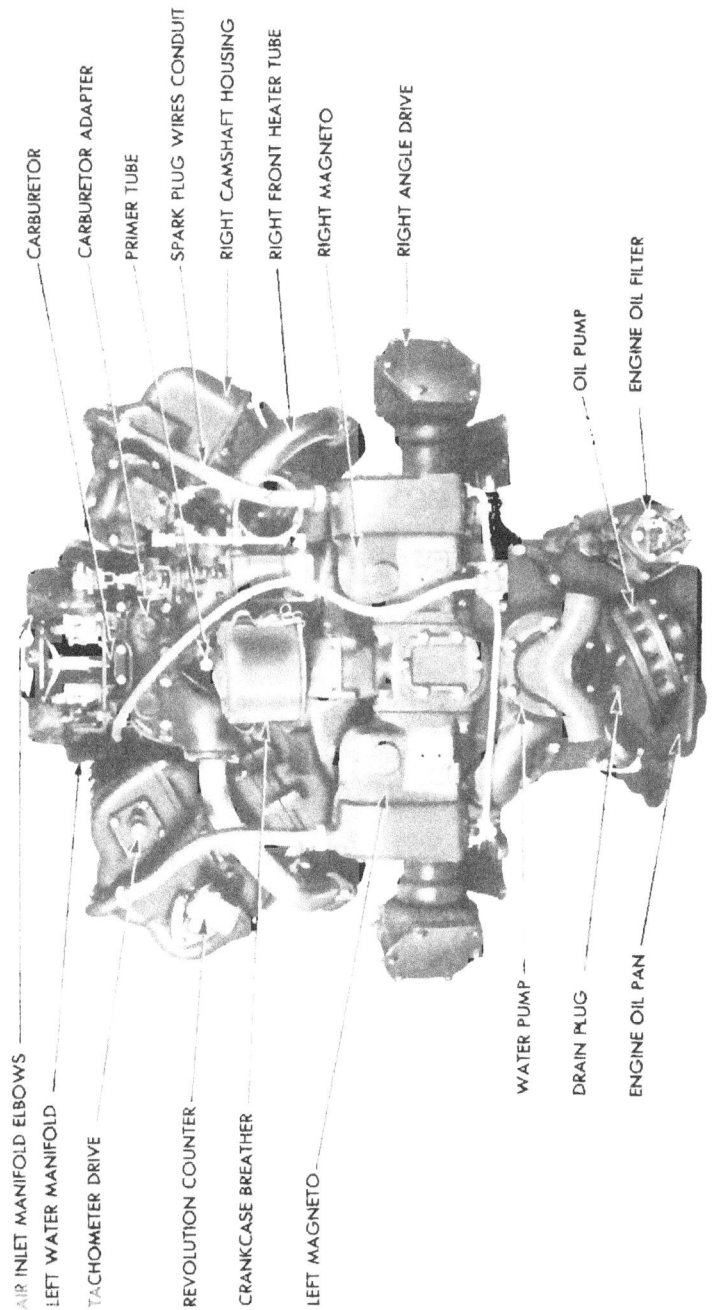

Figure 52 — Engine, Viewed From Magneto End

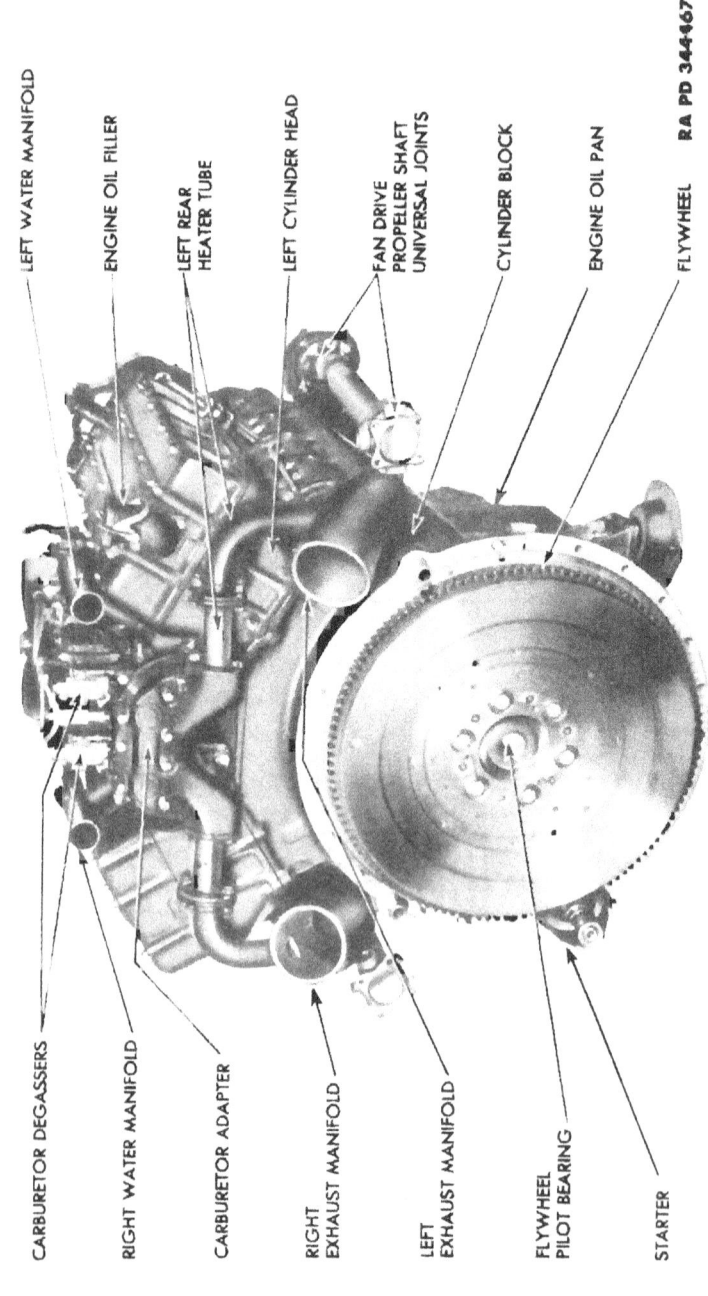

Figure 53 — Engine, Viewed From Flywheel End

TM 9-735
67

Engine Description and Maintenance in Vehicle

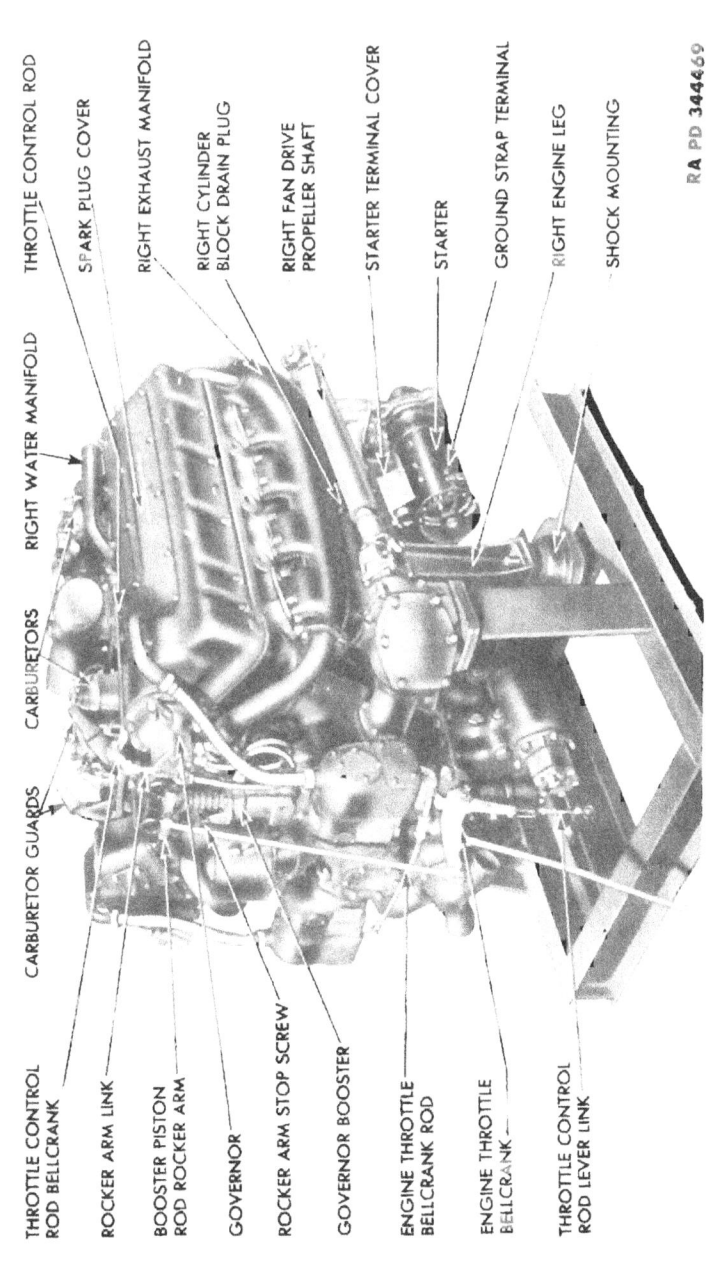

Figure 54 — Right Side of Engine Showing Starter

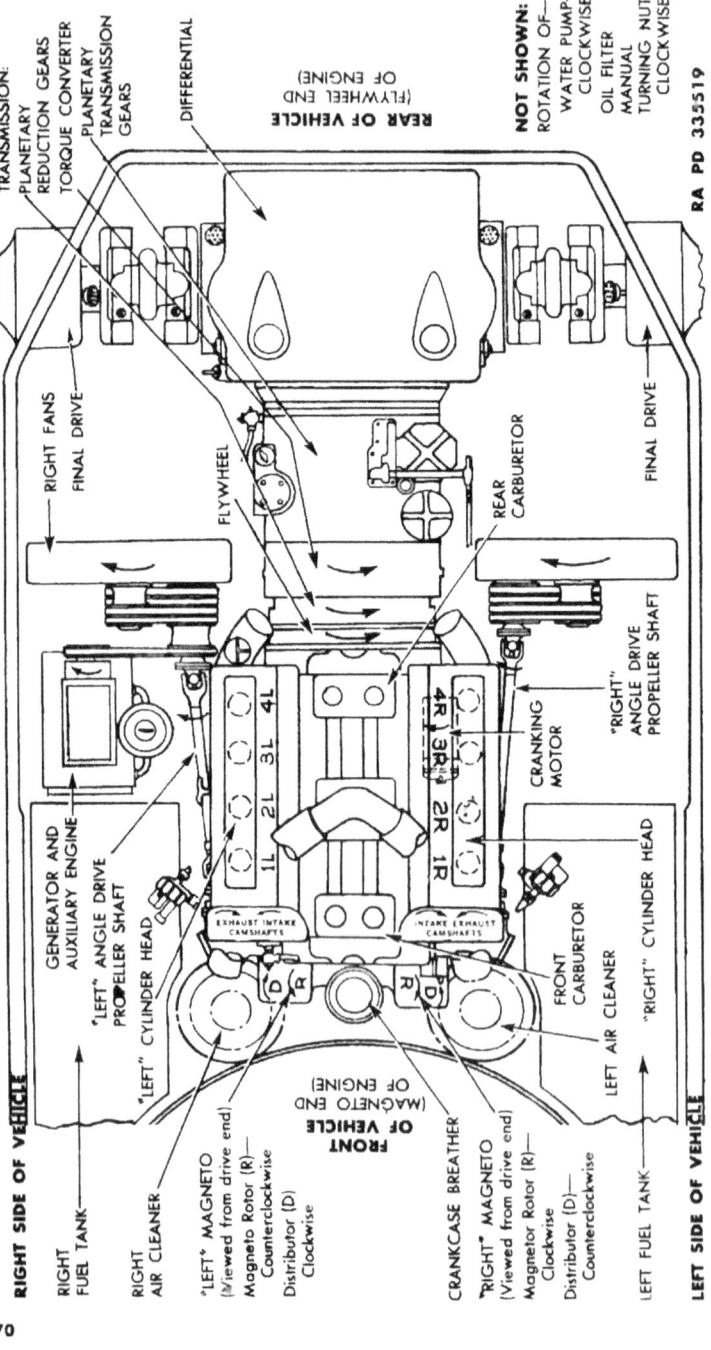

Figure 55 — Power Unit Identification Diagram

is the "right" cylinder bank but it is on the *left* side of the vehicle. The cylinder bank on the left is the "left" cylinder bank but it is on the *right* side of the vehicle (fig. 55). Cylinders are numbered from the magneto end and the number and cylinder bank designating letter "R" or "L" is stamped in each fuel intake manifold, 1 R, 2 R, 3 R, 4 R... 1 L, 2 L, 3 L, 4 L. See figure 55.

c. Direction of Rotation. In referring to the direction of rotation, the right-hand direction (the direction in which the hands of a clock turn) is clockwise, and the left-hand, or opposite direction, is counterclockwise. NOTE: *Rotational directions of engine parts should always be described and considered as being viewed from the magneto end of the engine.* Reference to figure 55 will show direction of rotation of the engine parts, as well as nomenclature of other parts of the engine.

68. TABULATED DATA.

a. General. The following data includes the general information and engine characteristics which are frequently required for reference:

b. Engine Tabulated Data.

Type	60-deg V-type, 8-cyl. 4-cycle, gasoline, tank engine.
Model and make	GAF-Ford
Octane rating of fuel	80 or higher
Horsepower	500 at 2,600 rpm
Bore and stroke	5.4 x 6 in.
Piston displacement	1,100 cu in.
Compression ratio	7.5 to 1
Maximum governed speed	2,800 rpm
Maximum warm-up speed	1,200 rpm
Minimum idling speed	500 rpm
Weight (less angle drive shafts, engine mountings, exhaust pipes, air cleaners and tubes)	1,416 lb
Capacity of cooling system	22 gal
Capacity of engine oil pan	8 gal
Capacity of each air cleaner	3½ qt
Firing order	1R, 2L, 3R, 1L, 4R, 3L, 2R, 4L
Magneto breaker point gap	0.014 to 0.016 in.
Spark plug gap (nonadjustable)	0.011 to 0.014 in.
Valve clearance (nonadjustable)	0.027 to 0.030 in.

69. ENGINE TUNE-UP.

a. General. Many factors contribute to loss of engine power, rough idling, and hard starting. These faults can usually be corrected by restoring the original clearances, fits, and adjustments in the various engine units. In many cases loose, faulty, or dirty connections are the cause. Usually the symptom will suggest the proper operation to be performed from those listed in subparagraph b following, since in each case it may not be necessary to perform them all to restore the engine to peak operating efficiency.

b. Tune-up Operations. Listed below are tests, checks, and operations which are frequently necessary to perform when engine has lost power, does not operate smoothly, or is hard to start:

(1) Remove spark plugs, clean, and inspect gap (par. 76).
(2) Test engine compression (par. 70 c).
(3) Clean and adjust magneto points (par. 75 d).
(4) Check and adjust magneto timing (par. 75 c).
(5) Clean and tighten ignition wires.
(6) Check carburetor float levels (par. 123 b (1)).
(7) Test fuel pump pressure (par. 70 e).
(8) Test fuel pump vacuum (par. 70 d)..
(9) Adjust throttle and governor booster linkage on engine (par. 126 b).
(10) Adjust engine idling speed (par. 123 b).
(11) Adjust carburetor idling fuel mixture (par. 123 b).
(12) Clean air cleaners and tighten all air connections (par. 128 b and d).
(13) Clean crankcase breather (par. 146 b).
(14) Test battery, fill, clean and tighten terminals (par. 92 b).
(15) Tighten starter cable connections and engine ground strap.
(16) Test operation of and clean oil filter (par. 147 b and c).
(17) Inspect and tighten cooling system connections.
(18) Fill or clean cooling system (par. 134 d and f).
(19) Check cooling system thermostat for proper operation (par. 137).
(20) Clean radiator and oil cooler air passages and cores.
(21) Test and adjust fan belt tension (par. 138 b).
(22) Road-test the vehicle (par. 53).

70. OPERATIONS PERFORMED WITH ENGINE IN VEHICLE.

a. General. Listed in subparagraph b following are operations and tests which can be performed with the engine in the vehicle.

TM 9-735
70

Engine Description and Maintenance in Vehicle

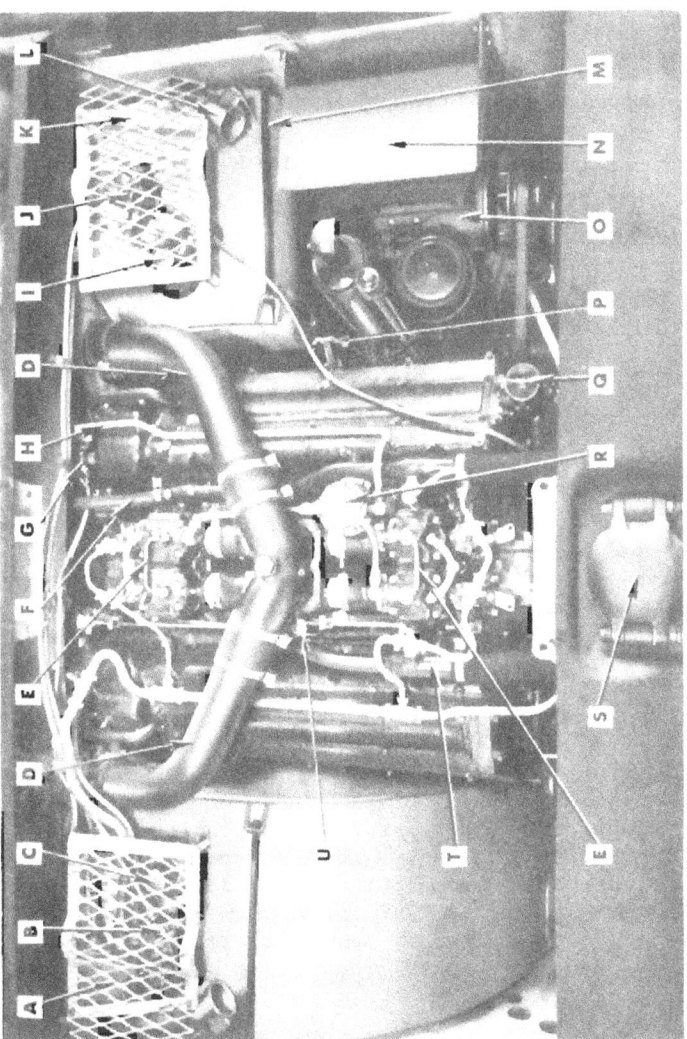

A—ENGINE COMPARTMENT TERMINAL BOX
B—TACHOMETER SENDING UNIT
C—LEFT FUEL TANK GAGE SENDING UNIT
D—AIR INTAKE TUBE
E—CARBURETOR
F—HEATER WATER HOSE
G—TACHOMETER ADAPTER
H—FUEL TUBE
I—RIGHT FUEL TANK GAGE SENDING UNIT
J—SPEEDOMETER SENDING UNIT
K—GUARD
L—FIRE-EXTINGUISHER DISCHARGE NOZZLE
M—RIGHT FUEL TANK
N—REGULATOR GUARD
O—AUXILIARY ENGINE AND GENERATOR
P—ENGINE OIL LEVEL INDICATOR
Q—OIL FILLER
R—FUEL PUMP
S—WATER FILLER CAP COVER
T—WATER MANIFOLD
U—THROTTLE LINKAGE

RA PD 344426

Figure 56 — Engine In Vehicle, Top View

The majority of these operations are covered in the sections pertaining to the system of which a particular unit is a part. A few operations not covered elsewhere in the manual but which can be performed with the engine in the vehicle are given in detail in subparagraphs c, d, and e following.

  b. **Operations Possible to Perform With Engine in Vehicle.**
  (1) Removal of spark plugs (par. 76 b).
  (2) Removal of carburetors (par. 123 c).
  (3) Removal of carburetor adapters (par. 124 h).
  (4) Removal of air intake manifold (par. 133 b (8)).
  (5) Adjustment of carburetor throttle linkage (par. 127 b).
  (6) Removal of fuel pump (par. 121 b).
  (7) Test carburetor float levels (par. 123 b (1)).
  (8) Removal of water manifolds (par. 142 b).
  (9) Adjust either magneto timing (par. 75 c).
  (10) Adjust either magneto points (par. 75 d).
  (11) Removal of either magneto (par. 75 b).
  (12) Removal of water pump (par. 143 b).
  (13) Test operation and clean oil filter (par. 147 b and c).
  (14) Service engine lubrication system (par. 145).
  (15) Clean crankcase breather (par. 146 b).
  (16) Replace distributor points (par. 75 d).
  (17) Removal of oil pump (par. 149 b).

  c. **Testing Engine Compression.** Engine compression pressure indicates the condition of valves, pistons, piston rings, and cylinder head gaskets. The test is made to determine if any of these parts require servicing or replacement. To test engine compression, start and run engine until it is warmed up to 150° F. This is to make sure all engine parts have reached normal operating temperature and are thoroughly lubricated. Remove all eight spark plugs (par. 76 b). Set throttle at full-open position. Insert universal-type cylinder compression gage (41-G-124) into No. 1 spark plug hole. Hold gage firmly seated and operate starter until gage reading reaches maximum value. Record reading and release gage pressure. Repeat test on other seven cylinders in numerical rotation and note readings. Compression pressure may vary between 155 to 180 pounds, measured at cranking speed. This variation will not affect engine performance to the extent that parts replacement is required. However, normally high compression pressure in any one cylinder may indicate excessive carbon deposit in the combustion chamber of that cylinder. Abnormally low individual cylinder pressure may indicate improper seating of the valves, worn piston rings, or leaking cylinder head gasket. When the pressure in any one or more cylinders is more than 20 to 30 pounds less than the average of the pressure in the other cylinders, the cause must be determined and corrected. Notify higher authority. NOTE: *The above compression pressures are based on operation at sea level. At higher elevations normal pressures will be correspondingly lower.*

d. **Testing Fuel Pump Vacuum.** Fuel pump vacuum is tested to determine the condition of the inlet valves, pump diaphragm, and also to gage the general performance of the fuel pump. Remove front center deck plate. Close both fuel shutoff valves. Disconnect fuel tube from fuel pump. Connect combination vacuum and pressure gage (41-G-500) to inlet port on fuel pump. Start and run engine at 700 revolutions per minute. Observe gage, and stop engine when maximum reading is indicated. A fuel pump under normal operating condition will develop a vacuum of 10 inches of mercury or its equivalent. If gage reading is less than this value, or if gage hand is seen to rapidly return to zero, remove fuel pump (par. 121 b), and install a new one. Disconnect gage and connect fuel tube to fuel pump. Install front center deck plate.

e. **Testing Fuel Pump Pressure.** As a further check on condition of fuel pump diaphragm, valves, and rocker arm spring and linkage, test fuel pump pressure. Remove front center deck plate. Disconnect fuel hose to carburetor at magneto end at tee connections. Attach combination vacuum and pressure gage (41-G-500) to tee connection. Start and run engine at 700 revolutions per minute. Observe gage reading. If pump is operating correctly the gage will indicate a pressure of from 4½ to 6 pounds. Remove fuel pump, and install a new one if pressure is not within these limits. Stop engine and disconnect gage. Connect fuel hose to tee connection. Install front center deck plate.

Section XVIII

POWER UNIT REMOVAL AND INSTALLATION

71. POWER UNIT REMOVAL.

a. **General.** The power unit of the T26E3 consist of the engine, transmission (torque converter and planetary gears), and the differential (fig. 63). The power unit is removed as a complete unit after the complete cooling unit has been removed as an assembly (par. 139 b). Facilities for removing power unit include suitable heavy duty hoisting equipment, necessary tools, clean containers to hold drained liquids, cable or chain slings, power unit lifting sling (KM-J-4349), wood blocking, and clean receptacles for small subassemblies and attaching parts. Keep removed parts separate and

TM 9-735

Part Three—Maintenance Instructions

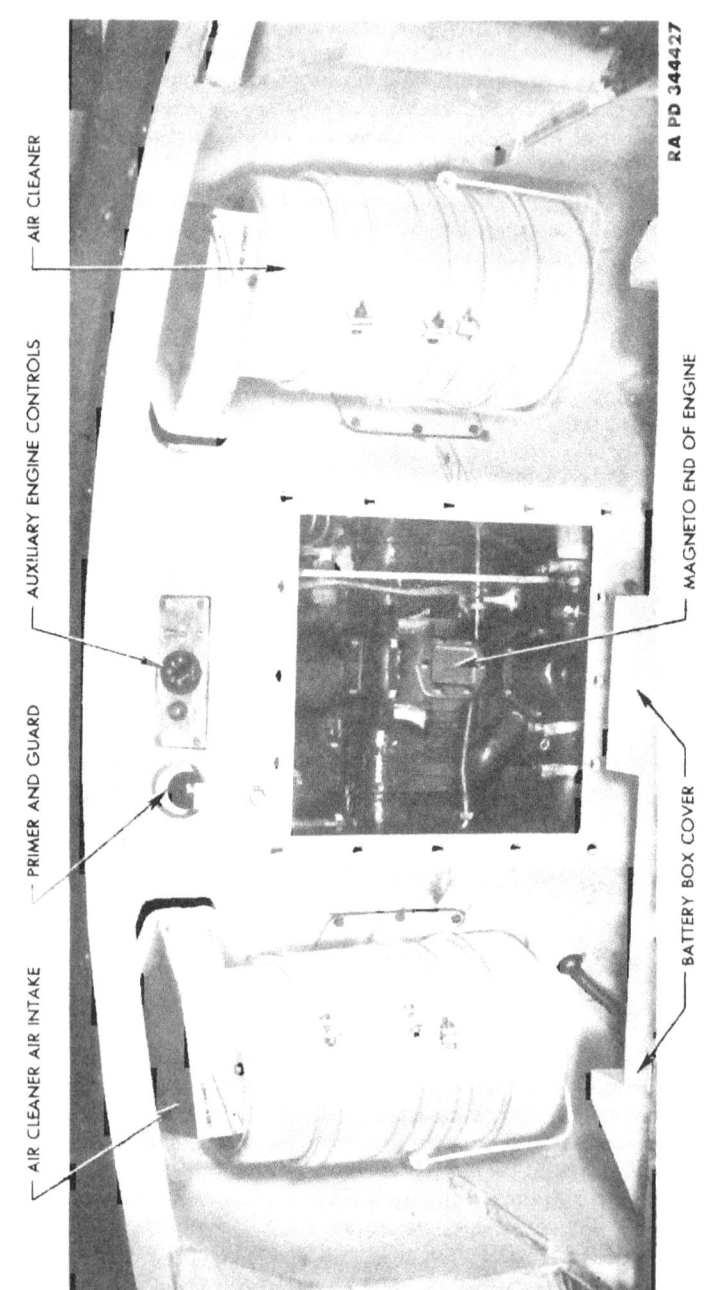

Figure 57 — Connections To Engine At Magneto End

TM 9-735

Power Unit Removal and Installation

RA PD 335522

Figure 58 — Lifting Power Unit From Engine Compartment

in related position to facilitate installation. Loosely reinstall all bolts, washers and nuts to ensure correct reassembly and to prevent loss. Tagging of conduits, wires, hoses, tubes and other parts aids in installation identification. WARNING: *Tape or cover all openings and ends of lines and tubes to prevent entrance of dirt. Protect all removed parts against dust, sand, and weather.*

b. **Distribution of Removal Procedure.** To facilitate removal of power unit, divide disassembly operations into groups. This permits dividing personnel into groups, each group proceeding simultaneously with its assigned operations:

(1) General preparatory.
(2) Disassembly in fighting compartment.
(3) Disassembly in engine compartment.
(4) Removal of power unit.

c. **Removal of Power Unit.**

(1) CLOSE FUEL TANK SHUT-OFF VALVES. Close both right and left fuel tanks, shut-off valves (par. 116 c).

TM 9-735

Part Three—Maintenance Instructions

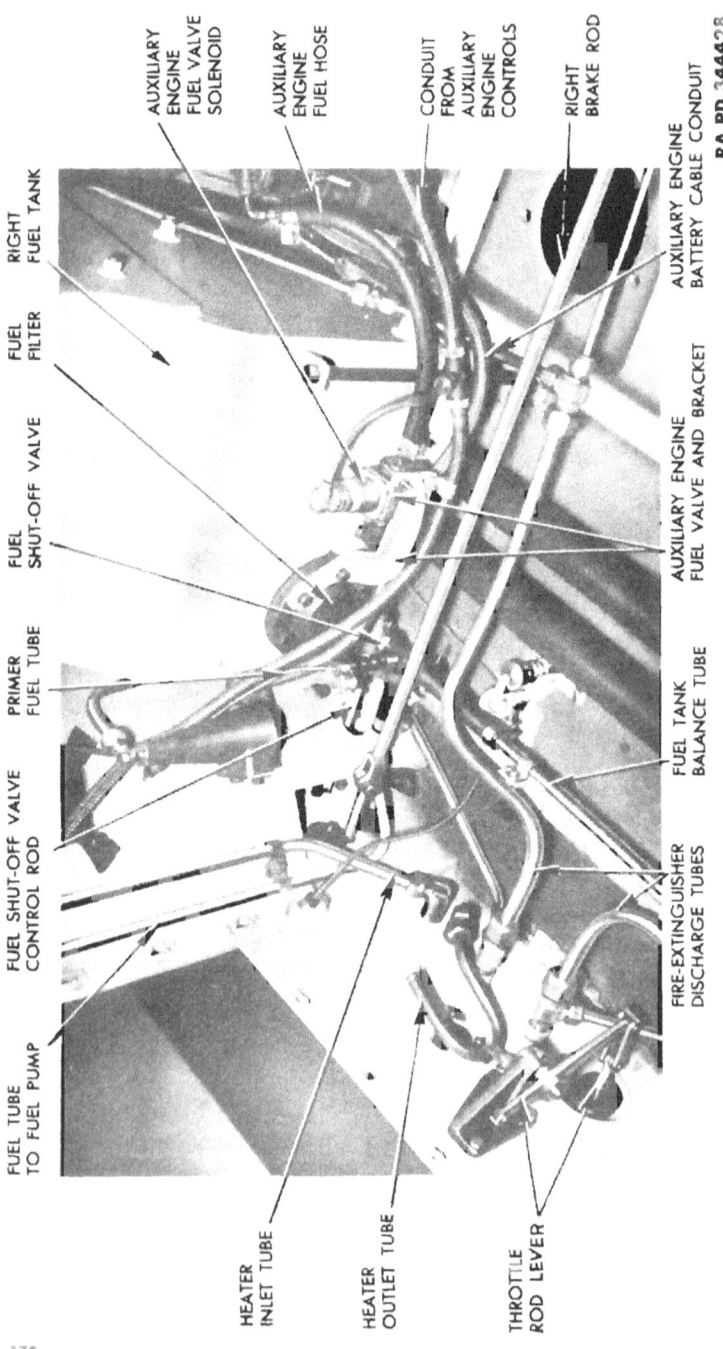

Figure 59 — Right Fuel Tank Connections

TM 9-735

Power Unit Removal and Installation

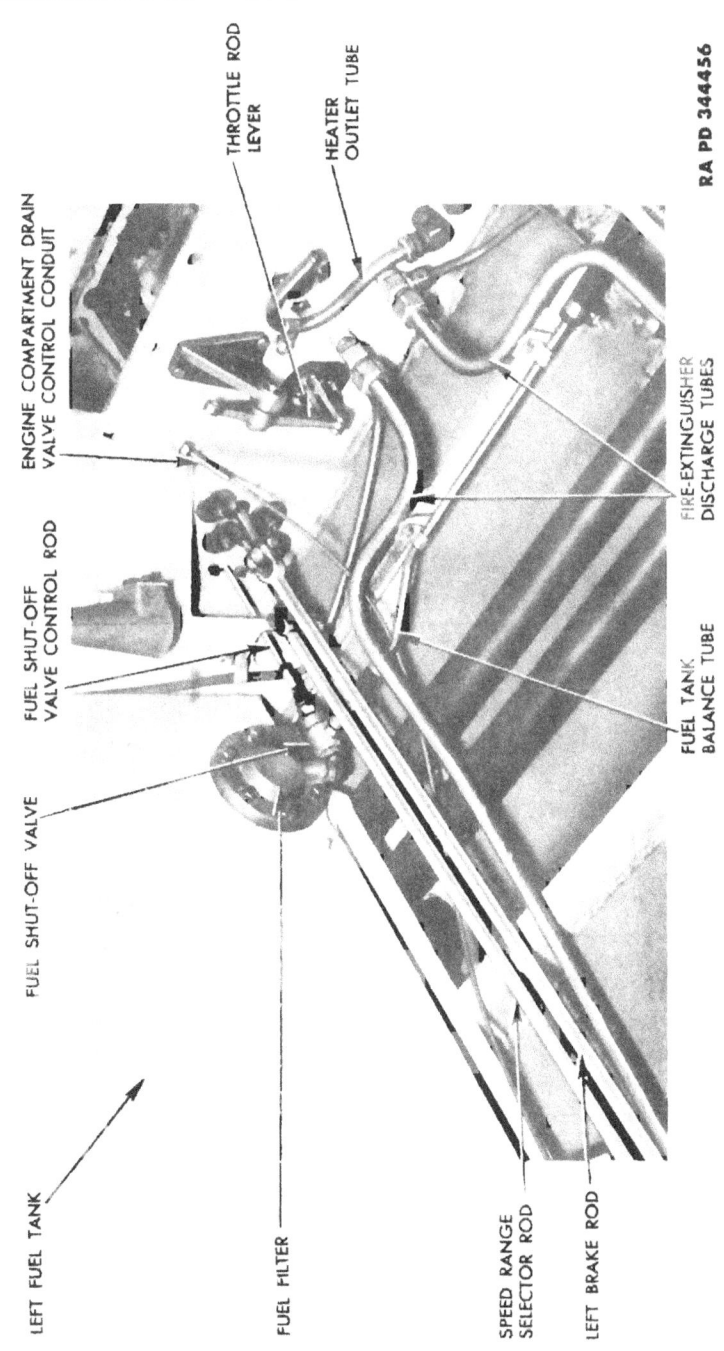

Figure 60 — Left Fuel Tank Connections

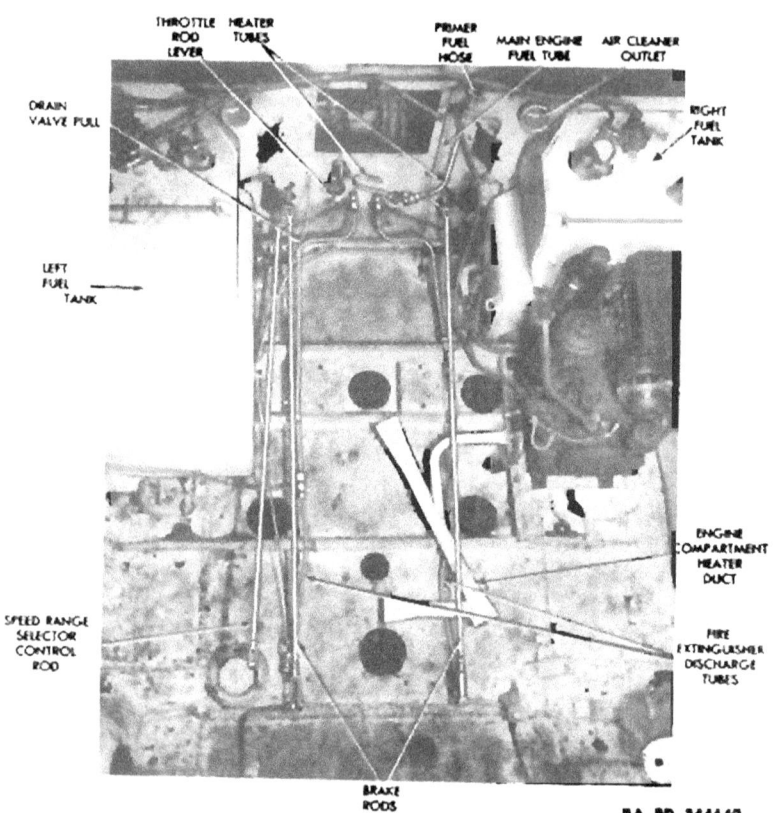

Figure 61 — Engine Compartment, Power Unit Removed

(2) DRAIN COOLING SYSTEM. Drain coolant into clean containers as described in paragraph 134 b.

(3) TRAVERSE TURRET. Traverse turret to left to clear engine compartment intake doors. Open all intake and exhaust doors.

(4) BATTERIES. Test batteries (par. 92 b). If batteries are not fully charged connect battery charger and recharge batteries or install new batteries. If batteries are fully charged disconnect ground cable only. Reinstall battery compartment cover, to provide a working surface when performing operations on magneto end of engine through bulkhead opening.

(5) REMOVE BULKHEAD COVER. Remove bulkhead cover (par. 181 b).

(6) REMOVE WATER PUMP HOSES. Loosen hose clamp on each

Power Unit Removal and Installation

water pump hose at pump intake connection and push clamp down on hose. Disconnect hoses from right and left pump intakes (fig. 114).

(7) DISCONNECT THROTTLE CONTROL ROD LEVER LINK ROD FROM LEVER. Remove clevis pin attaching throttle control rod lever link to lever (fig. 59) on rear side of engine compartment bulkhead.

(8) DISCONNECT HEATER TUBES. Loosen hose clamps on heater tube connecting hoses and move clamps down on tubes (fig. 60). Remove hoses from tubes. Tape ends of tubes and hoses.

(9) DISCONNECT PRIMER TO ENGINE FUEL HOSE. Loosen clamp attaching primer fuel tube hose to primer fuel tube (fig. 61). Push clamp down on hose and pull hose clear of tube. Tape ends of hose and tube.

(10) REMOVE FRONT AND REAR CENTER DECK PLATES AND DOORS. Using hoist remove intake and exhaust doors and front and rear center deck plates.

(11) DISCONNECT ALL WIRES AND CONDUITS. Disconnect all wires and conduits from power unit, sending units, and switches.

(12) REMOVE COOLING UNIT. Follow removal procedure described in paragraph 139 b.

(13) DISCONNECT ENGINE GROUND STRAP. Disconnect engine ground strap from vehicle connection. WARNING: *Do not disconnect strap from starter.*

(14) DISCONNECT STARTER CONDUIT. Remove cover from starter terminal box (fig. 72). Disconnect wire from terminal. Back off locking ring attaching starter cable conduit to terminal box. Pull conduit and wire clear of box and place on engine compartment floor.

(15) DISCONNECT SPEED RANGE SELECTOR ROD. Remove clevis pin attaching speed range selector rod to speed range selector shaft lever (figs. 127 and 133). Lower rod to engine compartment floor.

(16) DISCONNECT LEFT BRAKE REAR ROD. Remove clevis pin from clevis on end of left brake rear rod. Disconnect rod from left brake rear rod bell crank (fig. 133). Lay rod on engine compartment floor.

(17) DISCONNECT RIGHT TAILLIGHT CONDUIT. Follow removal procedure described in paragraph 95 c. Remove four clamps which attach conduit to rear hull plate and to differential housing. Lay conduit over stowage boxes on left side of vehicle to keep it clear of engine compartment.

(18) REMOVE FINAL DRIVE UNIVERSAL JOINTS. Remove locking wires from eight cap screws in each universal joint. Remove cap screws and lift out universal joints with rope sling (par. 163 b).

(19) REMOVE DIFFERENTIAL SHOCK MOUNTING PLATE BOLTS. Remove the two bolts in each differential shock mounting plate (fig. 126).

(20) DISCONNECT DIFFERENTIAL ANCHOR RODS. Remove clevis pins from both differential anchor rods at brackets on differential. Lower anchor rods to engine compartment floor.

(21) DISCONNECT RIGHT BRAKE REAR ROD. Remove clevis pin from clevis on end of right brake rear rod. Disconnect rod from right brake rear rod bell crank (fig. 132). Lay rod on engine compartment floor.

(22) DISCONNECT FUEL PUMP TUBE. Loosen upper hose clamp on fuel tube hose and slide clamp down over hose. Pull hose from fuel tube (fig. 56). Tape ends of tube and hose.

(23) DISCONNECT SPEEDOMETER SHAFT. Disconnect speedometer shaft from sending unit (fig. 87) (par. 99 f). Leave shaft attached to transmission adapter, but coil it up on top of engine out of the way.

(24) REMOVE CARBURETOR AIR INTAKE TUBES. Loosen and slide back clamps on air inlet connecting hoses (fig. 56). Remove hoses from carburetor air inlet elbows and from air cleaner outlets. Remove carburetor air intake tubes and clamps.

(25) DISCONNECT TACHOMETER SHAFT. Follow removal procedure described in paragraph 100 d.

(26) DISCONNECT CARBURETOR DEGASSER CONDUIT. Back off knurled locking ring and disconnect multiple pin plug from engine compartment terminal box connection (fig. 90). Lay conduit on top of engine out of the way.

(27) DISCONNECT CONDUITS AT ENGINE JUNCTION BOX. Back off knurled flange nuts and pull connectors out of receptacles at engine junction box.

(28) GROUND MAGNETOS. Remove cap screws and lock washers from magneto breaker inspection plate. Lift off inspection plate and gasket. Attach a short wire to the ground terminal of the magneto and attach other end of wire to one of the distributor cover screws on the magneto. Repeat operation on other magneto. NOTE: Grounding of the magnetos in this manner will prevent engine from starting in case it is accidentally cranked.

(29) INSTALL LIFTING EYES. Screw a lifting eye into the tapped hole in each differential shock mounting plate.

(30) REMOVE ENGINE MOUNTING BOLTS. Remove the two bolts in each engine mounting leg.

(31) INSPECT POWER UNIT. Make an inspection in fighting compartment and from both above and below in engine compartment to make sure all parts, conduits, wires, tubes, hoses and lines, which would interfere with the power unit removal, have been disconnected and swung aside or removed.

(32) REMOVE POWER UNIT. Attach power unit lifting sling (KM-J-4349) (fig. 63) to lifting eyes on engine and differential.

TM 9-735
71

Power Unit Removal and Installation

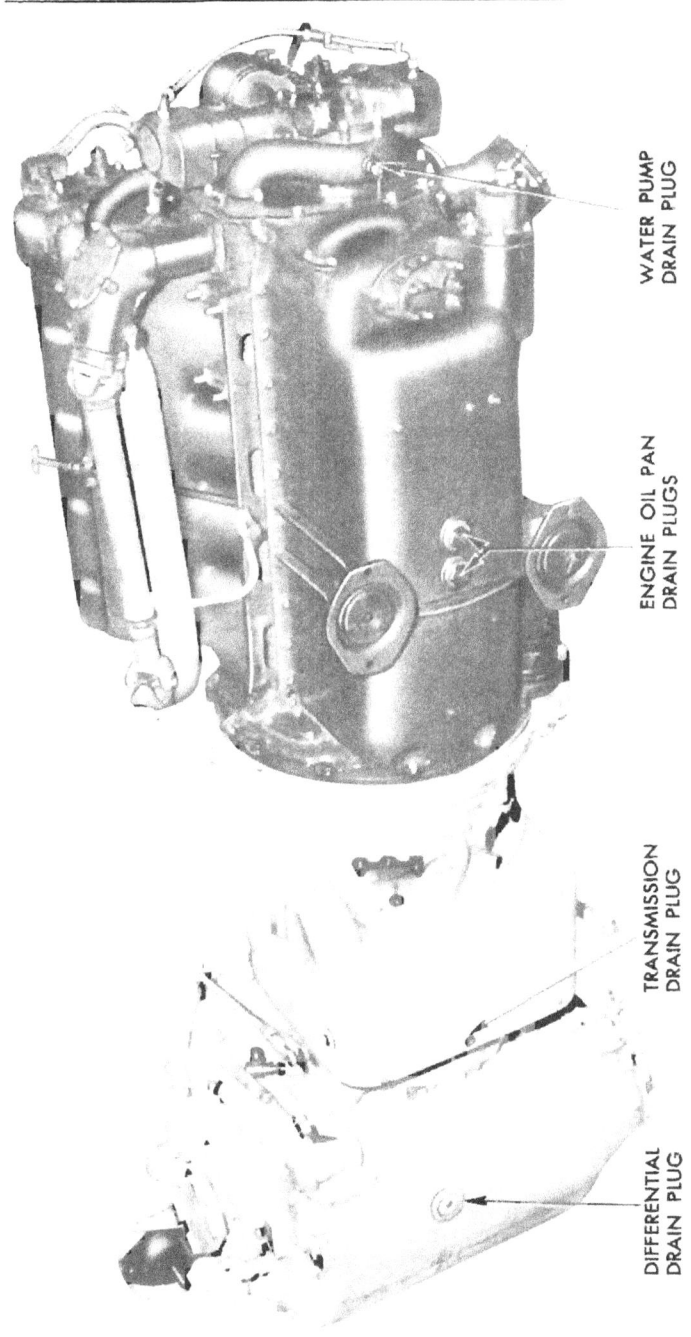

Figure 62 — Power Unit, Viewed From Below

With heavy-duty hoisting equipment slowly and carefully lift unit out of engine compartment. See figure 58 for correct angle at which to lift out power unit. WARNING: *Be sure both magnetos are grounded (par. 75 b) to prevent accidental starting of engine while out of vehicle.* Place unit on power unit stand or block up securely. NOTE: *When lifting power unit insert a long pinch bar between magneto end of engine and top front edge of engine compartment. Make sure that bar will not damage any engine part. Exert leverage on end of bar to force power unit toward rear of vehicle as unit hoisted out.* This is done to prevent the governor and/or the front carburetor from striking the front edge of the engine compartment or the bulkhead.

72. POWER UNIT INSTALLATION.

a. **General.** The power unit to be installed must be clean and in good operating condition. Make sure starter is lubricated (par. 46) and fuel filters are clean. Inspect engine, transmission, differential, and all component parts, units and assemblies for damage. Be sure conduits and hoses attached to the engine are so placed that they will not interfere with installation. Inspect engine compartment and make sure it is thoroughly clean and ready for the installation of the power unit. Facilities required for power unit installation are the same as required for removal (par. 71).

b. **Distribution of Installation Procedure.** To facilitate installation of the power unit, divide assembly operations into groups. This permits dividing personnel into groups, each group proceeding simultaneously with its assigned operations.

(1) General preparatory.
(2) Installation of power unit.
(3) Assembly in engine compartment.
(4) Assembly in fighting compartment.
(5) Road test.

c. **Power Unit Installation Procedure.**

(1) INSTALL LIFTING EYES. Screw a lifting eye into the tapped hole on each differential shock mounting plate.

(2) INSTALL POWER UNIT. Attach power unit lifting sling (KM-J-4349) (fig. 63) to lifting eyes on engine and differential. With heavy-duty hoisting equipment, lift power unit from power unit stand or blocking and position over engine compartment. See figure 58 for correct angle for lowering and installation. Slowly and carefully lower power unit into place in engine compartment. NOTE: *Insert a long pinch bar between magneto end of engine and top front edge of engine compartment. Make sure bar will not damage any engine part. Exert leverage on end of bar to force power unit toward the rear of vehicle until clears front edge of engine compartment.* This

TM 9-735
72

Power Unit Removal and Installation

will prevent governor or front carburetor from striking front edge of engine compartment or bulkhead. As engine is lowered, guide engine mounting bolt holes into alinement with bolt holes in engine mounting supports. Also line up shock mounting plates on differential with rear hull support plates.

(3) INSTALL ENGINE MOUNTING BOLTS. Install two bolts with lock washers, in each engine mount but do not tighten. Use long drift to line up bolt holes, if necessary.

(4) INSTALL DIFFERENTIAL SHOCK MOUNTING PLATE BOLTS. Install two bolts with lock washers in each differential mounting plate, but do not tighten.

(5) CONNECT DIFFERENTIAL ANCHOR RODS. Connect differential anchor rods. Tighten securely all engine and differential mounting bolts.

(6) INSTALL FINAL DRIVE UNIVERSAL JOINTS. Using rope sling, position universal joints between differential and final drive shafts (par. 163 c). Install eight cap screws in each universal joint loosely. When all screws are in place tighten securely, including engine and differential mounting bolts. Install new locking wires between each pair of screws.

(7) CONNECT RIGHT TAILLIGHT CONDUIT. Position right taillight conduit along rear hull plate. Install four conduit clamps. Connect conduit to right taillight as described in paragraph 95 d.

(8) ADJUST AND CONNECT RIGHT AND LEFT BRAKE REAR RODS. Back off locking nuts, adjust brake rod clevises (par. 164 e). Connect rods to right and left brake rear rod bell cranks. Install clevis pins.

(9) CONNECT FUEL PUMP FEED TUBE. Be sure hose clamp is in place on hose. Remove tape from ends of tube and hose. Install hose on fuel tube. Position hose clamp and tighten.

(10) CONNECT SPEEDOMETER SHAFT. Follow procedure described in paragraph 99 e.

(11) CONNECT TACHOMETER SHAFT. Follow procedure described in paragraph 100 e.

(12) INSTALL CARBURETOR AIR INLETS. Install hose clamps. Install hose connections of carburetor air inlets on the air inlet elbow and on carburetor outlets. Tighten hose clamps.

(13) CONNECT CARBURETOR DEGASSER CONDUIT. Connect multiple pin plug to socket on engine compartment terminal box. Tighten conduit locking ring.

(14) REMOVE TEMPORARY MAGNETO GROUND WIRE. Remove temporary ground wire, then tighten magneto ground terminal screw. Install gasket and magneto braker inspection plate.

(15) CONNECT CONDUITS AT ENGINE JUNCTION BOX. Plug connectors into receptacles at engine junction box, and secure with knurled flange nuts.

(16) CONNECT ENGINE GROUND STRAP. Be sure contacts are clean, dry, and bright. Position ground strap and install flat washer, strap, flat washer, and internal tooth lock washer, in order named. Tighten securely.

(17) CONNECT STARTER CONDUIT. Remove cover from starter terminal box. Connect wire in conduit to terminal on starter. Install terminal nut and tighten securely. Install terminal box cover.

(18) CONNECT SPEED RANGE SELECTOR ROD. Connect rod clevis to speed range selector shaft lever and install clevis pin.

(19) CONNECT ALL WIRES AND CONDUITS. Connect all wires and conduits to power unit, sending units, and switches.

(20) INSTALL COOLING UNIT. Follow procedure described in paragraph 139 c.

(21) ADJUST AND CONNECT THROTTLE CONTROL ROD LEVER LINK TO LEVER. Adjust and connect throttle control rod lever link to lever on rear side of engine compartment bulkhead. Adjust throttle linkage (par. 126 b).

(22) INSTALL FRONT AND REAR CENTER DECK PLATES AND DOORS. Position and install front and rear center deck plates and doors.

(23) CONNECT HEATER TUBES. Remove tape from hoses and tubes and connect hoses. Position hose clamps and tighten securely.

(24) CONNECT PRIMER TO ENGINE FUEL TUBE. Remove tape from primer fitting and hose connection on engine. Inspect hose and replace if not in good condition. Install hose on fuel tube. Position hose clamp and tighten securely.

(25) INSTALL WATER PUMP HOSES. Inspect water pump hoses and install new hoses if necessary. Connect hoses to water pump inlets. Position hose clamps and tighten securely.

(26) FILL COOLING SYSTEM. Follow procedure described in paragraph 134 d.

(27) INSTALL OR CONNECT BATTERIES. If batteries were removed follow installation procedure described in paragraph 92 e. If not removed, connect battery grounding strap securely. Turn off charger and disconnect wire if connected. Install battery compartment covers and stowage boxes.

(28) INSTALL BULKHEAD COVER. Place bulkhead cover in position and tighten wing nuts securely.

(29) OPEN FUEL TANK SHUT-OFF VALVES. Open both fuel tank shut-off valves.

Power Unit Removal and Installation

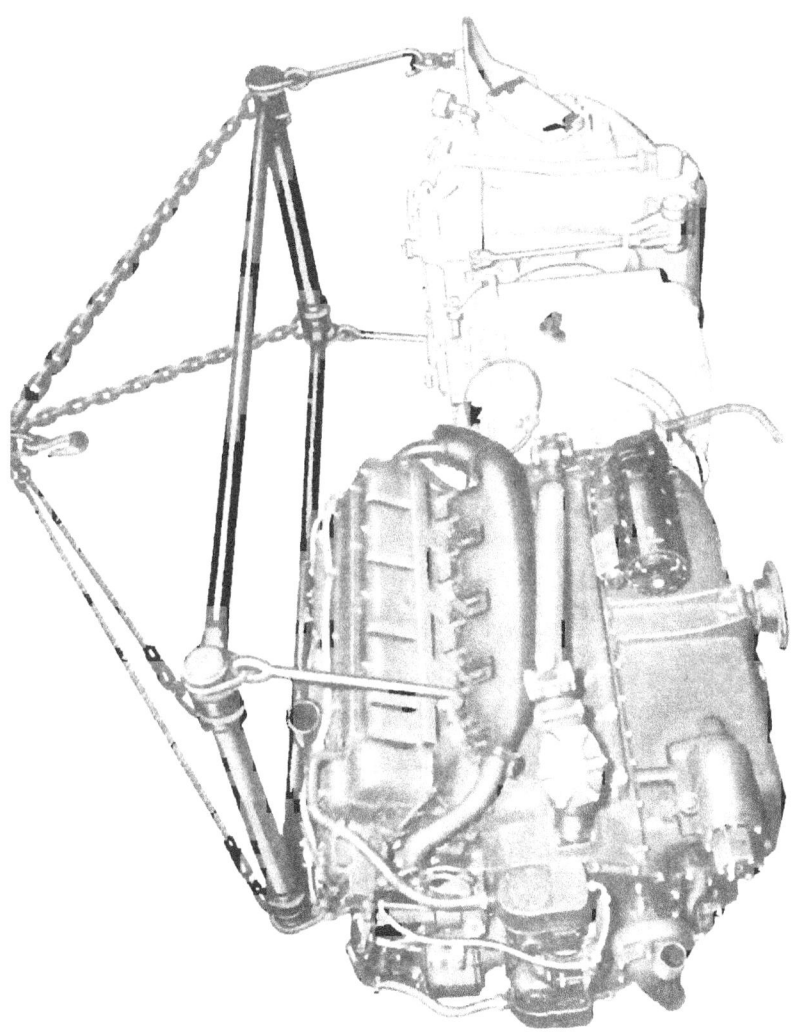

Figure 63 — Lifting Power Unit With Sling

(30) TUNE-UP ENGINE. Follow instructions outlined in paragraph 69.

(31) CLOSE INTAKE AND EXHAUST DOORS. Close and fasten all intake and exhaust doors. Position turret and gun for traveling.

(32) RECORD INSTALLATION OF POWER UNIT. Make proper entry on W.D., A.G.O. Form No. 478, "MWO and Major Unit Assembly Replacement Record."

(33) ROAD TEST VEHICLE. Follow instructions outlined in paragraph 53.

Section XIX
IGNITION SYSTEM

73. IGNITION SYSTEM.

a. **Description.** The ignition system (fig. 64) consists of two heavy-duty magnetos mounted on the magneto end of the engine, the magneto switch located on the instrument panel (fig. 16), and the necessary connecting wires and conduit.

74. MAGNETO SWITCH.

a. **Description.** The magneto switch is mounted on the instrument panel (fig. 16) in a unit which also contains the starter and booster switches and the speedometer-tachometer switch. Turning the magneto switch to "OFF" position grounds the magnetos and turns off the ignition. If the magnetos to magneto switch wires should become shorted, the ignition cannot be turned on with the magneto switch. If these same wires should break, the ignition cannot be turned off with the magneto switch. An electrically operated ON-OFF switch for the control of the speedometer and tachometer sending units is incorporated in the magneto switch. Operating the magneto switch automatically operates the speedometer-tachometer switch (par. 15 k). For a full description and operation of the magneto switch refer to paragraph 15 b.

b. **Removal of Magneto Switch.** Refer to paragraph 79 b.

c. **Installation of Magneto Switch.** Follow procedure as outlined in paragraph 79 c.

75. MAGNETOS.

a. **Description.** The two magnetos are induction type. Each has a fixed armature, the magnet rotating between the armature and field poles. The "right" magneto fires the "right" bank of four cylinders and the "left" magneto fires the "left" bank of four cylinders. A distributor is built into the end of each magneto. Access to dis-

Ignition System

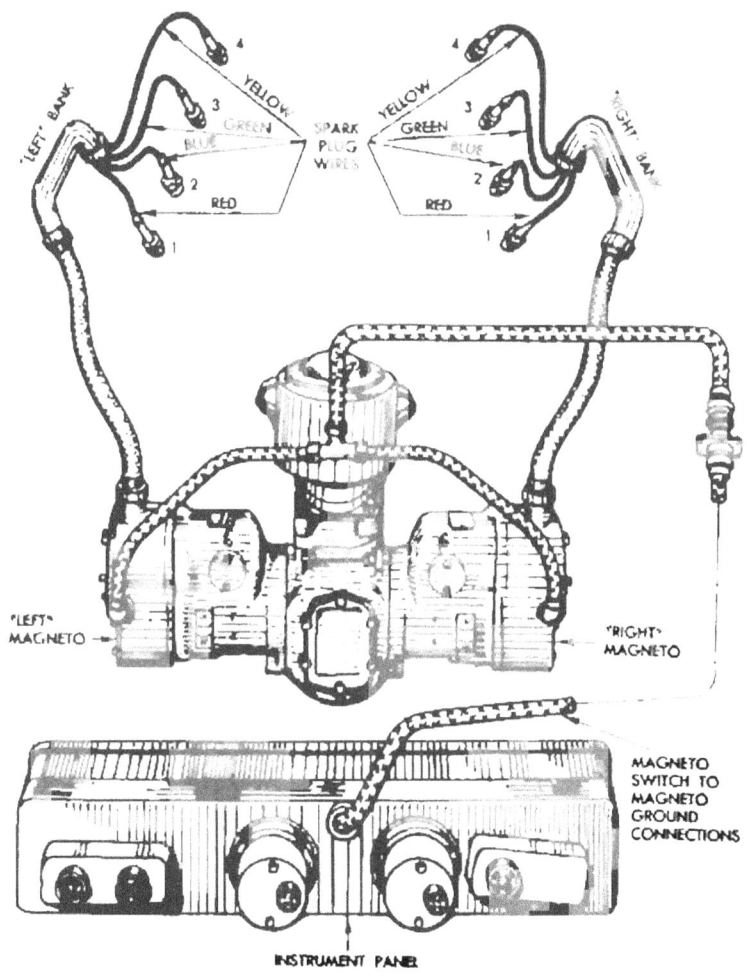

Figure 64 — Ignition System Diagram

tributor points is obtained by removing cover plates. An automatic spark advance mechanism is part of the engine and the one governor assembly advances the spark of both magnetos simultaneously. The numbering of the cylinders and the firing order are shown in figure 69.

b. **Removal of Magnetos.** Remove the circular plate at the end of the magneto which covers the spark plug wire terminals. Also remove the distributor cover plate located below the spark plug wires

TM 9-735

Part Three—Maintenance Instructions

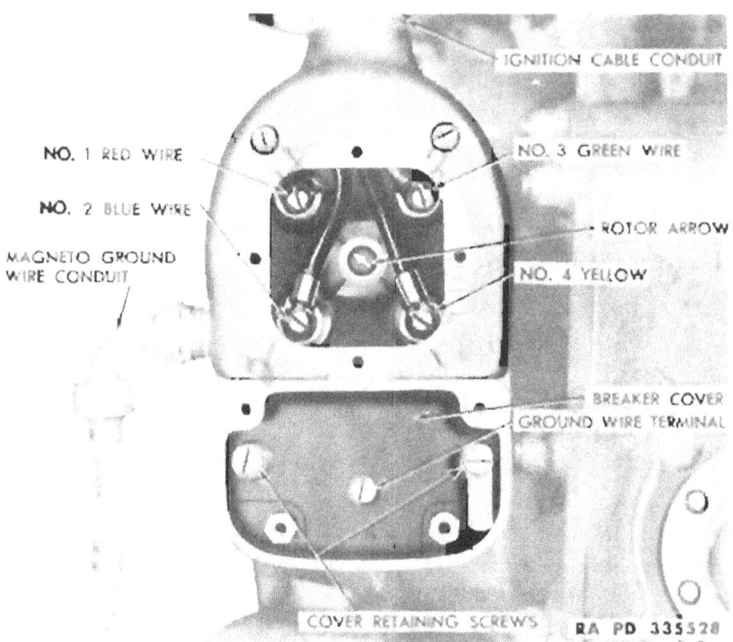

Figure 65 — Magneto (Covers Removed)

terminal cover plate. Remove ground wire terminal screw (fig. 65). Disconnect ground wire conduit and remove ground wire and conduit from magneto housing. CAUTION: *Do not turn engine over with ground wire removed and ignition cables connected to magneto. Magneto switch is inoperative with ground wire disconnected and engine may fire if cranked.* Remove four screws securing the ignition cables to distributor plate and lift ignition cables from recesses. Disconnect ignition cable conduit from magneto. Disconnect ignition cable from terminals on distributor plate. Remove conduit and cables. Tag cables for identification. Remove magneto from accessory gear cover. CAUTION: *Note position of rotor arrow (fig. 65) (center of distributor plate) before removing magneto.* Magneto must be installed with arrow pointing to same terminal as when removed. If the engine has been turned over with magneto removed, proceed as outlined in subparagraph c following.

c. **Installation of Magnetos.** Be sure rotor arrow in center of distributor plate (fig. 65) points to same terminal as when magneto

190

Ignition System

RA PD 335529

Figure 66 — Turning Propeller Shaft to Aline Timing Marks

was removed. If a new magneto is used, follow procedure as outlined in subparagraph c (4) following.

(1) TURN ENGINE BY HAND AND SET FLYWHEEL TIMING MARK. WARNING: *Be sure magneto switch is OFF before turning engine over.* Remove flywheel inspection plate (fig. 67) on top side of the flywheel housing toward left side of vehicle. Remove all spark plugs (par. 76 h) and in right cylinder bank install compression gage in No. 1 cylinder spark plug hole. Insert a short bar in the U-shaped coupling of the "right" fan drive shaft (fig. 66) and turn engine until No. 1 piston on the right-hand side is on its compression stroke. This is indicated when maximum compression is reached on gage. Stop turning when timing mark "SPARK SETTING R H" appears in the inspection hole opening opposite the pointer. Remove compression gage.

191

TM 9-735

Part Three—Maintenance Instructions

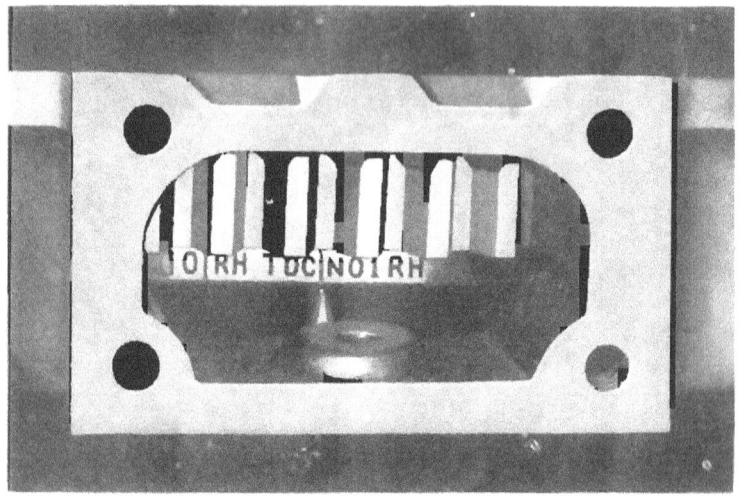

Figure 67 — Flywheel Timing Marks

(2) REMOVE MAGNETO DRIVE FLANGE. If a new magneto is to be installed, it will be necessary to remove the drive flange from the old magneto and install it on the new magneto. Install magneto drive flange holder on the old magneto mounting stud. Install the holder nut and tighten. Remove the drive flange lock nut and washer. Remove the drive flange holder nut and holder. Install the drive flange puller and remove the drive flange from the old magneto. Remove key from magneto drive shaft.

(3) INSTALL MAGNETO DRIVE FLANGE. Install key on magneto drive shaft and place drive flange on the shaft. Secure the flange on the shaft with a flat washer and a lock nut. Install flange holder on magneto drive flange and tighten drive flange nut until flange is seated on shaft. Remove drive flange holder nut and holder.

(4) INSTALL RIGHT-HAND MAGNETO AND SET TIMING. Install a new gasket if old one is unserviceable. Turn magneto in direction of rotation (clockwise) by means of driving flange on end of shaft until rotor arrow points to No. 1 (red wire) terminal (fig. 65). Install right-hand magneto, nuts and lock washers. Shift magneto so mounting studs are centrally positioned in timing slots. Turn nuts finger-tight. Remove distributor point covers (fig. 68) and place a narrow strip of cellophane between the points. Exert a slight pull on the cellophane while slowly turning the magneto. The cellophane

TM 9-735
75

Ignition System

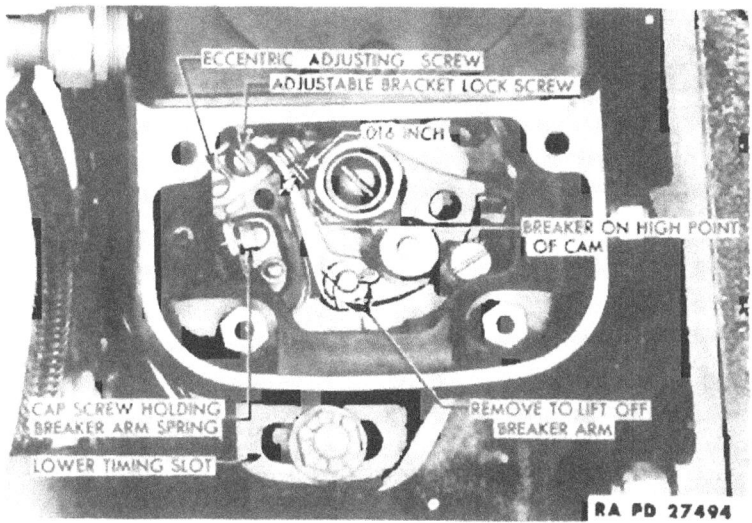

Figure 68 — Magneto-Distributor Point Cover Removed

will be released the instant points start to separate. Hold magneto in this position and turn attaching lock nuts up tight.

(5) RECHECK MAGNETO TIMING. Turn flywheel backward approximately one-quarter revolution. Then turn it in the direction of rotation until magneto points just begin to separate. Note whether timing mark on flywheel is opposite pointer (fig. 67). If a minor correction is necessary, tilt magneto forward or backward as required. Tilting magneto toward flywheel end of engine advances spark; tilting magneto toward magneto end of engine retards spark. If a major correction is to be made, loosen nut securing left-hand driving flange, and move flange and shaft forward or backward as necessary. Tighten nut and recheck timing. NOTE: *One serration on flange and driven gear is equal to five degrees at flywheel.*

(6) INSTALL LEFT-HAND MAGNETO AND SET TIMING. Turn flywheel (subpar. *c* (1) preceding) in the direction of rotation 300 degrees (five-sixths of a turn) until timing mark "SPARK SETTING L.H." appears opposite pointer. Turn magneto by means of driving flange until arrow in rotor window (fig. 65) is opposite No. 1 (red wire) terminal. Install left-hand magneto. Install a new gasket if necessary. Shift magneto until mounting studs are midway to timing slots and turn nuts finger-tight only. Place a narrow strip of cellophane between magneto points and exert a slight pull on the cellophane. Turn flywheel backward approximately one-quarter revolu-

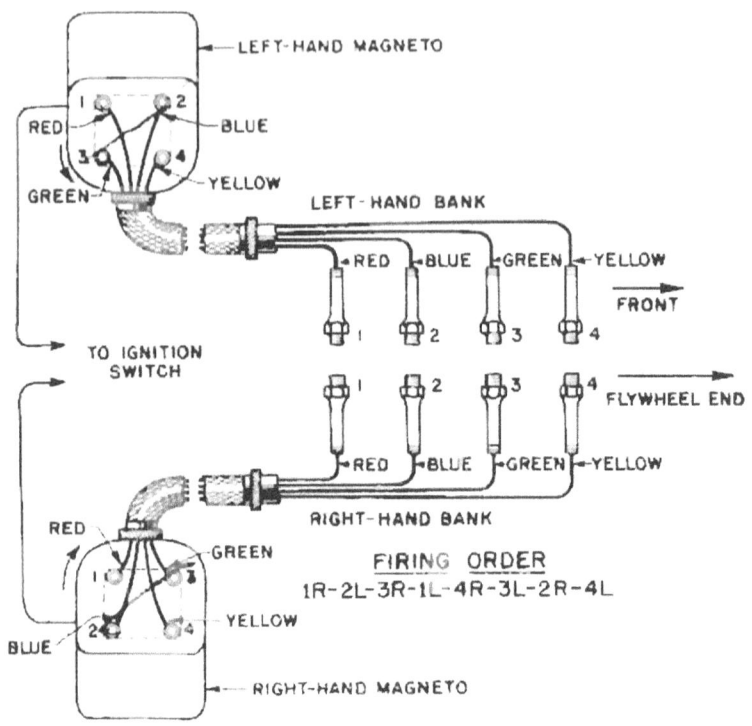

Figure 69 — Cylinder Numbering and Firing Order

tion, then turn it in the direction of rotation until magneto points just begin to separate and cellophane is released. Note whether timing mark on flywheel is opposite pointer. The reason for turning the engine backward, before bringing it up to the timing mark, is to compensate for any backlash in the magneto drive mechanism. Any correction in the timing of the left-hand magneto can then be made by tilting the magneto forward or backward.

(7) LOCK MAGNETOS AND CONNECT WIRING. After both magnetos have been properly set in mounting slots and correct timing established, lock the mounting nuts with wire. Install spark plug wires and attach to correct terminals (figs. 64 and 69). Tighten terminal screws and ignition cable conduit lock nut. Install magneto ground wire and conduit and tighten terminals screw and conduit lock nut. Install cover plates, tightening screws securely. Install spark plugs and connect wires.

(8) RECORD INSTALLATION OF MAGNETOS. Make proper entry

TM 9-735
75—76

Ignition System

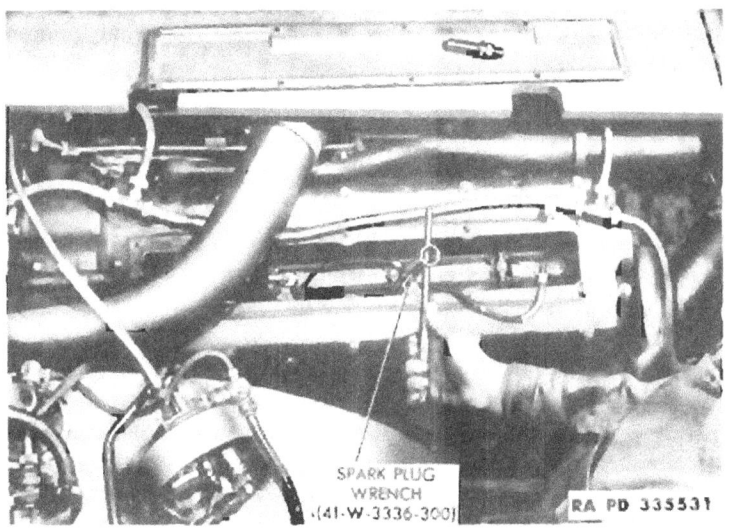

Figure 70 — Removing Spark Plug

on W.D., A.G.O., Form No. 478, "MWO and Major Unit Assembly Replacement Record."

d. **Adjust Distributor Points.** Remove magneto (subpar. b above). Adjust points to an opening of 0.016-inch when the breaker arm rests on the high point of the cam (fig. 68). Loosen lock screw of the adjustable bracket. Turn eccentric adjusting screw (fig. 68) until a 0.016-inch feeler gage can be moved between the points with a slight drag, then tighten lock screw to hold the adjustment. Adjust points so that they are in alinement with each other and that the full surfaces of both contacts meet squarely. If contact points are pitted, replace magneto (subpar. b and c above). In emergencies, pitted points can be cleaned with an extremely fine hone. Do not use a file at any time.

76. SPARK PLUGS.

a. **Description.** Aircraft type, radio-shielded spark plugs are used. When spark plugs are new, gap between electrodes is from 0.011 to 0.014 inch. Too wide a gap increases electrical resistance and interferes with efficient operation of engine. Replace plugs when gap increases to 0.030 inch. NOTE: *Do not attempt to adjust electrodes in this type of plug.*

Part Three—Maintenance Instructions

b. **Removal.** To remove a spark plug (fig. 70) first remove the spark plug cover plate on the cylinder head. Remove retainer nuts, spark plug wire and spark plug with special spark plug wrench (41-W-3336-300) (fig. 70).

c. **Install.** Clean plugs and check for proper gap (0.014-in.). Replace plugs if electrode gap is 0.030 inch or greater. Replace spark plug gaskets if unserviceable. Install spark plug and tighten with a torque wrench, to 25 to 30 foot-pounds torque. CAUTION: *Do not tighten spark plugs excessively.* Install spark plug wire and retainer nut and tighten moderately with spark plug wrench (41-W-3336-300). Install spark plug cover plate and tighten nuts securely.

(1) TEST SPARK PLUGS. With the engine running, remove spark plug cover plate and disconnect spark plug wire to be tested (subpar. b preceding). Hold wire approximately ¼ inch from engine. Note whether a strong, hot spark is produced. If spark fails to jump the gap or is weak and small, ignition cable is faulty and must be replaced. Remove spark plug from cylinder and connect it to the ignition wire just removed. Lay plug on camshaft housing and note condition of spark produced. If spark is weak and small, plug is faulty and must be replaced. NOTE: *A spark plug with a "thready" spark may not fire when under compression.* Test each spark plug and wire, and correct failures. Install spark plug cover plate.

77. IGNITION SYSTEM WIRING.

a. **Description.** Wires leading from the magnetos to the spark plugs (fig. 69) are identified for both "right" and "left" magnetos by wire colors on ends of wires as follows: No. 1 red, No. 2 blue, No. 3 green, No. 4 yellow. Replace spark plug wires, with cracked or damaged insulation. A wire in such condition can cause misfiring of spark plug to which it is attached. Replace damaged, frayed, or crushed flexible conduit.

b. **Removal (Harness).** To remove spark plug wires, remove spark plug cover plate on engine and the circular spark plug wires terminal cover plate on the magneto. Remove spark plug retainer nuts and disconnect wires at spark plugs. Disconnect conduit and wires at magneto. Remove ignition system wiring harness as a complete unit.

c. **Install. (Harness).** Install spark plug wiring harness as a unit. Attach wires to spark plugs and attach other ends to proper terminals on the magneto (figs. 64 and 69).

Section XX
STARTING SYSTEM

78. STARTING SYSTEM.

   a. Description. The starting system (fig. 71) consists of the starter, starter switch on the instrument panel, starter relay switch, batteries (par. 92) and wires and conduit connecting these units. NOTE: *Before removing or installing any part of this system, place magneto switch and 24-volt master switch in "OFF" positions.*

79. STARTER SWITCH.

   a. Description. The starter switch is mounted in the same unit as the magneto and booster (not operative) switches and is located in the upper center of the instrument panel (fig. 16). Operation of the starter switch is given in paragraph 6 c. When the switch lever is held down to "ON" position the starter relay switch closes the circuit, the starter pinion engages the flywheel ring-gear and cranks the engine.

   b. Removal. Remove instrument panel from case (par. 97 d). Remove starter switch support bracket. Disconnect wires from auxiliary circuit terminals. Move support bracket clear of assembly. Remove switch unit cover. Unscrew coupling nut on flexible conduit and carefully pull wires in conduit out of switch assembly as far as possible. Note numbers of wires for later identification. Lift cover off and from rear partially push unit out of panel. Disconnect wire from starter terminal. Disconnect wire from coil terminal. Disconnect wire from "LT" magneto terminal and wire from "RT" terminal. Pull unit cover and flexible conduit clear of panel, and remove gasket. Disconnect wires from the marked terminals in the unit and remove switch.

   c. Installation. Inspect old gasket and replace if necessary. Attach wires securely to proper terminals on the unit, reversing removal procedure described in subparagraph h preceding. Install switch unit in panel from the front and install knurled lock nut on conduit. Install unit cover and tighten nuts. Test operation of starter switch. Install switch support bracket. Install instrument panel in case (par. 97 e).

80. STARTER RELAY SWITCH.

   a. Description. A relay switch in the starter circuit is mounted in the battery compartment (fig. 79). The switch closes the circuit magnetically when the starter switch lever on the instrument panel is held down to "ON" position.

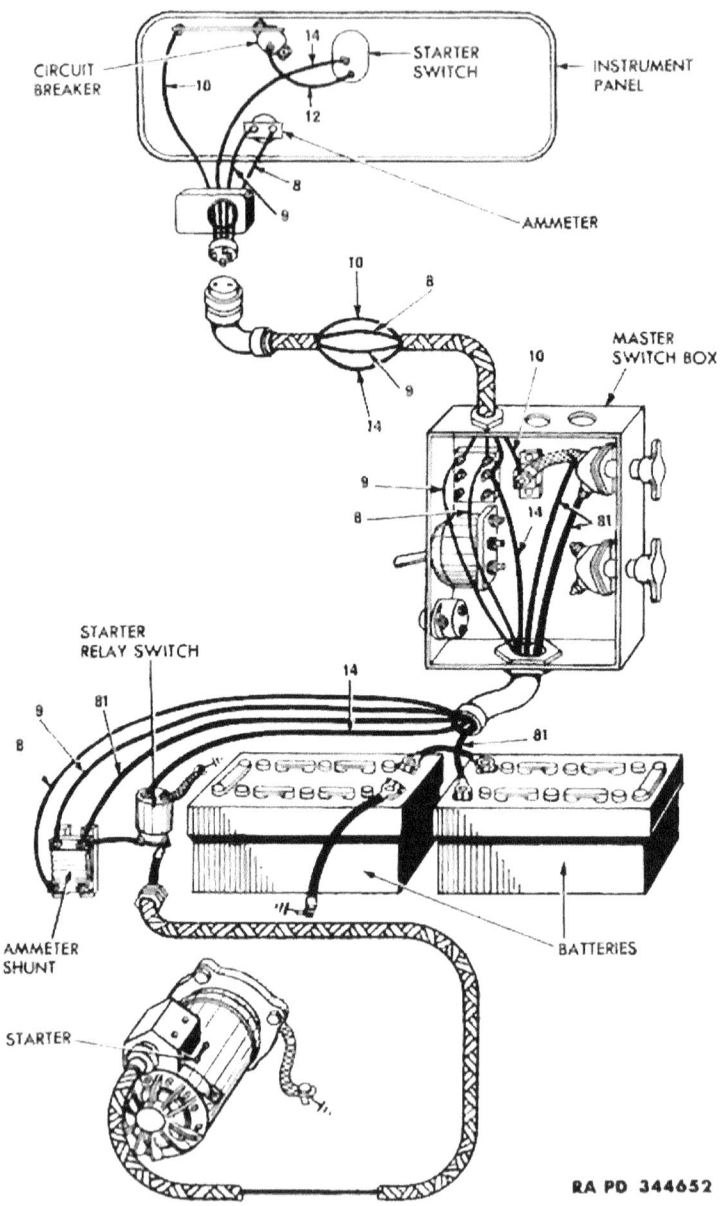

Figure 71 — Starting System Diagram

Figure 72 — Starter

b. **Removal.** Remove ground cable from battery terminal. Disconnect remaining cables from battery terminals and lift out right battery. Disconnect starter switch to starter relay switch wire from the relay switch terminal. Disconnect relay switch and remove switch from bracket.

c. **Installation.** Clean terminal posts and terminals before connecting wires to relay switch. Install switch on bracket in battery compartment. Connect starter switch to starter relay switch wire to relay switch terminal. Connect wires and ground to proper terminals (fig. 71). Install right battery in battery compartment and connect battery cables to battery terminals (fig. 79). Connect battery ground cable. Install battery box cover. Turn battery master switch and magneto switches on, and test relay switch by starting engine.

81. **STARTER.**

a. **Description.** The starter is a heavy-duty, 24-volt type, mounted on the right side of the flywheel housing (left side of vehicle) (fig. 72). Its power is transmitted to the engine through an automatic drive. Rotation of the starter drive shaft causes the pinion of the automatic drive to advance and mesh with the flywheel ring-gear. After engine starts and flywheel speed exceeds that of starter, the pinion releases from the flywheel automatically.

b. Removal. Turn 24-volt master switch off. Drain and remove left fuel tank (par. 116 d). Disconnect rear fan drive propeller shaft universal joint (par. 158 b). Tie shaft up out of the way. Remove terminal cover from starter. Disconnect cable from starter. Disconnect ground strap from ground terminal. Remove starter from engine, supporting it with a rope sling as last nut at top is removed. Lift starter out of engine compartment with sling.

c. Installation. If same starter is being installed, lubricate it according to instructions in Lubrication Order (par. 46). Inspect gasket and install new one if required. Use rope sling to lower and support starter while installing bolts. Install top bolt first, setting nut only finger-tight. Remove rope sling, aline starter and install other bolts, lock washers and nuts. Tighten the three nuts alternately until secure. Connect starting cable to starter. Install starter terminal box cover. Connect ground strap to ground terminal. Connect universal joint of fan drive propeller shaft (par. 158 c). Install and fill left fuel tank (par. 116 e). Turn 24-volt master switch on and test operation of starter by cranking engine.

d. Record Installation of Starter. Make proper entry on W.D., A.G.O. Form No. 478, "MWO and Major Unit Assembly Replacement Record."

82. STARTING SYSTEM WIRING.

a. General. The starting system wiring consists of the cables and wires leading from the batteries through the relay switch to the starter, and from the batteries to the battery master switch to the instrument panel (fig. 71). Replace damaged wires or conduits. Effective shielding of wires depends upon clean, tight coupling nuts and connections. Terminals must be clean and connections tight for efficient operation of electrical units.

b. Removal. Turn off 24-volt master switch. Disconnect wire at both terminals, loosen all holding clips, and carefully pull wire from vehicle, observing its relative position to other wires and parts of vehicles as an aid to installation. In case of doubt, note wire number and terminal connections before wire is removed.

c. Installation. Install wire in same relative position as old wire. Clean all terminal posts and connect wire securely. Attach all holding clips securely. Turn on 24-volt master switch and operate starter to test installation.

Generating System

Section XXI

GENERATING SYSTEM

83. GENERATOR.

a. *Description.* The generating system consists of a generator, generator regulator, auxiliary and/or main engine, auxiliary engine fuel shut-off valve, auxiliary engine controls, batteries, master switch box, ammeter shunt, circuit breaker, and ammeter (fig. 73). The 150-ampere, 24-volt generator (fig. 32) is mounted on top of the auxiliary engine by bands around the frame, and is supported by a hinge-type mounting. It furnishes 28.5 volts DC (direct current) for charging the batteries or operating the traversing mechanism and other electrical units. The generator is driven from either end V-belts. Over-running clutches and pulleys at each end permit operation by either the auxiliary or main engine. NOTE: *Do not operate both engines at the same time.* The generator is grounded by a strap attached to the hull. A spring-loaded device and an adjusting bolt for maintaining proper belt tension are provided (fig. 75). The generator regulator is mounted on top of the generator. The generator shunt and generator circuit breaker are mounted inside the battery box on the right side.

b. *Lubrication.* The generator bearings are packed with grease at assembly and no external means are provided for their lubrication. Do not attempt to clean or flush bearings with any solvent or light oil.

c. *Inspection.* Examine the generator brushes and commutator at every 100-hour Preventive Maintenance Service (par. 53) or whenever generator is removed (fig. 74). If brushes are oily, dirty, or worn so that the top of brush is even with or below the brush holder, install new brushes (subpars. d and e following). New brushes must move freely in holders and must be fitted to the contour of the commutator before operating generator. If commutator is dirty, clean it by holding a piece of flint paper, grade 2/0 (42-P-1210) against it with a piece of soft wood while turning armature by hand. CAUTION: *Do not use emery cloth to clean commutator.* Blow out dust with compressed air. Replace generator if commutator is out-of-round, badly worn, or scored.

d. *Removal.* Turn 24-volt master switch off. Remove generator regulator box shield. Disconnect generator ground strap at generator. Disconnect wires and conduit and remove generator regulator (par. 85 b). Back off generator drive belt tension adjuster. Fasten a sling around generator and take up slack. On side of generator, loosen jam nut on auxiliary engine drive belt tension adjusting screw and turn screw out of its mounting. Lower generator to relieve belt

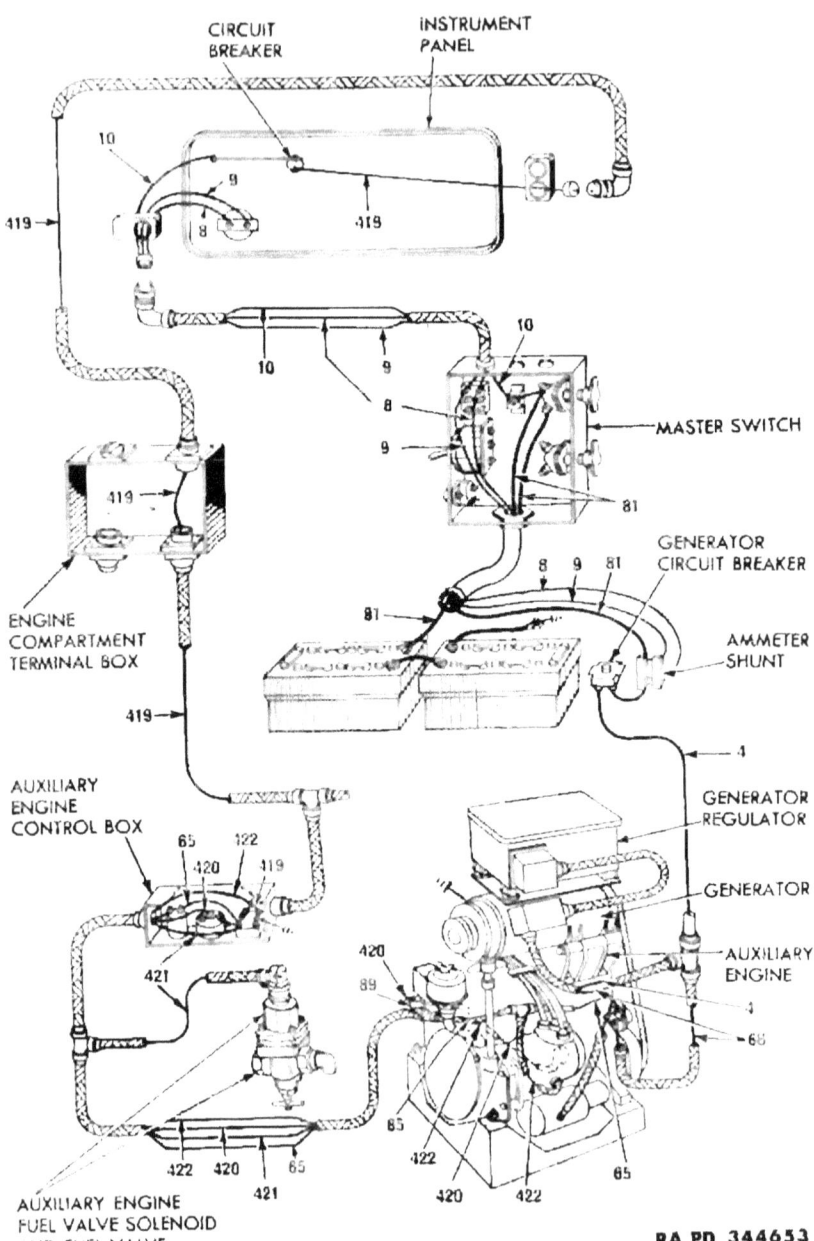

Figure 73 — Generating System Diagram

TM 9-735
83

Generating System

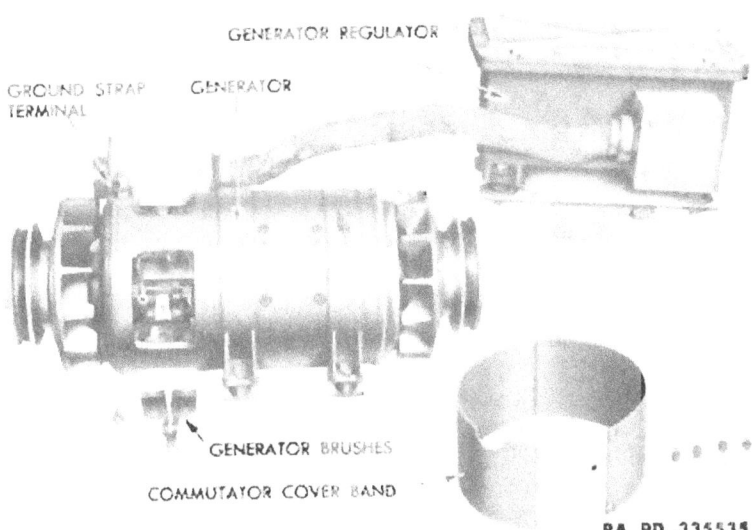

Figure 74 — Generator (Cover Band Removed)

tension and remove drive belts. Remove the two cap screws and lock washers from generator support and lift out generator with sling.

e. **Installation.** Be sure cables and conduits are out of the way. Position generator on support and install hinge lock washers and screws. Install generator drive belts (par. 86 b). Install auxiliary engine drive belt adjusting screw with jam nut, on side of generator. Adjust belt tension (par. 86 c). Install generator regulator and connect wires (fig. 73) and conduit. Connect generator ground strap to generator and install generator regulator box shield. Turn 24-volt master switch on. Start auxiliary engine (par. 29 b). Note polarity of generator by observing ammeter. If ammeter shows discharge, stop engine at once and polarize generator (subpar. f following).

(1) RECORD INSTALLATION OF GENERATOR. Make proper entry on W.D., A.G.O. Form No. 478, "MWO and Major Unit Assembly Replacement Record."

f. **Polarizing the Generator.** Whenever a new generator is installed it must be polarized before it will charge. The operation must be performed with ground strap and battery cables connected. Turn 24-volt master switch off. Remove terminal box cover, disconnect wire tagged "F"— (generator field wire) and connect it to the battery terminal which is the lower of the two terminals. Turn 24-volt master switch on for two or three seconds to "flash" the generator

203

field and correct the polarity. Turn 24-volt master switch off. Disconnect generator field wire from battery terminal post and reconnect it to generator field terminal. Install terminal box cover. Start engine and observe ammeter reading.

84. GENERATOR BRUSHES.

a. **Removal.** The generator brushes cannot be removed with generator installed in vehicle. To provide access to the brushes, first remove the generator regulator (par. 85 b) and then remove the generator (par. 83 d). Remove the generator brush cover bands (fig. 74). Remove each brush terminal screw and lift out the four pairs of brushes. Note position of brushes in holder and position of wires to brush terminals.

b. **Installation.** Before installing new brushes and operating the generator, fit the brushes to the contour of the commutator. To fit the brushes, place them in the brush holders in the normal position. Slip a piece of flint paper, grade 2/0 (42-P-1210) between the commutator and the brush with sand side toward brushes. CAUTION: *Do not use emery cloth.* Move the flint paper back and forth around the commutator until brushes have taken the contour of the commutator. Blow out all dust with compressed air. To install new brushes in the generator, reverse the removal procedure as outlined in subparagraph a preceding. Make certain that spring clips are in proper position on brushes and that terminal screws are tight.

85. GENERATOR REGULATOR.

a. **Description.** The generator regulator is mounted on rubber vibration dampers on top of the generator (fig. 32). The regulator consists of a voltage regular unit, a reverse current relay, and a voltage adjusting rheostat, all mounted in a water-tight cast aluminum box. The regulator is removed and installed as a complete unit and is protected from damage by a shield over the top. The function of the generator regulator is to hold the generator voltage to 28.5 volts over the entire speed range of the generator at no-load and to prevent the battery from discharging through the generator. No attempt should be made by the using arms to adjust the generator regulator. If the ammeter shows an incorrect reading and the trouble has been traced to the generator regulator, it must be replaced.

b. **Removal.** Turn 24-volt master switch off. Remove the regulator box shield supports and the shield. Remove the regulator terminal box cover. Disconnect the two heavy-duty cables. Note wire numbers for identification. Unscrew conduit coupling nuts and pull the wires clear of terminal box. Remove regulator mounting bracket and regulator box from generator frame. To remove regulator box from mounting bracket and plate, first disconnect the two ground wires from the mounting plate. Remove nuts and lock washers attaching the generator regulator rubber mountings and lift off the regulator.

c. **Installation.** Position generator regulator box on mounting plate. Connect the two ground wires to the mounting plate. Install generator regulator mounting bracket on generator frame. Connect the two heavy-duty cables (fig. 73). Connect conduit and install terminal box cover. Install generator regulator box shield and tighten screws securely.

(1) RECORD INSTALLATION OF GENERATOR REGULATOR. Make proper entry on W.D., A.G.O. Form No. 478, "MWO and Major Unit Assembly Replacement Record."

86. GENERATOR BELTS.

a. **Removal.** Disconnect fan drive propeller shaft from fan drive pulley (fig. 75). Loosen lock nut and back off generator belt tension adjuster spring nut to relieve spring pressure (fig. 75). Remove clevis pin to disconnect yoke from bracket on engine. Loosen jam nut on auxiliary engine drive belt adjusting screw on left side of generator. Turn adjusting screw until generator will swing in sufficiently to permit drive belt to be worked off fan drive pulley.

b. **Installation.** Fully release adjuster tension before attempting to install belts. Place belt in groove on generator drive pulley. Work belt into groove on fan drive pulley while turning fan. Connect fan drive propeller shaft to fan drive pulley (fig. 75). Adjust drive belt tension (subpar. c following). NOTE: *Always install all new belts in matched pairs. Do not install belts singly.* Use care in installing belts as all belts have a wire core. Install belts with arrows pointing in direction of rotation.

c. **Adjust Generator Drive Belt Tension.** Back off generator adjuster spring nut until nut is flush with end of spring stud (fig. 75). Loosen adjuster rod yoke lock nut and adjust length of rod so holes in yoke are alined with hole in bracket on engine. Install clevis pin and tighten yoke jam nut. Turn adjuster spring nut to compress spring and increase generator drive belt tension until a finger force of eight to ten pounds applied at center of belt will deflect belt $1/4$-inch. Tighten adjuster spring lock nut.

d. **Adjust Auxiliary Engine Drive Belt Tension.** Turn adjusting screw at left of generator to increase auxiliary engine drive belt tension until a finger force of 8 to 10 pounds, applied at center of belt, will deflect belt $3/8$-inch. CAUTION: *Exercise extreme care in tightening V-belts. Excessive tension will cause undue wear of the generator bearings.*

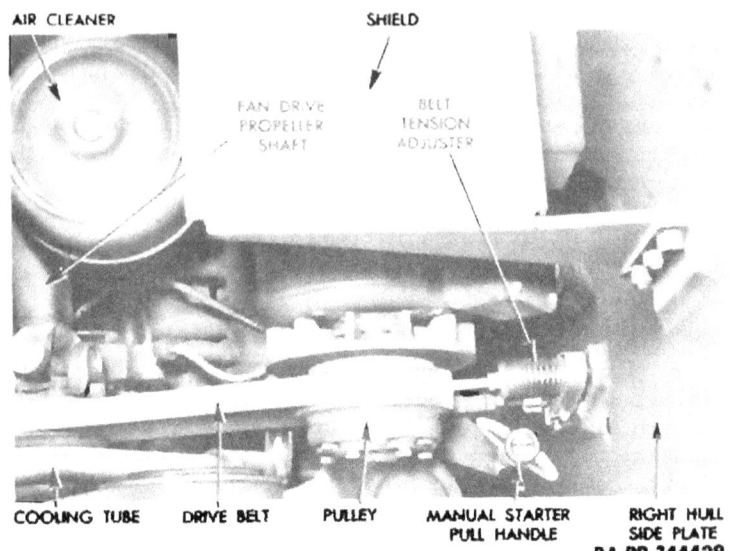

Figure 75 — Generator Belt Tension Adjuster

87. GENERATOR BELT TENSION ADJUSTER.

a. **Removal.** Release spring pressure by unscrewing jam nuts from the spring adjuster stud. Remove spring seat and spring. Unscrew the adjuster spring stud from the base (fig. 75). Remove clevis pin from adjuster yoke. Remove one connecting pin from hinge plate and remove adjuster assembly.

b. **Installation.** To install adjuster assembly, reverse removal procedure as outlined in subparagraph a preceding.

88. GENERATOR CIRCUIT BREAKER.

a. **Description.** The generator circuit breaker is mounted in the right end of the battery box (figs. 78 and 79). The circuit breaker cuts out automatically in case of an overload in the circuit and protects the wires in the generator circuit from burning. It must be reset manually by pushing in the red reset button after removing or correcting the cause of the trouble and allowing the circuit breaker to cool.

b. **Removal.** Turn 24-volt master switch off. Disconnect battery cables and remove right battery (par. 92 d). Note wire numbers for identification. Disconnect wire to shunt and wire leading to generator. Remove circuit breaker from bracket.

c. Installation. Place circuit breaker in position and install. Connect wires, install right battery and connect cables. Install battery box cover. Turn 24-volt master switch on and test circuit.

89. AMMETER.

a. Description. The ammeter is mounted on the instrument panel (fig. 16) and records the current output of the generator. Current used by electrical units does not flow through the ammeter. The ammeter shunt is mounted in the right end of the battery box, just beneath the generator circuit breaker. The shunt allows only a fraction of the current to pass through the ammeter, thus preventing damage to the instrument by the heavy electrical current from the generator.

b. Removal. Turn 24-volt master switch off. Remove instrument panel from the case (par. 97 d). Disconnect wires from instrument terminals and note numbers for identification. Remove ammeter support bracket, and fiber insulator. Push the ammeter out of the panel from the rear. Do not damage rubber gasket.

c. Installation. Replace rubber gasket if necessary. Install ammeter in position in the instrument panel and install support bracket. Connect proper wires to positive (+) and negative (−) terminals on ammeter. Install instrument panel in case (par. 97 e). Turn 24-volt master switch on. Test ammeter by starting auxiliary engine and note action of ammeter.

90. AMMETER SHUNT.

a. Removal of Ammeter Shunt. Turn 24-volt master switch off. Remove the right battery (par. 92 d). Disconnect the cables to the relay switch and circuit breaker. Remove shunt from battery box.

b. Install. Reverse the removal procedure as outlined in subparagraph a preceding. Use figure 73 for a wiring diagram.

91. GENERATING SYSTEM WIRING.

a. Description. The generating system wiring (fig. 73) consists in general, of the wires, cables and conduit leading from the generator circuit breaker to the generator regulator.

b. Removal. Follow procedure as outlined in paragraph 95 c.

c. Install. Follow procedure as outlined in paragraph 95 d.

TM 9-735
92

Part Three—Maintenance Instructions

Section XXII

BATTERY AND LIGHTING SYSTEM

92. BATTERIES.

a. **Description.** Two 12-volt storage batteries, connected in series, are located in the rear of the fighting compartment, just forward of the engine compartment bulkhead (fig. 79). The negative post of the battery on the right is grounded to the hull, and the positive post of the left battery is connected to the 24-volt master switch. When this switch is turned off, no current flows to any electrical units.

b. **Maintenance and Inspection.** Inspect batteries at least once a week and after every long run. Maintain level of battery fluid (electrolyte) ⅜-inch above tops of the plates by adding clean water (distilled water preferred) as required. Clean tops of batteries, battery terminals, and posts with a solution of baking soda and water (1 tablespoon soda to 1 gallon water). Clean and coat terminals and posts with petrolatum to prevent corrosion. Whenever batteries are removed, wash out battery compartment with soda solution. Inspect batteries for possible leaks and cracks. Take specific gravity or hydrometer reading every 25 hours, and recharge or exchange low charge batteries for fully charged ones if reading is 1.225 or less at normal temperatures. At temperatures below freezing the load on the battery becomes greater and the relative capacity of the batteries is reduced. For this reason, when low temperatures prevail, the specific gravity of battery electrolyte must be maintained at 1.250 or higher. The batteries must be replaced or recharged when specific gravity is below 1.250. The following table shows the capacity of the batteries and the relative freezing point of the electrolyte:

(1) FREEZING POINTS OF BATTERIES TABLE.

Capacity	Actual Specific Gravity	Freezing Temperature
Battery charged	1.285	−96° F
Battery ⅓ discharged	1.255	−60° F
Battery ½ discharged	1.220	−31° F
Battery ¾ discharged	1.185	− 8° F
Battery normally discharged	1.150	+ 5° F
Battery completely discharged	1.100	+18° F

c. **Specific Gravity Tests.** Specific gravity testing of battery electrolyte (battery fluid) determines the state of charge in each battery cell. Use storage battery testing hydrometer (18-H-1240), equipped with a thermometer and a temperature compensating table (fig. 77). Adjust reading to compensate for the temperature of the electrolyte. A specific gravity reading of 1.275 in each cell indicates

Battery and Lighting System

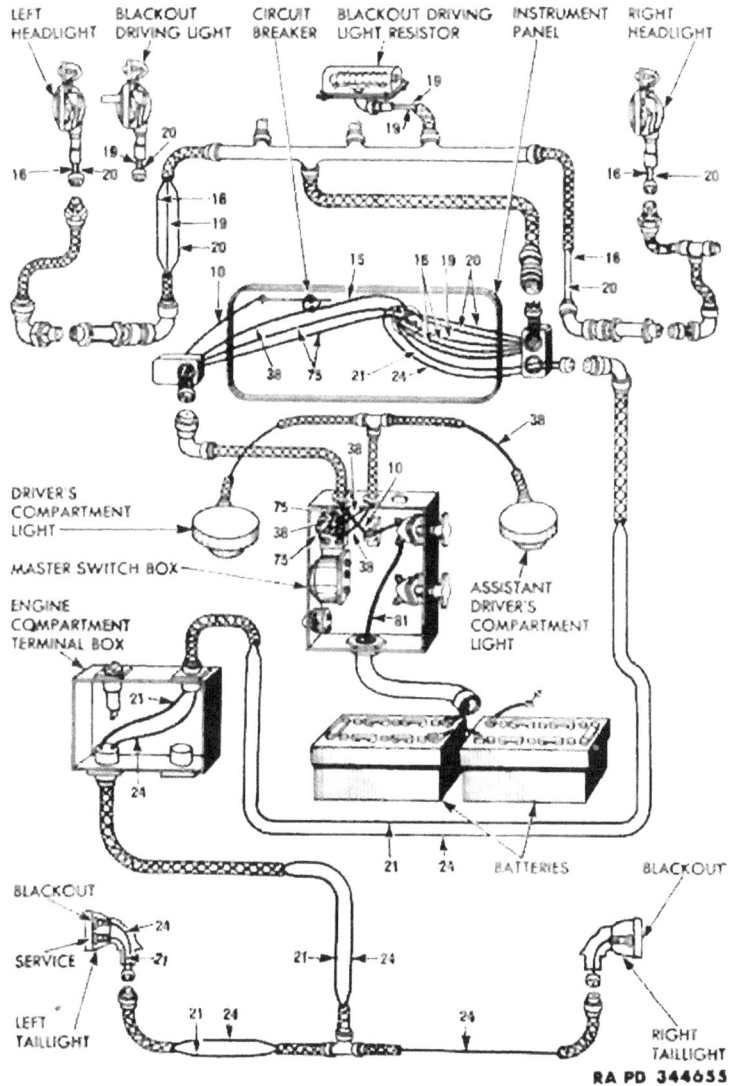

Figure 76 — Hull Lighting System Diagram

TM 9-735
92–93

Part Three—Maintenance Instructions

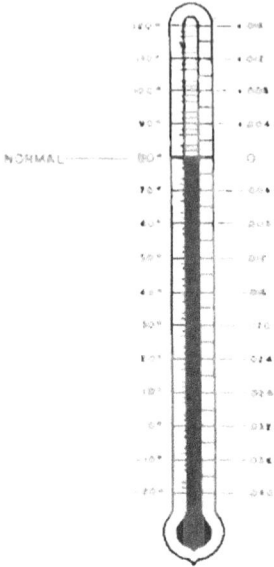

Figure 77 — Battery Electrolyte Temperature Correction Chart

a fully charged battery. A specific gravity reading of 1.225, or less, in each cell, indicates battery is discharged and must be recharged or replaced.

d. Removal. Remove battery box cover. Disconnect battery ground cable. Disconnect battery cables. Remove battery retaining frame. Remove batteries.

e. Installation. Install batteries in battery box. Install battery retaining frame. Connect battery cables and cover all terminals with a coating of petrolatum or general purpose grease. Install battery box cover.

f. Record Installation of Batteries. Make proper entry in W.D., A.G.O. Form No. 478, "MWO and Major Unit Assembly Replacement Record."

93. BATTERY BOX.

a. Description. The battery box (fig. 78) is located behind the 90-mm ammunition stowage compartment at the rear of the fighting compartment (fig. 79). The ammeter shunt, starter relay switch, and generator circuit breaker are mounted inside the battery box on

TM 9-735

Battery and Lighting System

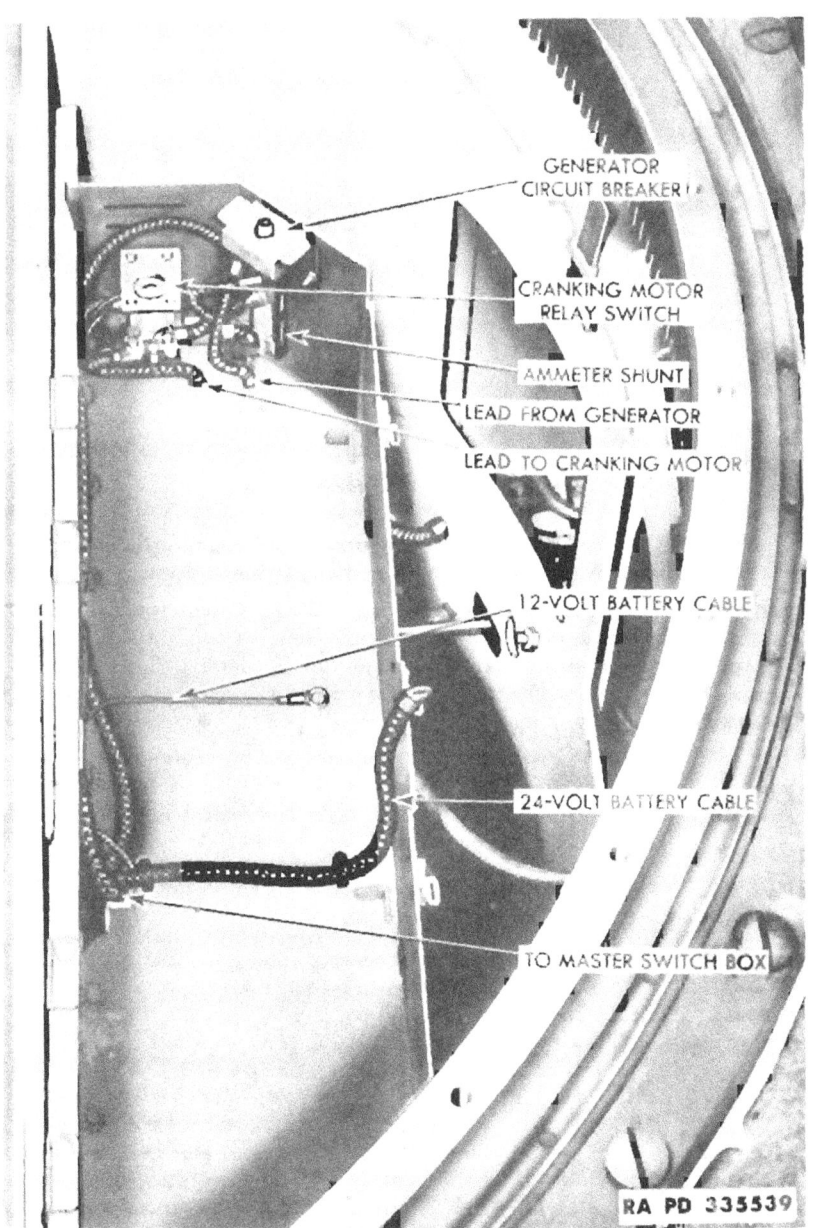

Figure 78 — Battery Box (Batteries Removed)

Part Three—Maintenance Instructions

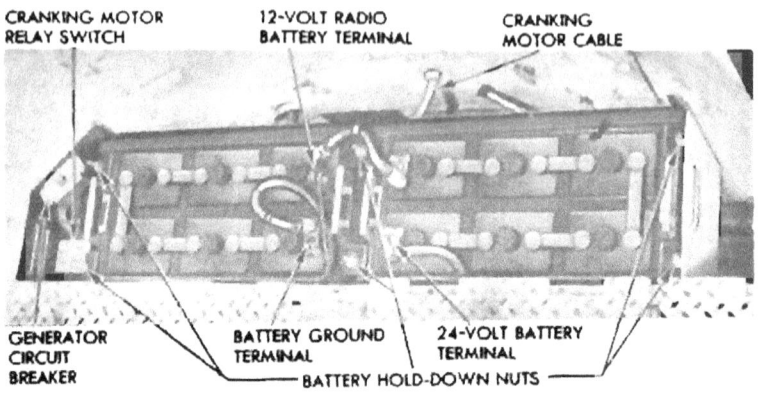

Figure 79 — Batteries

the right wall. The circuit breaker operates automatically in case of an overload, and is reset by pushing in the red reset button.

b. Removal. Remove batteries (par. 92 d). Disconnect the two cables leading into bottom of battery box. Remove conduit lock nuts and pull cables and conduit down and out of battery box. Note numbers of cables for later installation. Remove all attaching bolts, nuts, and lock washers, and lift out of box.

c. Installation. Install battery box and line up bolt holes. Install conduit and lock nuts in bottom of box and connect cables to ammeter shunt and relay switch according to numbers (fig. 76). Install and connect batteries (par. 92 e).

94. MASTER SWITCHES AND BOX.

a. Description. The master switch box is mounted on the front of the turret ring front support, between the driver and assistant driver (fig. 12). The 24-volt and 12-volt master switches are mounted vertically in the box. The upper switch controls the 24-volt main circuit. The lower switch controls the 12-volt circuit. For use and operation of master switches, see paragraph 14 i and j. A battery charging receptacle is also provided in the box which has a ground wire attached to the turret ring front support. An "ON-OFF" toggle type switch for controlling the operation of the ventilator blower and a circuit breaker controlling the ventilator circuit are also mounted in the master switch box (fig. 39). In case of an overload of current in the ventilator circuit, the circuit breaker will cut out automatically, and can be reset by pushing in the red reset button.

TM 9-735
94

Battery and Lighting System

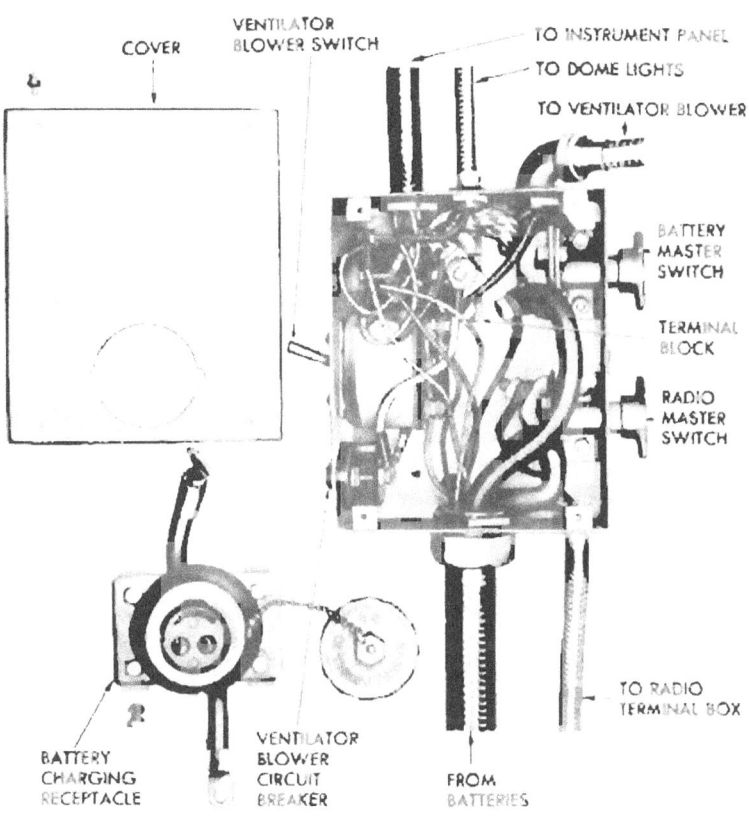

Figure 80 — Master Switch Box (Cover Removed)

b. **Removal.** Disconnect battery ground cable. Remove master switch box cover (fig. 80). Disconnect cables and wires. Note wire numbers for identification. Disconnect conduits, and pull ends of wires out of box. Disconnect the charging receptacle ground wire. Remove master switch box from turret front support. To remove master switches, first remove screws in handle knobs and remove knobs, then remove master switches from mounting brackets. Remaining ventilator switch, charging receptacle, and circuit breaker can be disassembled from the box by disconnecting ground strap from terminal block and removing units from box. NOTE: *Switches and wiring may be serviced or removed from box without removing master switch box from turret front support.*

TM 9-735
94–96

c. **Installation.** Install switches, charging receptacle, and circuit breaker in master switch box. Install ground strap to terminal block. NOTE: *Make certain ground cable is not frayed or in position to cause a short in the circuit.* Install master switch box on turret front support if removed. Install wires and cables through openings in box and attach conduit. Connect wires to proper terminals according to numbers (fig. 76). Connect charging receptacle ground wire to turret support. Install master switch box cover. Connect ground cable to battery terminal and tighten securely. NOTE: *Throughout sections on electrical equipment reference is made to turning 24-volt master switch off whenever any unit is to be disconnected, or wiring is to be exposed. This is important as a protection to both personnel and the electrical units.*

95. BATTERY AND LIGHTING WIRING SYSTEM.

a. **Description.** The battery and lighting wiring system (fig. 76) consists of the cables, conduit, and wires leading from the batteries to the master switch box and to the various lights and their switches.

b. **Inspection and Maintenance.** Inspect the wiring system for broken or crushed conduit, damaged insulation, dirty or corroded conduit and wire connections. Efficient functioning of electrical units and prevention of radio interference depends upon proper shielding of wires and clean, tight terminals and coupling nuts. Use a fine wire brush to clean corrosion from conduit and wire connections and terminal screws. Replace or repair any wires having damaged insulation, conduit or terminals.

c. **Removal.** Turn battery master switch off before making any repairs to the wiring system. Before disconnecting any wire from the circuit, fasten a tag to the terminal post being worked on, stating the number of the wire, and where the other end is connected. Disconnect the conduit couplings and the enclosed wires. Remove brackets or clips holding conduit. Attach a piece of strong twine to the end of each wire before removing the wire from the conduit. The twine can be used to pull a new wire into the conduit or for pulling a cleaning cloth through a dirty but otherwise serviceable conduit.

d. **Installation.** Use only wire of the same size and number as the wire removed. Pull a new wire through conduit by means of the heavy twine left in conduit when defective wire was withdrawn. Clean terminal posts and wire terminals before connecting any wire. Connect wires and conduit and tighten terminal screws and conduit couplings securely. Install conduit clips or brackets.

96. VEHICLE LIGHTS AND SWITCH.

a. **Service Headlights and Blackout Marker Lights.** The vehicle is equipped with two demountable service headlights (fig. 1).

TM 9-735

Battery and Lighting System

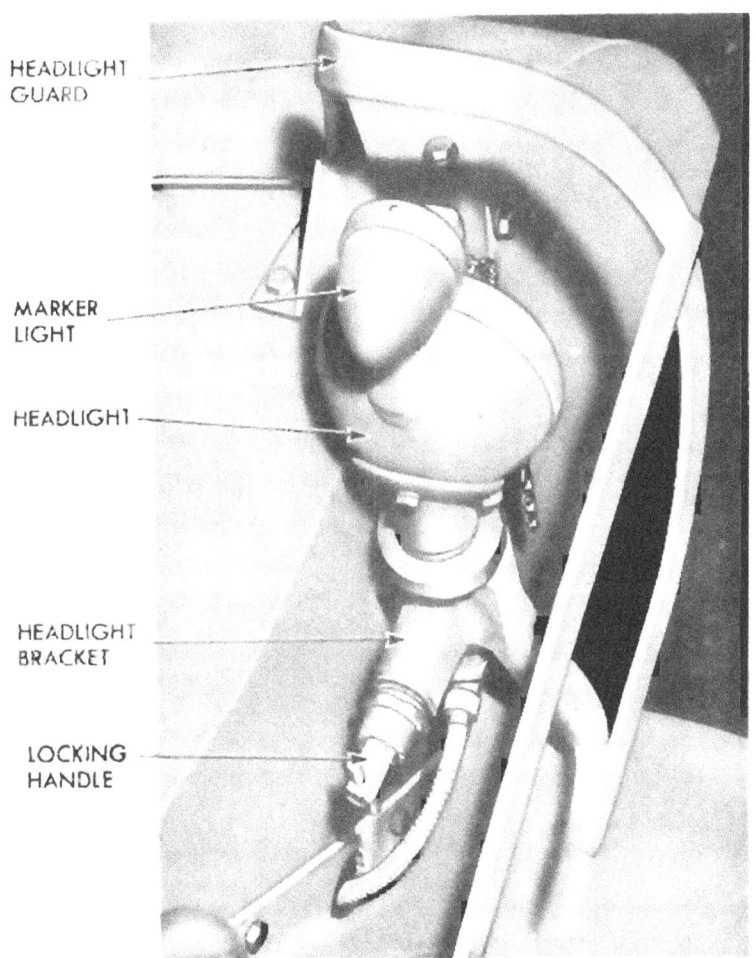

RA PD 335541

Figure 81 — Headlight — Installed

A headlight is mounted in a socket on each side of the front slope of the vehicle. Each light has a blackout marker light unit mounted on the top of the service headlight unit. Both light units are controlled by the driving lights switch on the instrument panel. For operation of driving lights switch see paragraph 15 a. The service headlights are held in position in the sockets by screw-and-ratchet type locking handles, operated from outside the vehicle (fig. 81). With the ser-

215

TM 9-735

Part Three—Maintenance Instructions

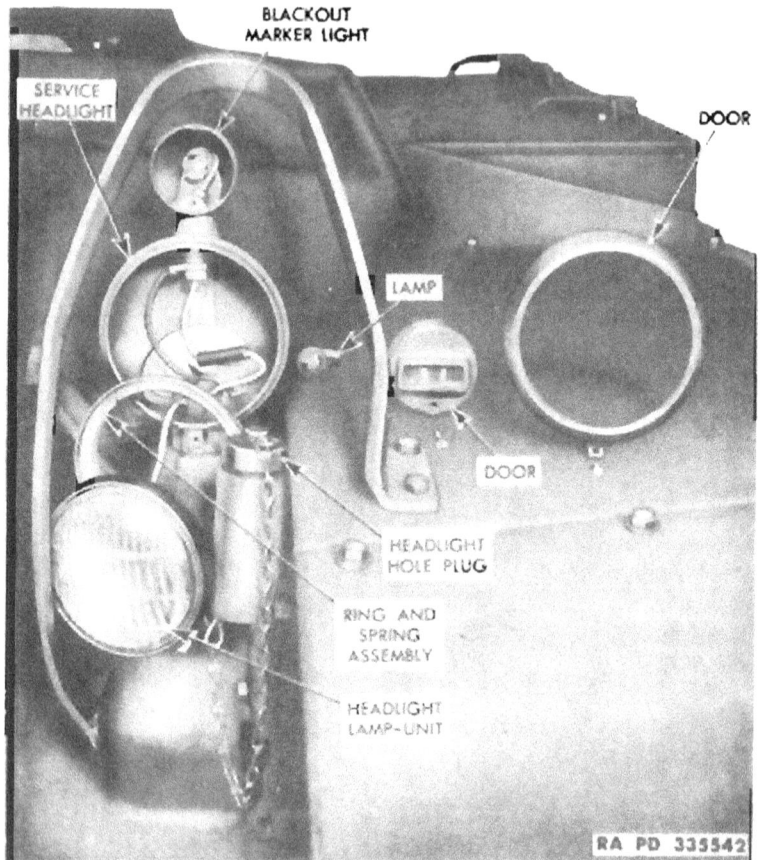

Figure 82 — Headlight — Disassembled

vice headlights in place, either the service head lamps or the blackout marker lamps can be used.

(1) CHANGE HEADLIGHTS. Remove or install service headlights as explained in par 14 l.

(2) REMOVE SEALED BEAM UNITS. Turn 24-volt master switch off. Remove the screw attaching the headlight door assembly to headlight body. Remove lamp-unit and door assembly. Disconnect wire. Release retainer springs. Remove sealed beam lamp-unit (fig. 82).

(3) INSTALL SEALED BEAM UNITS. To install sealed beam lamp-

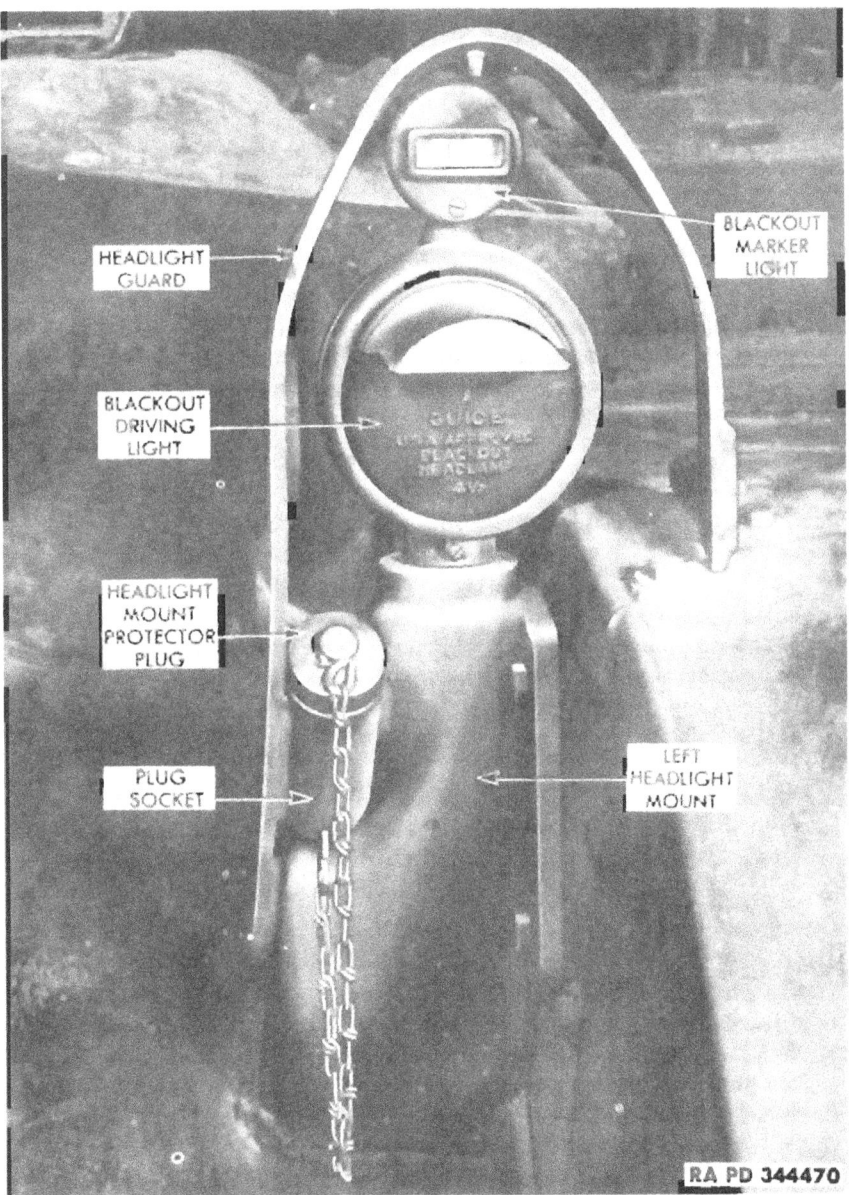

Figure 83 — Blackout Driving Light Installed

unit, reverse procedure as outlined in step (2) preceding. Test unit by turning on battery master switch and operating main lights switch.

(4) ADJUST HEADLIGHTS. Loosen three screws at rear of headlight, tilt light up or down to adjust beam, and tighten bolts.

(5) REMOVE BLACKOUT MARKER LAMPS. Remove door retainer screw and lift off door and gasket. Press lamp in, turn to left to release, and remove lamp.

(6) INSTALL BLACKOUT MARKER LAMPS. Insert the new lamp, press it in and turn to the right to lock. Install door with gasket and tighten screw. Test lamp by operating main light switch.

b. **Blackout Driving Light and Blackout Marker Light.** Mount the blackout driving light in the left headlight socket only (fig. 83). When not in use stow it in the bracket next to the driver's left foot (fig. 18) together with a service headlight. This light throws a thin, flat beam that cannot readily be seen from above because only a narrow strip of lens is exposed, yet sufficient illumination is provided to allow the driver to proceed with caution. The blackout driving light is also equipped with a blackout marker. Lamps are replaced as described in subparagraph *a* preceding.

c. **Taillights.** A taillight is mounted on the rear of the vehicle on each side (fig. 2). They are operated by the driving lights switch on the instrument panel, in conjunction with the headlights. Each taillight consists of a sealed beam unit. A blackout lamp-unit is mounted in the lower section of each taillight. NOTE: *There are no service or blackout stoplights on the vehicle although the service stoplight position is marked on the driving lights switch position plate.*

(1) REMOVE LAMP-UNIT. Remove two screws and pull door off. Remove defective lamp-unit.

(2) INSTALL LAMP-UNIT. Insert lamp-unit, install door and tighten screws. Test unit operating main lights switch. Be sure 24-volt master switch is on.

(3) REMOVE TAILLIGHTS. Turn conduit coupling nut off taillight bracket and pull wire and socket from the bracket. Remove the two brass lock nuts and pull the taillight assembly from the hull. Since the two taillights are not identical, they must be marked either "right" or "left" on removal, if removed together, to insure their installation on the proper side of vehicle. The right taillight contains only a blackout taillight. The left taillight contains a blackout taillight and the service taillight.

(4) INSTALL TAILLIGHTS. Before inserting taillight bracket into the hull, apply joint sealing compound around outside and inside of hole to prevent entrance of moisture. Install left and right taillights as marked and be sure they are adjusted to the proper angle as

viewed from the rear of the vehicle. Insert socket in taillight bracket and connect wires. Screw on the conduit coupling nut.

d. **Compartment Lights.** There are two compartment lights in the driver's compartment and three in the turret. They are connected in series with the 24-volt master switch which must be turned on before lights are operative. All compartment lights are equipped with two lamps. The lenses are divided into red and white sections. Each light is controlled by a toggle-type switch equipped with a spring-loaded safety latch on the toggle to prevent accidental turning on of the white light during blackout or night operations. The toggle cannot be moved from the "OFF" position to the white light "ON" position without first releasing the safety latch. To turn red light on simply move toggle in the direction in which it will move freely. To turn white light on pull out on toggle safety latch and then move toggle into slot in plate on switch shoulder and release safety latch. Turn off either red or white light by moving toggle to center position.

(1) REPLACE LAMPS. To replace the compartment light lamps remove the two screws holding the bezel in place, and remove the bezel and lens. Push lamp in, turn to the left and withdraw lamp. To install lamp, reverse the foregoing procedure.

(2) REMOVE COMPARTMENT LIGHT. Turn 24-volt master switch off before attempting any repairs to electrical system. Remove the two nuts attaching compartment light to mounting bracket. Remove bezel and lens as outlined in step (1) preceding. Disconnect wires and conduit. Remove light.

(3) INSTALL COMPARTMENT LIGHTS. To install compartment light, reverse procedure as outlined in step (2) preceding.

e. **Driving Light Switch (Main Lights).** The description and use of the driving light switch is given in paragraph 15 *s*.

(1) REMOVAL. Remove instrument panel from case (par. 97 d). Remove the sponge rubber gasket from the double-pin type terminal box. Remove terminal box from the three support posts. Note numbers of all wires as removed. Disconnect wire from the fuel gage control switch. Lift and tilt terminal box toward bottom of panel to clear light switch. Disconnect wire from fuel cut-off switch. Disconnect wires from vehicle light switch terminals. Remove set screw and pull switch lever off shaft. Lift off switch position plate. Lift driving light switch out of panel. Do not remove safety lock button. It will slide through opening in panel as switch is removed.

(2) INSTALL. Position vehicle light switch in instrument panel. Place switch position plate on shaft. Install lock washer, flat washer and nut. Install switch lever on shaft. Connect wire to fuel cut-off switch. Connect wire to fuel gage control switch. Connect wires to

vehicle light switch terminals. Install switch in instrument panel. Install position plate and switch lever. Install terminal box on support posts. Install sponge rubber gasket on terminal box. Install instrument panel in case (par. 97 e). Turn on 24-volt master switch and test vehicle light switch by operating through full-position range.

Section XXIII

INSTRUMENTS

97. INSTRUMENT PANEL.

a. General. The instrument panel consists of the instrument panel face plate assembly (fig. 85) and the case in which it is mounted. The face plate assembly contains the instruments, lights, gages, switches, and circuit breakers necessary for the operation and control of the power unit and vehicle. The face plate assembly is secured in the case by screws. The complete instrument panel assembly is shock mounted on vertical supports between and ahead of the driver and assistant driver (figs. 13, 18 and 19). Full descriptions on the use of the various instruments and units on the instrument panel are given in section VI. Units discussed in this section are those which have no direct relation to any one system. Units which are parts of systems, such as engine temperature gage, oil pressure gage, etc., are discussed in the system of which they are a part.

b. Removal of Instrument Panel and Case. Turn off 24-volt master switch. Disconnect the six electrical conduit connections and yellow ground wire from the back of the instrument panel case (fig. 84). Leave ground wire attached to instrument panel support. Tag wires and conduits for later identification. Remove screws which attach rubber insulated mounting brackets to panel supports. Support panel and case as screws are removed to prevent damage.

c. Installation of Instrument Panel and Case. Install instrument panel and case by reversing removal procedure described in subparagraph b preceding. Remove identification tags from wires and conduits. Turn on 24-volt master switch and test panel.

d. Removal of Instrument Face Plate. Turn off 24-volt master switch. Remove the two top mounting bolts from the rubber shock mounting brackets. Loosen, but do not remove, the two lower bracket mounting bolts (fig. 13). NOTE: *It is not necessary to loosen or remove bolts attaching mounting brackets to instrument panel supports.* Tilt top of the panel toward rear of vehicle as far as it will go. Disconnect the six electrical conduit connections and tag wires and conduits, for later identification. Leave yellow ground wire con-

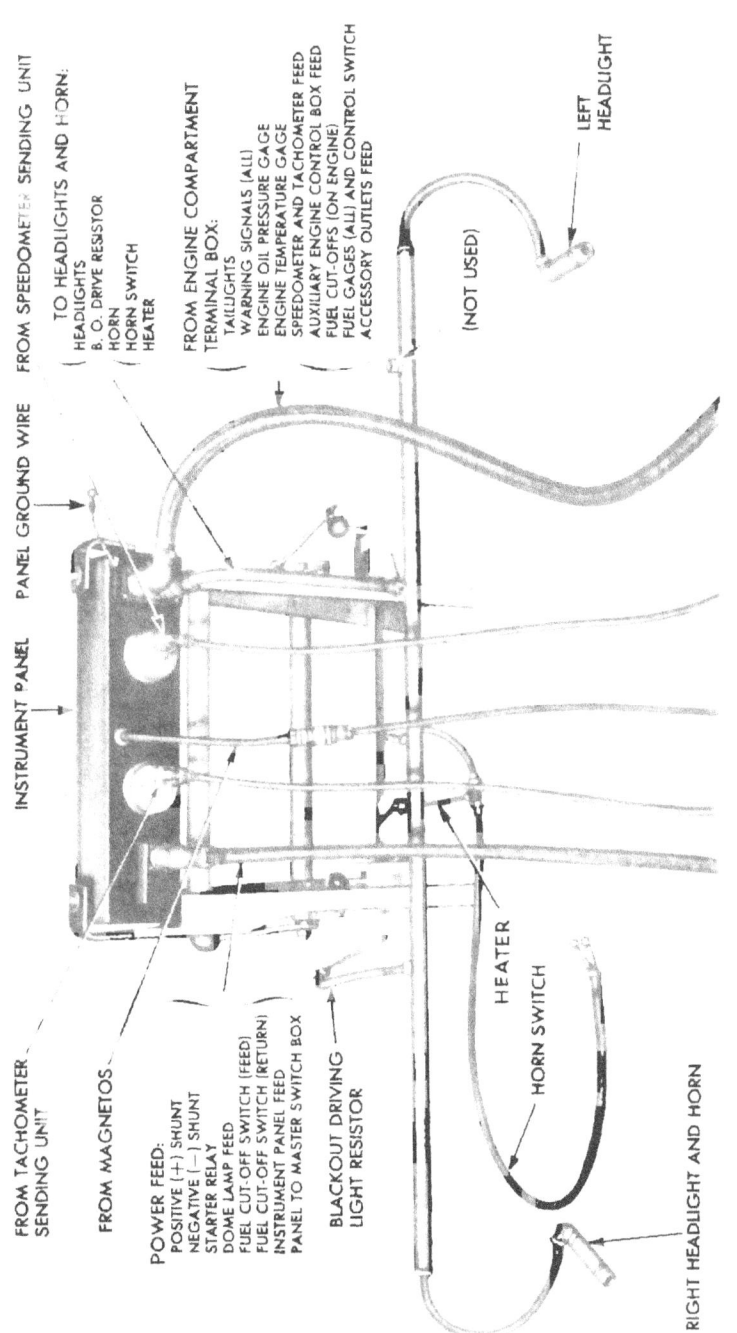

Figure 84 — Instrument Panel and Controls, Rear View

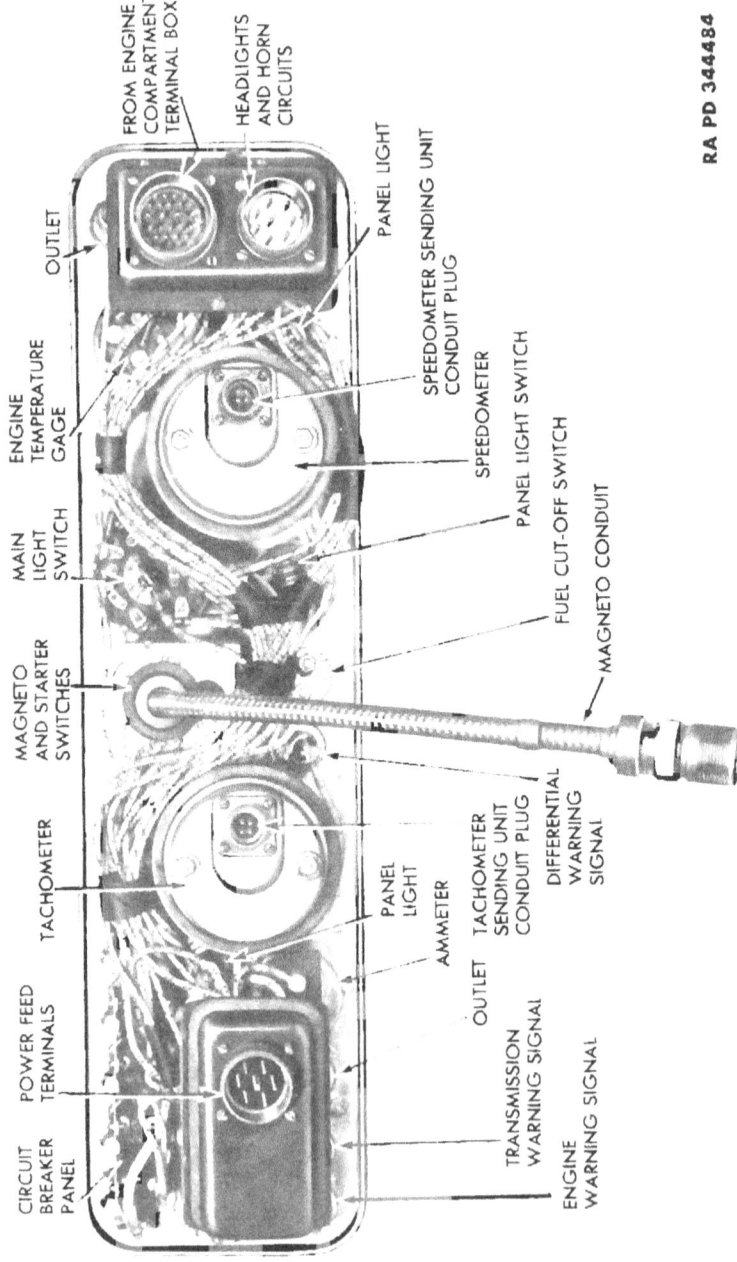

Figure 85 — Instrument Panel Face Plate, Rear View

nected to back of case. Return panel and case to vertical position. Remove screws which hold instrument panel in case and lift out panel face plate.

e. **Installation of the Instrument Panel Face Plate.** Place instrument panel case in vertical position. Install panel face plate in case and install attaching screws. Tilt panel and case toward rear of vehicle. Connect the six electrical conduit connections. Connect ground wire. Return panel and case to vertical position. Install top mounting bolts and tighten all bolts securely. Remove identification tags. Turn on 24-volt master switch and test panel.

98. INSTRUMENTS.

a. **General.** Whenever a unit in the panel becomes defective, it must be replaced with a new unit. Before removing any of the instruments, be sure 24-volt master switch is turned off. When removing wires from terminals note wire numbers and tag wires and terminals to identify them in making connections (fig. 86). When connecting wires from 12-volt or 24-volt circuits to instruments, be sure correct wires are connected to proper terminals (fig. 86).

b. **Driving Lights Switch (Main Lights).** The description and use of the driving light switch is given in paragraph 15 s. See paragraph 96 e for removal and installation.

c. **Accessory Outlets and Lamp Sockets.** The description and use of the accessory outlets and lamp sockets is given in paragraph 15 v. NOTE: *These sockets are welded to the panel and are not removable.*

d. **Speedometer and Tachometer Switch.** The description of the speedometer and tachometer switch is covered in paragraph 15 k.

e. **Speedometer.** The description and use of the speedometer are given in paragraph 15 l.

(1) REMOVAL. Remove instrument panel from case (par. 97 d). Remove speedometer case. Lift speedometer and built-in receiving unit out of position in panel. Remove rubber gasket.

(2) INSTALLATION. Install new speedometer and receiving unit by reversing the removal procedure described in step (1) above. Install new rubber gasket if necessary. Install panel in case (par. 97 e). Test unit.

f. **Tachometer.** The description and use of the tachometer are given in paragraph 15 m.

(1) REMOVAL. To remove the tachometer and its built-in receiving unit, repeat the operations described in subparagraph e (1) above for the removal of the speedometer.

(2) INSTALLATION. To install a new tachometer and receiving unit, repeat the speedometer installation procedure described in sub-

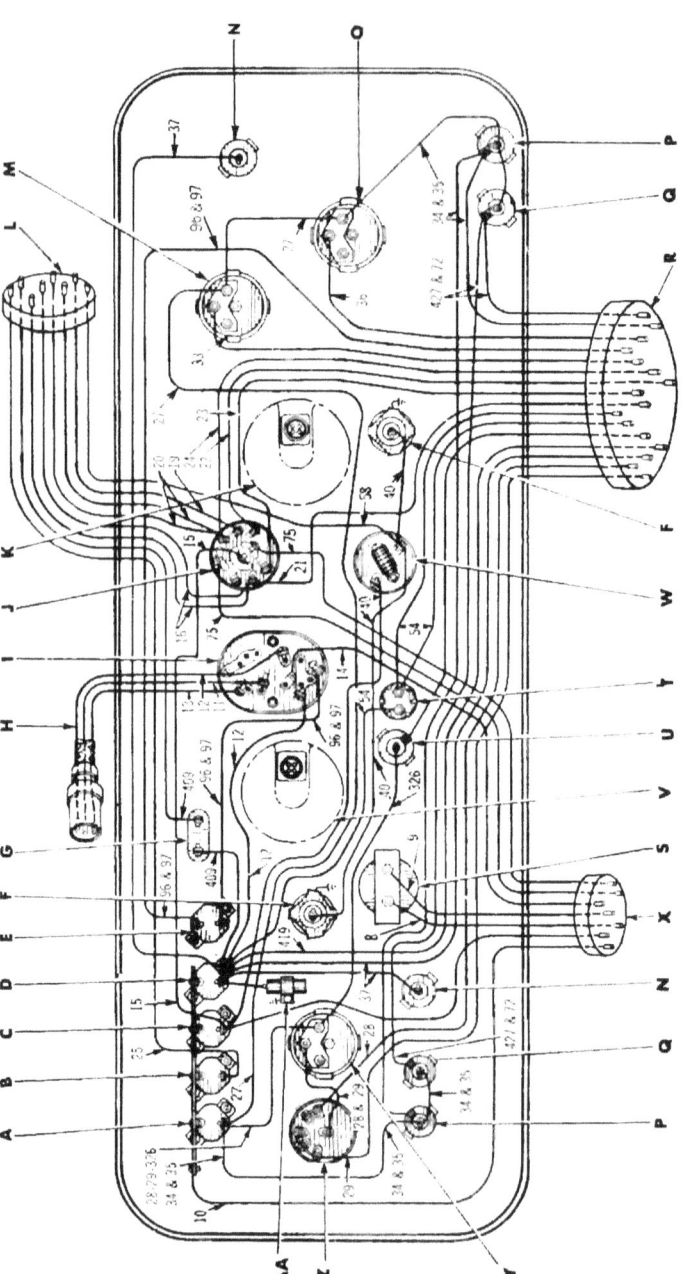

Figure 86 — Instrument Panel Schematic Wiring Diagram

A—CIRCUIT BREAKER—ALL GAGES AND WARNING SIGNALS
B—CIRCUIT BREAKER—HORN ONLY
C—CIRCUIT BREAKER—HULL LIGHTS AND SWITCHES
D—CIRCUIT BREAKER—ALL ACCESSORIES
E—CIRCUIT BREAKER—SPEEDOMETER AND TACHOMETER ONLY
F—PANEL LIGHT
G—HEATER SWITCH
H—MAGNETO CONDUIT
I—MAGNETO AND STARTER SWITCHES
J—MAIN LIGHTS SWITCH (DRIVING LIGHTS)
K—SPEEDOMETER
L—HEADLIGHTS AND HORN CONDUIT PLUG
M—ENGINE TEMPERATURE GAGE
N—ACCESSORY OUTLET
O—ENGINE OIL PRESSURE GAGE
P—ENGINE LOW OIL PRESSURE—HIGH WATER TEMPERATURE WARNING SIGNAL
Q—TRANSMISSION LOW OIL PRESSURE—HIGH OIL TEMPERATURE WARNING SIGNAL
R—ENGINE COMPARTMENT TERMINAL BOX CONDUIT PLUG
S—AMMETER
T—FUEL CUT-OFF SWITCH
U—DIFFERENTIAL LOW OIL PRESSURE WARNING SIGNAL
V—TACHOMETER
W—INSTRUMENT PANEL LIGHT SWITCH
X—POWER FEED CONDUIT PLUG
Y—FUEL GAGE
Z—FUEL GAGE CONTROL SWITCH
AA—CONDENSER

RA PD 344657B

Legend for Figure 86

paragraph e (2) preceding. Install panel in case (par. 97 e). Test unit.

g. **Heater Switch.** The description and use of the heater switch are given in paragraph 15 g.

(1) REMOVAL. Remove instrument panel from case (par. 97 d). Note wire numbers and disconnect wires from switch terminals. Remove switch from panel.

(2) INSTALLATION. Install a new heater switch by reversing the removal procedure (subpar. g (1) preceding). Test switch by operating.

h. **Panel Light Switch.** The description and use of the panel light switch are given in paragraph 15 t.

(1) REMOVAL. Remove instrument panel from case (par. 15 t). Disconnect wires from switch terminals. Pull switch lever off shaft. Remove switch position plate. Lift switch out of panel.

(2) INSTALLATION. Install a new panel light switch by reversing the removal procedure described in step (1) preceding. Connect wires to proper terminals (fig. 86). Be sure dowel on switch seats in dowel hole in panel. Install panel in case (par. 97 e). Test switch.

i. **Circuit Breakers.** The description and use of circuit breakers are given in paragraph 15 w.

(1) REMOVAL. Remove instrument panel from case (par. 97 d). Remove single-connection pin-type terminal box from support posts. Disconnect wire leading from capacitor in terminal box to horn circuit breaker at circuit breaker. This will also disconnect wire from the same circuit breaker terminal. Leave wires disconnected. Tip terminal box toward bottom of panel. Disconnect brass bus bar from all circuit breaker terminals. Disconnect all wires from the circuit breaker terminals (fig. 86). Disconnect circuit breaker mounting plate from panel. Lift out mounting plate with all circuit breakers attached. Remove screws attaching faulty circuit breaker to mounting plate, and lift off circuit breaker.

(2) INSTALLATION. Install a new circuit breaker and circuit breaker plate by reversing removal procedure described in step i (1) preceding. Connect wires to horn circuit breaker terminal (fig. 86).

99. SPEEDOMETER SYSTEM.

a. **Description.** The speedometer system consists of the speedometer adapter mounted on the right side of the transmission housing, a flexible drive shaft and casing, the sending unit (mounted on right fuel tank) (fig. 87), the wiring (flexible conduit), and combination receiving unit and speedometer mounted in the instrument panel.

TM 9-735
99

Instruments

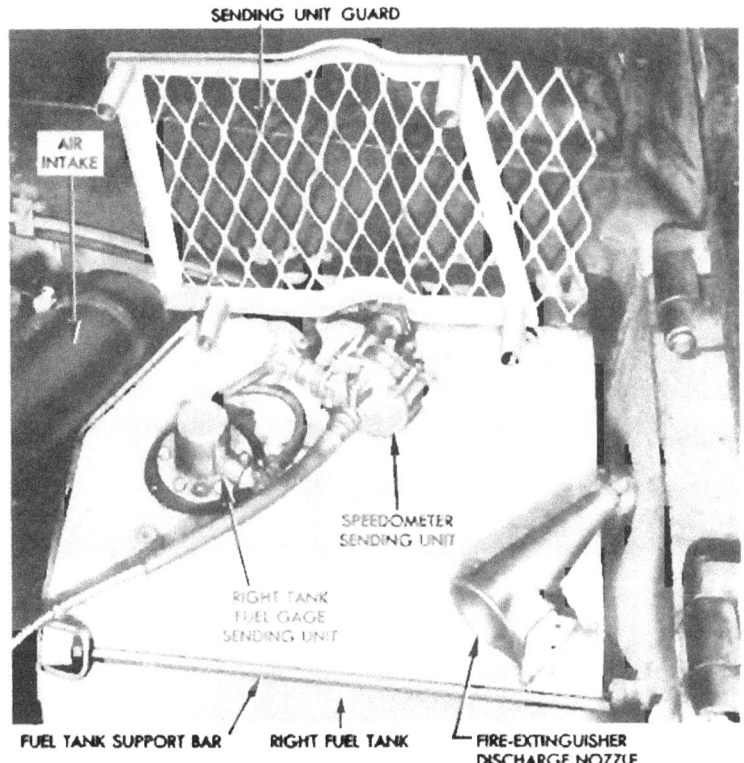

Figure 87 — Speedometer Sending Unit

b. **Removal of Speedometer Adapter.** Disconnect speedometer drive shaft conduit from adapter in transmission housing (right side) by backing off knurled coupling ring and pulling drive shaft from adapter. Remove adapter by unscrewing from transmission housing. Remove drive quill.

c. **Installation of Speedometer Adapter.** Install drive quill. Install speedometer adapter by reversing removal procedure (subpar. b preceding).

d. **Removal of Speedometer Drive Shaft.** Disconnect speedometer shaft casing from adapter on right side of transmission housing. Back off knurled coupling ring and pull drive shaft from adapter. Place strip of tape across open end of adapter to hold drive quill in

227

TM 9-735

Part Three—Maintenance Instructions

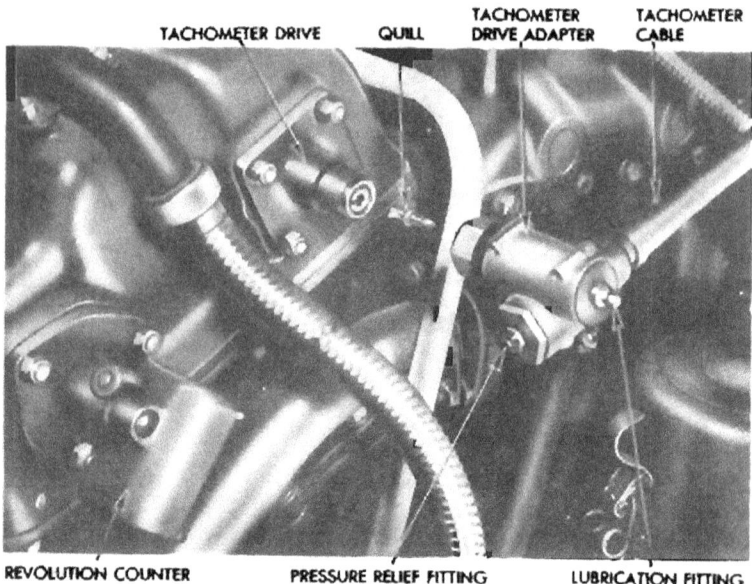

Figure 88 — Tachometer Drive Adapter

place. Remove clamp from cylinder head. Reinstall lock nut for safekeeping. Disconnect casing from sending unit. Where speedometer casing passes between fan shroud and transmission, push casing down toward bottom of engine compartment until adapter end of casing can be pulled clear of baffle. Pull speedometer drive shaft out of casing.

e. **Installation of Speedometer Drive Shaft.** Place speedometer drive shaft in position. Install square end of shaft in sending unit. Lead casing under fan shroud on right side of transmission, and then raise casing between shroud and transmission housing until end of drive shaft can be inserted in adapter. Turn transmission end of shaft with fingers to be sure it turns freely. Remove tape from open end of adapter. Insert driving lug on shaft into slot in adapter shaft. Push casing against adapter, and tighten knurled locking ring. Install cylinder head clamp. Test operation of unit.

f. **Removal of Speedometer Sending Unit.** Remove guard screen. Disconnect speedometer drive shaft from sending unit mounted on the right fuel tank (fig. 87). Remove conduit clamps, and disconnect the two electrical conduits from the sending unit. Remove unit from mounting bracket and guard. NOTE: *Ground* screw is attached to one cap screw. Tag ground strap so that it can be connected to correct screw. Lift out unit.

g. **Installation of Speedometer Sending Unit.** Install speedometer sending unit by reversing the removal procedure described in subparagraph f above. NOTE: *Be sure ground strap is attached to proper mounting screw.* Remove identification tags. Install unit guard.

h. **Removal of Speedometer System Wiring.** Remove guard screen and disconnect flexible electrical conduit from speedometer sending unit. Remove clamps attaching conduit to bottom of turret ring splash guard. Pull conduit through bulkhead seal into fighting compartment. Remove clamps attaching conduit to top of hull side plate on left side of fighting compartment and plates in drivers' compartment. Disconnect plug from speedometer receiving unit on rear of instrument panel.

i. **Installation of Speedometer System Wiring.** Install speedometer system wiring by reversing removal procedure described in subparagraph g preceding.

j. **Removal and Installation of Speedometer Receiving Unit and Speedometer.** See paragraph 98 e.

100. TACHOMETER SYSTEM.

a. **Description.** The tachometer system consists of the tachometer adapter in the magneto end of the "left" camshaft housing (fig. 88), a flexible drive shaft and casing, a sending unit mounted on the left fuel tank (fig. 90), the wiring (flexible conduit), and combination receiving unit and tachometer mounted in the instrument panel. For further description see paragraph 15 m.

b. **Removal of Tachometer Adapter.** Disconnect tachometer drive shaft from adapter. Unscrew adapter from magneto end of "left" camshaft housing (fig. 88). Use care not to lose short drive quill which may pull out when removing drive shaft. Remove adapter and gasket from camshaft housing. If adapter drive quill comes out, reinstall it in camshaft. Tape open end of adapter to hold drive quill in place.

c. **Installation of Tachometer Adapter.** Install tachometer adapter by reversing removal procedure described in subparagraph b preceding. Be sure adapter drive quill is installed in camshaft.

d. **Removal of Tachometer Drive Shaft.** Disconnect casing from adapter and pull drive shaft out of unit. Be careful not to lose drive quill as shaft is pulled from adapter. See that drive quill is in place in adapter and then tape open end of adapter. Disconnect casing from sending unit and pull drive shaft out of casing.

TM 9-735

Part Three—Maintenance Instructions

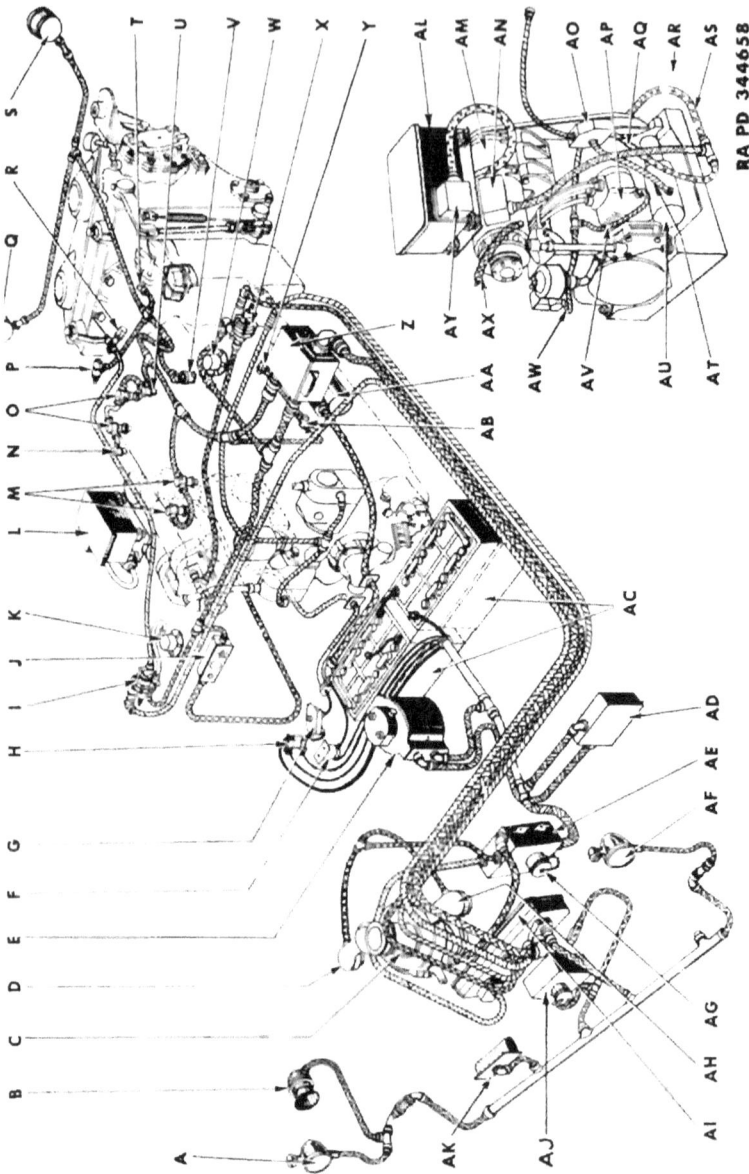

Figure 89 — Hull Electrical System Diagram

Legend for Figure 89

A—RIGHT HEADLIGHT
B—HORN
C—BLOWER AND MOTOR
D—ASSISTANT DRIVER'S COMPARTMENT LIGHT
E—COLLECTOR RING BOX
F—STARTER RELAY
G—CIRCUIT BREAKER
H—AMMETER SHUNT
I—SPEEDOMETER SENDING UNIT
J—AUXILIARY ENGINE CONTROL BOX
K—RIGHT FUEL GAGE
L—AUXILIARY ENGINE GENERATOR REGULATOR
M—FRONT CARBURETOR DEGASSERS
N—ENGINE HIGH WATER TEMPERATURE WARNING SIGNAL SWITCH
O—REAR CARBURETOR DEGASSERS
P—DIFFERENTIAL LOW OIL PRESSURE WARNING SIGNAL SWITCH
Q—RIGHT TAILLIGHT
R—SPEEDOMETER ADAPTER
S—LEFT TAILLIGHT
T—TRANSMISSION LOW OIL PRESSURE WARNING LIGHTS SWITCH
U—ENGINE WATER TEMPERATURE SENDING UNIT
V—TRANSMISSION HIGH OIL TEMPERATURE WARNING SIGNAL SWITCH
W—LEFT FUEL GAGE
X—TACHOMETER SENDING UNIT
Y—ENGINE LOW OIL PRESSURE SENDING UNIT
Z—ENGINE COMPARTMENT TERMINAL BOX
AA—STARTER
AB—ENGINE LOW OIL PRESSURE WARNING SIGNAL SWITCH
AC—BATTERIES
AD—TURRET RADIO TERMINAL BOX
AE—MASTER SWITCH BOX
AF—LEFT HEADLIGHT
AG—BATTERY CHARGING RECEPTACLE
AH—DRIVER'S COMPARTMENT LIGHT
AI—INSTRUMENT PANEL
AJ—DRIVER'S COMPARTMENT HEATER
AK—BLACKOUT DRIVING LIGHT RESISTOR
AL—GENERATOR REGULATOR
AM—GENERATOR
AN—GENERATOR TERMINAL BOX
AO—STARTER RELAY
AP—MAGNETO
AQ—GENERATOR TO STARTER RELAY
AR—AUXILIARY ENGINE AND GENERATOR
AS—BATTERY TO GENERATOR
AT—STARTER RELAY TO STARTER
AU—STARTER
AV—CONTROL BOX TO MAGNETO
AW—CONTROL BOX STARTER SWITCH TO STARTER RELAY

RA PD 344658B

TM 9-735
100

Part Three—Maintenance Instructions

Figure 90 — Tachometer Sending Unit

e. **Installation of Tachometer Drive Shaft.** Push new drive shaft into casing. Insert sending unit end of drive shaft into sending unit, and tighten knurled coupling nut on casing. Lead adapter end of drive shaft across magneto end of the engine to adapter in "left" camshaft housing. Remove tape from adapter. Be sure drive quill is in place, "short end" of quill first. Connect drive shaft to quill and screw knurled locking ring onto adapter. Start engine and check operation of unit.

f. **Removal and Installation of Tachometer Sending Unit.** The tachometer sending unit is mounted at the top of the left fuel

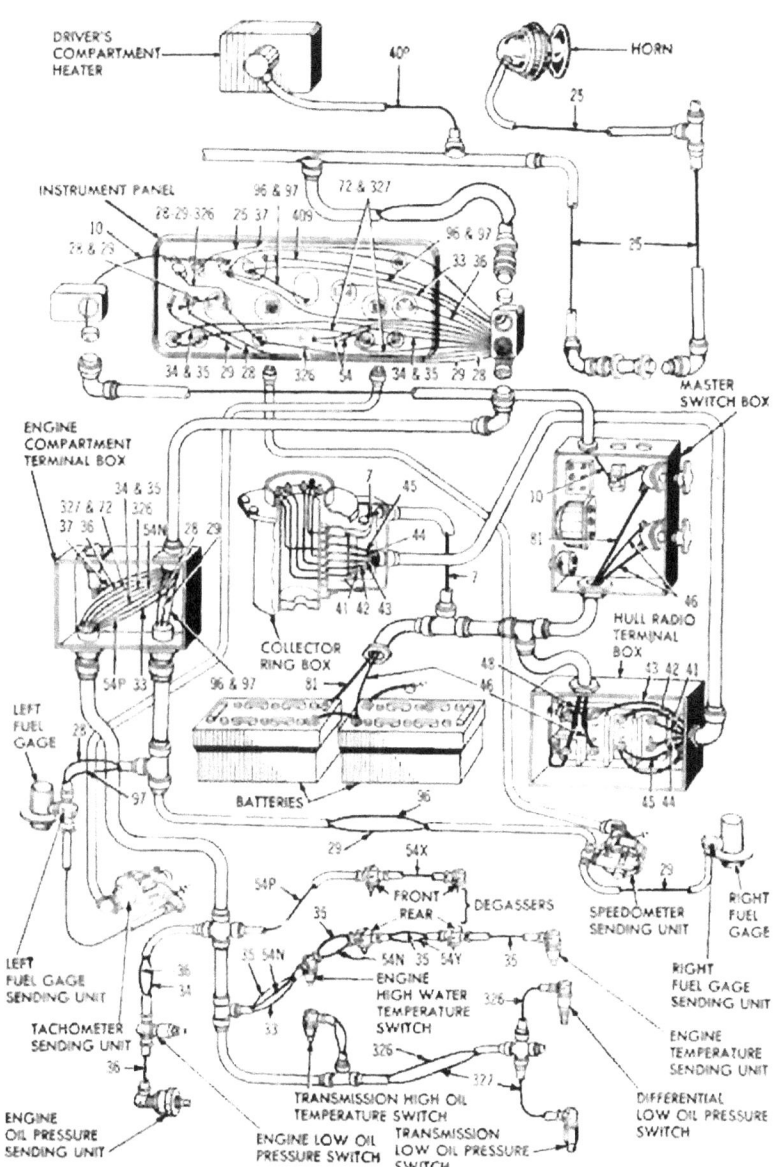

Figure 91 — Miscellaneous Hull Electrical System Diagram

tank and is removed and installed as described in paragraph 99 preceding.

g. Removal of Tachometer System Wiring. Disconnect flexible electrical conduit from tachometer sending unit and then follow removal procedure described in paragraph 95 c.

h. Installation of Tachometer System Wiring. Follow procedure described in paragraph 95 d and connect conduit to tachometer sending unit.

i. Removal and Installation of Tachometer Receiving Unit and Tachometer. See paragraph 98 f.

Section XXIV

TURRET ELECTRICAL SYSTEM

101. TURRET ELECTRICAL SYSTEM.

a. Description. The turret electrical system (fig. 92) consists of the wires, conduits, switches, and electrically operated units located in the turret, and the turret collector ring. The collector ring feeds both 12 and 24-volts to the turret switch box, which is the distribution point for the turret electrical system. The electric traversing motor, lights, solenoids, radios and interphone system operate on the 24-volt circuit. The British radio and interphone system operate on the 12-volt circuit. All wiring is shielded for protection from oil and dirt, and to prevent radio interference. The radio terminal box in the turret is the distribution point for the radio system.

b. Inspection and Maintenance. Inspect wiring and conduit frequently, and replace any damaged wires. Clean corrosion from terminal connections, nuts, and coupling threads with a small wire brush. Replace broken, crushed or frayed shielding conduits. Clean dirty or corroded conduit connections with dry-cleaning solvent. Examine and securely tighten coupling nuts and holding clips.

102. WIRING AND CONDUITS.

a. Description. Wires in the turret electrical system are encased in metal conduit to provide both protection and radio shielding. Conduit and fittings are treated to prevent corrosion and soldered to conduits. Conduits are attached to the vehicle structure by removable clips which also serve as bonds in radio shielding. Service all wiring harnesses and conduits as complete units, except as outlined in subparagraph d following.

TURRET ELECTRICAL SYSTEM

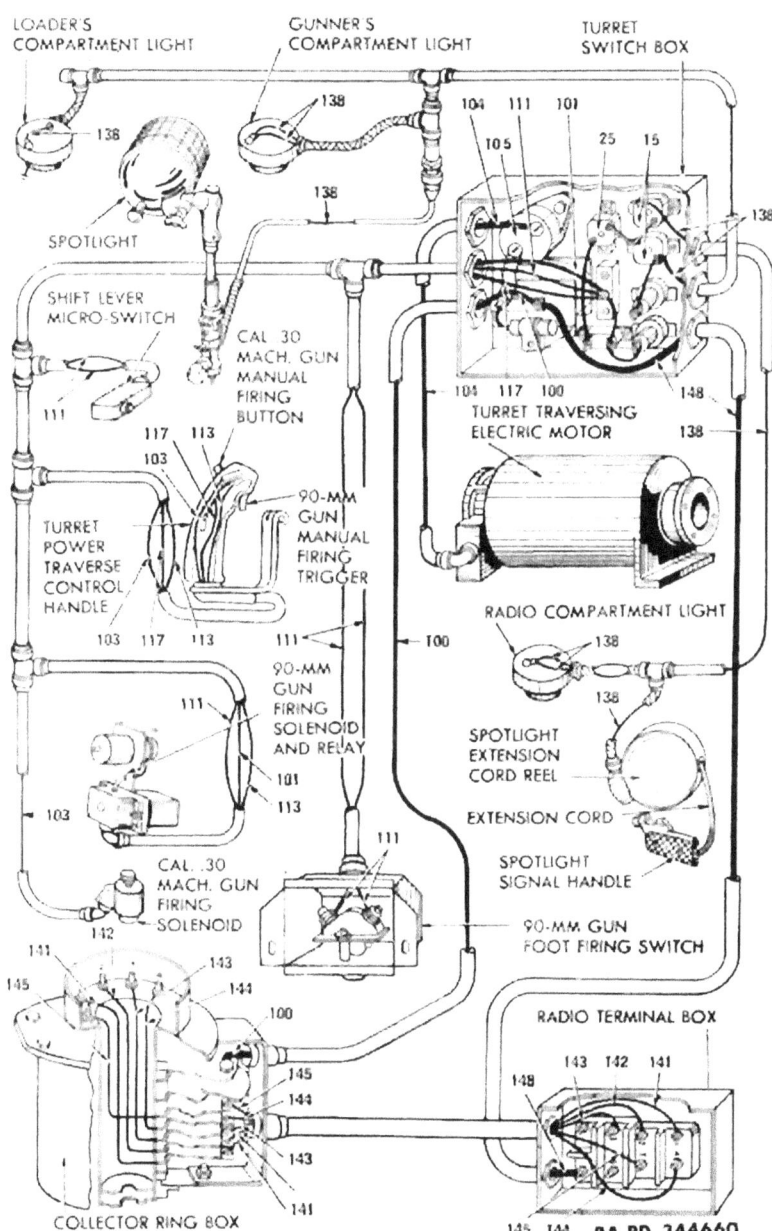

Figure 92 — Turret Electrical System Diagram

Part Three—Maintenance Instructions

b. **Removal of Conduit.** Turn 24-volt master switch off, and disconnect conduit and wires at terminals. Tag wires and terminals for identification when installing. Remove conduit holding clips, and lift the wiring harness and conduit out as a complete unit.

c. **Installation of Conduit.** Install wiring harness and conduit as a complete unit. Connect wires to terminals. Tighten coupling nuts securely. Remove identification tags. Install holding clips. Turn 24-volt master switch on. Test circuit by operating electrical units serviced by new wiring harness.

d. **Replace Damaged Conduit.** Replace any damaged section of a conduit with a new section, only if the damaged piece can be readily removed from the wiring system. Otherwise replace entire conduit harness in the damaged circuit. Disconnect conduit and wires where necessary. Tag wires for identification. Remove damaged section of conduit and install new section, or remove and install complete harness. Connect wires to correct terminals and tighten conduit coupling nuts securely.

e. **Removal of Wires.** After disconnecting any wire in a circuit, fasten tags, stating wire(s) number, to terminal post(s). Tag the wires as they are removed, marking terminal identification on tag. (If this method is followed, it will not be necessary to waste time tracing circuits.) Disconnect conduit couplings, and enclosed wires. Remove brackets or clips supporting conduit. Attach a piece of strong twine to end of each wire before removing wire from conduit. Leave twine in conduit to pull new wire(s) through, or to pull cleaning cloth through a dirty but serviceable conduit.

f. **Installation of Wires.** Attach new wire to the twine left in conduit (subpar. e preceding), and pull wire through conduit. Clean terminal posts and wire terminals before connecting wire. Connect wires according to identification tags, and tighten terminal screws securely.

103. RADIO TERMINAL BOX.

a. **Description.** The radio terminal box is mounted at the rear of the fighting compartment (fig. 205). It serves as the distribution point for the radio system wiring, and contains a terminal block for connecting the electrical units of the radio system. It is fed by wires and conduit from the collector ring (fig. 92).

b. **Removal.** Turn 24-volt battery master switch off. Remove terminal box cover. Disconnect wires from terminal posts, and note numbers for installation identification. Unscrew conduit coupling nuts, and pull wires clear of box. Remove terminal box.

c. **Installation.** Install terminal box. Insert wires through the openings ir the box. Attach conduit by tightening knurled coupling

TM 9-735
103

TURRET ELECTRICAL SYSTEM

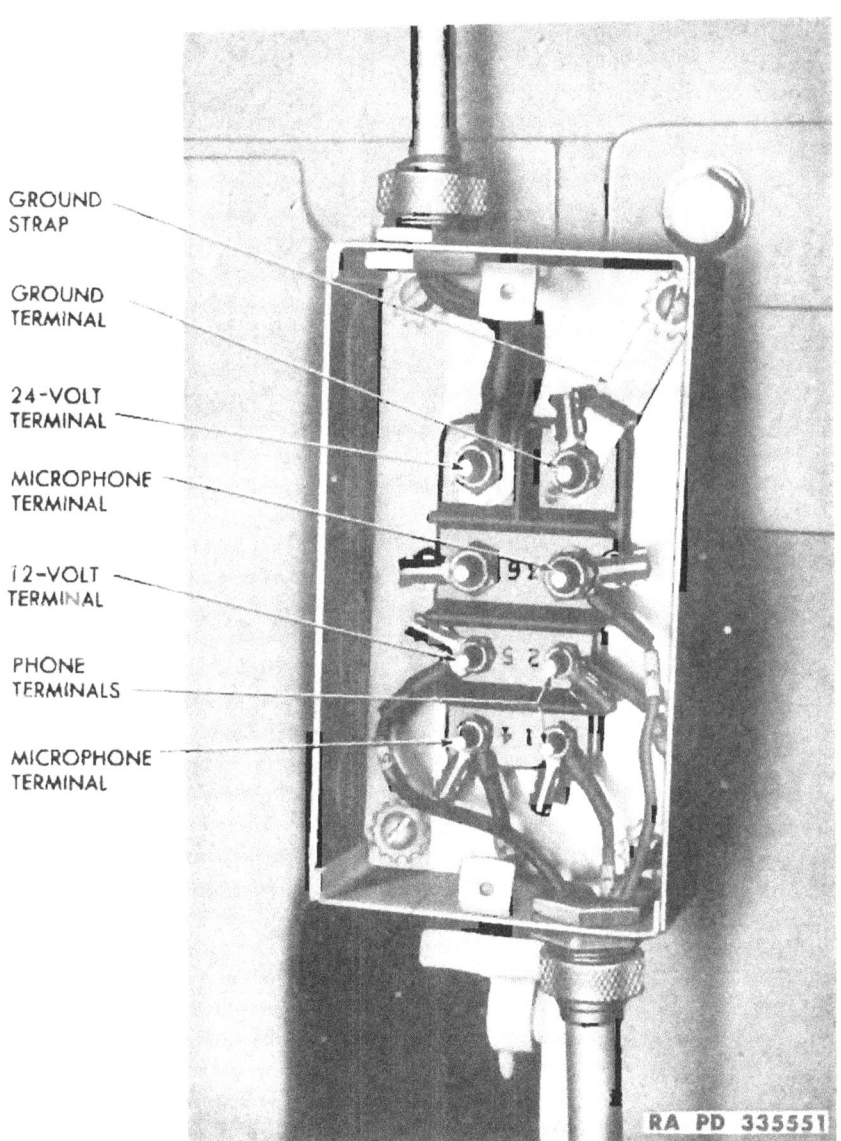

Figure 93 — Turret Radio Terminal Box (Cover Removed)

TM 9-735
103

Part Three—Maintenance Instructions

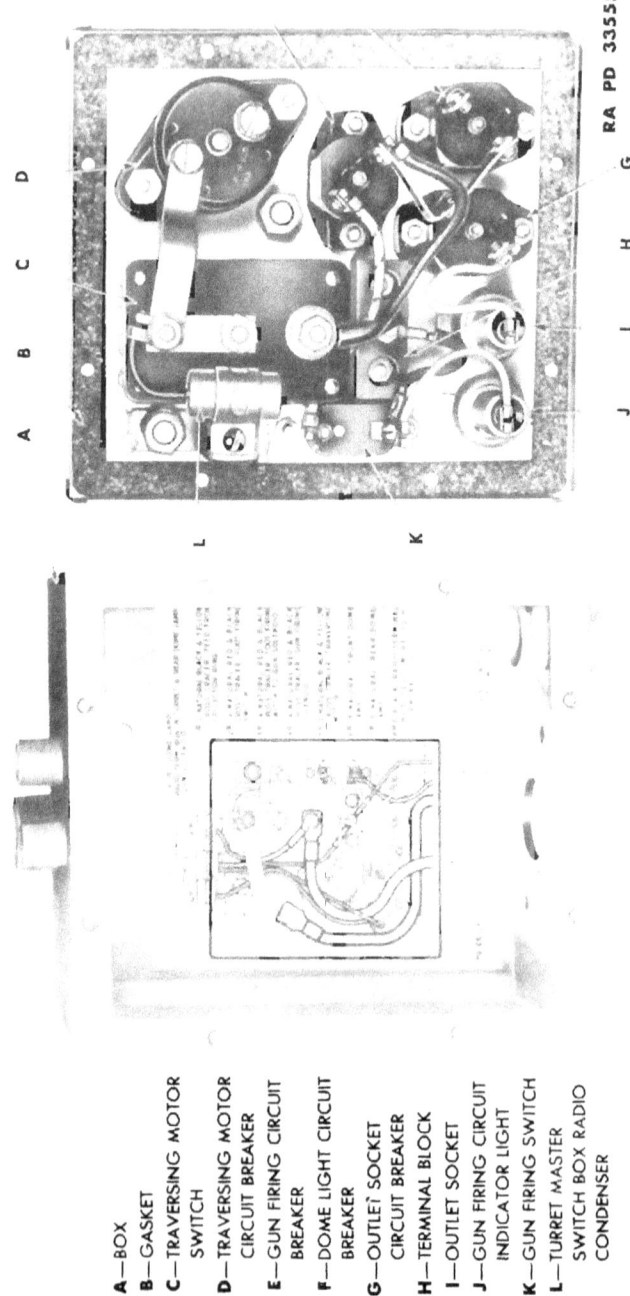

Figure 94 — Turret Switch Box (Face Plate Removed)

A—BOX
B—GASKET
C—TRAVERSING MOTOR SWITCH
D—TRAVERSING MOTOR CIRCUIT BREAKER
E—GUN FIRING CIRCUIT BREAKER
F—DOME LIGHT CIRCUIT BREAKER
G—OUTLET SOCKET CIRCUIT BREAKER
H—TERMINAL BLOCK
I—OUTLET SOCKET
J—GUN FIRING CIRCUIT INDICATOR LIGHT
K—GUN FIRING SWITCH
L—TURRET MASTER SWITCH BOX RADIO CONDENSER

nuts. Connect wires to correct terminal posts according to numbers (fig. 92). Install terminal box cover. Test circuits by operating radio.

104. COLLECTOR RING BOX.

a. **Description.** The collector ring is located in the center of the turret. A plate on top of the box supports the bracket which contains the electric traversing motor, hydraulic pump, and stowage boxes (fig. 27). An arm extends from the bracket and plate upward to the turret rim to revolve the turret. It provides a 360-degree rotary contact for the wiring between turret and hull. The collector ring consists of an assembly of rings, brushes, and terminals contained within a dustproof housing. The brush-holding part of the assembly revolves with the turret while the ring part remains stationary. Both the 12-volt and the 24-volt turret circuits are supplied through the collector ring assembly, with 12-volt and 24-volt circuits going to the radio terminal box and a 24-volt circuit to the turret switch box. The upper rings and brushes carry the 12-volt current. The lower rings and brushes carry the 24-volt current.

b. **Inspection and Maintenance.** The collector ring seldom requires attention. At regular inspections remove cover and inspect for dirt, worn brushes, and loose connections. Clean assembly by blowing out with compressed air. Turn battery master switch off and wash collector ring parts with dry-cleaning solvent. Keep collector ring tightly covered at all other times.

c. **Removal.** Turn 24-volt and 12-volt master switches off. Remove both front and rear sections of fighting compartment floor. Remove stowage boxes. Remove bolts attaching stowage box bracket to plate on top of collector ring box. Remove bolts attaching turret arm to collector ring box plate. Remove plate from top of box. Loosen knurled thumb screws holding cover to collector ring mounting flange, and lift off cover (fig. 27). Disconnect the five wires leading from the radio terminal box to collector ring brush terminals at the collector ring. Note wire number for installation identification. Disconnect the 24-volt cable and conduit leading to the turret switch box at the collector ring terminal. Disconnect 24-volt and 12-volt power feed cables and conduits. Note wire numbers for installation identification. Disconnect collector ring box ground strap. Remove bolts attaching box to support bracket and remove collector ring box (fig. 27).

d. **Installation.** Install collector ring box on support bracket. Connect 24-volt and 12-volt power feed wires to terminals (fig. 92). Correct ground strap. Tighten conduit lock nuts. Install 24-volt cable from turret switch box through opening in collector ring box and connect to terminal post (fig. 92). Insert the five 12-volt feed wires through fitting in collector ring box and connect to proper terminals (fig. 92). Install collector ring box cover. Install collector ring box mounting plate and bolts stowage box bracket and turret arm to plate. Turn 24-volt and 12-volt master switches on and operate turret traversing mechanism to test operation of collector ring assembly.

105. TURRET SWITCH BOX.

a. **Description.** The turret switch box is mounted on the right side wall of the turret behind the gunner's right shoulder (figs. 7 and 94). It is the control center of the turret electrical system and contains the traversing motor and gun firing switches, indicator light for the gun firing circuit, and circuit breakers for these units.

b. **Removal of Turret Switch Box.** Turn both 24-volt and 12-volt master switches off. Remove switch box cover and cover out as far as possible and disconnect wires from their terminals. Note each wire number for installation identification. Remove switch box cover. Unscrew conduit lock nuts and pull wires clear of box. Remove turret switch box.

c. **Installation of Turret Switch Box.** Install switch box on mounting bracket. Connect wires to proper terminals according to wiring diagram inside box cover (fig. 94). Connect conduits and tighten lock nuts securely. Install switch box cover. Turn 24-volt and 12-volt master switches on. Test traversing motor switch by operating traversing motor. Test gun firing circuits. Note if solenoids operate gun firing mechanisms. WARNING: *Before testing firing mechanism open gun breech to be sure breech is clear and a round has not been left in the gun. Fully depress gun before testing mechanism.* Observe if firing indicator light operates when gun firing switch is turned on.

d. **Removal of Traversing Motor Switch.** Remove switch box cover (fig. 94). Remove condenser. Remove bus bar. Remove condenser with wire attached. Disconnect circuit breaker connecting wire. Remove switch from box cover and remove traversing motor switch.

e. **Installation of Traversing Motor Switch.** Install traversing motor switch in turret switch box. Install switch face plate with rubber boot to front of turret switch box, and tighten mounting bolts. Connect direct breaker wire. Install condenser and bus bar according to wiring diagram shown in figure 94.

f. **Removal of Circuit Breakers, Gun Firing Switch, and Terminal Block.** Disconnect wire. Lift unit from turret switch box cover.

Turret Electrical System

g. **Installation of Circuit Breakers, Gun Firing Switch and Terminal Block.** Reverse removal procedure as outlined in subparagraph f preceding. Connect wires to correct terminals according to wiring diagram shown in figure 94.

h. **Remove Indicator Lamp.** Pry indicator light cap from switch cover, being careful not to damage rubber gasket. Disconnect wire, press lamp in, turn left to release locking lugs, and remove lamp.

i. **Install Indicator Lamp.** Insert lamp, press in and turn right to lock in place. Connect wire and tighten terminal nut securely. Install light cap and new rubber gasket if necessary.

106. GUN FIRING SOLENOIDS.

a. **Removal of 90-mm Gun Firing Solenoid.** Remove firing solenoid from bracket. Remove clamp screw and nut from push-pull cable. Remove cable from solenoid (fig. 95). Unscrew conduit lock nut and pull cover back far enough to disconnect wire at solenoid. Remove wire, conduit, and cover. Remove solenoid.

b. **Installation of 90-mm Gun Firing Solenoid.** Connect wire to relay terminal. Install conduit and cover. Install push-pull cable to solenoid, and tighten clamp screw. Install solenoid assembly in position on gun. Tighten nuts securely.

c. **Removal of Cal. .30 Gun Firing Solenoid.** Release solenoid from bracket (fig. 95). Through opening provided in solenoid bracket, remove wire terminal screw and lock washer. Unscrew conduit lock nut, and pull wire, and conduit from solenoid.

d. **Installation of Cal. .30 Gun Firing Solenoid.** Connect wire to solenoid terminal, install conduit lock nut, and tighten securely. Position solenoid to bracket, install two screws and lock washers, and tighten.

107. COMPARTMENT LIGHTS.

a. **Description.** The turret is equipped with three combination red-white lens type compartment lights mounted on the turret roof. See paragraph 23 a for full description.

b. **Changing Lamps.** For instructions on changing lamps refer to paragraph 96 d (1).

c. **Removal of Compartment Lights.** For removal of compartment lights use procedure in paragraph 96 d (2).

d. **Installation of Compartment Lights.** To install compartment lights follow procedure in paragraph 96 d (3).

TM 9-735
108

Part Three—Maintenance Instructions

Figure 95 — 90-mm Gun Firing Solenoid

108. ELECTRIC TRAVERSING MOTOR.

a. Description. A 24-volt, 1¾-horsepower electric motor furnishes power to operate the turret traversing mechanism (fig. 167). It is located ahead of the stowage boxes mounted above the turret collector ring box beneath the 90-mm gun. The turret traversing hydraulic pump is mounted on the right end of the motor.

b. Removal. Turn the 24-volt master switch off. Remove feed wire terminal cover. Unscrew conduit coupling nut and disconnect wire at motor terminal. Pull wire clear of terminal and move conduit clear. Attach a sling to support hydraulic pump as motor is removed. Remove locking wire and screws from pump flange connection. Support motor with a rope sling. Move motor carefully to left to clear pump coupling. Be careful not to damage oil tubes. Lower motor to fighting compartment floor.

Turret Electrical System

c. **Installation.** Install motor on mounting bracket. Aline coupling on armature shaft to mate with pump drive shaft. Install screws and locking wire in pump flange connection and remove supporting sling. Connect wire to motor terminal. Install terminal cover and connect conduit coupling nut. Turn 24-volt master switch on and test electric motor by operating turret traversing mechanism.

109. GUN FIRING SWITCHES.

a. **Description.** The 90-mm gun firing foot switch box (fig. 27) is located below the gunner's seat. A trigger type switch in the power traversing handle can also be used to fire the 90-mm gun. A switch button located at the top of the handle is used to fire the cal. .30 coaxial mounted machine gun.

b. **Removal of 90-mm Gun Firing Foot Switch.** Turn 24-volt master switch off. Remove the gunner's platform. Remove the switch box from below gunner's seat. Unscrew conduit coupling nut and pull switch box down to reach switch and wires. Disconnect incoming wire, note wire number for installation identification, and pull wire clear of box. Remove mounting screws and push switch out through top of box.

c. **Installation of 90-mm Gun Firing Foot Switch.** Insert switch through top of switch box and install. Insert wires through opening in bottom of the box and connect them to correct terminals. Tighten conduit coupling nut. Install switch box on gunner's seat. Install platform. Turn 24-volt master switch and gun firing switch on. Test gun firing foot switch by operating with heel, and note if gun firing solenoid operates properly. Observe action of striker plate. WARNING: *Open gun breech to be sure breech is clear and a round has not been left in the gun before testing firing switches. Depress gun fully.*

d. **Removal of Gun Hand Firing Switch.** Turn 24-volt master switch off. Remove screws and washers from cap at base of pistol grip (fig. 27). Remove two handle mounting screws. Remove screws from the top of the grip and compress the flexible spring on the conduit. Pull top and bottom parts of handle with switch and wires attached out far enough to expose switch terminals. Remove terminal nuts, pull wires through, and remove remaining parts of pistol grip.

e. **Installation of Gun Hand Firing Switches.** Insert wires into the pistol grip, and connect to switch terminals. Install screw to top of the pistol grip. Position the assembly on the handle bracket and install remaining screws. Install cap at base of pistol grip. Press trigger switch, and switch button to determine if gun firing solenoids operate correctly. WARNING: *Open gun breech to be sure breech is clear and a round has not been left in the gun before testing firing switches. Fully depress gun.*

Part Three—Maintenance Instructions

Section XXV

RADIO INTERFERENCE SUPPRESSION SYSTEM

110. INTRODUCTION.

a. **Purpose.** Radio interference suppression is the elimination or minimizing of electrical disturbance which would interfere with radio reception, or would disclose the location of the vehicle to enemy detection equipment. It is important that all vehicles, with or without radio equipment, be properly suppressed to prevent interference with reception of neighboring radio equipment.

b. **Description and Data.** Suppression is accomplished in this vehicle by use of bond straps and toothed washers, resistor-suppressors, and single condensers (capacitors) in all circuits or systems where shielding of wires and units does not completely confine or dissipate electrical disturbances which would otherwise cause radio interference. Suppression units used in specific assemblies or units of the vehicle are listed below:

(1) GENERATOR. One 0.1 microfarad capacitor is mounted internally from the positive brush to ground. The generator ground cable, mounted from the generator housing to the hull, serves as a bond strap for radio suppression (fig. 172).

(2) GENERATOR REGULATOR (fig. 74). Two grounding wires, mounted from adjacent corners of the regulator to the regulator mounting plate, serve to ground the regulator and are very important for suppression.

(3) ENGINE (fig. 72). One ground strap, mounted from the starter housing to the hull, serves as a radio suppression bond.

(4) INSTRUMENT PANEL. One 0.5 microfarad capacitor is mounted inside the instrument panel box and connected from the accessory outlet lead at the socket to ground (fig. 86). One ground strap, mounted from the instrument panel to the panel mounting bracket, serves as a radio suppression bond (fig. 84).

(5) WINDSHIELD WIPER MOTOR (fig. 33). One 0.1 microfarad capacitor is mounted under the windshield wiper motor on each hatch hood and connected from the positive lead of the wiper motors to ground.

(6) BLOWER. One 0.5 microfarad dual capacitor is mounted inside the motor terminal box and connected one from the positive lead and one from the negative lead to ground.

(7) HULL RADIO TERMINAL BOX. Two 0.1 microfarad capacitors are mounted in the hull radio terminal box, and are connected one from the 12-volt battery supply terminal to ground and one from the 24-volt battery supply terminal to ground.

(8) TURRET RADIO TERMINAL BOX. Two 0.1 microfarad capacitors are mounted in the turret radio terminal box and connected from the 24-volt battery supply terminal to ground.

(9) TURRET SWITCH BOX (fig. 94). One 0.1 microfarad capacitor is mounted in the turret switch box and connected from the battery side of the traversing motor switch to ground.

(10) TURRET TRAVERSING MOTOR. The 0.1 microfarad capacitors are mounted internally on the traversing motor frame and connected one from each positive brush to ground.

(11) TURRET COLLECTOR RING (fig. 165). One ground strap which acts as a suppression bond strap is mounted on "legs" welded to the fighting compartment floor and connected to the turret collector ring ground terminal.

(12) TACHOMETER AND SPEEDOMETER SENDING UNITS. Two 0.15 microfarad feed-through capacitors are mounted internally in sending units. One bond strap is mounted from each sending unit to gas tanks (fig. 90).

(13) RADIO SHELF. One bond strap is mounted from radio shelf in turret bulge to right wall of turret.

(14) TAILLIGHTS. One bond strap is mounted from taillights right and left sides to hull.

(15) MISCELLANEOUS MOUNTINGS AND FASTENINGS. Toothed lock washers are used throughout the vehicle for the purpose of insuring a good electrical connection between components or parts of components. Cable clamps are used to secure conduits and wiring harnesses of the vehicle in place and at the same time serve as a bonding connection between the conduit and the hull or other part of the vehicle to which it is secured. These clamps are located every two feet along any shielded conduit or wiring cable.

c. Maintenance. If radio interference resulting from the operation of this vehicle is reported or encountered, faulty equipment will have to be traced down, by referring to the tests described in the trouble shooting section of this manual (par. 66), then replaced as described in the following paragraphs. If replacement of suspected faulty radio suppression equipment is not described in this section, its replacement is the responsibility of a higher echelon of maintenance or Signal Corps personnel and will not be attempted by using personnel, without specific permission of the authority responsible.

111. ACCESSORIES.

a. Instrument Panel Utility Outlet Capacitor (fig. 86).

(1) REMOVAL. Remove instrument panel cover to gain access to capacitor (par. 97 d). Disconnect capacitor lead from accessory outlet terminal. Remove capacitor by unscrewing mounting screw.

(2) INSTALLATION. Secure new capacitor in place with mounting screw using toothed lock washer under head of screw. Connect capacitor lead to accessory outlet terminal. Install instrument panel cover (par. 97 e).

b. Windshield Wiper Motor Capacitors.

(1) REMOVAL. Remove two windshield wiper motor attaching screws and lock washers. Disconnect capacitor lead from positive terminal of wiper motor. Cut conduit shield from capacitor clamp and remove capacitor.

(2) INSTALLATION. Solder the conduit shield to capacitor clamp of new capacitor and connect capacitor lead to wiper motor positive terminal. Position capacitor and wiper motor against windshield frame and install attaching lock washers and cap screws.

112. RADIO AND INTERPHONE SYSTEM.

a. Hull or Turret Radio Terminal Box Capacitors.

(1) REMOVAL. Remove terminal box cover retaining screws and lock washers and remove cover. Disconnect capacitor lead of capacitor to be replaced from terminal block. Remove capacitor mounting screw and toothed lock washer and remove capacitor from terminal box.

(2) INSTALLATION. Secure new capacitor and place in terminal box with mounting screw, using toothed lock washer under head of screw. Connect capacitor lead to terminal on terminal block from which lead of defective capacitor was removed. Install terminal box cover with cover screws, using toothed lock washer under head of each screw.

113. TURRET ELECTRICAL SYSTEM.

a. Turret Switch Box Capacitor (fig. 94).

(1) REMOVAL. Remove turret switch box cover retaining screws and toothed washers and pull cover out to a position where the capacitor is accessible as shown in (fig. 94). Disconnect capacitor lead from terminal on battery side of traverse motor switch. Remove capacitor mounting screw and lock washer, and remove capacitor.

(2) INSTALLATION. Install new capacitor by reversing procedure in step (1), immediately above, making sure toothed lock washers are installed under the heads of capacitor and cover mounting screws.

114. BOND STRAPS, MOUNTINGS, AND FASTENERS.

a. Replacement. When replacing components or accessories of the vehicle, be sure to replace toothed lock washers in exactly the same order to "pile-up" in which they were found at the time of removal. When replacing bond or ground straps or when replacing units which are grounded by means of braided wire straps, be sure the contact surfaces of the strap as well as the surface to which it is to be secured are clean and free from paint so that a very good electrical connection can be made. When securing conduits or cables in place be sure to effect a good ground or bond at least every two feet along the conduit or cable by use of cable clamps or solder.

Section XXVI

FUEL AND AIR INTAKE SYSTEM, EXHAUST SYSTEM, AND CYLINDER HEAD

115. FUEL SYSTEM.

a. **Description.** Fuel (gasoline) is stored in two tanks, one on each side of the engine compartment (fig. 96). The fuel tanks are connected by tubing to the fuel pump on top of the engine (figs. 59, 60 and 61). Fuel is drawn from the fuel tanks by the fuel pump and distributed to the dual carburetors simultaneously. A balance tube between the two fuel tanks permits automatic balancing of the fuel to provide expansion space when both fuel tanks are full and both fuel valves are open. The fuel flow to the main engine is controlled by two manual-type, remote control operated shut-off valves, one in each fuel tank (figs. 59 and 60). A valve operating lever is located at the base of each instrument panel support (fig. 15). Each shut-off valve is connected by a pipe nipple to an edge-type fuel filter installed in the bottom of each fuel tank (figs. 59 and 60). The level of the fuel in each fuel tank is measured by electrically operated fuel tank gages which register the individual levels on the fuel gage on the instrument panel. The auxiliary engine draws its fuel from the right fuel tank through an electrically controlled shut-off valve (fig. 59). The primer is connected to the outlet tube from both fuel tanks, and is used to inject "raw" fuel directly into the cylinders for quick starting in cold weather. Dual throttles and accelerators permit either the driver or assistant driver to control fuel feed to the engine (fig. 127). Engine speed is limited by the governor which is driven by the "left" intake camshaft (fig. 54).

116. FUEL TANKS.

a. **Description.** Fuel is carried in the two armor-steel fuel tanks, one on each side of the engine at the front of the engine compartment (fig. 96). They are filled through filler openings located on each side of the rear deck behind the turret. Filler openings are protected by hinged armor steel covers. Drain plugs in the bottom of the fuel tanks are reached through holes covered by steel plates bolt-

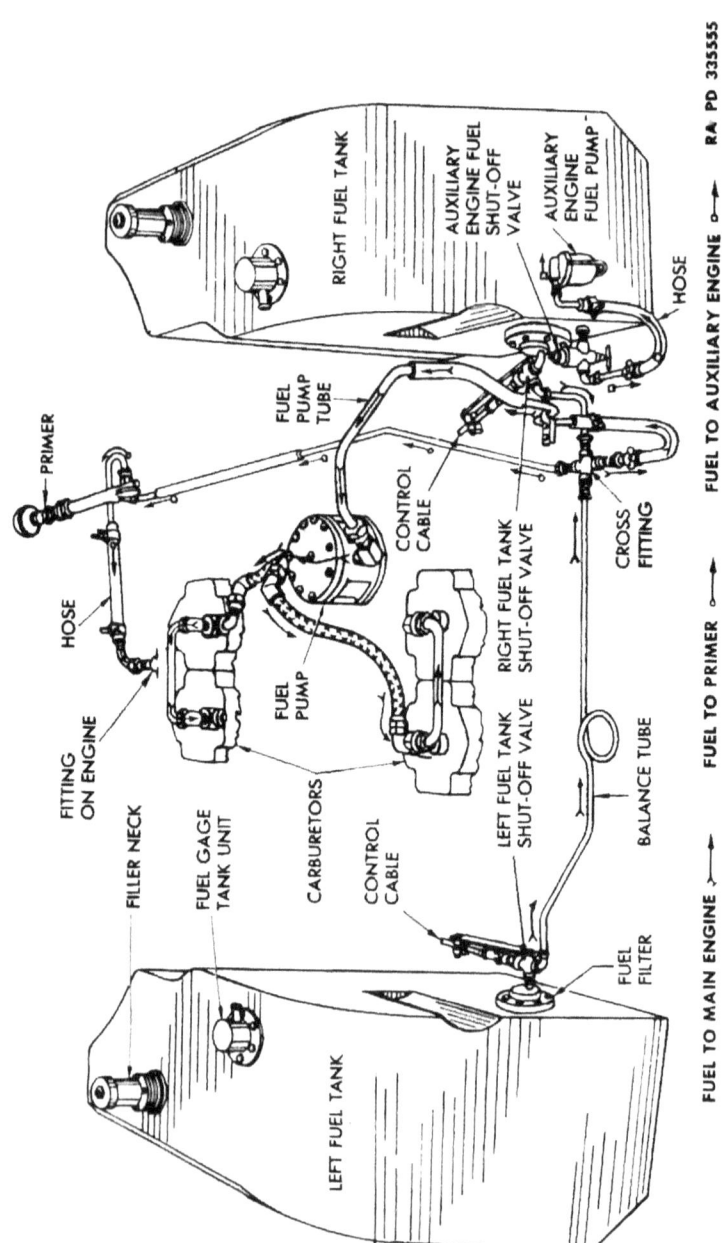

Figure 96 — Fuel System Diagram

Fuel and Air Intake System, Exhaust System, and Cylinder Head

RA PD 344645

Figure 97 — Fuel Tank Mountings in Fighting Compartment

ed to the hull floor (fig. 162). The left tank has a capacity of 116 gallons but is filled with only 111½ gallons. The left tank capacity is 75½ gallons but is filled with only 71½ gallons. This difference between capacity and actual is to permit sufficient expansion space in each fuel tank.

b. **To Fill Fuel Tanks.** Traverse turret as required to reach and open fuel filler covers at rear of turret. Wipe off dirt around filler openings and caps. WARNING: *When filling fuel tanks make sure hose nozzle or container is clean, and that nozzle or container contacts filler neck to carry off static electricity.* Fill tanks until fuel level is approximately 6½ inches below top of filler neck. CAUTION: *Do not overfill as space for expansion must be provided.* Inspect filler cap gasket to make sure it is sealing tightly. Install filler pipe cap. Close and lock filler opening covers.

c. **To Drain Fuel Tanks.** From below vehicle remove drain plug cover (fig. 162), and wipe off drain plug and surrounding surface. Provide suitable clean containers with a total capacity of 112 gallons for draining left tank, and 72 gallons for right tank. The balance tube permits draining the entire fuel system dry by removing either drain plug, and opening both fuel shut-off valves. If only one fuel tank and fuel tubes are to be drained, open the fuel shut-off valve on fuel tank to be drained, and close valve on other fuel tank. Position containers under drain opening and remove drain plug. After tank has drained dry, install drain plug. Inspect drain plug cover gasket, use new gasket if required, and install drain plug cover.

d. **Removal.** If a removed fuel tank is to be discarded, it will be necessary to remove the speedometer sending unit from the right fuel tank (par. 99 f); or tachometer sending unit and engine compartment terminal box from the left fuel tank. The upper fire extinguisher tubes and nozzles (par. 204 e), fuel gage tank sending unit (par. 119 f), fuel shut-off valve (par. 118 h), auxiliary engine fuel shut-off valve (par. 118 h), (right fuel tank only) and fuel filter will also have to be removed after the fuel tank has been taken out of the vehicle. However, if the removed fuel tank is to be reinstalled, speedometer and/or tachometer sending units and other units and fittings except shut-off valves can be left attached. To remove either fuel tank proceed as described in the following paragraphs:

(1) REMOVE COOLING UNIT. Follow procedure in paragraph 139 h. Move fan drive propeller shaft in against engine and wire in this position.

(2) REMOVE AUXILIARY ENGINE. If right fuel tank is to be removed, follow procedure in paragraph 202 e for removing auxiliary engine. Also remove drive belt adjuster mechanism, generator regulator guard and mounting bracket.

(3) DISCONNECT ELECTRICAL CONDUITS AND WIRES (LEFT FUEL TANK). On left fuel tank disconnect all conduits from terminal box (fig. 90). Disconnect tachometer sending unit to instrument panel conduit. Disconnect conduit and wire from fuel gage tank unit.

(4) DISCONNECT ELECTRICAL CONDUITS AND WIRES (RIGHT FUEL TANK). On right fuel tank disconnect conduits and wires from speedometer sending unit and fuel gage tank unit (fig. 87). Disconnect wire from auxiliary engine fuel shut-off valve. From either tank remove clips which attach any disconnected conduits to the tank.

(5) DRAIN FUEL TANK. Close fuel shut-off valve on opposite fuel tank and drain fuel from fuel tank to be removed. Unscrew drain pipe from fuel tank. Do not install drain plug cover plate.

(6) DISCONNECT FUEL TUBE AND SHUT-OFF VALVE. Disconnect fuel tube and fuel shut-off valve control rod. Disconnect primer fuel hose. Remove fuel shut-off valve (par. 118 h).

(7) REMOVE AUXILIARY ENGINE FUEL SHUT-OFF VALVE. When removing right fuel tank disconnect wire and auxiliary engine fuel tube from auxiliary engine fuel shut-off valve. Remove valve from bracket and remove bracket from fuel tank.

(8) REMOVE FIRE EXTINGUISHER TUBE AND NOZZLE. From top of either fuel tank disconnect and remove fire extinguisher nozzle

Fuel and Air Intake System, Exhaust System, and Cylinder Head

from tube. Disconnect tube from tee connection at main tube on engine compartment floor. Remove tube, and upper nozzle.

(9) REMOVE FUEL TANK FILLER NECK. Raise fuel tank filler cover. Remove filler cap. Remove screws from filler neck grommet assembly. Work grommet up and out of rear hull plate opening. Lift out lower grommet ring. Remove screws which attach filler neck to adapter, and lift out filler neck and adapter upper washer. Remove countersunk screws which attach adapter to fuel tank flange. Lift out screen, adapter, and lower washer. Cover fuel tank opening.

(10) LOOSEN FUEL TANK FROM HULL. Remove tie bolts which attach upper and lower engine compartment ends of fuel tank (fig. 61). Remove lower hold-down bolt and upper clamp from inside fighting compartment (fig. 97).

(11) REMOVE FUEL TANK FROM VEHICLE. Make sure all conduits, tubes and attaching clips are disconnected or removed. Work fuel tank straight towards rear of vehicle by prying with a pinch bar. As front end is withdrawn from bulkhead recess, swing fuel tank slightly inward. Continue working fuel tank toward rear of engine compartment until front end of fuel tank is clear of upper hull plate. Lift fuel tank out of vehicle with suitable hoist.

e. *Installation of Fuel Tank.* If spacers were loosened when fuel tank was removed, cement spacers firmly into position. Proceed as described in following paragraphs to install fuel tank in vehicles:

(1) POSITION FUEL TANK IN VEHICLE. Lower tank into engine compartment. Slide it forward into position, keeping it as close as possible to hull side plate. Make sure spacers under tank have not shifted.

(2) ATTACH FUEL TANK TO HULL. Install upper and lower tie bolts. Tighten bolts to draw tank into position. In fighting compartment install lower hold-down bolt and upper fuel tank clamp.

(3) INSTALL FUEL TANK FILLER NECK. Position lower washer and adapter on tank, and insert screen. Install countersunk screws in adapter. Place upper washer on filler neck and attach filler neck to adapter. Insert grommet assembly on filler neck. Shift upper and lower sections of grommet as required to completely seal hull plate opening. Install and tighten attaching screws in grommet assembly. Install filler cap.

(4) CONNECT FIRE EXTINGUISHER TUBES AND NOZZLES. Connect fire extinguisher tube to main tube tee connection on engine compartment floor. Install upper discharge nozzle.

(5) CONNECT FUEL TUBES AND INSTALL SHUT-OFF VALVES. Install fuel shut-off valve (par. 118 c). Connect shut-off valve control rod to valve (par. 118 c (1)). Connect fuel tube to shut-off valve. Connect primer fuel hose.

(6) INSTALL AUXILIARY ENGINE FUEL SHUT-OFF VALVE. Install bracket on fuel tank. Install shut-off valve on bracket. Connect wire and fuel tube.

(7) CONNECT ELECTRICAL CONDUITS AND WIRES (RIGHT OF LEFT FUEL TANK). Connect wires and conduits to units from which they were removed (left fuel tank, see subpar. d (3) preceding—right fuel tank subpar. d (4) preceding). Install conduit clips.

(8) INSTALL AUXILIARY ENGINE. Install belt adjuster mechanism. Install auxiliary engine (par. 202 f). Install generator regulator bracket and guard.

(9) INSTALL COOLING UNIT. Follow instructions given in paragraph 139 c.

(10) FILL FUEL TANK. Coat threads of drain pipe with liquid joint and thread compound (52-C-3122). Install drain pipe, drain plug, and drain plug cover. Fill fuel tank (subpar. b preceding). Start engine. Inspect for leaks. Stop engine when inspection is completed.

117. FUEL FILTERS.

a. *Description.* A disk-type fuel filter is installed in each fuel tank (figs. 59, 60, and 98) to filter the fuel as it flows to the carburetors. A removable edge-type element is attached to the inner side of the filter cover. As the fuel flows out of the fuel tank it passes between the closely stacked disks, depositing any dirt or foreign matter on their outer edges (fig. 98). Clean filter element whenever power unit is removed.

b. *Removal.* Remove power unit (par. 71). Disconnect brake throttle control rods, and speed range selector rod (left fuel filter only). Disconnect fuel valve control rod at valve plunger (figs. 59 and 60). Disconnect balance tube hose from valve outlet elbow. Unscrew shut-off valve. Remove screws attaching filter to fuel tank. On right side disconnect primer tube hose and remove auxiliary engine fuel shut-off valve bracket. Carefully pull filter out of tank so as not to damage disk edges. Hold filter cover, and remove safety nut. Lift filter element off cover stud (fig. 98).

c. *Cleaning.* Thoroughly agitate filter element in dry-cleaning solvent. CAUTION: *Do not use brush to clean element.* Carefully blow element dry with reduced air pressure so as not to crimp disks.

d. *Installation.* Position filter element on cover stud and install safety nut. Coat one side of new gasket with liquid joint and thread compound and position gasket on filter cover flange. Carefully insert filter into fuel tank so as not to damage disk edges and install. Tighten screws alternately to compress gasket evenly. Install fuel shut-off valve. Connect fuel balance tube hose to valve outlet elbow.

Fuel and Air Intake System, Exhaust System, and Cylinder Head

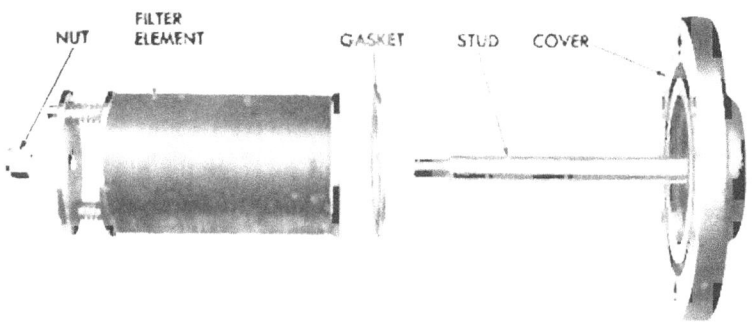

Figure 98 — Fuel Filter — Disassembled

Install control rod. Install power unit (par. 72). Be sure fuel tank drain plug is installed. Fill fuel tank. Open fuel shut-off valve, and inspect all connections for leaks. Close fuel shut-off valve. On right side connect primer tube hose and auxiliary engine fuel shut-off valve bracket.

118. FUEL SHUT-OFF VALVES.

a. Description. The flow of fuel from each fuel tank is controlled by a manual-type, remote control operated shut-off valve attached to a pipe nipple in each filter cover (figs. 59 and 60). Each valve is operated by a rod connected to a lever at the base of each instrument panel support (figs. 59 and 60). An electrically operated fuel shut-off valve connected to a pipe nipple in the right-hand fuel tank (fig. 59) controls fuel flow to auxiliary engine. This valve opens and closes with "on-off" operation of auxiliary engine ignition switch.

b. Removal. Procedure for removing main and auxiliary engine fuel shut-off valves follows:

(1) REMOVE POWER UNIT. Follow instructions in paragraph 71.

(2) REMOVE MAIN FUEL SHUT-OFF VALVES. On right side disconnect right brake rear rod. On left side disconnect left brake rear rod and speed range selector control rod. CAUTION: *Do not remove fuel filter.* Loosen outlet connection at fuel filter, and unscrew valve and nipple from filter cover.

(3) REMOVE AUXILIARY ENGINE FUEL SHUT-OFF VALVE. Remove power unit (par. 71). Disconnect auxiliary engine fuel tube. Remove valve from bracket. To clear torsion bar spring when unscrewing valve, remove solenoid from valve body. Lift diaphragm and needle valve out of valve body. Wrap solenoid in a clean cloth

and lower it to hull floor. Leave wire connected to solenoid. If new valve is to be installed, detach conduit shielding cap and disconnect wire from solenoid removed from old valve.

c. **Installation of Fuel Shut-off Valves.** The procedure to install the main and auxiliary engine fuel shut-off valves is given in the following steps:

(1) INSTALL MAIN FUEL SHUT-OFF VALVES. Screw valve onto nipple and hold nipple with wrench while tightening valve, to place valve in proper position to connect fuel and balance tube. On right side connect right brake rear rod. On left side connect left brake rear rod and speed range selector control rod. Connect control rods to valves. Install power unit (par. 72).

(2) INSTALL AUXILIARY ENGINE FUEL SHUT-OFF VALVE. Make sure nipple is tight in fuel tank. Screw valve body onto nipple and tighten until it is vertical. Insert needle valve, and place diaphragm in valve body. Make sure gasket is serviceable, and position gasket on valve body. Attach solenoid to valve body. Install valve on bracket. Connect fuel tube to valve outlet elbow. Turn 24-volt master switch on. Turn auxiliary engine ignition switch on, and listen for click to indicate valve is operating. Inspect all fuel connections for leaks. Turn switches off. Turn off 24-volt master switch and install power unit (par. 72).

119. FUEL GAGE.

a. **Description.** The fuel gage on the instrument panel is connected through the fuel gage control switch (fig. 16) to the fuel gage sending unit in each fuel tank (figs. 87 and 90). The level of fuel in the tank varies the resistance of the fuel gage sending unit, which changes the position of the pointer on the instrument panel fuel gage. The control switch connects the fuel gage to either right or left fuel gage sending units so the fuel level in either tank can be determined.

b. **Removal of Fuel Gage.** Remove instrument panel face plate (par. 97 d). Note wire numbers and disconnect wires and fiber insulator from terminals. Remove nuts and fiber insulator from terminals. Lift off mounting bracket. Withdraw gage and gasket from front of panel face plate.

c. **Installation of Fuel Gage.** Inspect gasket, and install new gasket if necessary. Place gasket on gage, and insert gage through front of panel face plate. Install mounting bracket and fiber insulator, but do not tighten nuts. Make sure gasket is evenly spaced under rim of gage, and that gage is properly positioned. Tighten mounting bracket nuts. Attach wires to terminals, (fig. 86) and install fiber insulator. Install instrument panel face plate (par.

Fuel and Air Intake System, Exhaust System, and Cylinder Head

97 e). Turn 24-volt master switch on and test action of gage by moving fuel gage control switch lever to right and left fuel tank positions. Turn 24-volt master switch off.

d. **Removal of Fuel Gage Control Switch.** Remove instrument panel face plate (par. 97 d). Note wire numbers and terminal identifying numerals, and disconnect wires from terminals. At front of face plate remove set screw from lever and remove lever from shaft. Remove position plate from shaft. Withdraw switch from rear of panel ace plate.

e. **Installation of Fuel Gage Control Switch.** Insert switch through panel face plate from the rear. Install position plate on shaft. Install lever and set screw. Connect wires to proper terminals (fig. 86). Install instrument panel face plate (par. 97 e). Turn 24-volt master switch on. Operate fuel tank control switch to make sure fuel gage indicates fuel lever in each fuel tank. Turn 24-volt master switch off.

f. **Removal of Fuel Gage Sending Unit.** Raise front intake door. Turn 24-volt master switch off. Remove guard above the sending unit. Detach conduit shielding cap, and disconnect wire from fuel gage sending unit. Wipe off top of fuel tank and fuel gage sending unit. Remove attaching cap screws. Carefully withdraw fuel gage tank unit from fuel tank. Cover opening in fuel tank.

g. **Installation of Fuel Gage Sending Unit.** Inspect sending unit gasket, and use new gasket if required. Make sure surface of tank is clean to provide positive ground contact. Position gasket on tank and insert sending unit into tank, making certain lower end is centered in guide in bottom of tank. Tighten cap screws alternately to compress gasket evenly. Connect wire and attach conduit shielding cap to sending unit. Turn 24-volt master switch on. Operate fuel tank control switch to test action of fuel tank gauge unit. Turn 24-volt master switch off. Install guard previously removed.

120. FUEL TUBES.

a. **Description.** Preformed steel tubing is used throughout the fuel system. Special fireproof rubber hose is used at points where rigid steel tubing would be subjected to vibration. The tubes are insulated against heat where they pass close to hot parts of the engine.

b. **Removal of Fuel Tubes.** Before disconnecting any fuel tube or hose, first make sure that fuel supply to the tube is shut off, or the system drained dry, to avoid loss of fuel and the creation of a fire hazard. When disconnecting fuel tubes, use wrench with snug fit on couplings while holding elbows or nipples from turning with second wrench. Disconnect both ends of tube, if necessary, to avoid excessive bending or springing. Tape ends of tubes and cover openings in units to prevent entrance of dirt. Place tubes removed from vehicle in safe, clean place where they will not be damaged or exposed to dust or dirt.

c. **Installation of Fuel Tubes.** Carefully inspect all tubes and hose for evidence of leaks before installing. If preformed tubes are not available for replacement, cut tubes to length of original tubes from bulk tubing. Make sure seats in elbows or nipples and ends of tubes are perfectly clean. Coat threads with liquid joint and thread compound. Use wrench with snug fit on couplings while holding elbows or nipples from turning with second wrench. Tighten couplings until snug, but avoid using excessive force. When installing a tube with hose on one end, first position hose and then connect tube fitting finger-tight. Hold tube in proper installed position and tighten coupling, then tighten hose clamps. Position hoses and clamps on tubes so ends of tubes extend through hose beyond the clamps. Position clamps at least ½ inch from ends of hose.

121. FUEL PUMP.

a. **Description.** The diaphragm-type fuel pump on top of the engine (fig. 56) is mechanically operated by the "left" intake camshaft. When the carburetor needle valves close and shut off the flow of fuel to the carburetors, the resulting back pressure on the fuel pump diaphragm stops the pumping action. The upper part of the pump acts as a surge chamber to absorb pulsations, and to insure an even flow of fuel to the carburetors.

b. **Removal.** Remove front center deck plate. Make sure both fuel shut-off valves are closed. Disconnect fuel tube at fuel pump. Disconnect the two flexible hoses at tee connection on fuel pump (fig. 56). Remove tee connection from pump. Compress hose running from air inlet elbow to carburetor, as necessary, to permit fuel pump to be pulled off attaching studs. Remove fuel pump.

c. **Installation of Fuel Pump.** Make sure gasket on camshaft housing is serviceable. Screw tee connection into outlet of fuel pump. Compress air box to carburetor hose sufficiently to allow fuel pump to be positioned on studs. Install fuel pump. Connect fuel tube to fuel pump. Connect the two flexible hoses to the tee connection. Open either fuel shut-off valve. Turn magneto switch on. Turn starter switch on. NOTE: *It may be necessary to operate starter several times at 20- to 30-second intervals to fill hoses and pump, before engine fires and runs smoothly.* Inspect fuel connections for leaks while engine is running. Stop engine. Install front center deck plate.

Fuel and Air Intake System, Exhaust System, and Cylinder Head

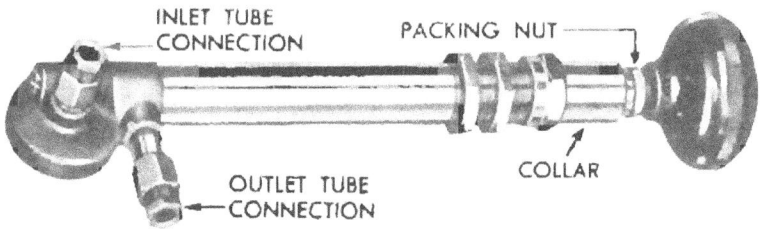

Figure 99 — Engine Primer Pump

122. PRIMER.

a. **Description.** The primer is mounted at the rear of the fighting compartment on the bulkhead, immediately to the right of the auxiliary engine control panel (fig. 173). It is operated manually from the fighting compartment. It takes the place of the conventional choke and is used to facilitate cold-weather starting. The primer draws fuel from the cross-connection in the fuel tank outlet tubes (fig. 59), and then forces it through spray jets into the intake ports in the cylinder heads. A spring-loaded guard prevents the pump handle from being accidentaly pulled out when primer is not in use.

b. **Removal.** Close both fuel shut-off valves. Working from on top of engine disconnect both fuel tubes from primer pump body. Inside fighting compartment, swing pump handle guard down and hold in this position. Unscrew packing nut from pump body, and pull plunger out of pump (fig. 99). Remove front lock nut from pump body. In engine compartment withdraw primer pump body from hole in bulkhead.

c. **Install.** Install rear lock nut on primer pump body. From engine compartment insert pump body through hole in bulkhead. In fighting compartment install front lock nut finger-tight. Connect the two fuel tubes to pump body, and tighten connections. Hold pump body from turning, and tighten rear lock nut. Insert plunger in pump body, taking care not to cut or curl leather washer. Screw packing nut on pump body snugly, but without excessive force. Release pump handle guard.

123. CARBURETORS.

a. **Description.** Two dual, double-barrel, downdraft carburetors are used. They are located at each end of the cylinder heads (fig. 54), in the "V" between the cylinder banks. They are mounted on carburetor adapters which are connected to the intake passages in cylinder heads. There are two degassers on each carburetor (fig. 100)

TM 9-735
123

Part Three—Maintenance Instructions

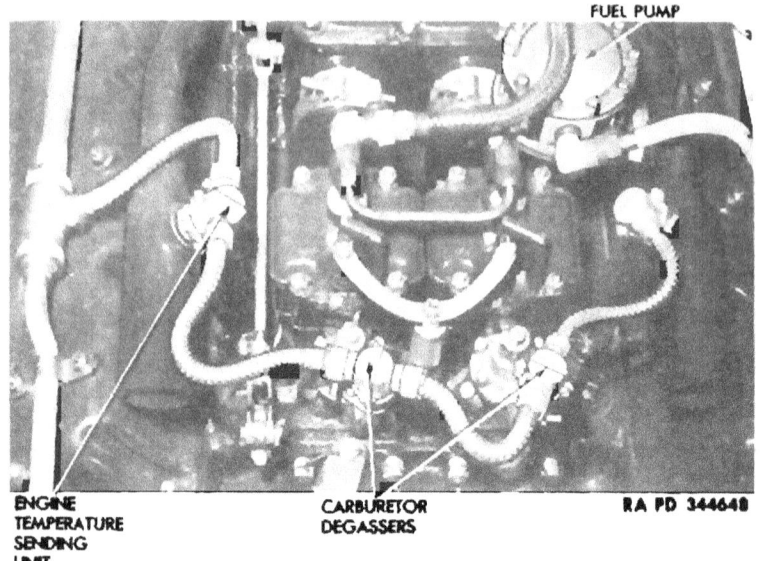

ENGINE TEMPERATURE SENDING UNIT

CARBURETOR DEGASSERS

RA PD 344648

Figure 100 — Carburetor Degassers

which control the flow of fuel to the idling fuel wells in the carburetors. When high intake vacuum builds up on deceleration, the degassers automatically shut off the flow of fuel through the idling well and thus prevent flooding. The degassers are also electrically operated by pressing the fuel cut-off button on the instrument panel (fig. 16) to stop the engine.

b. Adjustment of Carburetors. The carburetor idling mixture and engine idling speed can be adjusted and the float level checked by second echelon personnel, according to the following procedure. When repairs to a carburetor, which include correcting float level, are necessary the carburetor must be removed and a new one installed.

(1) CHECK CARBURETOR FLOAT LEVEL. With vehicle on level ground, remove front center deck plate. Start engine and run at idling speed and check float level in each bowl of the two carburetors as follows: Remove locking wire from float level plug (fig. 101), and unscrew plug. If fuel is not visible at the bottom of the threads, the float level is too low. If fuel flows freely from hole, the float level is too high. Install and tighten plug, and secure with locking wire. Repeat tests on all four float bowls. Top engine. If a float level is found to be incorrect, mark bowl that has incorrect level. Remove

Fuel and Air Intake System, Exhaust System, and Cylinder Head

carburetor, and install a new one (subpar. c following). Install front center deck plate.

(2) ADJUST ENGINE IDLING SPEED. Engine idling speed can be adjusted at the two carburetors, providing accelerator linkage is not out of adjustment (par. 127 h). Start and run engine until it is warmed up to normal operating temperature, at least 100° F. Return throttle linkage to fully closed position and observe tachometer reading, which must be 500 revolutions per minute. If speed varies more than 50 revolutions per minute from the recommended idling speed, adjust carburetor idling speed screws (fig. 101). Adjust idling screws on both carburetors in unison by turning so that each screw rests on stop pad when adjustment is completed. To increase idling speed turn the screws clockwise. To decrease idling speed turn screws counterclockwise. If engine does not idle smoothly and float level has been tested and found to be correct, adjust carburetor idling mixture. Stop engine.

(3) ADJUST CARBURETOR IDLING MIXTURE. Adjust basic carburetor idling mixture with engine stopped. Remove front center deck plate. Turn each of the idling mixture adjusting screws on both carburetors (fig. 101) clockwise until they are felt to seat lightly. CAUTION: *Do not turn screws down tight.* Back off each screw one-quarter turn. Start and run engine until it is warmed up to 100 F. Return throttle linkage to closed position. If engine still does not idle smoothly, it may be necessary to vary the adjustment by turning the screws slowly in the direction which improves the idling. Stop engine. Install front center deck plate.

c. **Removal of Carburetors.** Either front or rear carburetor can be removed with the power unit in the vehicle by the following procedure: Remove front center deck plate. Remove screws which attach conduit shielding caps to degassers (fig. 100). Remove caps from degassers, and pull wires out of connectors. Disconnect flexible fuel hose from carburetor. On carburetor at flywheel end unhook throttle pull-back spring from bracket on adapter. Remove cotter pin from clevis pin, and remove clevis pin to disconnect throttle operating rod from throttle lever (fig. 101). On carburetor at magneto end, remove nut and washer from bell crank pivot post (fig. 101). Remove guard from carburetor adapter. Loosen clamps on the two carburetors to air inlet elbow hose connections. Slide hose connections off carburetor and air inlet elbow. Remove carburetor by lifting straight up (fig. 102). Remove carburetor insulator from adapter.

d. **Installation of Carburetors.** Make sure carburetor insulator is in serviceable condition. Insert two outer upper attaching cap screws through adapter, and position insulator on these screws. Lower carburetor into position, and screw two attaching screws into carburetor finger-tight. Install the other attaching screws, and tighten screws alternately to compress insulator evenly. Form loop of locking wire through each pair of adjacent screws, and twist loop until tight. Slide hose connections over carburetor, and tighten clamp screws securely. On carburetor at magneto end, place guard on top of adapter with pivot post entered in governor throttle operating lever. Form loop of locking wire through each pair of adjacent screws, and twist loop until tight. Install plain washer and safety nut on pivot post. Connect throttle operating rod to throttle lever with clevis pin. On carburetor, at flywheel end, hook end of throttle return spring into hole in bracket on adapter. Connect flexible fuel hose to carburetor. Insert wire in connectors on degassers. Install conduit shielding caps on degassers, and tighten screws. Start and run engine to test carburetor adjustment. Adjust carburetor (subpar. b preceding) or throttle linkage (par. 127 h) as required. Inspect fuel connections for leaks and tighten as required. Stop engine. Install front center deck plate.

124. CARBURETOR ADAPTERS.

a. Description. Carburetors are attached to carburetor adapters at each end of the cylinder heads (figs. 52 and 53). These adapters are connected to the fuel intake passages at the front and rear ends of both cylinder heads. Exhaust gases are directed through the adapters to preheat the ingoing fuel mixture.

b. Removal. To remove the carburetor adapter at the magneto end remove power unit (par. 71). Then remove adapter (par. 133 b (9)). The adapter at flywheel end can be removed with power unit in the vehicle as follows: Remove carburetor (par. 123 c). Disconnect exhaust manifold extension flange from adapter and slide flanges over extension. Remove safety nuts which attach adapter to cylinder heads. Slide adapter off cylinder head studs.

c. Install. Make sure adapter to cylinder head gaskets are serviceable. Position gasket on studs on end of cylinder head. Install adapter on cylinder heads. Make sure adapter to exhaust manifold extension flange gaskets are serviceable. Attach flanges to adapter and tighten cap screws securely. Form loop of locking wire through heads of cap screws and twist loops until tight. Install carburetor (par. 123 d). If adapter at magneto end was removed, install power unit in vehicle (par. 72).

125. GOVERNOR AND GOVERNOR BOOSTER.

a. Description. The engine speed is limited by the action of the governor and governor booster which is connected to the carburetor linkage (fig. 54). The hydraulic booster is used to supply the power required to overcome the friction in the carburetor linkage. The mechanical flyball type governor is driven by the "right" intake camshaft. As engine speed increases, the mechanical governor moves a valve which admits engine lubricating oil under pressure to the tube leading to the top of the booster cylinder. The oil pressure in the top cylinder moves piston down. The downward movement of the piston rod lowers the rear end of the rocker arm (fig. 54), and partially closes the throttle valves to limit engine speed below 2,800 revolutions per minute. When engine speed decreases to within the normal operating range, the governor control valve admits oil pressure to the bottom of the booster, and the piston rod returns to its normal raised position.

126. THROTTLE, ACCELERATOR, AND CONTROLS.

a. **Throttle.** Dual throttle levers are provided for manual control of engine speeds (fig. 127). A throttle is mounted at each end of a cross shaft on which the speed range selector lever cross shaft also rotates (fig. 13). The throttle cross shaft is mounted in bearings which are supported by caps mounted on brackets bolted to the instrument panel supports. A friction device at the left end holds throttles in any position between closed and open. The throttle assembly is connected by a rod on the left-hand side with the dual accelerator assembly. Throttle control of engine speeds is obtained through the accelerator linkage. The throttles are used to run the engine during tests, but not for driving the vehicle. When accelerator and throttle are closed, the engine must idle at 500 revolutions per minute. The length of the throttle to accelerator rod must be adjusted so that when the throttle is fully closed the accelerator will be in the fully raised position.

b. **Accelerator and Controls.** Dual accelerator pedals are mounted, one on each end of cross shaft, on the floor at the front of the driver's and assistant driver's compartments (fig. 127). The accelerator cross-shaft is connected to a control rod, the rear end of which is connected to the throttle rod lever on the rear side of the engine compartment bulkhead (fig. 60). The throttle rod lever is connected to the engine throttle linkage by the throttle rod lever link, engine throttle bell crank and engine throttle bell crank rod (fig. 54). This last rod is connected to the governor booster piston rod rocker arm (fig. 54). When either accelerator is depressed the front end of the governor booster rocker arm is pulled down, and, through the throttle linkage, both carburetor throttle valves are opened. The front carburetor opens immediately and the rear carburetor opens at 1,500 revolutions per minute. Both carburetors operate at idling speed. A return spring on the left accelerator pedal assists in closing the throttle when the accelerator is released. The return spring connected to the rear end of the throttle control rod also helps close the throttle, and takes up any slack in the linkage.

TM 9-735

Part Three—Maintenance Instructions

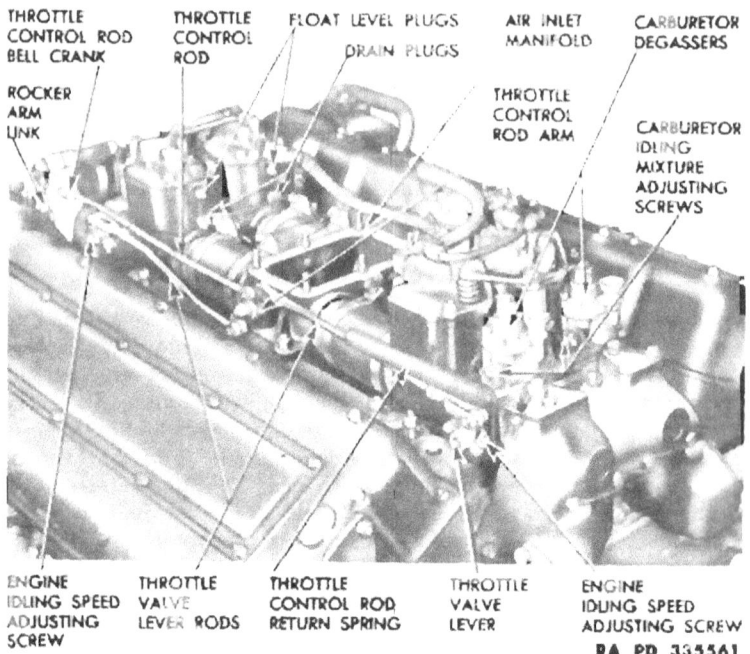

Figure 101 — Carburetor Control Linkage

c. **Adjustment of Accelerator Controls.** When any of the accelerator controls have been removed, or when it is impossible to obtain correct engine idling speed adjustment at the carburetors (par. 123 b), it will be necessary to completely adjust accelerator controls.

(1) ADJUST CARBURETOR CONTROL LINKAGE. Remove front center deck plate. Adjust governor booster linkage to provide full operating range from closed to open. Unhook throttle control rod return spring from bracket. Disconnect rod from rear throttle valve lever (fig. 60). NOTE: *Front carburetor linkage is fixed.* Disconnect rocker arm link from throttle control rod bell crank (fig. 54). Adjust front carburetor first. Back off carburetor idling speed adjusting screw (fig. 101). While holding throttle valve in closed position by pressing lightly on throttle valve lever, insert a 0.002-inch feeler gage between idling speed adjusting screw and its stop. Turn screw clockwise until slight drag is felt when feeler gage is withdrawn from under screw. Repeat adjustment on rear carburetor, then turn each screw clockwise one full turn to open throttles sufficiently to prevent

engine stalling. With idling speed adjusting screws held against stops, adjust length of rear throttle valve lever rod so rear carburetor starts to open at 1500 revolutions per minute. NOTE: *Both carburetors are in operation when engine is idling.* Connect rear throttle rod and install spring. Connect rocker arm link to throttle control rod bell crank. Test for full throttle which is 2,750 revolutions per minute at no load.

(2) ADJUST ACCELERATOR CONTROL LINKAGE IN ENGINE COMPARTMENT. The only adjustment necessary, in addition to throttle linkage and governor booster, is the throttle rod lever link. Remove bulkhead cover. Disconnect throttle control rod from lever on rear side of bulkhead. Disconnect throttle rod lever link from lever. Set lever so that center of clevis hole in lever upper arm measures $1\frac{7}{16}$ inches from bulkhead. Adjust length of throttle rod lever link so clevis pin will enter hole in lever when governor booster rocker arm is down against stop (fig. 54). Tighten jam nut on link and connect link to lever with clevis pin. Make necessary adjustments in driver's compartment (step (3) following). Adjust throttle control rod and connect rod to front bell crank and to bulkhead lever. Install crankcase breather and bulkhead cover.

(3) ADJUST ACCELERATOR CONTROL LINKAGE IN DRIVER'S COMPARTMENT. In driver's compartment disconnect accelerator return spring from bracket. Loosen jam nut on control rod clevis at accelerator cross shaft. Set front bell crank upper lever so that center of clevis hole measures $2\frac{25}{32}$ inches from cross brace. Set accelerator so bottom edge measures 5 inches from hull floor. Adjust control rod length, tighten jam nut and connect rod to cross shaft lever. Attach return spring to accelerator. Adjust and connect main throttle control rod if disconnected (step (2) preceding). Test action of accelerator for full travel. In closed position throttle valve lever on each carburetor must be firmly seated against stops with engine idling at 500 revolutions per minute. In open position rocker arm must be against stop on governor-booster (fig. 54).*

(4) ADJUST THROTTLE CONTROL LINKAGE IN DRIVERS' COMPARTMENT. With accelerator in closed position, adjust length of throttle pull rod (fig. 101) so that when connected to accelerator pedal lever there is not less than $\frac{1}{16}$-inch, nor more than $\frac{1}{8}$-inch, clearance between clevis pin and top of slot in lever. Connect throttle pull rod to accelerator with clevis pin. Tighten jam nut at clevis. Operate throttle to make sure friction devices at each throttle will hold levers in any position throughout entire length of travel. Adjust friction device if necessary.

(5) TEST ACTION OF THROTTLE AND ACCELERATOR. Start and run engine until it is warmed up to at least 100° F. With throttle in closed position observe tachometer to make sure engine is idling at

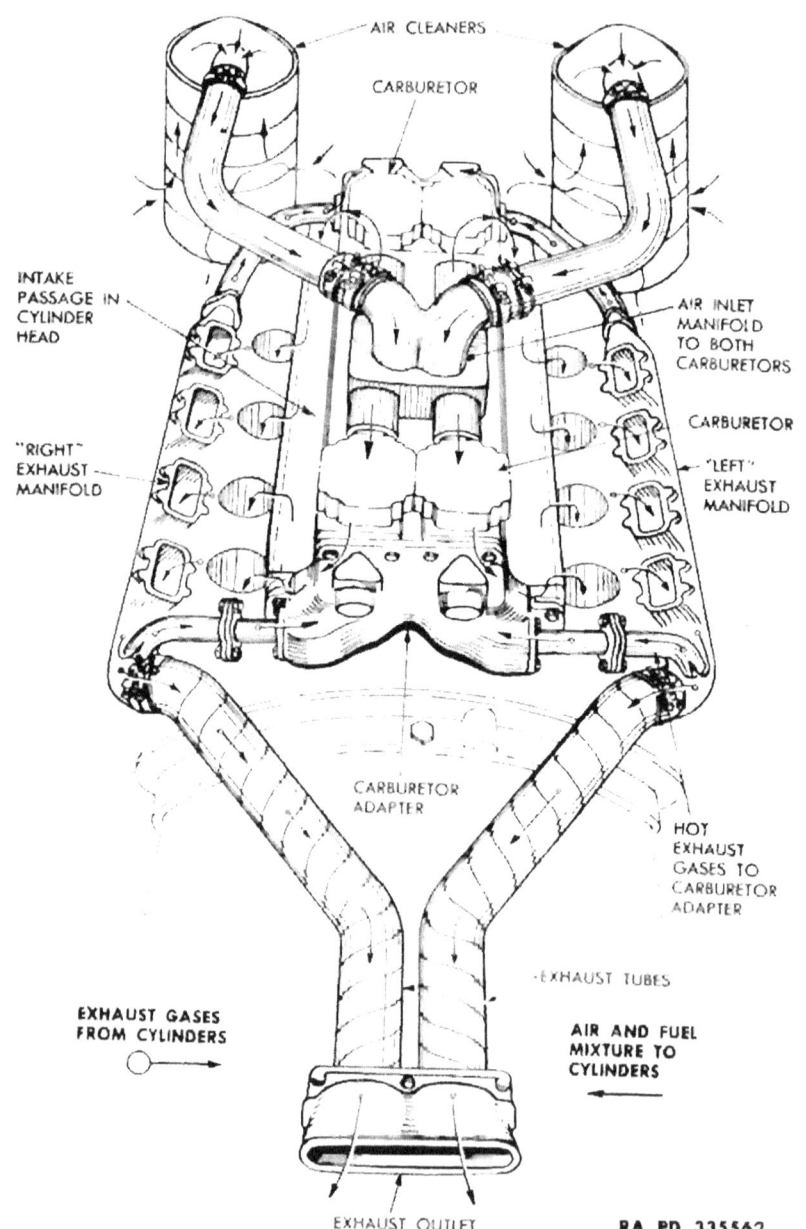

Figure 102 — Air Intake and Exhaust System Diagram

recommended speed of 500 revolutions per minute. Adjust engine idling speed screws (par. 123 h (2)) or carburetor idling mixture screws (par. 123 h (3)) as required. When engine is idling smoothly at 500 revolutions per minute test action of accelerator and throttle through full limit of travel. Make sure there is no bind in the linkage and that full throttle is 2,750 revolutions per minute at no load. Stop engine. Install front center deck plate.

127. AIR INTAKE SYSTEM.

a. **Description.** Air for combustion in the engine enters the air intake system through two air cleaners, one in each rear corner of the fighting compartment (figs. 102 and 103). From the air cleaners air is drawn through tubes to the air inlet manifold between the carburetors. From the air inlet manifold the air passes to the dual carburetors where it is mixed with the fuel. Each carburetor is connected to the fuel inlet passages at the ends of both cylinder heads by a carburetor adapter. These adapters are warmed by exhaust gases circulating within them. The fuel mixture passes through the adapters, where it is pre-heated, and then into the intake passages in the cylinder heads to the combustion chambers.

128. AIR CLEANERS.

a. **Description.** Air for combustion is filtered and cleaned as it passes through the two oil-bath air cleaners (fig. 103). Air is drawn into the air cleaner through an intake duct at the top. A manually controlled valve permits taking air from either the fighting or engine compartments. From the duct the air passes down into the body of the air cleaner (fig. 104) across the top of the oil in the reservoir, and then up through the metallic gimp element into the air inlet manifold tubes. Most of the larger particles of dirt are trapped in the oil bath. The air as it moves upward carries an oil mist into the gimp element which further filters out the impurities. Most of the dirt which collects on the element is carried back into the oil reservoir as the oil drains out of the cleaner.

b. **Servicing Air Cleaners.** The air cleaners must be serviced regularly in accordance with the following procedure:

(1) GENERAL. Air cleaners must be serviced at intervals specified in paragraph 46 or more frequently as conditions require. If allowed to become clogged, they will not effectively filter the incoming air. The moving parts of the engine will then be subjected to the abrasive action of lubricating oil mixed with dirt, dust particles and other foreign matter. Serious engine damage will result from this abrasive action. Clogged air cleaners, or air cleaners with too high an oil level due to overfilling, or accumulation of dirt in the reservoirs, will cause excessive fuel consumption and loss of power. Air cleaner

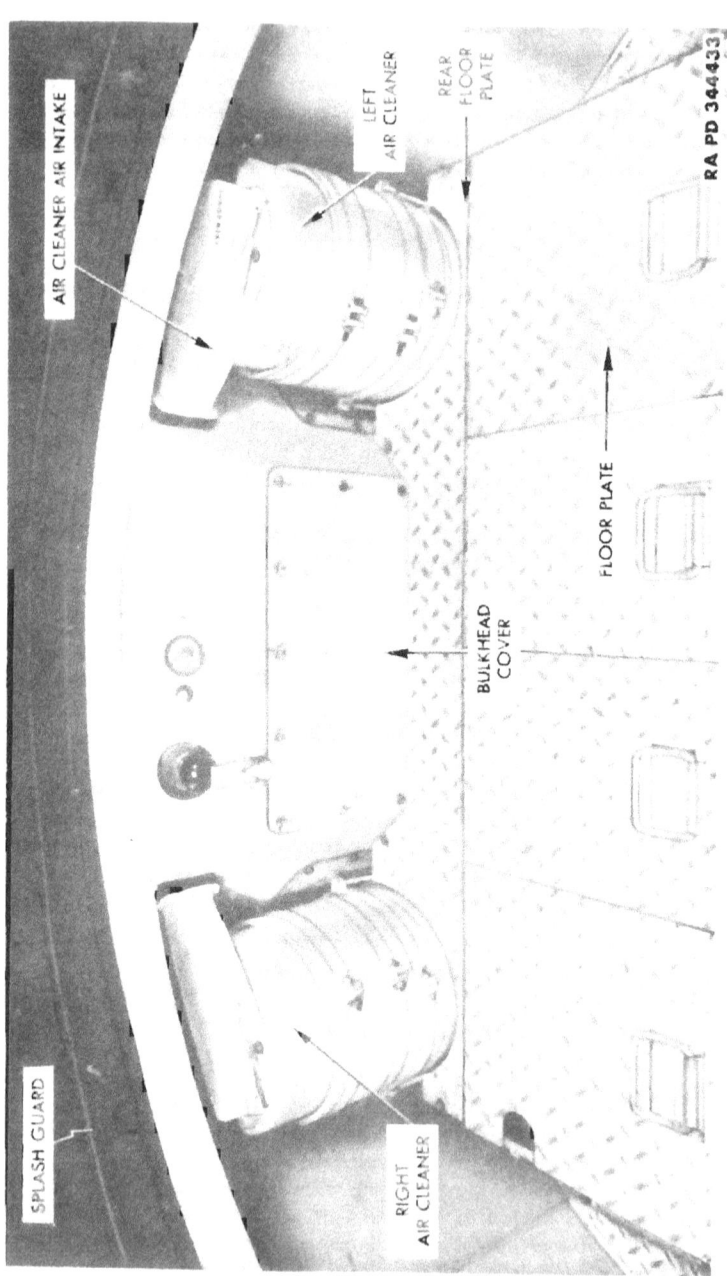

Figure 103 — Air Cleaner — Installed

TM 9-735
128

Fuel and Air Intake System, Exhaust System, and Cylinder Head

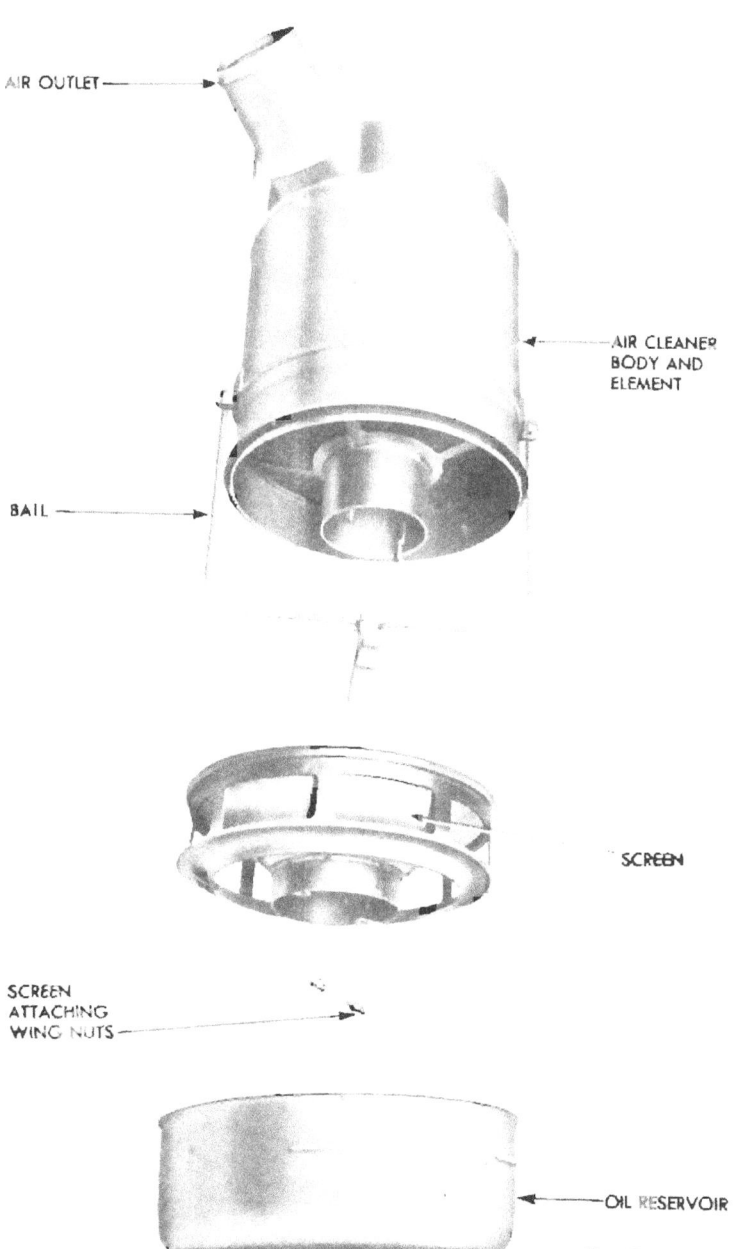

Figure 104 — Air Cleaner — Disassembled

gimp elements must be cleaned, and oil reservoirs cleaned and refilled to proper level with correct grade of oil, at regular intervals.

(2) To CLEAN AND REFILL AIR CLEANER OIL RESERVOIR. Remove air cleaner oil reservoir and screen (subpar. c (1) following). Remove wing nuts and lift out screen. Pour out old oil. Scrape accumulated dirt out of bottom and sides of oil reservoir. Wipe reservoir clean inside and out with cloth soaked in dry-cleaning solvent. Fill reservoir to level mark on side of reservoir with used crankcase oil or seasonal grade of fresh engine oil. Inspect element and clean if required (step (3) following). Install element and reservoir (subpar. d (2) following).

(3) To CLEAN AIR CLEANER SCREEN. While screen is removed from oil reservoir, wipe off dirt. Slush unit up and down in container partially filled with dry-cleaning solvent until it is washed clean. Allow it to drain, and then blow dry with compressed air.

c. **Removal of Air Cleaners.** Procedure for removal of the air cleaners is given in the following steps:

(1) REMOVE AIR CLEANER OIL RESERVOIR AND/OR ELEMENT. Support oil reservoir and release locking bail handle (fig. 104). Push bail out of the way. Lower oil reservoir with screen, and move clear of air cleaner body. Remove two wing nuts and lift screen out of oil reservoir.

(2) REMOVE AIR CLEANER BODY. Loosen clamps on hose connection at front end of air tube. Work hose off air cleaner body outlet and over tube. Disconnect air cleaner clamps while supporting air cleaner. Lower air cleaner to fighting compartment floor.

d. **Installation of Air Cleaners.** Procedure for installation of the air cleaners is given in the following steps:

(1) INSTALL AIR CLEANER BODY. Place air cleaner body in position. Attach clamps. Rotate body until center of outlet is in line with engine air intake tube. Tighten clamp bolts. Slide air intake tube hose onto air cleaner outlet and tighten hose clamps.

(2) INSTALL AIR CLEANER OIL RESERVOIR AND SCREEN. Make sure reservoir is filled to level mark. Install screen in reservoir. Raise reservoir into position under air cleaner body. Swing bail under reservoir while supporting reservoir. Pull up on locking lever to secure reservoir in position.

129. EXHAUST SYSTEM.

a. **Description.** The exhaust system includes two exhaust manifolds, heater tubes, carburetor adapters, two exhaust tubes, and an exhaust outlet (figs. 53 and 54). The exhaust manifolds (fig. 54) connect directly with the exhaust ports in each cylinder head. Heater tubes at each end of the exhaust manifolds connect both manifolds with each carburetor adapter (fig. 53). The adapters are heated by the hot exhaust gases passing through them. Exhaust tubes connect the manifolds to the rear center deck plate exhaust outlet flange. The exhaust gases pass out through the exhaust outlet, which is attached to the rear face of the deck plate flange (fig. 3).

130. EXHAUST MANIFOLDS.

a. **Description.** The two exhaust manifolds are made of stampings welded together. They are attached to the cylinder heads at each exhaust port (fig. 54). A heater tube at each end of the manifold is connected to each carburetor adapter (fig. 53). Four metal gaskets are used to seal the joints between each manifold and the cylinder head.

b. **Removal of Exhaust Manifold or Gaskets.** Remove power unit (par. 71). Disconnect manifold from cylinder head (fig. 54) and carburetor adapter heater tubes. Remove heater tubes (fig. 53). Support manifold as last nuts are removed. Lower manifold off studs. Remove the used gaskets from cylinder block and manifold. Scrape all gasket surfaces clean.

c. **Installation of Exhaust Manifold or Gaskets.** Coat one side of new gaskets lightly with liquid joint and thread compound. Position new gasket on each of the four cylinder head flanges and on magneto end of manifold. Carefully raise manifold into position on studs, making sure end gasket is positioned. Screw on two nuts flush with ends of lower studs. Insert the two bolts at magneto end. At flywheel end insert new gasket between heater tube flange, and insert the two bolts. Install the remaining manifold to cylinder head nuts. Tighten nuts alternately to compress gaskets evenly. Install heater tubes. Install power unit (par. 72).

131. EXHAUST TUBES.

a. **Description.** Two large-diameter steel tubes connect the exhaust manifolds with the rear center deck plate flange. The tubes are wrapped with sheet asbestos for heat insulation.

b. **Removal of Exhaust Tubes.** Remove rear center deck plate. Loosen clamps which attach tubes to exhaust manifolds. Rotate rear end of tube upward and to the outside to clear cut-out in rear hull plate. Withdraw tube through hole in upper baffle plate.

c. **Installation of Exhaust Tubes.** Position clamp on end of exhaust manifold. Insert exhaust tube through hole in upper baffle plate with rear end raised and rotated to outside. Guide front end into clamp while swinging rear end down into cut-out in rear hull plate. Install rear center deck plate.

Part Three—Maintenance Instructions

132. EXHAUST OUTLET.

a. Description. The exhaust outlet is a casting attached to the rear face of the rear center deck plate flange (fig. 181). The joint between the outlet and deck plate is sealed by a copper-asbestos gasket. The gun traveling lock is mounted on the exhaust outlet.

b. Removal. Remove nuts from studs and pull outlet to rear so studs can be withdrawn. Lift off outlet. If required, remove gun traveling lock from exhaust outlet (par. 188 b).

c. Install. Install gun traveling lock on exhaust outlet (par. 188 c) if previously removed. Inspect exhaust outlet gasket and install new gasket, if required, on studs. Install outlet on rear center deck plate.

133. CYLINDER HEADS.

a. Description. The engine has two valve-in-head cast aluminum cylinder heads with two overhead camshafts in each head. Each cylinder has two intake and two exhaust valves with alloy steel valve inserts as valve seats. The two camshafts are bearing mounted in supports on the top of each cylinder head. The valve clearance of 0.025 to 0.030 inch is established during manufacturing or overhaul, and is not adjustable. A flat, copper-asbestos cylinder head gasket and a cork oil seal are used between the cylinder head and cylinder block. Gasket and seal must be removed and new ones installed in *both* cylinder heads each time either or both cylinder heads are removed. Notify higher authority when it is necessary to remove the cylinder head and install a new one. A passage cast in the inner side of each cylinder head serves as a fuel intake manifold.

b. Removal of Cylinder Head Gasket and Compression Seals. When it is necessary to replace the cylinder head gasket and/or seal on one bank of cylinders always install a new gasket and seal on the opposite cylinder bank. CAUTION: *This is important.* The same operations will be performed on both cylinder heads, except when otherwise directed. The procedure for removing the cylinder head or gasket is given in the following steps. Remove and install locking wires wherever necessary.

(1) DISCONNECT SPARK PLUG WIRES FROM MAGNETO. Follow procedure described in paragraph 75 b.

(2) REMOVE WATER MANIFOLDS. Disconnect sending unit. Disconnect vehicle heater water tube from end of "left" cylinder head water manifold (fig. 53 and 54). Remove manifolds and two gaskets. Cover openings in cylinder head with tape.

(3) REMOVE THROTTLE ROD ASSEMBLY. Disconnect rocker arm link from throttle control rod bell crank. Disconnect throttle

Fuel and Air Intake System, Exhaust System, and Cylinder Head

control rod from bell crank (fig. 101). Remove throttle control rod return spring. Disconnect throttle rods at each carburetor (fig. 101). Remove safety nut attaching throttle control rod arm support bracket to air intake manifold base stud. Remove throttle rods and support bracket as a complete assembly.

(4) DISCONNECT CARBURETOR METALLIC FUEL HOSES. Back off connector nuts attaching metallic fuel hoses to each carburetor (fig. 101). Disconnect fuel hoses from each carburetor.

(5) DISCONNECT CARBURETOR AIR INTAKE HOSES. Remove locking wires and loosen hose clamps (fig. 102). Work hoses away from carburetors back against carburetor air intake manifold base.

(6) REMOVE CARBURETOR. Loosen conduit clamps. Disconnect carburetor degasser wires from degasser terminals by pulling wires out of sockets. Remove insulator and two gaskets between each carburetor and adapter. Lift off carburetors, and remove intake manifold gaskets.

(7) REMOVE FUEL PUMP AND FUEL TUBE. Disconnect fuel tube from fuel pump (fig. 100) and swing clear of left camshaft housing if left cylinder head is being removed. Pry open tube clamp on engine oil pan but leave clamp attached to pan. Remove fuel pump from "left" camshaft housing. Remove gasket between pump and camshaft housing. NOTE: *When removing "left" cylinder head, disconnect fuel tube from pump and leave fuel pump attached to "left" camshaft housing. Also disconnect speedometer cable.*

(8) REMOVE CARBURETOR AIR INTAKE MANIFOLD. Remove nuts from air intake manifold base studs on each intake manifold (fig. 102). Leave hoses attached and lift off manifold. Do not remove air intake elbow from base.

(9) REMOVE CARBURETOR ADAPTERS. Remove nuts from the four adapter studs. Remove nuts and "through" bolts attaching lower section of adapter to intake manifolds, using carburetor heat-box nut special offset wrench (41-W-639-850). NOTE: *"Through" bolts on magneto end adapter are ⅛-inch longer than the "through" bolts on the flywheel end carburetor adapter. The magneto end "through" bolts are longer so as to support primer fuel tube bracket.* At flywheel end of engine disconnect both short air heater tubes at connections (flanges) nearest exhaust manifolds. At magneto end disconnect long, curved air heater tubes from exhaust manifolds. Remove gaskets between flanges. Leave tubes attached to adapters, and remove assemblies.

(10) REMOVE PRIMER FUEL TUBE AND BRACKET. Disconnect primer fuel tube at first tee-connection (from magneto end) on each cylinder head (fig. 52). Remove U-shaped section of tube and tube support bracket.

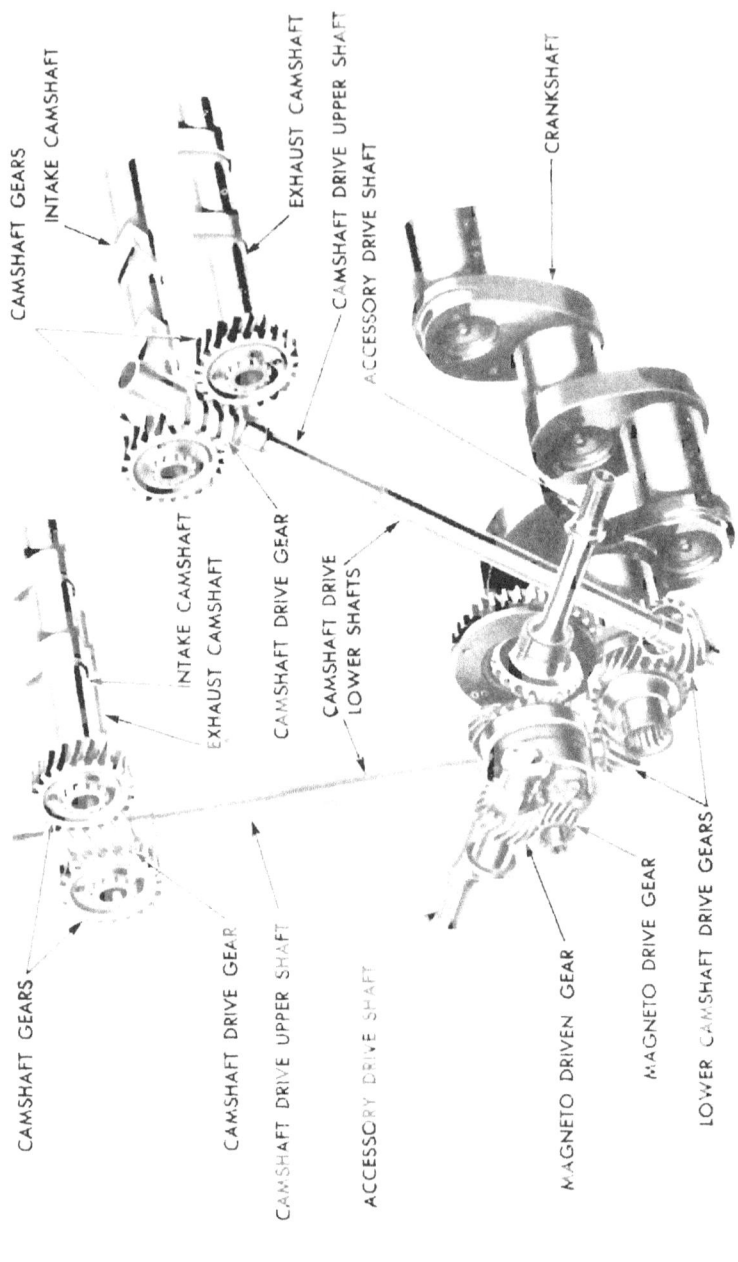

Figure 105 — Accessory Drives

Fuel and Air Intake System, Exhaust System, and Cylinder Head

(11) REMOVE GOVERNOR BOOSTER. When removing "right" camshaft housing and/or cylinder head, remove governor booster. Remove gasket and at the same time disconnect the three oil tubes from both the governor and the booster (fig. 54).

(12) REMOVE GOVERNOR. When removing "right" camshaft housing and or cylinder head, remove governor. Remove nuts attaching governor to magneto end of camshaft housing. Carefully pull governor away from camshaft housing so, as not to damage splines on governor drive shaft. Remove gasket.

(13) REMOVE TACHOMETER DRIVE. When removing "left" camshaft housing and/or cylinder head, remove tachometer drive located in magneto end of housing. Remove nuts attaching drive to camshaft housing (fig. 88). Lift off drive and remove gasket. After drive has been removed pull out short tachometer drive quill. Tape opening.

(14) REMOVE REVOLUTION COUNTER. When removing "left" camshaft housing and/or cylinder head, remove the revolution counter located on magneto end of housing. Remove nuts and carefully pull counter off camshaft housing. Remove gasket. After counter has been removed pull out short revolution counter drive quill and tape opening.

(15) REMOVE ENGINE OIL LEVEL INDICATOR HOUSING. When removing "left" camshaft housing and or cylinder head, remove engine oil level indicator housing. Pull indicator from housing. Remove nuts attaching housing clamp to cylinder head. Remove both sections of clamp from studs. Remove cylinder block drain plug locking wire from around housing. Remove nuts attaching lower part of housing to cylinder block studs. Pull housing off studs and up between exhaust manifold and cylinder head. Remove gasket. Tape opening.

(16) REMOVE ELECTRICAL CONDUIT (RIGHT CAMSHAFT HOUSING). Disconnect wires from carburetor degassers and water manifold engine temperature sending units and move tube clear of camshaft housing.

(17) REMOVE SPARK PLUG COVER AND SPARK PLUGS. Remove spark plug cover and spark plugs (par. 76 b).

(18) REMOVE SPARK PLUG CHAMBER STUD SAFETY NUTS. Remove four safety nuts and copper washers from studs in bottom of spark plug chamber.

(19) REMOVE CAMSHAFT HOUSING SAFETY NUTS. Remove 10 safety nuts and copper washers from top of camshaft housing. Remove 10 safety nuts and copper washers from bottom edge of camshaft housing.

(20) REMOVE CAMSHAFT HOUSING. Lift camshaft housing evenly and carefully off long cylinder head studs so that studs will not be bent. Remove inner and outer gaskets.

TM 9-735
133

Part Three—Maintenance Instructions

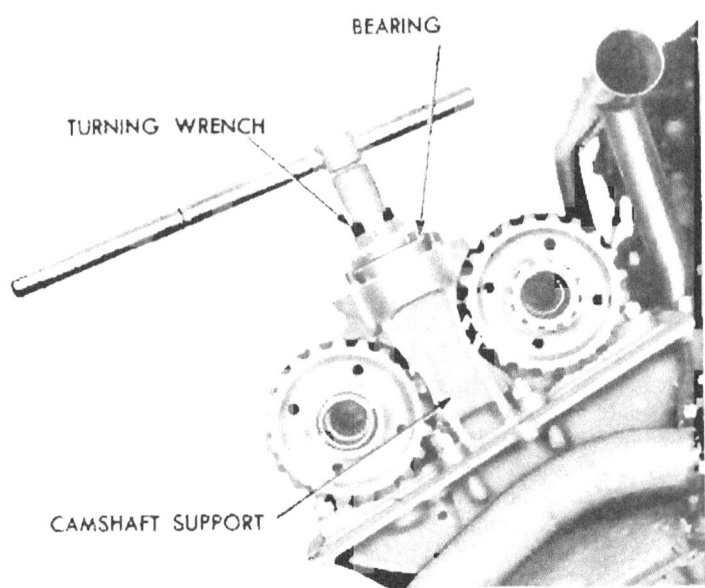

Figure 106 — Camshaft Turning Wrench (41-W-2964-300), Installed in Drive Gear

(21) REMOVE CAMSHAFT UPPER DRIVE SHAFT. Turn flywheel until timing marks on the top camshaft gears are lined up with the reference marks on the camshaft support bracket (fig. 112). If flywheel is not turned again the camshafts can be removed and reinstalled without disturbing the camshaft timing in relation to the crankshaft. CAUTION: *Do not turn flywheel after camshaft upper drive shaft has been removed.* Remove snap ring from top of the camshaft drive worm gear, and lift the upper drive shaft from the gear (fig. 105). Install camshaft drive shaft puller (41-P-2905-75) by screwing puller into end of drive shaft. Pull shaft out carefully so as not to damage splines.

(22) INSPECT PUSH ROD CLEARANCE. Before removing camshafts check push rod clearance with a feeler gage (fig. 107). Correct clearance is 0.025 inch to 0.030 inch, and is taken between top of push rod and heel of cam. To check clearance install camshaft turning wrench (41-W-2964-300) in top of worm gear (fig. 106). Turn camshafts as required to position heel of each cam at bottom of camshaft. Insert feeler gage. As feeler gage is inserted a very slight drag should be felt. If any push rod clearances are not within recommended meas-

TM 9-735
133

Fuel and Air Intake System, Exhaust System, and Cylinder Head

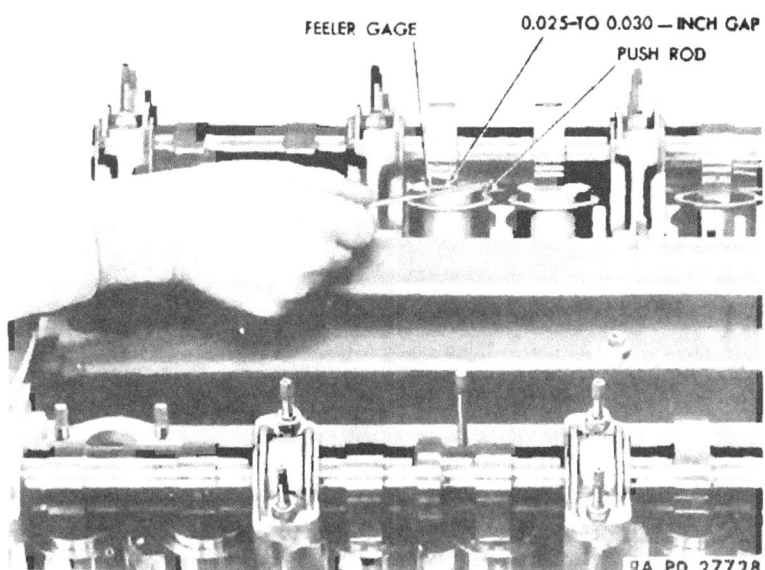

Figure 107 — Checking Push Rod Clearance

urements, notify higher authority as it will be necessary to install a new cylinder head or push rod.

(23) REMOVE CAMSHAFT BEARING CAPS. Remove upper camshaft gears and support bracket assembly nuts. To remove camshaft bearing caps start at center and alternately work toward each end of cylinder head, loosening each nut a turn at a time until pressure of all nuts is full released. CAUTION: *Remove nuts evenly to prevent undue strain on camshaft due to valve spring pressure.* Remove nuts from camshaft bearing studs, being careful not to drop nuts or washers into cylinder head openings. Lift off camshaft bearing caps. NOTE: *Each bearing cap is indexed with a letter and numeral which corresponds to the same markings on the cylinder head, to insure correct reassembly.*

(24) REMOVE CAMSHAFTS, UPPER CAMSHAFT GEARS AND SUPPORT BRACKET ASSEMBLY. Carefully lift off camshaft upper camshaft gears, and bracket assembly as a unit (fig. 108).

(25) REMOVE CAMSHAFT BEARINGS. Tag camshaft bearings for reassembly identification and lift from bearing studs.

(26) REMOVE INNER ROWS OF CAMSHAFT BEARING STUDS AND LONG CYLINDER HEAD NUTS. Slide long cylinder head nut wrench (41-W-866-200) down on stud. Seat key on end of wrench securely

TM 9-735
133

Part Three—Maintenance Instructions

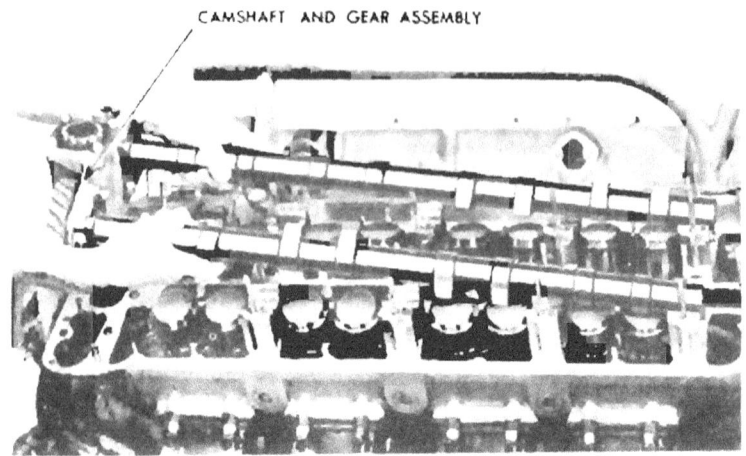

Figure 108 — Removing Camshaft and Gear Assemblies

in keyway on nut, and remove long cylinder head nut and stud together (fig. 110). Upper and lower outside rows of camshaft bearing studs may be left in place. Leave washers in place.

(27) REMOVE CYLINDER HEAD END NUTS. Remove two nuts from magneto end of cylinder head and three nuts from flywheel end of head.

(28) REMOVE SHORT CYLINDER HEAD NUTS. The short cylinder head nuts are also recessed in the cylinder head. Remove nuts by using short cylinder head nut wrench (41-W-866-250). Be sure key on end of wrench seats securely in keyway in nut. Leave washers in place.

(29) REMOVE CYLINDER HEAD. Lift cylinder head off cylinder block evenly to prevent binding on studs. Leave exhaust manifold attached to cylinder head. WARNING: *Do not drop or remove valve lifters from guides. Lifters are individually fitted and changing installation will cause incorrect valve clearance.*

(30) REMOVE CYLINDER HEAD GASKET. Lift off copper-asbestos cylinder head gasket (fig. 109).

(31) REMOVE CYLINDER COMPRESSION SEALS. Remove steel compression seal from top of each cylinder.

(32) REMOVE CARBON. Remove carbon from cylinder head and from tops of pistons. Tag and remove each push rod, or remove one rod at a time, and place them on a bench in their respective installed positions. Wipe top of cylinder block and combustion chambers. In-

spect top of cylinder block and cylinder head for cracks, burned or pitted valves, loosened spark plug inserts, or other damage. Examine push rods, camshafts, camshaft drive gears and bearings for excessive wear or damage. Notify higher authority if any of the above parts are to be replaced. After cleaning and inspection, install push rods in original locations.

c. **Installation of Cylinder Head Compression Seals and Gaskets.** The procedure for installing cylinder head or gasket is given in the following steps. Be sure to install locking wires wherever required. When elastic stop nuts are installed and tightened, bolt must extend through fiber at least three full threads. Do not install elastic stop nuts on bolts with cotter pin holes.

(1) INSTALL CYLINDER COMPRESSION SEALS. Install a *new* steel compression seal, lip side down, on the top of each cylinder sleeve (fig. 109).

(2) INSTALL CYLINDER HEAD GASKET. Wipe off top of cylinder block and install new copper-asbestos cylinder head gasket. Use gasket 0.014 inch thick, identified by two small circles stamped in gasket face. Always replace gaskets in both cylinder heads.

(3) INSTALL CYLINDER HEAD. Wipe off lower side of cylinder head. Lower cylinder head evenly and carefully onto cylinder block so that it does not bind on studs.

(4) INSTALL CYLINDER HEAD NUTS. Place a flat washer on each cylinder head stud. CAUTION: *Before placing new washers on studs be sure no washers are in holes in cylinder head.* Unscrew camshaft bearing studs from long nuts, otherwise studs will "bottom" on cylinder head, preventing cylinder head nuts from being securely tightened. Screw long nuts finger-tight on inner rows of cylinder head studs. Screw short cylinder head nuts finger-tight on outer rows of studs. Use torque wrench with long cylinder head nut wrench (41-W-866-200) on inner stud nuts (fig. 110) and short cylinder head nut wrench (41-W-866-250) on the outer rows of studs. Start at the center and work alternately in each direction. Tighten nuts with one turn each to uniformly compress gasket. When gasket has been uniformly compressed, tighten each nut a turn at a time with torque wrench, using a pull of 60 foot-pounds.

(5) INSTALL CYLINDER HEAD END NUTS. Install two nuts on magneto edge of cylinder head, and three nuts on flywheel edge. CAUTION: *Do not tighten these nuts until long and short cylinder head nuts have been tightened.*

(6) INSTALL CAMSHAFT BEARING STUDS. Install camshaft bearing studs in inner rows of long cylinder head nuts (fig. 111). CAUTION: *Do not tighten studs with so much pressure that replacer binds on studs.*

Part Three—Maintenance Instructions

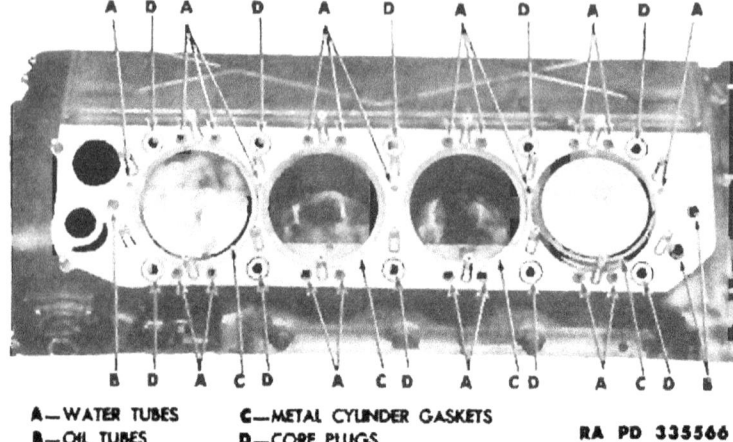

A—WATER TUBES
B—OIL TUBES
C—METAL CYLINDER GASKETS
D—CORE PLUGS

RA PD 335566

Figure 109 — Top of Cylinder Block — Gasket Removed

(7) INSTALL CAMSHAFT BEARINGS. Install camshaft bearings on camshaft bearing studs, making sure each bearing is installed in its original location.

(8) INSTALL CAMSHAFTS, UPPER CAMSHAFT GEARS AND SUPPORT BRACKET ASSEMBLY. Position assembly on top of cylinder head with camshafts resting in camshaft bearings (fig. 108). Install five nuts attaching upper camshaft gears and bracket assembly to cylinder head studs. CAUTION: *Do not tighten nuts.* Oil bearings lightly.

(9) INSTALL CAMSHAFT BEARING CAPS. Install camshaft bearing caps on camshaft bearing studs. Be sure index mark on cap corresponds to and registers with index mark on camshaft bearing. Install stud nuts. Tighten nuts evenly. Work alternately from center of cylinder head toward each end to prevent strain on camshaft from thrust of valve springs. After nuts have been uniformly tightened use a torque wrench and tighten all camshaft bearing stud nuts with 16 foot-pounds torque. Tighten five nuts attaching upper camshaft gear and bracket assembly to cylinder head. Check push rod and cam clearance (subpar. b (22) preceding).

(10) CHECK CAMSHAFT GEAR INDEX MARKS. Install camshaft turning wrench (41-W-2964-300) in top of worm gear (fig. 112). Turn camshafts in direction of normal rotation (fig. 113) until No. 1 intake valve is just starting to open. The valve opening point can be determined by noting when toe of the cam contacts push rod with sufficient force to make it difficult to rotate rod with thumb and forefinger. Inspect timing gear marks. Be sure they are alined with

Fuel and Air Intake System, Exhaust System, and Cylinder Head

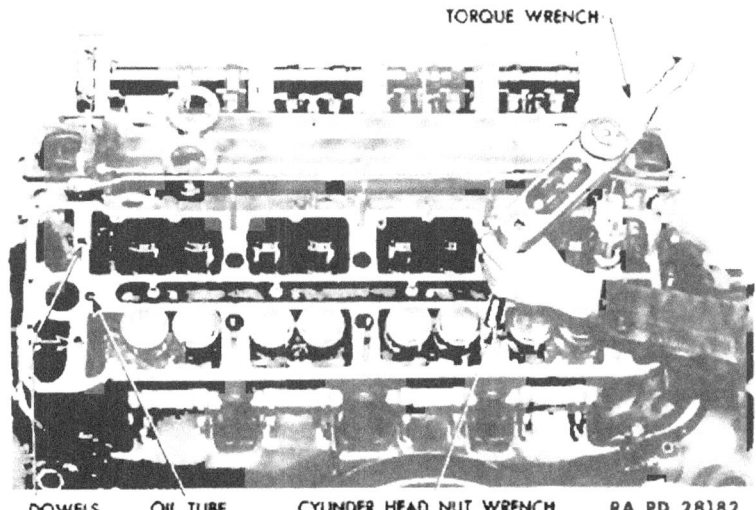

Figure 110 — Tightening Long Cylinder Head Nuts, Using Wrench (41-W-866-200)

reference marks on camshaft support (fig. 112). If marks are not alined, camshafts are not in proper timing with one another. Notify higher authority to make corrections.

(11) INSPECT PUSH ROD CLEARANCE. Follow procedure in subparagraph b (22) preceding.

(12) TIME CAMSHAFTS WITH CRANKSHAFT. Turn camshaft on "right" cylinder head until No. 1 intake valve starts to open (subpar. c (10) preceding). NOTE: *Remove spark plugs from both cylinder heads before attempting to turn flywheel.* Remove cap screws which attach inspection cover to flywheel housing, and remove cover and gasket. To turn flywheel use heavy screwdriver to engage flywheel ring gear teeth. Turn flywheel counterclockwise at least one complete revolution. Stop turning when flywheel timing mark "IO-RH" (intake opens—right-hand) is centered under pointer (fig. 67). CAUTION: *If fan drive propeller shafts are connected to angle drives, make sure they do not strike or bind when flywheel is turned.* Remove camshaft turning wrench (41-W-2964-300) and insert upper camshaft drive shaft without disturbing position of either camshafts or flywheel. Make sure shaft is fully entered into lower camshaft drive shaft. NOTE: *It may be necessary to try the shaft in several positions to aline splines in lower camshaft drive shaft at bottom and drive gear at top.* Install snap ring in top of camshaft

TM 9-735

Part Three—Maintenance Instructions

Figure 111 — Camshaft Bearing Studs

drive gear. Repeat timing operation for "left" cylinder head. Turn camshaft until No. 1 intake valve starts to open. Turn flywheel until timing marks "IO-LH" (intake opens—left-hand) center under pointer. Install camshaft upper drive shaft. Recheck timing on both cylinder banks by again turning flywheel counterclockwise to set each timing mark under pointer. Check to see that No. 1 intake valve is just starting to open. Install inspection cover, using new gasket.

(13) INSTALL CAMSHAFT HOUSING. Coat one side of gaskets with liquid joint and thread compound and place new camshaft inner and outer gaskets in position on cylinder head. Lower housing evenly into place on cylinder head. Place copper washer on each of attaching studs and install nuts, but do not tighten. NOTE: *Make sure nuts are installed on four studs in spark plug chamber.* Tighten nuts alternately to compress gaskets uniformly.

(14) INSTALL SPARK PLUGS. Remove protective covering from spark plug holes and install spark plugs and connect spark plug wires (par. 76 c). Install new spark plug chamber cover gasket. Position cover on camshaft housing, louver opening at top. On "left" camshaft housing attach speedometer cable housing and fuel tube clamps. On "right" camshaft housing attach clamp holding degasser conduits to end studs in upper row.

(15) CONNECT SPARK PLUG WIRES TO MAGNETOS. Follow procedure in paragraph 75 c (7).

(16) INSTALL REVOLUTION COUNTER. Insert revolution counter drive quill in magneto end of "left" exhaust camshaft. Place new

Fuel and Air Intake System, Exhaust System, and Cylinder Head

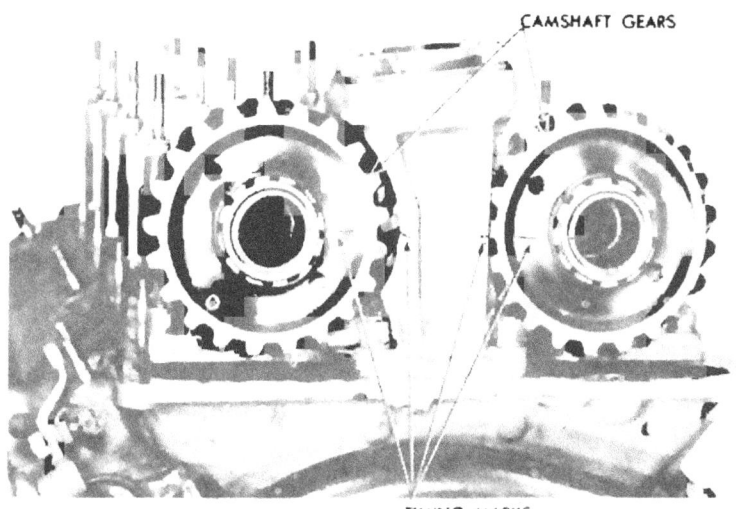

Figure 112 — Camshaft Gear Index Marks

gasket on studs. Aline splines on drive quill and install revolution counter, face down (fig. 88). Tighten nuts evenly.

(17) INSTALL TACHOMETER DRIVE. Insert tachometer drive quill in magneto end of "left" intake camshaft (fig. 88). Place new gasket on studs. Turn drive to aline splines, and install tachometer drive. Tighten nuts evenly. Install tachometer drive adapter (par. 100 c), making sure adapter quill is properly positioned. Do not tighten adapter coupling nut on drive.

(18) INSTALL GOVERNOR. Coat one side of new gasket with liquid joint and thread compound and place gasket on governor mounting studs located at magneto end of "right" camshaft housing (fig. 54). Rotate governor shaft to aline splines, and install governor on housing. Tighten nuts evenly.

(19) INSTALL PRIMER FUEL TUBE AND BRACKET. Position primer tube with bracket on magneto end of engine. Connect fuel tube to each nozzle tube.

(20) INSTALL CARBURETOR ADAPTER AT MAGNETO END. Place new adapter to cylinder head inlet passage gasket on each cylinder head. Position adapter, with air heater tubes attached, on cylinder head, and attach with long bolts through the primer tube bracket. Do not tighten bolts. Insert new gaskets between air heater tubes and exhaust manifolds. Insert bolts and screw on nuts to hold bolts in

Part Three—Maintenance Instructions

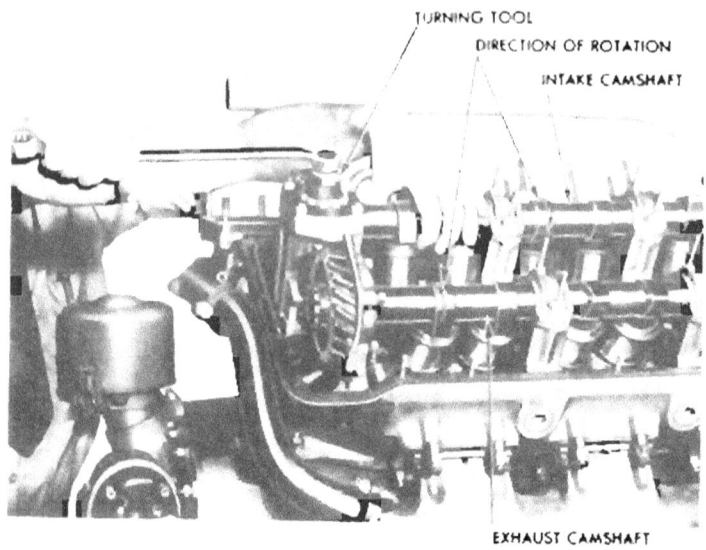

Figure 113 — Timing Camshafts, Using Wrench (41-W-2964-300)

place. Install remaining adapter bolts and nuts. Using carburetor heat box special offset wrench (41-W-639-850), tighten bolts and nuts alternately to compress gaskets uniformly. Tighten heater tube flange nuts. Install locking wires as required.

(21) INSTALL CARBURETOR ADAPTER AT FLYWHEEL END. Follow procedure in step (20) preceding as it applies. Use new gaskets and install locking wires as required.

(22) INSTALL CARBURETOR AIR INTAKE MANIFOLD. With hoses and hose clamps attached to air intake manifold, position manifold on attaching studs (fig. 101). Install and tighten nuts.

(23) INSTALL CARBURETORS. Insert carburetor attaching screws through center of adapter. Place a new gasket, insulator, and second gasket on the screws. Position carburetor and install remaining screws. Tighten cap screws alternately to compress gaskets uniformly, and secure with locking wires as required. Connect degasser wires, and attach conduit shielding caps (fig. 100).

(24) INSTALL GOVERNOR BOOSTER. Connect oil supply tube to governor and cylinder block. Install booster on cylinder block and tighten nuts. Connect oil tubes to governor and to booster.

(25) INSTALL FUEL PUMP AND FUEL TUBE. Place new gasket on studs. Slide fuel pump into position on studs, and install nuts.

Position fuel pump tube on camshaft housing and connect tube loosely to pump. Install tube bracket on cylinder block. Tighten connection at pump. Connect and tighten fuel hoses at both carburetors.

(26) CONNECT AIR INTAKE HOSES. Coat inside of hose ends with liquid joint and thread compound. Slide hoses onto carburetors, and tighten hose clamps. Install locking wires in clamp screws.

(27) INSTALL WATER MANIFOLDS. Place new gaskets on studs at each outlet. Install manifold, using washers and safety nuts. Tighten nuts alternately to compress gaskets uniformly. Connect heater rear inlet tube to magneto end of "left" manifold. Connect wires and conduits to sending unit and switch (fig. 100).

(28) INSTALL THROTTLE CONTROL RODS. Place throttle control rods in position. Install throttle control rod arm on stud on air intake manifold. Follow procedure in paragraph 126 for adjusting linkage.

(29) TIGHTEN NUTS. Test ALL installation nuts with a wrench to be sure they are tight.

(30) INSTALL POWER UNIT IN VEHICLE. Follow procedure in paragraph 72.

(31) INSTALL ELECTRICAL CONDUIT. Install tube on right camshaft housing and connect wires to carburetor degassers and engine temperature sending units in water manifolds.

(32) INSTALL SPEEDOMETER CABLE. Install speedometer cable on left camshaft housing.

(33) RECORD INSTALLATION OF CYLINDER HEAD. Make proper entry on W.D., A.G.O. Form No. 478, "MWO and Major Unit Assembly Replacement Record."

(34) ROAD TEST VEHICLE. Perform the operations and make such required inspections as listed in paragraph 53.

Section XXVII

COOLING SYSTEM

134. COOLING SYSTEM.

a. **Description.** The cooling system is a pressure-sealed, thermostatically controlled, liquid-type, 22-gallon capacity system (fig. 114). The auxiliary engine and the driver's compartment heater are connected to the cooling system. The water pump at the magneto end of the engine (fig. 52) draws water from the bottom of each of the two radiators. It forces the water through the water jackets, passages, and water manifolds on each cylinder head to the expansion tank (fig. 114). From the expansion tank the water passes directly to each radiator when the two thermostats are open. When the two thermostats are closed, the hot water is bypassed back through the water pump to the engine for quick engine warm-up. A sending unit and switch for the water temperature gage and engine high temperature warning signals are mounted in the water manifolds (fig. 100). The water is cooled by passing through two cross-flow radiators directly behind the engine, one on each side of the engine compartment. Two five-bladed fans, mounted in each fan shroud, provide air circulation for each radiator (fig. 115). They are driven by belts from a drive pulley on the end of each fan drive propeller shaft. The fans draw air in through the intake doors in front of the transverse housing to which the radiators are attached. The air is then forced through the radiators and oil coolers, mounted on the rear sides of the radiators, and out the exhaust doors at the rear of the vehicle. Baffle plates between the two radiators and around the transmission housing prevent recirculation of air within the engine compartment. The fans, shrouds, radiators, oil cooler, and expansion tank are suspended from the transverse housing, and are all removed as one unit (fig. 115).

b. **Drain Cooling System.** When draining the cooling system to disconnect the tubes, hoses, or radiators, it is not necessary to drain the cylinder block water pump. The pump and block, however, must be drained separately when the system is to be cleaned, flushed, filled with antifreeze solution, or completely drained for storage or shipment. From below vehicle remove cooling system drain plug covers (fig. 162). If coolant is to be saved, place a clean container of 12-gallon capacity under each drain plug. Remove the drain plug from each radiator outlet elbow (fig. 115). At top center of transverse housing raise filler cap cover (fig. 115) and remove filler cap from expansion tank. CAUTION: *If system is hot, vent expansion tank (subpar. d following) before removing filler cap, or serious burns will result.* To completely drain the cooling system, drain the heater, cylinder block, and water pump after radiators have drained dry. Open engine compartment drain valves. Remove locking wire and remove drain plug on each side at the base of the cylinder block below No. 3 exhaust port (fig. 54). Remove bulkhead cover. Remove locking wire and drain plug from bottom of water pump (fig. 52). Open driver's compartment drain valve and remove drain plug from bottom of heater. Install and tighten drain plugs which were removed, and secure with locking wire as required. Close driver's and engine compartment drain valves. If cooling system is to be left drained, place suitable warning tag on 24-volt master switch. If engine is to be operated, fill cooling system (subpars. d and e following) and inspect (subpar. c following).

c. **Inspect Cooling System.** Be sure cooling system is full. Start and run engine at idling speed for 3 minutes while inspecting all hoses, tubes, connections, and drain plugs for leaks, replacing any

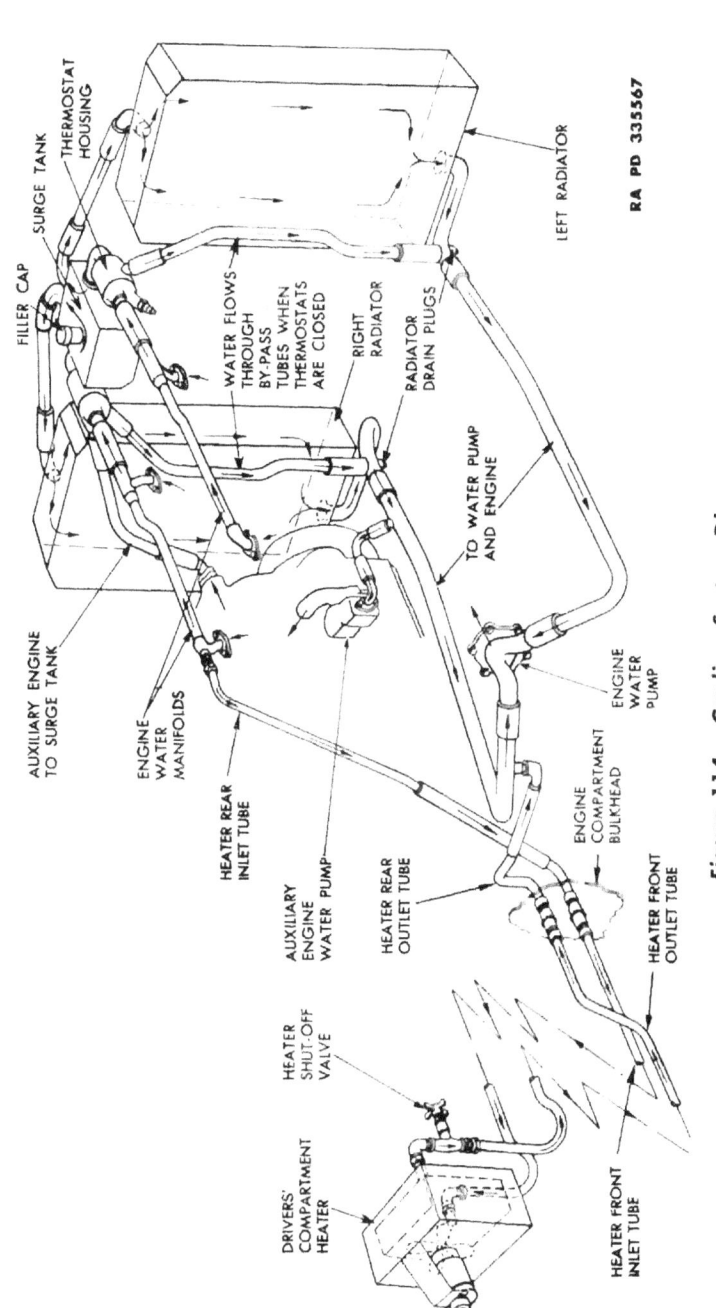

Figure 114 — Cooling System Diagram

defective items. Tighten connections as required or, if necessary, report to higher authority. Note temperature gage indications, and examine radiator cores for dirt or insects. Clean, if necessary. Test pressure relief valve in filler cap to see that it opens freely. Inspect vent at side of filler neck to see that it is not plugged. Inspect filler cap gasket, and install new gasket if required. Inspect condition and tension of fan belts. If belts are frayed or cracked, replace them. Examine all drain plugs and gaskets, and close or install drain plug covers.

d. **Fill Cooling System.** Maintain coolant at bottom of screen in filler neck. Daily, or more frequently as required, check level. At temperatures above 32° F use clean water (suitable for drinking). When adding water or filling the cooling system, add one container of corrosion inhibitor compound to each 4 gallons of water. To check level of coolant or to add water, raise cover and turn expansion tank filler cap one-quarter turn to left to first vent system. WARNING: *If system is hot, wait and allow steam pressure to escape before removing cap.* After system has been vented press down on cap, press and turn until cap can be lifted off. Lift screen out of filler neck to see if water is visible in bottom of surge tank. If water temperature is above 200° F and coolant is not visible in bottom of tank, allow coolant and engine to cool to at least 180° F before adding water. CAUTION: *Adding any considerable quantity of water when engine is overheated will cause permanent damage to cylinder heads or blocks.* Install screen in filler neck. Start and run engine at idling speed. WARNING: *Do not operate at speeds above 500 revolutions per minute with cap removed, as coolant will be forced from expansion tank.* Add water which has been treated with corrosion inhibitor compound until coolant is visible above bottom of filler neck screen, or can be felt with finger tip. Make necessary inspection (subpar. c preceding). Install filler cap and turn to right until tight. Start and idle engine for 3 minutes to thoroughly fill system. Recheck level of coolant and add if required. Keep heater shut-off valve closed if heater is not to be used.

e. **Flush Cooling System.** When coolant becomes discolored, or there is evidence of rust and scale deposits, flush cooling system. If flushing does not remove rust and scale, system will have to be cleaned (subpar. f following). Flush cooling system only after engine has been operated for at least 15 minutes with water temperature about 180° F. Open heater shut-off valve (fig. 35) and completely drain cooling system (subpar. h preceding). Refill cooling system with clean, fresh water. Run engine until it has operated at least 5 minutes with water temperature between 180° F and 200° F. Stop engine and again completely drain the system. If the drained water is not reasonably clear, repeat flushing operation as necessary, or until

TM 9-735
134

Cooling System

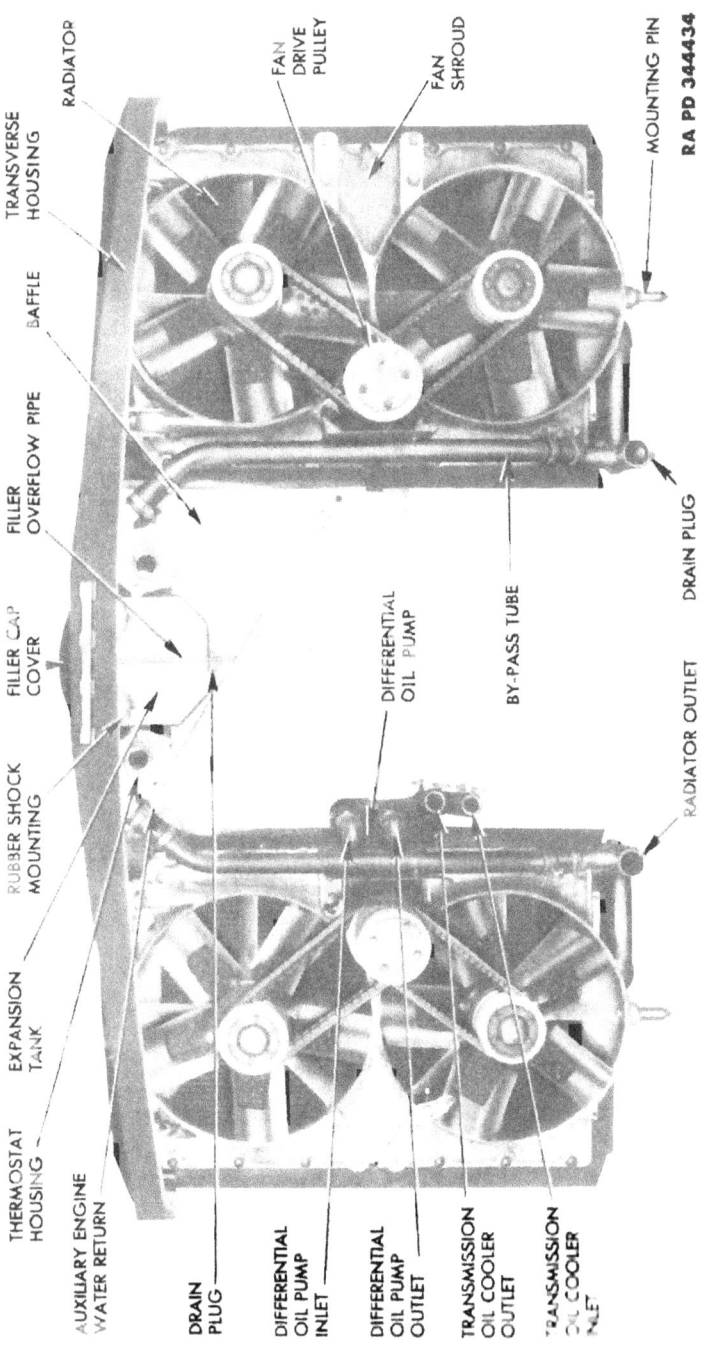

Figure 115 — Cooling Unit, Front View

drained water is clear. CAUTION: *Do not run engine with cold water supply hose running into expansion tank.* Refill cooling system with clean water (subpar. h preceding) or with antifreeze solution (subpar. g following).

f. **Clean Cooling System.** Completely clean the cooling system with cleaning compound before installing, after removing antifreeze compound, or when system becomes clogged causing overheating. Notify higher authority if overheating continues after system has been cleaned. Before cleaning, operate engine until it warms up to 180° F but stop engine before temperature reaches 200 F. Completely drain cooling system (subpar. h preceding). Open heater shut-off valve. Allow engine to cool to 180° F. Slowly fill system with 10 gallons of water. Empty 5½ containers of cleaning compound into expansion tank. Start and run engine at idling speed to completely fill system. CAUTION: *Cleaning compound must be mixed only with clear water WITHIN the cooling system.* Do not allow cleaning solution to come in contact with skin, clothing, or painted surfaces. Cover intake doors if required and run engine for at least ½ hour at a speed sufficient to keep engine temperature between 180° F and 200° F.

CAUTION: *Do not allow the engine temperature to exceed 200° F.* After cleaning solution has been heated and circulated for 30 minutes, neutralize it with the neutralizing compound found in the cleaning compound kit. Stop engine and again completely drain cooling system. Allow engine to cool to 180° F while installing drain plugs. Slowly fill system with 10 gallons of water. Empty 5½ containers of neutralizing compound into expansion tank. Start and run engine at idling speed to completely fill system. CAUTION: *Do not mix neutralizing compound with water before pouring it into the system.* Run engine to circulate the neutralizing solution for 5 minutes or more at 180° F to 200° F. Completely drain system (subpar. h preceding). Flush system (subpar. e preceding) until drained water is clear. Fill cooling system with fresh water (subpar. d preceding) or with antifreeze solution (subpar. g following). Start and run engine. While engine is operating, inspect cooling system for leaks (subpar. e preceding) which may have developed during cleaning process. Tighten connections or, if necessary, stop engine, drain system, and install new gaskets or parts as required to stop leaks, or notify higher authority if necessary. Close heater shut-off valve if heater is not to be used.

g. **Protect Cooling System with Antifreeze Compound.** Complete instructions for preparing and protecting the cooling system with antifreeze compound are given in paragraph 38 f.

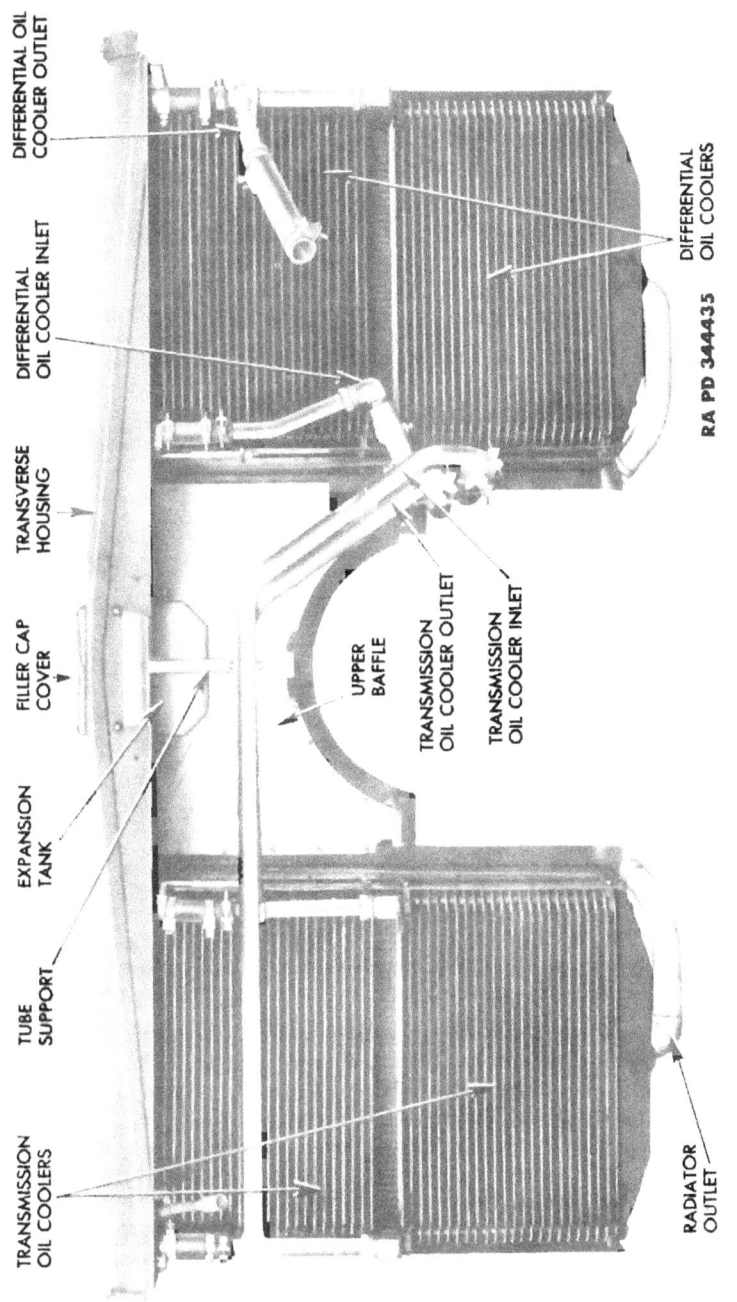

Figure 116 — Cooling Unit, Rear View

TM 9-735
135

Part Three—Maintenance Instructions

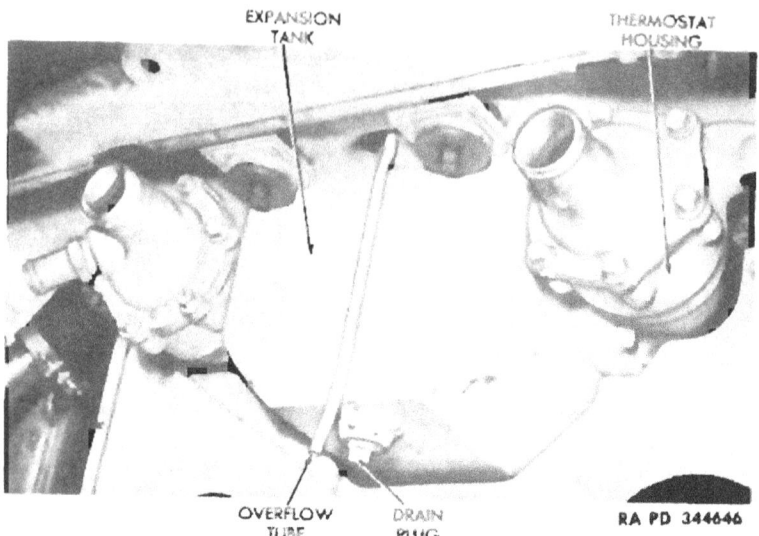

Figure 117 — Water Expansion Tank and Thermostat Housings

135. EXPANSION TANK.

a. **Description.** The expansion tank, mounted on shock rubber supports on the center underside of the transverse casting, provides the necessary space for expansion of the coolant, condensation of vapor, and convenient means for filling both radiators through the expansion tank filler opening (figs. 115 and 117). A thermostat housing is connected by a hose with each engine water manifold. At outlet on each side the expansion tank is connected by a hose to each radiator. A pressure-type filler cap is used to maintain a slight pressure in the cooling system which permits operation at higher temperatures without boiling of the coolant.

b. **Removal.** Turn battery master switch off. Remove front center deck plate. Remove exhaust tubes (par. 131 b). Remove filler cap from expansion tank. Drain at least 3 gallons from each radiator (par. 134 b). Disconnect and remove all hoses. Remove hoses from expansion tank outlets and from thermostat housings. Remove lifting eyes from transmission. Remove bolts which attach upper baffle plate to transverse casting. Tip baffle plate to the rear. Remove four screws which attach expansion tank to transverse casting. Lower expansion tank and slide it toward rear from under transverse casting.

Cooling System

c. **Installation.** Slide expansion tank into position under transverse casting from the rear. Position upper baffle plate and install. Install transmission lifting eyes. Connect hoses to expansion tank outlets. Connect hoses to thermostat housings. Fill cooling system (par. 134 d). Start and run engine. Carefully inspect all water and oil connections for leaks. Stop engine. Install front center deck plate.

136. ENGINE TEMPERATURE GAGE AND DUAL ENGINE HIGH WATER TEMPERATURE WARNING SIGNALS.

a. **Description.** Engine water temperature is recorded on an electric-type gage on the instrument panel. The engine high water temperature warning signals on the instrument panel (fig. 16) light when the temperature exceeds 220° F. The gage and signals are operated by a thermostatic switch and sending unit mounted in the water manifold (fig. 100).

b. **Removal of Engine Temperature Gage.** Remove instrument panel face plate (par. 97 d). Note wire numbers. Remove wires No. 27 and 33 from gage terminals, and lift off fiber insulator. Remove nuts from terminals which attach bracket. Lift off support bracket and fiber insulator. Carefully withdraw gage and gasket from front of panel.

c. **Installation of Engine Temperature Gage.** Inspect gasket and use new gasket if required. Position gage in panel. Install support bracket and fiber insulator, but do not tighten nuts. Make sure gasket is evenly spaced under edge of gage and gage is properly positioned, then tighten bracket attaching nuts. Place fiber insulator on terminals and connect wires No. 27 and 33 to proper terminals. Install instrument panel face plate (par. 97 e). Start and run engine, and observe action of gage. Stop engine.

d. **Removal of Engine Temperature Gage Sending Unit and Switch.** Turn 24-volt master switch off. Raise rear intake doors. Disconnect conduit shielding cap and wire from sending unit and switch at bottom of expansion tank. Remove unit.

e. **Installation of Engine Temperature Gage Sending Unit and Switch.** Screw sending unit and switch into water manifold. Connect wire and attach conduit shielding cap. Start and run engine while observing action of engine temperature gage. Inspect unit for leaks and tighten as required. Stop engine.

f. **Removal of Engine Temperature Warning Signal Lamp.** With 24-volt master switch turned off, use screwdriver and raise edge of red signal jewel rim on instrument panel. Pry out, remove jewel. Press in and turn left to remove lamp.

g. **Installation of Engine Temperature Warning Signal Lamp.** Install new lamp by pressing in and turning to right. Snap red jewel into place.

h. **Removal of Engine Temperature Warning Signal Switch.** Follow procedure outlined in subparagraph d preceding, disconnecting wire.

i. **Installation of Engine Temperature Warning Signal Switch.** Follow procedure in subparagraph e preceding, as it would apply to high water temperature warning signals sending unit. Connect wire.

137. THERMOSTATS.

a. **Description.** The circulation of water through the radiators is controlled by two thermostats, one in each thermostat housing on each side of the expansion tank (fig. 117). When thermostats are open, water flows from the engine into the expansion tank, and then into the radiators. During warm-up period, or when thermostats are closed, the flow of water into the expansion tank is shut off and the water directed down the by-pass tube (fig. 115) to the water pump from which it is recirculated through the engine. The thermostats start to open at 150° F to 155° F and are fully opened at 185° F.

b. **Removal.** Remove front center deck plate. Drain expansion tank. Disconnect hoses from thermostat housings. Take out screws and withdraw thermostat and two gaskets from housing.

c. **Installation.** Make sure thermostat to be installed is serviceable by testing action in water heated to at least 160° F. Inspect both thermostat gaskets and cover gasket to make sure they are serviceable. Lightly coat one side of each of the three gaskets with general purpose grease. Insert two gaskets in housing, smaller gasket first. Place cover gasket in position on cover. Install thermostat. Slide hose connections into place on cover and tighten hose clamps. Fill cooling system. Install front center deck plate.

138. FAN BELTS.

a. **Description.** The two fans in each radiator shroud (fig. 115) are each driven by a matched pair of fan belts. These belts are driven by a drive pulley attached to the rear end of each fan drive propeller shaft. A fifth belt, on the right side, drives the generator. Fan belts are tested and run-in under load and are serviced in matched pairs only.

b. **Adjust Fan Belt Tension.** Fan belt tension is adjusted by shifting the position of the drive pulley bracket on the fan shroud. Raise intake doors and loosen four drive pulley bracket stud nuts (fig. 115). Loosen jam nut on adjusting screw. Turn adjusting screw

Cooling System

in to increase tension until a finger force of 8 to 10 pounds, applied at the center of each belt, will deflect belt ¼ inch. It may be necessary to twist pulley support slightly on studs to equalize tension of upper and lower belts. Tighten stud nuts securely. Tighten adjusting screw jam nut. Recheck belt tension, and repeat adjustment procedure if required.

c. **Removal.** Disconnect fan drive propeller shaft from drive pulley (par. 158 b (1)). Lower shaft into out-of-way position. If removed belts are to be reinstalled, mark or tag all of them so they can be reinstalled in their original location. On right side, remove generator belt (par. 86 a). On right side, loosen or disconnect oil tubes at oil pump elbows (fig. 115) as required to prevent damaging tubes when pulley is tipped outward. Loosen jam nut on adjusting screw, and back off adjusting screw. Back off stud nuts until flush with ends of the two outer studs which attach pulley support to fan shroud. Remove nuts and fan belt shield from two inner studs. Pull pulley part way off studs and tip outward. Work belts off drive and fan pulley. NOTE: *Each belt has a wire core, so fully release belt tension before attempting to remove belts.*

d. **Installation.** When only one of a pair of fan belts is worn or damaged, the other belt in the set must be discarded also and two new fan belts installed. All belts removed and which are serviceable must be reinstalled in original locations. Loosen pulley and bracket. Place belts in grooves in fan pulleys and work belts into proper grooves in drive pulleys. Install all belts with arrows pointing in direction of belt travel. Place shield on inner studs (fig. 115), and slide drive pulley bracket against fan shroud. Tighten stud nuts finger-tight. Adjust fan belt tension (subpar. b preceding). On right side install generator drive belt (par. 86 b). On right side tighten compression fittings on oil tubes at oil pump elbows. Connect fan drive propeller shaft to drive pulley (par. 158 c (2)). Start and run engine; inspect all water and oil connections for leaks, and fan and generator belts for proper tension. Stop engine. Tighten stud nuts securely. NOTE: *Fan drive belts have wire cores. Do not attempt to stretch or force belts over pulleys when installing as this will break wires. Pulley brackets must be fully loosened and belts installed without stretching.*

139. COOLING UNIT.

a. **Description.** The cooling unit is a removable assembly consisting of the transverse housing expansion tank, baffle plates, fans, fan shrouds, radiators, and oil coolers (figs. 115 and 116). These units are suspended from the transverse casting and are removed as a complete unit from vehicle with all connecting tubes assembled.

TM 9-735
139

Part Three—Maintenance Instructions

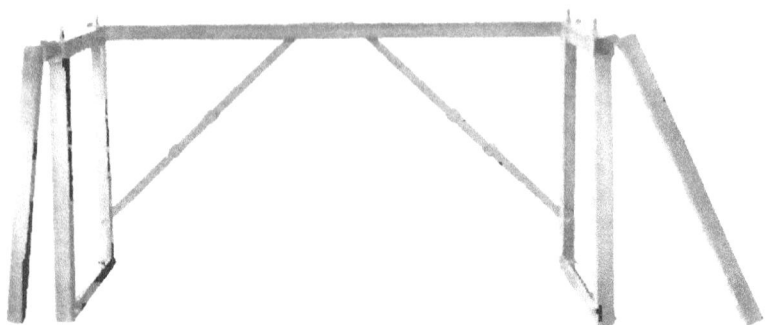

Figure 118 — Cooling Unit Stand (41-S-4942-300)

b. **Removal.** Procedure for removing the cooling unit from the vehicle as an assembly is given in the following steps:

(1) DRAIN RADIATORS. Traverse turret until gun is pointing 90 degrees left. Turn 24-volt master switch off. Drain engine cooling system (par. 134 b). Install drain plugs, but do not install drain plug covers.

(2) REMOVE CENTER DECK PLATES. Remove front and rear center deck plates with hoist, and lower to ground.

(3) REMOVE EXHAUST TUBES. Follow procedure in paragraph 131 b.

(4) DISCONNECT FAN DRIVE SHAFTS. Detach universal joint flanges from fan drive pulleys and lower shafts into out-of-way position (par. 158 b (1)). Remove generator drive belt from fan drive pulley (par. 86 a).

(5) DISCONNECT WATER AND OIL TUBES. Working through radiator drain plug openings, loosen hose clamps and slide hose connection off each radiator outlet elbow (fig. 115). At thermostat housings disconnect engine water manifold and auxiliary engine outlet hoses (fig. 115). Disconnect differential to oil pump tube at connector (fig. 115). Disconnect oil cooler to differential tube at connector at oil cooler. Unscrew connector at each side of transmission to disconnect oil tubes.

(6) REMOVE COOLING UNIT FROM VEHICLE. Disconnect both ends of transverse housing section of taillight wiring conduit. Remove two fillister-head screws from top of transverse housing, and install lifting eyes. Attach sling and suitable hoisting equipment to lifting eyes. Remove cap screws which attach each end of transverse housing to hull side plates. Operate hoist, and guide cooling unit up out of engine compartment. Carefully lower unit onto cooling unit

TM 9-735
139

Cooling System

stand (41-S-4942-300) (fig. 118). CAUTION: *Do not support cooling unit by resting it on fan shroud guide pins or radiator outlet elbows.*

c. **Installation.** Procedure for installing complete cooling unit in the vehicle is given in the following steps:

(1) CLEAN AND INSPECT COOLING UNIT. Clean surfaces of oil coolers and radiators by blowing out with compressed air. Wipe off tubes and hoses which show evidence of leaks to determine if they are badly damaged. Test fan pulley hubs for loose or worn bearings by working them from side to side and revolving fans by hand. Inspect all fan belts for wear or damage, and replace in matched pairs as required. Make sure all connections and gaskets are tight. NOTE: *Inspect other units and connections in the engine compartment while the cooling unit is removed.*

(2) POSITION COOLING UNIT IN VEHICLE. Raise cooling unit into position so it is centered above its installed position in the vehicle. Lower unit, slowly, with a man on each side as a guide, until fan shroud guide pins enter bushings on hull floor and upper baffle plate lines up with lower baffle plate. NOTE: *It may be necessary to pry transverse housing to rear to aline it with dowel pins in hull side plates.* Attach transverse housing to hull side plates. Detach hoist, remove lifting eyes, and install fillister-head screws. Make sure all seals are turned in right direction and are not damaged. Back off radiator tie rod nuts until rubber cushions are only slightly compressed, and tighten jam nuts. Connect both ends of transverse casting section of taillight wiring conduit.

(3) CONNECT WATER AND OIL TUBES. At expansion tank, slide hose connections into position on thermostat housings and outlet elbows. Tighten hose clamps. Working through drain plug openings, slide hose connections into position on radiator outlet elbows, and tighten clamps. Make sure drain plugs are tight, but do not install covers. Connect differential to oil pump tube, and tighten connector. Attach oil cooler to differential oil tube to oil cooler connector, and tighten connector. Connect transmission oil tubes to transmission housing, and tighten connector.

(4) CONNECT FAN DRIVE SHAFTS. Install and adjust generator drive belts (par. 86 b). Connect fan drive shafts to fan drive pulleys.

(5) INSTALL EXHAUST TUBES. Follow procedure in paragraph 131 c and install rear center deck plate.

(6) INSTALL FRONT CENTER DECK PLATE. Hoist front center deck plate from ground and lower into position on vehicle.

(7) COMPLETE AND INSPECT INSTALLATION OF COOLING UNIT. Fill cooling system (par. 134 d). Start and warm up engine. Carefully inspect all water, oil and exhaust connections and tighten as

required. Install radiator drain plug covers. Observe operation of fans and generator to make sure belts are properly installed. Make sure engine temperature gage indicates proper temperature. Stop engine. Test operation of taillights. Turn 24-volt master switch off. Check oil level in transmission and differential, and add oil as required. Fill cooling unit.

140. RADIATORS.

a. Description. Two cross-flow type radiators, identical in construction, are mounted on each side of the transmission (fig. 115). The radiator support or frame also supports two oil coolers mounted at the rear of each radiator. Each radiator with its two oil coolers can be removed from the fan shroud as an assembly, but only after the complete cooling unit has been removed from the vehicle. Coolant flows from the expansion tank into the tops of the radiators, down through the radiator cores, and is discharged from the outlet elbows at the bottom of the radiators to the water pump. Air is forced through the radiators by dual fans mounted in shrouds (fig. 115).

b. Removal. The procedure for removing radiators is given in the following steps:

(1) REMOVE COOLING UNIT. Follow procedure in paragraph 139 b.

(2) DISCONNECT WATER AND OIL TUBES. Disconnect top and bottom by-pass tube hose connections (fig. 115). Loosen hose clamps and slide hose connections over tubes. On right side disconnect differential oil pump hoses from oil pump connections (fig. 115). Remove fan drive belt shields. Disconnect water by-pass tube hose connections. Disconnect all oil cooler hoses and tubes from oil coolers. When removing either radiator, remove oil cooler elbows and install pipe plugs to prevent spilling oil from the oil coolers.

(3) DETACH RADIATOR ASSEMBLY FROM TRANSVERSE HOUSING. Block fan shroud from swinging outward. Remove locking wire bolts attaching fan shrouds to radiators. Remove clevis pins from upper right and left fan shroud mounting brackets. Remove both mounting brackets. Remove right and left tie rods from top of oil cooler side of assembly. Remove fan shroud, fans, pulley and fan drive belts as a complete assembly. Remove bolts attaching oil cooler to radiator mounting side plates. Remove oil coolers. NOTE: *Handle oil coolers carefully, so as not to spill oil trapped in coolers.* Remove drain plugs and drain oil coolers. Carefully remove rubber baffles.

(4) REMOVE FAN SHROUD FROM RADIATOR. See step (3) preceding.

(5) REMOVE OIL COOLERS FROM RADIATOR. Turn radiator over on blocks. Take out the cap screws which attach oil coolers to frame. Lift out oil coolers.

Cooling System

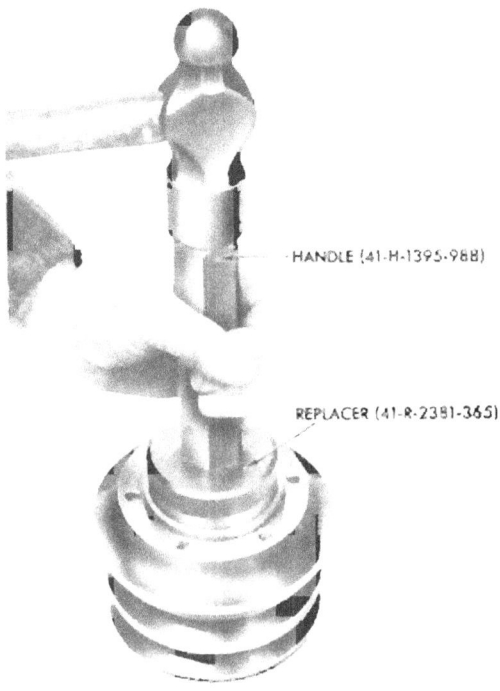

Figure 119 — Installing Fan Hub Inner Bearing

c. **Installation.** The radiators are installed according to procedure given in the following steps:

(1) ASSEMBLE OIL COOLERS TO RADIATORS. Place radiator, oil cooler side up, on blocks spaced to avoid damage to core. Position oil coolers in frame to aline bolt holes. Install cap screws, with lock washers, and tighten.

(2) ATTACH FAN SHROUD TO RADIATOR. Turn radiator over with oil coolers resting on blocks. Position rubber baffles. Place fan shroud on radiator. Install bolts and washers. Using new gasket, attach radiator outlet elbow with by-pass tube connection to radiator lower tank if removed. Tighten all fan shroud bolts evenly, and install locking wire.

(3) ATTACH RADIATOR ASSEMBLY TO TRANSVERSE HOUSING. Install support brackets on transverse housing. Raise radiator assembly into position under transverse housing, and insert support pins in brackets. Install both tie rods on upper right and left sides of oil

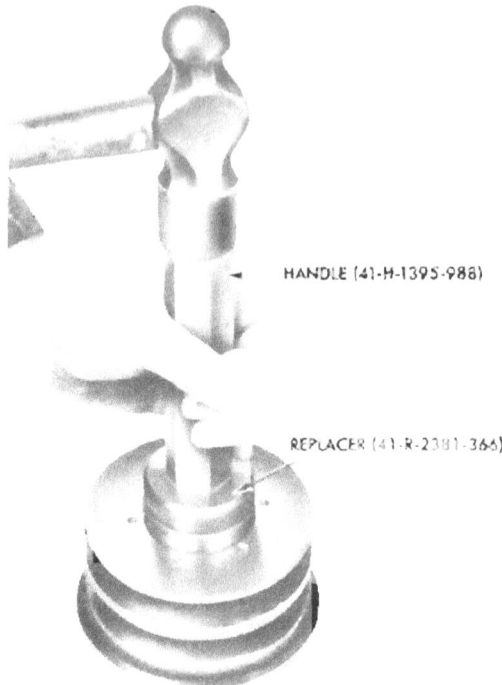

Figure 120 — Installing Fan Hub Outer Bearing

cooler. Adjust tie rod rubber bushings so radiator will hang in vertical position, and tighten jam nut.

(4) CONNECT WATER AND OIL TUBES. Connect all tubes and hoses to oil coolers. Install differential outlet and inlet oil pump connection hoses through hole in upper baffle plate, and connect to oil pump. Connect water by-pass tube hose connections to thermostat housings, and tighten all hose clamps.

(5) INSTALL COOLING UNIT. To install cooling unit follow procedure in paragraph 139 c.

(6) RECORD INSTALLATION OF COOLING UNIT. Make proper entry on W.D., A.G.O. Form No. 478, "M.W.O. and Major Unit Assembly Replacement Record."

141. FANS AND SHROUDS.

a. **Description.** Two fans, each driven by two matched belts from the drive pulley, force the air through each radiator (fig. 115).

TM 9-735
141

Cooling System

The fans, identical in construction, are attached to hubs which turn on ball bearings mounted on shafts in bosses in the fan shroud. Fan shrouds are rigid aluminum castings with air flow straightening vanes centering at the hub boss in each of the two fan openings. Air is drawn down through the intake doors, forced through the radiators, and out through the exhaust doors. Drive pulleys are mounted on adjustable supports attached to the inner edge of each fan shroud.

b. **Removal of Fans.** Fans and hubs can be removed with the cooling unit in the vehicle. Second echelon is permitted to remove bearings from fan hubs according to procedure in steps (2) through (4) below. If inspection reveals that the shaft is damaged, a new hub with shaft assembly will be installed according to procedure in subparagraphs d (1) through (3).

(1) REMOVE FAN BELTS. Remove fan belts from the hub (par. 138 c).

(2) REMOVE FAN HUB FROM SHAFT. Remove cover from hub. Release locking washer and using lug wrench (41-W-3734-200), unscrew nut from end of shaft. Lift off lock washer. Pull hub with fan off shaft. NOTE: *If fan blades strike shroud, remove fan from hub and lift fan out after hub has been removed. Use care to see that fan blades do not damage radiator cores.* Remove fan from hub.

(3) INSPECT FAN HUB BEARINGS AND SHAFTS. Carefully examine fan hub shft for damage or wear. If a new shaft is required it will be necessary to remove the fan shroud to remove the shaft (subpar. d (1) through (3) following) and install a complete new fan hub with shaft assembly. Inspect shroud and fan hub for evidence of grease leaks. If oil retainers are leaking, install new bearings. Inspect bearings to see that they are not worn or damaged. If new bearings are required, remove damaged bearings according to procedure in step (4) following.

(4) REMOVE FAN HUB BEARINGS. Place hub, outer face down on blocks spaced to clear bearing bore. Use drift with small flat point to drive out the outer bearing by driving on outer race. Reverse hub and drive out inner bearing. Clean interior of hub and wipe off shaft.

c. **Installation of Fans.** The procedure to install the fans when hub shafts have not been removed from the shrouds is covered in steps (1) through (4) following. When shafts have been removed from the shrouds refer to subparagraph e (2) through (4).

(1) INSTALL FAN HUB BEARINGS. Pack new bearings with grease and place approximately 1 cubic inch of grease in hub bore as instructed on Lubrication Order (par. 46). Start outer bearing squarely into bore with oil retainer to the top. Use replacer (41-R-2381-366) with handle (41-H-1395-988) (fig. 120), and drive bear-

ing in until it seats against shoulder in hub. Turn hub over and install inner bearing (fig. 119) in the same manner, using replacer (41-R-2381-365) with handle (41-H-1395-988).

(2) ATTACH FAN TO HUB. Place fan on hub so that, when hub is installed, right-hand or leading edge of the top blade, as viewed from the rear, will be closer to hub and away from radiator.

(3) INSTALL FAN HUB ON SHAFT. Slide hub onto shaft, taking care not to damage oil retainers as they pass onto ground surfaces on shaft. Place new lock washer on shaft with lug entered in keyway. Screw on nut and tighten securely with lug wrench (41-W-3734-200). Bend lip of washer down against flat of nut. Place new gasket on hub, and install cover.

(4) INSTALL FAN BELTS. Install fan belts according to procedure in paragraph 138 d.

d. **Removal of Fan Shrouds.** The procedure for removing the fan shrouds is given in the following steps:

(1) REMOVE RADIATORS. Follow procedure in subparagraph b (1) through (4) to remove radiators.

(2) REMOVE FAN BELTS. Loosen jam nut on fan drive pulley adjusting screw. Back off adjusting screw. Back off nuts until flush with ends of two outer studs which attach drive pulley bracket to shroud. Remove nuts and shields from the other two studs. Pull pulley part way off studs and tip outward. Work fan belts off drive pulley and fan hub.

(3) REMOVE FAN HUB AND SHAFT FROM SHROUD. From rear of fan shroud, remove cotter pin from fan shaft nut. Unscrew nut and remove plain washer. Remove fan shaft from shroud by tapping on end of shaft.

(4) REMOVE FAN DRIVE PULLEYS FROM SHROUD. Remove the two remaining nuts from studs, and lift off drive pulley assembly.

e. **Installation of Fan Shroud.** The procedure to install the fan shrouds is given in the following steps:

(1) INSTALL FAN DRIVE PULLEY. Place fan drive pulley assembly on attaching studs. Place plain washer and lock washer on the two outer studs, and screw nuts on until flush with ends of studs.

(2) INSTALL FAN HUB IN SHROUD. Insert fan hub shaft in shroud boss. From rear install plain washer and nut. Tighten nut securely while tapping lightly on center of hub cover to firmly seat shaft in shroud boss. Insert and spread new cotter pin.

(3) INSTALL FAN BELTS. Tip fan pulley outward and work fan belts over hub and into proper groove in pulley. Do not force belts. Install shield and nuts with lock washers on the other two studs, and tighten all nuts to hold, but not to lock pulley in place. Adjust fan belts acocrding to procedure in paragraph 138 h.

(4) INSTALL RADIATOR. To install radiators follow procedure in paragraph 140 c (1) through (3).

142. WATER MANIFOLDS.

a. Description. A tubular water manifold on each cylinder head carries the water from the cylinder head through tubes to the expansion tank (fig. 53). A tube connected to the magneto end of the "left" water manifold delivers hot water to the driver's compartment heater. A sending unit and switch is mounted in each manifold.

b. Removal. Drain approximately 5 gallons from radiator on side from which manifold is to be removed (par. 134 b). Remove front center deck plate. Remove hose clamp connection from air intake tube at air intake manifold (fig. 56). Loosen hose and swing end out of the way. When removing "left" manifold close both fuel tank shut-off valves and disconnect fuel tube at fuel pump (fig. 56). Disconnect driver's compartment heater water tube hose from end of manifold. Remove expansion tank hose from end of manifold. Remove manifold stud nuts. Disconnect sending and switch unit wires. Work manifold clear of cylinder head. Lift manifold off studs, and work it out to rear over transverse casting. Remove gaskets. Scrape gasket surfaces clean if gasket has been damaged in removing manifold.

c. Installation. Inspect water manifold to cylinder head gaskets. Use new gasket. Coat both sides of gasket with liquid joint and thread compound. Position gasket on studs. Position water manifold on studs (fig. 56). Install and tighten manifold nuts alternately to compress gaskets evenly. Install manifold to expansion tank hose. Position air intake tube, and install air box hose connection. Tighten all hose clamps. If "left" manifold was removed, connect fuel tube to fuel pump and driver's compartment heater water tube to end of manifold. Install sending and switch units. Fill cooling system (par. 134 d). Open both fuel shut-off valves. Start and run engine. Inspect all fuel, water, and air intake connections for leaks. Tighten as required. Stop engine. Install front center deck plate.

143. WATER PUMP.

a. Description. The water pump is mounted on the magneto end of the engine (fig. 52). It is a centrifugal-type pump driven from the crankshaft by a splined shaft.

b. Removal. To remove water pump drain both radiators (par. 134 b). Remove bulkhead cover. Remove drain plug, and drain water pump dry. Install water pump drain plug. Loosen hose clamps on both hose connections at water pump inlet-elbows (fig. 52). Pull hose connections off elbows. Remove the two safety nuts which attach the pump outlet elbow to cylinder block (fig. 52). Remove safety nuts which attach water pump housing to cylinder block. Pull pump straight forward until impeller shaft clears crankshaft. Lift pump out through bulkhead opening. Remove seal from cylinder block inlet opening.

c. **Installation.** Examine cylinder block inlet seal and water pump to cylinder block gasket. Use new seal and gasket if required. Coat both sides of seal and gasket with liquid joint and thread compound. Position seal and gasket on studs. Rotate impeller shaft as required to aline splines and install pump on attaching studs. Install and tighten safety nuts alternately to compress gasket and seal evenly. Fill cooling system (par. 134 d). Start and run engine, and inspect water pump connections for leaks. Tighten as required. Stop engine.

144. WATER TUBES AND HOSE CONNECTIONS.

a. **Description.** Preformed, seamless steel tubing of various diameters is used in the cooling system. The tubes have a raised projection around each end to provide more positive sealing for hose connections. Water tube hoses are made of three-ply fireproof hose. Hose connections are used instead of rigid steel tubes at locations where vibration or movement might cause tubes to fracture, and also to facilitate removal and installation of various units.

b. **Removal.** Loosen hose clamps and slide them down over hose. If joint is tightly sealed carefully raise edge of hose with screwdriver to loosen it before attempting to twist tube or hose loose. Remove all hose connections which have deteriorated, or show evidence of leaking, and install new hoses. Reseal or tighten clamps on hose connections which are in good condition but are leaking at the joints or connections. When disconnecting tubes, loosen both clamps and slide hose connection off one tube or fitting and over the other. Remove crushed or bent tubes which will restrict flow of water, or possibly develop leaks, and install new tubes.

c. **Installation.** Clean ends of tubes and inside of hoses to remove sealing compound if used parts are reinstalled. Coat contact surfaces on ends of tubes or inside of hoses with liquid joint and thread compound. Connect tubes by positioning hose connections with clamps in place so that each end of hose extends an equal distance over ends of tubes or fittings. Position hose clamps so they are centered between end of hose and raised ring on end of tube or fitting which is within the hose. Turn clamps so screws will be accessible when all units are installed. Position tubes so they will not interfere with or chafe against other parts. Tighten hose clamps securely but without excessive force which would cause them to cut into hose or crush tubes.

Section XXVIII

LUBRICATION SYSTEMS

145. ENGINE LUBRICATION SYSTEM.

a. **Description.** The engine is lubricated by a wet sump pressure lubrication system. The oil pump is located at the magneto end of the oil pan (fig. 52) and is accessible through the bulkhead opening. It draws oil from the sump in the engine oil pan and forces it under pressure through the oil filter and then to the internal distribution system, and to the moving parts of the engine. The oil returns by gravity to the oil pan sump. An electric low oil pressure signal switch and an oil pressure gage sending unit are installed on the lower right side of the right cylinder block. These units are connected to the low oil pressure warning signals and oil pressure gage on the instrument panel.

b. **Check Engine Oil Level.** Daily, or more frequently as required (par. 46) check engine oil level. Raise right rear intake door to reach oil level indicator. If engine has been stopped for 2 hours or more, start and run engine for 3 minutes for oil to heat. Stop engine and wait 5 minutes for oil to drain back into sump. Remove indicator, wipe dry and fully insert indicator into housing. Remove indicator and read oil level. Add engine oil, seasonal grade, as required to bring level up to "FULL" mark on gage as instructed in subparagraph d following.

c. **Drain Engine Lubrication System.** When draining to refill with new oil, drain while oil is hot. From below vehicle remove drain plug cover from hull floor (fig. 162). Wipe off oil pan around the drain plug. Position clean 8-gallon capacity receptacle under drain plug. Remove drain plug and drain oil pan dry. Thoroughly clean, reinstall, and tighten drain plug securely. Inspect drain plug gasket and use a new gasket if required. Make sure cover plate gasket is serviceable and install cover plate. NOTE: *If vehicle is to be left dry after engine oil has been drained, place tag to that effect on 24-volt master switch so engine will not be started.*

d. **Fill Engine Lubrication System.** Raise right rear intake door. Unlock filler cap (fig. 56) and pour 8 gallons of engine oil, seasonal grade, into filler neck. Check oil level (subpar. b preceding). Inspect filler cap gasket. Install new gasket if required. Install filler cap. Close and lock filler cap cover.

146. ENGINE CRANKCASE BREATHER.

a. **Description.** The crankcase breather is located above and between the two magnetos (fig. 52). Air drawn into the crankcase

TM 9-735
146

Part Three—Maintenance Instructions

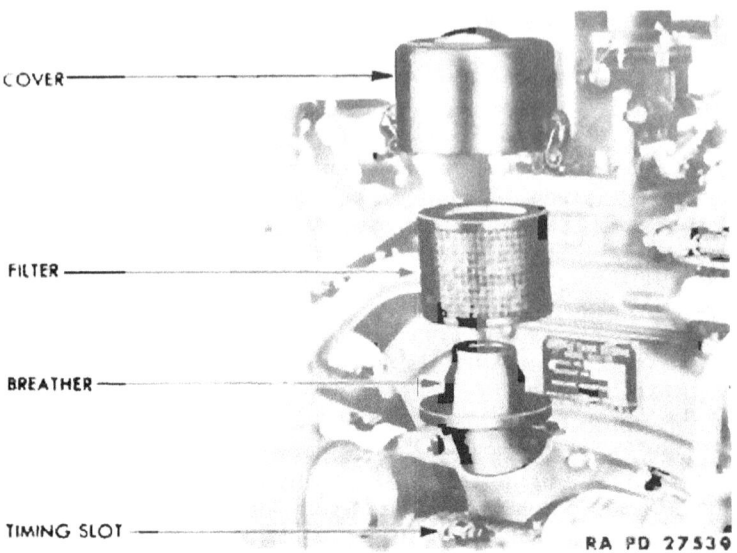

Figure 121 — Crankcase Breather

through the breather is filtered as it passes through the oil-soaked, metallic-wool element.

b. Service Engine Crankcase Breather.

(1) TO CLEAN ENGINE CRANKCASE BREATHER ELEMENT. Remove breather element (step (2) following). Slosh element in dry-cleaning solvent until clean. Drain until dry, then blow out with compressed air. Oil element (step (2) following) and install.

(2) TO OIL ENGINE CRANKCASE BREATHER ELEMENT. Remove bulkhead cover. Release the two catches (fig. 52) and lift breather cover off and remove through bulkhead opening. Lift off filter element (fig. 121). Inspect element and, if dirty or clogged, clean (step (1) preceding). To oil element immerse in clean or used engine oil, crankcase grade, for several minutes. Remove element and allow it to drain thoroughly, then install. Install breather cover, and securely fasten both catches.

c. Removal of Engine Crankcase Breather. Remove bulkhead cover. Remove nuts which attach breather base to accessory drive gear cover. Lift breather off studs, and remove through bulkhead opening.

Lubrication Systems

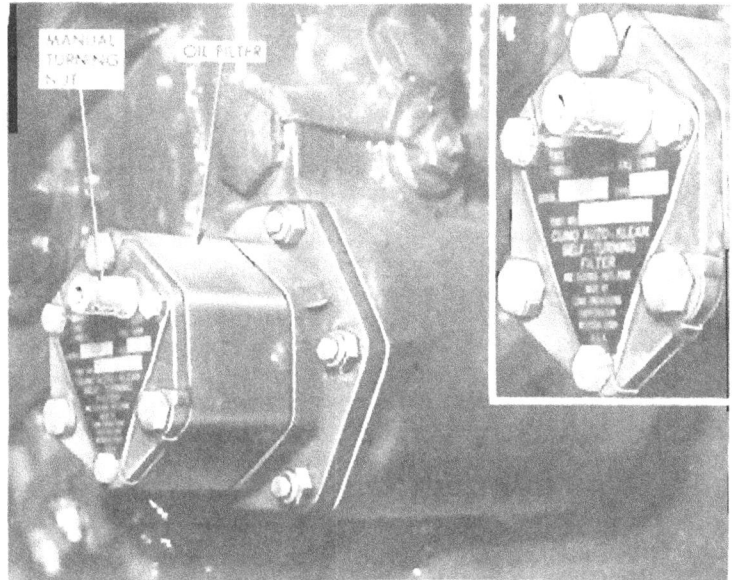

Figure 122 — Oil Filter

d. **Installation of Engine Crankcase Breather.** Inspect breather cover seal, and install new seal if required. Position seal in cover opening. Position breather on cover studs, install and tighten nuts.

147. ENGINE OIL FILTER.

a. **Description.** The engine oil filter is enclosed in a housing in the magneto end of the oil pan on the right side (figs. 52 and 122). It is a self-cleaning type operated by a small built-in hydraulic motor which is driven by engine lubricating oil pressure. Filter disks are mounted on a spindle which is driven by the filter motor. The disks pass between cleaning blades, which remove dirt and solids deposited on the outer edges of the disks. The flow of oil through the oil filter is controlled by a pressure relief valve and pressure regulator, or valve, in the oil filter housing inside the oil pan.

b. **Test Engine Oil Filter Operation.** Test engine oil filter each day to make sure it is operating properly. Remove bulkhead cover. Remove locking wire and remove manual turning nut (fig. 122). Reverse nut and install it finger-tight on shaft. Make pencil index mark on top flat of nut. Start and run engine at speed required to provide 30 pounds oil pressure. Observe manual turning nut. If it makes one revolution in five minutes it indicates filter is operating properly. If nut does not turn, or does not make a complete revolution within the five-minute period, filter is not operating properly. Remove filter (subpar. d following). Clean, inspect and install new filter if necessary. If filter is operating properly, stop engine. Remove manual turning nut, reverse nut and install in original position. Secure nut with locking wire through nut and clip.

c. **Clean Engine Oil Filter.** Remove oil filter (subpar. d following). Remove locking wire and remove manual turning nut. Reverse nut and install it on shaft. Slosh element in dry-cleaning solvent and rotate disks by turning manual turning nut with fingers. Submerge and agitate element until all dirt and solids are removed. When thoroughly clean allow element to drain dry. CAUTION: *Do not use wire brush on element and do not blow element dry with compressed air as doing so will crimp disks.* Thoroughly clean filter housing by wiping inside with cloth soaked in dry-cleaning solvent. Remove manual turning nut and install in original position. Secure nut with locking wire through nut and clip. Install oil filter (subpar. e following).

d. **Removal of Engine Oil Filter.** Remove bulkhead cover. Remove engine oil filter from oil pan. Carefully withdraw filter from housing so as not to damage disks or blades. Wrap it in cloth to prevent oil dripping as it is removed from engine.

e. **Installation of Engine Oil Filter.** Inspect gasket and install new one if required. Position gasket on studs. Install oil filter. Tighten nuts alternately to compress gasket evenly. Start and run engine and inspect oil filter for leaks. Reverse manual turning nut to test action of filter. Upon completion of test, stop engine and install nut in correct position. Secure nut with locking wire.

148. ENGINE OIL PAN.

a. **Description.** The oil pan is a heavy aluminum casting attached to the base of the crankcase. Legs, cast in the sides of the pan (fig. 62), support the engine when mounted in the engine compartment. The oil pump and oil filter are mounted at the magneto end of the oil pan (fig. 52), and the starter at the flywheel end of the "right" side (fig. 54). Gaskets seal the joints between oil pan, cylinder block and flywheel housing. Remove power unit (par. 71) to remove oil pan.

149. ENGINE OIL PUMP.

a. **Description.** A gear-type oil pump is mounted on the "right" side of the oil pan at the magneto end (fig. 52). It is driven from the lower camshaft worm gears by means of a splined quill. Oil is

Lubrication Systems

drawn by the pump, through a screen, from the sump in the base of the oil pan. From the pump it is forced into the oil filter where the flow is controlled by a pressure relief valve and pressure regulator valve in the oil filter housing.

b. **Removal.** Drain oil pan (par. 145 c). Remove engine compartment bulkhead cover (par. 181 b). Remove the four oil pump mounting nuts and remove the oil pump and gasket. NOTE: *It may be necessary to lightly tap the body of the pump to release it from the gasket and driving quill.*

c. **Installation.** Place new gasket on pump. Insert pump in oil pan. If necessary, turn engine by hand (fig. 66) to aline driving quill with pump shaft and lightly tap pump into place on studs. Install the four attaching nuts and tighten them evenly to compress gasket uniformly. Fill engine oil pan (par. 145 d). Start and run engine and inspect for leaks. Stop engine. Install engine compartment bulkhead cover (par. 181 c).

150. ENGINE OIL PRESSURE GAGE AND WARNING SIGNALS.

a. **Description.** The electric-type engine oil pressure gage on the instrument panel (fig. 16) is connected to a pressure-operated sending unit on the lower right side of the right cylinder block (fig. 89). A hose leading from the engine lubrication system supplies oil to the electrical sending unit which operates the gage on the instrument panel. The low oil pressure warning signals, which light when engine lubricating oil pressure falls below normal safe operating pressure (11 pounds), are operated in the same way by the low oil pressure switch mounted on the lower right side of the right cylinder block.

b. **Removal of Oil Pressure Gage.** Remove instrument panel face plate (par. 97 d). Note wire numbers and remove them from terminals. Lift off fiber insulator. Remove bracket nuts from terminals and lift off support bracket and black fiber insulator. Remove gage with rubber gasket from front of panel.

c. **Installation of Oil Pressure Gage.** Install gage in instrument panel by reversal of removal procedure in subparagraph b preceding. Connect wires to proper terminals (fig. 86). Make sure gasket is centered under face of gage before tightening bracket. Install instrument panel face plate (par. 97 e). Start and run engine to test action of gage.

d. **Removal of Oil Pressure Gage Sending Unit.** Turn 24-volt master switch off. Remove conduit shielding cap and wire from sending unit. Cover opening in tee to prevent entrance of dirt and loss of oil.

e. **Removal of Low Engine Oil Pressure Warning Signal Switch.** Follow procedure which applies to removal of switch in subparagraph d preceding.

f. **Installation of Low Engine Oil Pressure Warning Signal Switch.** Follow procedure in subparagraph g following as applies to installation of switch. See figure 89 for wiring diagram.

g. **Installation of Oil Pressure Gage Sending Unit.** Screw unit into fitting. Connect wire to unit and attach conduit shielding cap. Start and run engine, and observe action of instrument. Inspect all oil connections for leaks, and tighten as required. Stop engine.

h. **Removal of Low Engine Oil Pressure Warning Signal Lamps.** Turn 24-volt master switch off. Use screwdriver to raise edge of red jewel (fig. 16), and remove jewel. Press in and turn to left to remove lamp.

i. **Installation of Low Engine Oil Pressure Warning Signal Lamps.** Install new lamp, and test by turning 24-volt master switch on. Turn 24-volt master switch off and snap red jewel into place.

151. TRANSMISSION LUBRICATION SYSTEM.

a. **Description.** The transmission, including the torque converter and planetary reduction gears, is lubricated by oil pumped from a sump in the transmission housing. It is delivered under pressure and/or by gravity to the moving parts of the unit. A screen in the right side of the transmission housing filters the oil before it is drawn into the built-in oil pump. The oil is forced under pressure through the two oil coolers mounted at the rear of the left-hand radiators (fig. 116) to dissipate the heat generated by torque converter and transmission. Dual combination low oil pressure and high oil temperature warning signals on the instrument panel (fig. 16) are connected to two sending units on the transmission (fig. 132).

b. **Checking Transmission Oil Level.** Place vehicle on level ground before checking oil level. If possible, check oil level while oil is at operating temperature of 180° F (hot) (step (1) following). If transmission has not been previously operated, check oil level as outlined in step (2) following.

(1) CHECK OIL LEVEL (HOT). Run engine at idling speed with transmission in neutral. Remove filler cap with oil level indicator, wipe indicator, insert fully and secure cap. Remove indicator and read oil level at "HOT FULL" mark. Add engine oil SAE 10 to bring level to "HOT FULL" mark if necessary. Install cap with oil level indicator.

(2) CHECK OIL LEVEL (COLD). Remove oil level indicator and note if transmission is supplied with oil. If transmission has not been operated within the previous 30 minutes, run engine for a few min-

utes with transmission in neutral to fill torque converter. Stop engine and wait 3 minutes for oil to settle into sump. Wipe indicator before inserting in transmission. Read oil level at the "COLD FULL" mark and add engine oil SAE 10 to bring oil level to "COLD FULL" mark if necessary. Install cap with indicator.

c. **Drain Transmission Lubrication System.** If drained oil is to be discarded and new oil used, drain only when oil is hot. If drained oil is to be used again oil can be drained cold. From below vehicle remove drain plug cover (fig. 162). Wipe off transmission case around drain plug. Position clean 8-gallon capacity receptacle under drain. Remove drain plug and allow transmission to drain dry. NOTE: *Since approximately 23½ quarts of oil will be trapped in the flywheel housing, oil coolers and tubes, only 31 quarts of oil can be drained off. Trapped oil can be drained only by removing and draining individual units when the cooling unit is removed.* Clean drain plug. Install drain plug, making sure it is tight. Inspect drain plug cover plate gasket; use new gasket if required. Install drain plug cover.

d. **Fill Transmission Lubrication System.** Fill transmission with 31 quarts (or more as required) of new engine oil, after screen has been cleaned (subpar. e following). Remove filler cap with bayonet gage attached (C, fig. 46). Pour oil into filler opening, and then check oil level with bayonet gage (subpar. b preceding).

e. **Clean Transmission Oil Screen.** Clean transmission oil screen after transmission has been drained, and before filling. Remove screws and carefully lift screen out of transmission case (fig. 124). Wipe off accumulated sediment, and thoroughly wash screen by sloshing it in dry-cleaning solvent until clean. CAUTION: *Do not use wire brush on screen.* Blow screen dry with compressed air at reduced pressure. Wipe out inside of screen cover gasket, using new gasket if required. Position gasket on transmission case and insert screen. Install and tighten evenly to compress gasket uniformly. Fill transmission with oil (subpar. d preceding).

152. DUAL TRANSMISSION OIL PRESSURE AND TEMPERATURE WARNING SIGNALS.

a. **Description.** Dual red warning signals on the instrument panel (fig. 16) are connected to two sending units on the transmission (fig. 132). These combination warning signals light when the transmission oil pressure drops below safe operating pressure, or when oil temperature rises above normal safe operating temperature. The oil *temperature* warning signal switch is mounted in the outlet elbow at the right side of the transmission (fig. 132). The oil *pressure* warning signal switch is mounted on top of the transmission near the shifter lever (fig. 132).

b. **Removal of Transmission Oil Pressure and Temperature Warning Signal Lamp.** Follow same procedure as in paragraph 150 h.

c. **Installation of Transmission Oil Pressure and Temperature Warning Signal Lamp.** Follow procedure in paragraph 150 i.

d. **Removal of Transmission Oil Pressure Warning Signal Switch.** Turn 24-volt master switch off. Raise left front exhaust door. Remove shield and conduit shielding cap and disconnect wire from switch at top of transmission (fig. 132). Unscrew switch unit and cover opening to prevent entrance of dirt.

e. **Installation of Transmission Oil Pressure Warning Signal Switch.** Screw switch into transmission case until tight. Connect wire and attach conduit shielding cap to sending unit. Install shield. Start and run engine with transmission in neutral. Observe action of warning signal, and inspect switch for oil leaks. Tighten switch as required. Stop engine.

f. **Removal of Transmission High Oil Temperature Warning Signal Switch.** Follow procedure in subparagraph d preceding (fig. 132).

g. **Installation of Transmission High Oil Temperature Warning Signal Switch.** Follow procedure in subparagraph e preceding (fig. 132).

153. DIFFERENTIAL LUBRICATION SYSTEM.

a. **Description.** The oil pump is built into the right-hand fan drive pulley and is attached to the inside of the differential cooling unit shroud (fig. 115). It lubricates the moving parts of the differential by either splash and or gravity lubrication systems. Oil is drawn through screens from the sump in the base of the differential housing and forced through the right-hand oil coolers. It then returns to the top of the differential, from which point it is delivered through oil manifolds to the moving parts of the differential.

b. **Check Differential Oil Level.** Place vehicle on as level ground as possible. Lift out bayonet gage at front of differential (E, fig. 46). Wipe dipstick clean and fully insert in housing. Withdraw it and read oil level. Add engine oil, seasonal grade as required, to bring level up to "FULL" mark. Reinstall gage.

c. **Draining Differential Lubrication System.** If oil drained is to be discarded and new oil used, drain only when oil is hot. If drained oil is to be used again oil can be drained cold. From below vehicle remove drain plug cover from hull floor (fig. 162). Wipe off differential case around drain plug. Position clean 15-gallon capacity receptacle under drain plug. Remove drain plug, and allow differ-

Lubrication Systems

ential to drain dry. Thoroughly clean drain plug and install making sure plug is tight. Inspect drain plug cover gasket, using new gasket if required. Install drain plug cover.

d. **Fill Differential Lubrication System.** Fill differential lubrication system with 60 quarts of new engine oil, seasonal grade, after screens have been cleaned (subpar. e following). Pour oil into right brake adjusting nut port in differential cover. Check oil level (subpar. h preceding).

e. **Clean Differential Oil Screens.** Clean both differential oil screens after oil has been drained, and before filling. On right side disconnect differential oil pump suction tube. Unscrew compression fitting at lower end of outlet tube. Remove oil pump suction tube elbow fitting from differential case. Carefully withdraw screens. Wipe accumulated dirt from screens. Clean screens by sloshing them in dry-cleaning solvent. CAUTION: *Do not use wire brush on screens.* Blow screens dry with compressed air at reduced pressure. Wipe inside of screen housing clean with cloth soaked in dry-cleaning solvent. Inspect screen flange gaskets, and use new gaskets if required. Coat one side of gasket with liquid joint and thread compound. Position gasket on screen flange. Install screens in housing. Install oil pump suction tube elbow fitting. Tighten screws alternately to compress gasket evenly. Install oil pump suction tube in elbow. Fill differential lubrication system (subpar. d preceding). Start engine and inspect system for leaks. Tighten connections if necessary. Stop engine.

154. DIFFERENTIAL OIL PUMP.

a. **Description.** Lubricating oil for the differential is circulated by a gear type pump built into the right fan drive pulley hub (fig. 115). The pump is driven by the left fan drive propeller shaft. The pump draws oil from a sump in the base of the differential housing, and forces it through the differential oil coolers (fig. 116) and back to the differential. Pump and drive pulley are removed and serviced as an assembly.

b. **Removal.** Remove all four fan belts and generator drive belt from right fan drive pulley (par. 86 a). Remove outer nuts attaching drive pulley bracket to fan shroud studs. Disconnect oil tubes from pump. If it is necessary, disconnect radiator by-pass tube at top (fig. 115), so it can be swung inward to clear pump elbows, first drain approximately 5 gallons of coolant from right radiator (par. 134 h). Slide pump and pulley assembly off studs and out from behind by-pass tube.

c. **Installation.** Position oil pump and pulley assembly on fan shroud studs, making sure oil tubes squarely enter elbows. Screw on the two outer stud nuts with flat washers and lock washers until nuts are flush with end of studs. Install other nuts, with lock washers. finger-tight. Install the four fan belts (par. 86 b), and adjust fan belt tension (par. 86 c). Tighten all nuts securely. If radiator by-pass tube was disconnected, connect tube and fill cooling system. Connect oil tubes to pump.

155. OIL COOLERS.

a. Description. Two cross-flow type oil coolers are mounted at the rear of each radiator (fig. 116). They are cooled by the air forced through the radiators by the fans. The transmission oil circulates through the two left-hand oil coolers which are connected together by a system of oil tubes and hoses. The differential oil circulates through the two right-hand oil coolers (fig. 116).

b. Removal. Remove cooling unit (par. 139 b). Disconnect and or remove oil tubes as required to permit the individual oil cooler to be removed. Support cooler and remove attaching cap screws. Lift cooler out of radiator frame.

c. Installation. Position oil cooler in radiator frame, and aline bolt holes. Install attaching screws. Connect oil tubes previously removed or loosened. Install cooling unit (par. 139 c). Start engine, and operate with transmission in neutral. Examine all oil and water connections for leaks. Tighten connections as required. Stop engine.

156. OIL TUBES AND HOSES.

a. Description. The external oil tubes are made of preformed seamless steel tubing with straight ends to receive the compression fittings or hose connections. Flexible hoses and hose connections used in the lubrication system are fireproof and oil-resistant. Flexible hoses with fittings permanently attached are used instead of rigid steel tubes where vibration or movement might cause a tube to fail. Hose connections are also used for the same purpose, and to facilitate removal and installation of the steel tubes.

b. Removal of Oil Tubes and Hoses. The general instructions for removing oil tubes and hoses are given in the following steps:

(1) REMOVE OIL TUBES AND FLEXIBLE HOSES. Use wrenches with snug fit to hold coupling or elbow from turning while loosening fittings at both ends. CAUTION: *Never use a pipe wrench on these fittings.* Unscrew compression fitting and pull hose or tube straight out of coupling or elbow. In many cases it is advisable to tag one end of hose or tube to facilitate installation in original position, and thus assure more positive leakproof connections. Cover ends of tubes or hoses, and openings or fittings from which they are removed, with tape to prevent entrance of dirt and loss of compression fittings.

Lubrication Systems

Tubes which are bent, crushed, dented, or squeezed so as to either restrict oil flow or develop leaks, must be removed and new tubes installed.

(2) REMOVE HOSE CONNECTIONS. To disconnect hose connections, loosen both hose clamps and slide hose connection off one of the tubes and over the other. To remove hose connections disconnect hose and either swing tube to one side, or remove it to permit hose connection to be pulled off tube. Tighten hose connections at joints which show evidence of leaking, or reseal joint. Remove all hose connections which have deteriorated or have developed leaks, and install new connections of same diameter and length.

c. **Installation of Oil Tubes and Hoses.** The general instructions for installing oil tubes and hoses are given in the following steps, which apply to all similar lines and connections in the vehicle.

(1) INSTALL OIL TUBES AND FLEXIBLE HOSES. Before installing oil tubes or flexible hoses, examine them for evidence of leaks and then wipe them clean. Make sure couplings or fittings are clean and threads are not damaged. Coat threads of fittings with liquid joint and thread compound. With couplings on tube, enter both ends of tube squarely in fitting or elbow. Screw couplings on finger-tight. Position tube and attach clips so tube does not contact or chafe against any object. Use wrenches with snug fit to hold fittings or elbows from turning. Tighten coupling nuts securely but without excessive force. Follow same procedure to install flexible hoses. Test connections for leaks by operating units, and tighten connections as required.

(2) INSTALL HOSE CONNECTIONS. If replacement hose connections are not available make sure bulk hose has same characteristics and specifications, and is of the same diameter. When connecting or installing hose connections, remove connection and clean all sealing compound from ends of tubes and from the inside of hose connections to be reinstalled. Coat inside of hose connections along length in contact with tubes or fittings with liquid joint and thread compound. Inspect clamps and install new clamps as required on hose connection. Position hose connection on ends of tubes or fittings so they extend an equal distance inside hose connection. Position clamps on hose connections so they are equally spaced between end of hose and end of tube or fitting extending into hose. Turn screws so they are accessible when units have been installed. Tighten screws securely but without excessive force which would cause clamp to cut into hose or crush tube within hose. Test all connections for leaks by operating units and tightening clamps as required.

Section XXIX
ANGLE DRIVES AND FAN DRIVE PROPELLER SHAFTS

157. ANGLE DRIVES.

a. **Description.** Power to drive the fans, generator, and differential oil pump is supplied by the main engine through the angle drives (figs. 52 and 53) and fan drive propeller shafts on each side of the engine. The angle drives are driven from the crankshaft through the accessory drive gear train inside the engine at the magneto end (fig. 52). The angle drives transmit power to the fan drive propeller shafts at the correct angle so that the shafts are connected in straight lines with the fan drive pulley hubs.

b. **Removal.** The angle drives are removed by the procedure outlined in the following steps, in the order given below:

(1) REMOVE POWER UNIT. Follow procedure in paragraph 71.

(2) REMOVE FAN DRIVE PROPELLER SHAFT. Remove bolts which attach propeller shaft to angle drive flange and lift off shaft.

(3) REMOVE ANGLE DRIVE. Support angle drive housing, and remove nuts which attach angle drive to engine. Pull angle drive straight out until drive shaft has cleared internal bevel gear.

c. **Installation.** The installation of the angle drives is described in the following steps:

(1) INSTALL ANGLE DRIVES ON ENGINE. Place new gasket on engine studs, if inspection shows old gasket is not serviceable. Raise angle drive into position. Rotate drive flange to aline splines on drive shaft with internal bevel gears. Slide assembly into position on studs. Install and tighten nuts alternately to compress gasket uniformly.

(2) CONNECT FAN DRIVE PROPELLER SHAFT TO ANGLE DRIVE. Follow procedure in paragraph 158 c (1) following.

(3) INSTALL POWER UNIT. Follow procedure in paragraph 72.

158. FAN DRIVE PROPELLER SHAFTS.

a. **Description.** A tubular propeller shaft (fig. 53) connects each angle drive with the fan drive pulley. A universal joint at each end and a slip joint at the rear end of each shaft are provided to accommodate difference in angularity and any variation in distance between angle drive and fan drive pulley. NOTE: *Fan drive propeller shafts are not to be disassembled. If a universal joint becomes defective, replace complete shaft assembly.*

b. **Removal.** The removal of fan drive propeller shafts is performed as described in the following steps:

(1) DISCONNECT FAN DRIVE PROPELLER SHAFTS. With power unit in the vehicle the fan drive propeller shafts can be disconnected at the rear ends only. Remove locking wire from heads of attaching bolts. Remove nuts, bolts and/or cap screws, and tie or support propeller shaft in out-of-way position.

(2) REMOVE FAN DRIVE PROPELLER SHAFT. To remove either or both fan drive propeller shafts, first remove power unit from vehicle (par. 71). Disconnect both ends of shaft by following procedure in step (1) preceding.

c. **Installation.** The fan drive propeller shafts are installed as outlined in the following steps, in the order given:

(1) CONNECTING FAN DRIVE PROPELLER SHAFT TO ANGLE DRIVE. Make sure that arrow on rear universal joint sleeve is alined with arrow on propeller shaft. Thoroughly lubricate universal joints and slip joint (par. 46). Raise shaft to aline bolt holes in flange with angle drive flange. Insert bolts with heads to vehicle rear. Install nuts and tighten alternately and securely. Install power unit (par. 72).

(2) CONNECTING PROPELLER SHAFT TO FAN DRIVE PULLEY. Raise shaft and turn fan to aline bolt holes in drive pulley. Install bolts and tighten securely. Thread locking wire through each bolt head, and form a tight loop by twisting ends of wire together.

Section XXX

TRANSMISSION

159. **TRANSMISSION.**

a. **Description.** The transmission housing is bolted with the front to the engine and the rear to the differential case (figs. 62, 124 and 125). The transmission consists of the planetary reduction drive gears which are driven by the flywheel, the torque converter which serves as a fluid clutch and a torque multiplier, and the transmission planetary gears which provide the three forward and one reverse gear ratios. Oil for the converter, complete lubrication of transmission, and hydraulic control of bands and clutches is furnished by two pumps—one driven by the engine and the other driven by the transmission output shaft. Either pump is capable of furnishing a sufficient volume of oil to operate the transmission under certain conditions. The rear pump, driven by the output shaft, does *not* furnish the volume necessary to operate the transmission (with the engine dead) until a vehicle speed of approximately 18 miles per hour is attained. With engine idling, the front pump furnishes sufficient volume of oil to operate the transmission. Under normal conditions and with the vehicle in motion, both pumps function in unison to operate the transmission. Pistons operated by hydraulic pressure actuate the transmission bands and clutches to obtain the desired transmission gear ratios. The action of these pistons is manually controlled by the speed range selector lever (fig. 127) connected by a control rod to the control shaft lever and valve mechanism on top of the transmission. A bevel drive pinion at the rear end of the transmission output shaft meshes with the differential ring gear. Since proper adjustment of the pinion and ring gear requires special gages and tools not available to the second echelon, it will be necessary to notify higher authority if the transmission is to be separated from the differential. Only in emergencies is the second echelon permitted to separate the transmission from the engine and then only on the order of higher authority.

b. Adjust Transmission Bands. The procedure for adjusting the second speed band and the third and reverse band is given in the following steps:

(1) GENERAL. The *first* band adjustment should be at 1,500,000 engine revolutions and every 15,000,000 engine revolutions thereafter. However, the bands should be adjusted regardless of engine revolutions if malfunction of the transmission indicates an adjustment is necessary. NOTE: *A band adjustment should be performed on all new or rebuilt transmissions after being operated three or four miles in the vehicle.*

(2) ADJUSTMENT OF TRANSMISSION BANDS. Torque both band adjusting screws to approximately 50 foot-pounds, then back off each screw three-quarters of one turn (fig. 124). CAUTION: *To avoid disconnecting either band adjusting screw from band strut, do not back out screw more than two turns. Otherwise the entire transmission will have to be disassembled in order to insert the band strut in adjusting screw socket.* NOTE: *No adjustments to the transmission, other than the adjustment of the bands, and manual control linkage can be made by the using arms. Should any abnormal conditions develop, notify higher authority.*

c. Test Transmission Oil Pressure. Operate transmission in neutral for five minutes to warm oil. Remove either pipe plug from transmission oil pump fitting located on top of transmission end cover directly behind the valve body assembly (fig. 123). Install adapter in opening. Install copper tube in end of adapter and connect pressure gage (41-G-446) to other end of tube (fig. 123). Set parking brake and run engine at 500 revolutions per minute with speed range selector lever in first speed range while pressure is being checked. (Gage should read at least 85 pounds. Operating range is 85 to 105 pounds.) If pressure is less than 85 pounds, notify higher authority. Stop engine. Remove gage and adapter. Install and tighten plug.

TM 9-735
160

Transmission

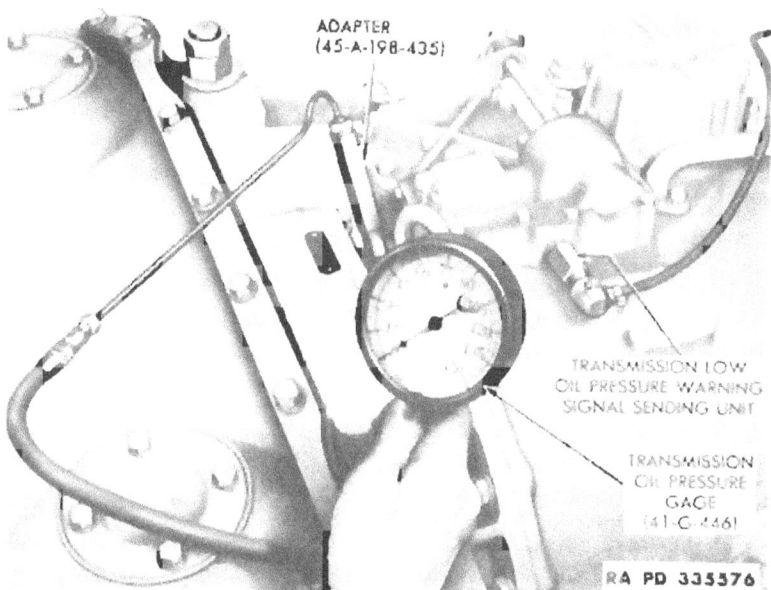

Figure 123 — Testing Transmission Oil Pressure

160. SPEED RANGE SELECTOR CONTROLS.

a. **Description.** The speed range selector controls (fig. 127) consist of the dual speed range selector levers and linkage in the drivers' compartment, speed range lever control rod, brackets, and a speed range shifter lever and a shifter lever shaft connected to the valve operating mechanism on top of the transmission housing (figs. 125 and 127). For a complete description of the operation of the speed range selector levers and controls, see paragraph 14 d.

b. **Adjustment of Speed Range Selector Controls.** When any part of the speed range selector control system has been removed and a new part installed, or when transmission does not respond properly to the movement of the selector lever, it will be necessary to adjust the speed range selector controls. In drivers' compartment disconnect speed range selector lever pull rod and front end of control rod (fig. 127). In engine compartment disconnect rear end of control rod from shifter lever (fig. 127) on transmission housing. Pull and push on end of rod to see that it does not bind and has full travel. Operate shifter lever by hand to make sure it seats properly in each speed range position and does not bind. Move shifter lever all the way to the rear into THIRD position. Adjust position of clevis on rod so threaded end of rod is flush with inside of clevis. Tighten jam nut. Attach clevis to shifter lever with lever still in THIRD position. In drivers' compartment, adjust position of clevis on front end of rod so rod can be connected to bell crank with clevis pin. Tighten jam nut. Loosen jam nut on clevis at end of selector lever pull rod. Place selector lever in THIRD position. Adjust and connect pull rod and control rod. Tighten clevis jam nut and oil linkage. Operate speed range selector lever in all speed ranges to test action of linkage.

c. **Removal of Speed Range Selector Controls and Linkage.** The procedure to remove the speed range selector controls and linkage is given in the following steps:

(1) REMOVE SPEED RANGE SELECTOR LEVER ASSEMBLY. Remove stowage boxes and speed range selector quadrant. Remove instrument panel (par. 97 b). Remove speed range selector lever retractor spring. Disconnect throttle pull rod from left throttle lever (fig. 127). Disconnect speed range selector lever linkage from levers on cross shaft. Lift out levers and linkage. Remove plunger and spring from center of left lever. NOTE: *It is not necessary to remove guide rod attaching the two levers.* Remove bolts attaching cross shaft, throttle levers, and linkage brackets to right and left instrument panel supports. Move assembly as necessary until an end clears inside edge of panel support. Pull assembly at an angle between panel supports into assistant driver's compartment. NOTE: *Escape doors can be dropped and personnel can stand in escape opening to facilitate handling of assembly.* Pass assembly up through assistant driver's opening. As assembly is lifted it may be necessary to move throttle levers to clear supports and cross brace. Remove assembly from vehicle. Instrument panel supports are attached to pads on hull floor by three bolts and lock washers on each side. To remove supports, remove bolts and lift supports into assistant driver's compartment. Pass supports into fighting compartment, and then out of vehicle through commanders' hatch.

(2) REMOVE SPEED RANGE SELECTOR LEVER CONTROL ROD. Remove ammunition stowage racks from left stowage compartment (fig. 166). Remove bulkhead cover. Disconnect control rod at clevis in engine compartment. Remove control rod bulkhead fitting or sleeve. Disconnect rod at front connection in driver's compartment. Pull rod back into engine compartment until front end of rod can be raised clear of front wall of 90-mm ammunition stowage box. Remove rod from vehicle.

d. **Installation of Speed Range Selector Controls and Linkage.** The procedure to install the speed range selector controls and linkage is given in the following steps:

(1) INSTALLATION OF SPEED RANGE SELECTOR LEVER ASSEMBLY. If instrument panel supports have been removed, enter sup-

ports and cross brace, less linkage, into vehicle through loader's hatch and then into position on pads on hull floor. Work speed range selector levers and throttle linkage into position by first passing left end of assembly at an angle between the panel supports from the assistant driver's side. Install spring and plunger in hole in left selector lever on cross shaft. Install speed range selector lever linkage in cross shaft levers. Remove knob from left speed range selector lever, and install speed range quadrant and oddment tray. Install selector lever knob. Connect and adjust selector lever linkage (subpar. h preceding). Connect and adjust throttle pull rod (par. 126 h). If desired, drop escape doors to provide additional working space.

(2) INSTALLATION OF SPEED RANGE SELECTOR LEVER CONTROL ROD. Insert rear end of rod through seal in bulkhead. Carefully work rod to rear until front end clears front wall of left 90-mm ammunition stowage compartment. Lower front end of rod and slide rod forward and connect with speed range selector lever linkage. Connect clevis on rear end of rod to speed range selector lever rod clevis in engine compartment. Install bulkhead fitting. Adjust linkage (par. 160 h) and install ammunition stowage racks in left ammunition stowage compartment. Install bulkhead cover.

161. TRANSMISSION EXTERNAL OIL TUBES.

a. Description. Three preformed steel oil tubes with flanges are attached to the outside of the transmission. The top tube (pressure regulator to valve body) connects the pressure regulator valve with the control valve body. The oil pump suction tube on the right side (fig. 124) connects the oil pump with the sump. The oil return tube at the bottom (reduction gear oil return tube) returns the oil from the reduction gear case to the oil sump. Remove damaged oil tubes and install new ones. If tubes leak at joints tighten attaching screws. If joints continue to leak, remove old gasket and install new one. If joints still leak, install new tube and new gasket.

b. Removal of Transmission External Oil Tubes. The procedure to remove the transmission external oil tubes is given in the following steps:

(1) REMOVAL OF PRESSURE REGULATOR OIL TUBE. Raise left rear intake door. Remove rear center deck plate. Remove attaching screws from both flanges. Swing rear end of tube to the right, and pull it back until front end strikes upper baffle plate. Pry up lower edge of upper baffle plate so tube can be pulled from under baffle plate, and lift out tube.

(2) INSTALLATION OF PRESSURE REGULATOR OIL TUBE. Clean gasket surfaces on flanges and transmission. Pry up lower edge of upper baffle plate, and insert front end of tube under baffle plate

from the rear. Slide tube into position on transmission and bend baffle plate back into place. Coat one side of gaskets with joint and thread compound and position a new gasket under each flange. Install attaching screws. Tighten screws alternately to compress gasket evenly. Install rear center deck plate. Check transmission oil level and fill if necessary. Start and run engine. Inspect connections for leaks and tighten if necessary. Stop engine. Close and lock intake door.

(3) REMOVAL OF OIL PUMP SUCTION TUBE. Remove cooling unit (par. 139 b). Remove screws which attach tube flanges to transmission, and lift off oil tube.

(4) INSTALLATION OF OIL PUMP SUCTION TUBE. Clean gasket surfaces on flanges and transmission. Coat one side of gaskets with joint and thread compound. Position new gaskets on flanges and install attaching screws. Tighten cap screws alternately to compress gasket evenly. Install and fill cooling unit (par. 139 c). Start engine, inspect connections for leaks and tighten if necessary. Stop engine.

(5) REMOVAL OF REDUCTION GEAR CASE OIL RETURN TUBE. Remove power unit (par. 71). Drain transmission (par. 151 c). Remove screws which attach tube to bottom of transmission housing and remove tube.

(6) INSTALLATION OF REDUCTION GEAR CASE OIL RETURN TUBE. Clean gasket surfaces on flanges and transmission. Coat one side of gaskets with joint and thread compound and position new gaskets on flanges. Attach tube to bottom of transmission. Tighten cap screws alternately to compress gaskets evenly. Install power unit (par. 72). Fill transmission (par. 151 d). Start engine, inspect connections for leaks, and tighten if necessary. Stop engine.

Section XXXI

DIFFERENTIAL AND FINAL DRIVES

162. DIFFERENTIAL.

a. Description. The controlled-type differential (figs. 62, 124 and 125) is located in the extreme rear of the engine compartment, and is bolted to the rear end cover of the transmission housing. It is supported on each side at the top by two rubber-insulated mounting brackets, which rest on two support brackets welded to the rear hull plate (figs. 124 and 125). Two anchor rods, one on each side at the bottom of the differential, help to stabilize the unit. The differential transmits the power from the transmission output drive pinion through a universal joint on each side (fig. 126) to the final drive units and drive sprockets. The controlled type differential permits the outer track to travel faster than the inner track on turns, or one track to travel faster than the other on uneven ground. It is called a controlled-type differential because it incorporates a brake drum and a three-shoe external-contracting type brake assembly on each side of the differential carrier which are used for steering and stopping the vehicle. The brake drums and shoes (fig. 130) are mounted within the differential housing, one assembly on each side of the differential carrier, where they receive lubrication and are protected from dirt. The brake mechanism is operated by dual steering levers in the drivers' compartment (figs. 18, 19, and 127). Except in an emergency, and only on authorization of higher authority, will second echelon be permitted to separate the differential from the transmission, because the installation requires the use of special third echelon tools. When it is necessary to remove or replace only the differential, notify higher authority. Engine, transmission, and differential units are removed or installed as a complete power unit (pars. 71 and 72).

(1) RECORD INSTALLATION OF DIFFERENTIAL (AND/OR POWER UNIT. Make proper entry on W.D., A.G.O. Form No. 48, "MWO and Major Unit Assembly Replacement Record."

163. UNIVERSAL JOINTS.

a. **Description.** Each final drive is connected to the differential by means of a universal joint (fig. 126) to allow for any movement between the units. The universal joint connects the end of the differential drive shaft with the end of the final drive pinion shaft. It is mounted between the companion flanges on the ends of the shafts, and is held in position by cap screws secured by locking wire.

b. **Removal.** Remove locking wires and screws attaching universal joint to differential and final drive shafts. Attach rope sling to universal joint for lifting. Pry final drive pinion shaft away from universal joint flange with a bar, and lift out the universal joint with the rope sling.

c. **Installation.** Place rope sling around universal joint, and lower it into position between the differential and final drive shaft flanges. Match the universal joint flanges with the flanges on the differential and final drive shafts. Install screws and tighten securely. Install locking wire.

164. STEERING LEVERS AND CONTROL RODS.

a. **Description.** Dual steering levers are provided for operating the brakes to steer and stop the vehicle. A set of right and left steering levers are mounted on the right and left front corners of the stationary frames of both the driver's and the assistant driver's seat

TM 9-735
164

Part Three—Maintenance Instructions

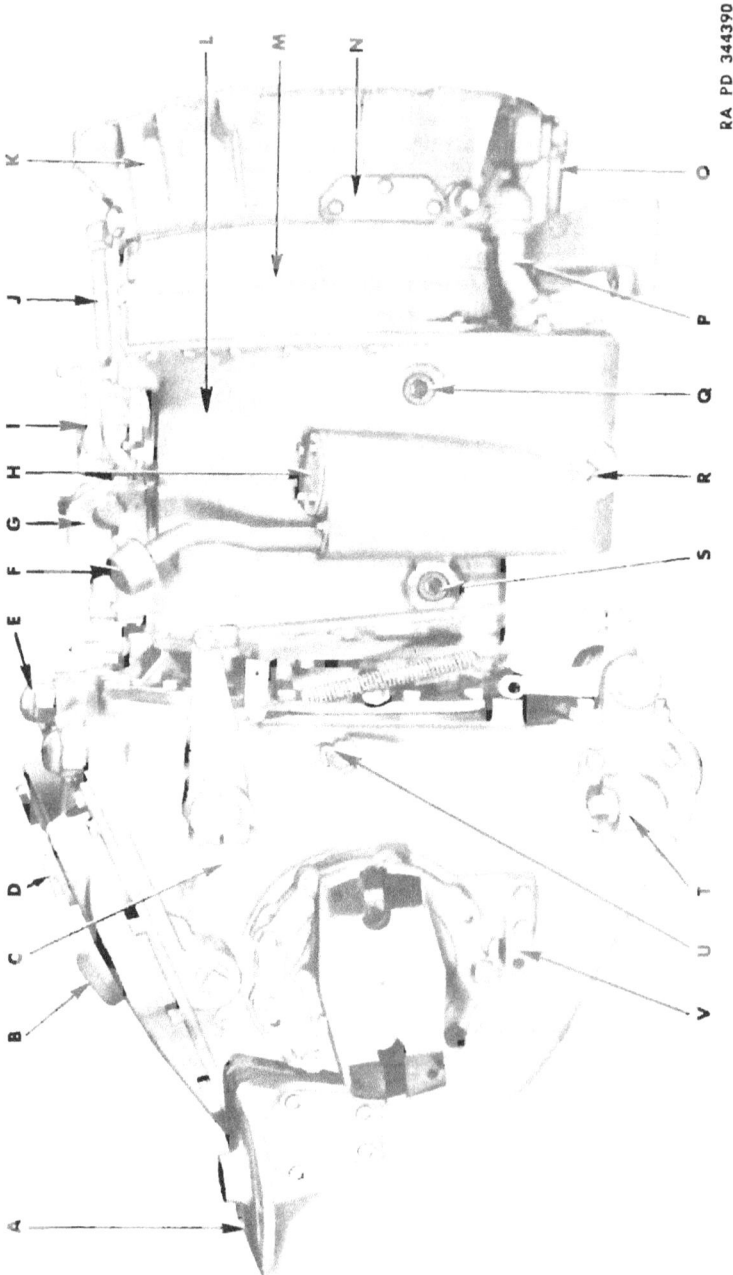

Figure 124 — Transmission and Differential, Right Side

Differential and Final Drives

A—MOUNTING
B—BREATHER
C—DIFFERENTIAL
D—BRAKE ADJUSTING NUT COVER
E—REVERSE ANCHOR COVER
F—OIL LEVEL INDICATOR
G—SECOND SPEED VALVE BODY AND SERVO COVER
H—OIL SCREEN AND COVER
I—THIRD AND REVERSE SERVO COVER
J—PRESSURE REGULATOR TO VALVE BODY OIL TUBE
K—FLYWHEEL AND PLANETARY REDUCTION DRIVE GEARS HOUSING
L—TRANSMISSION
M—TORQUE CONVERTER
N—OIL COOLER INLET AND OUTLET CONNECTIONS
O—REDUCTION DRIVE GEAR HOUSING OIL RETURN TUBE
P—FRONT PUMP SUCTION TUBE
Q—THIRD AND REVERSE SPEED BAND ADJUSTING SCREW
R—PLUG
S—SECOND SPEED BAND ADJUSTING SCREW
T—SCREEN COVER
U—OIL LEVEL INDICATOR
V—ANCHOR ROD BRACKET

RA PD 34439OB

Legend for Figure 124

TM 9-735

Part Three—Maintenance Instructions

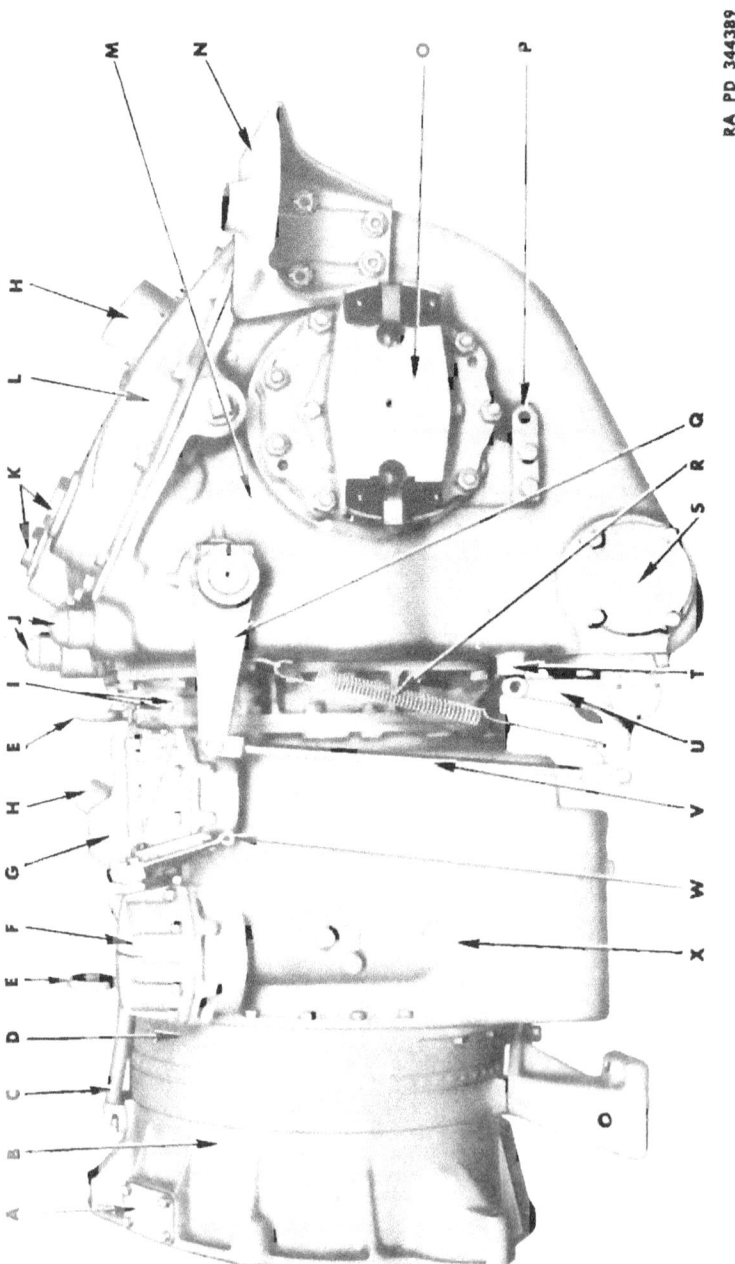

Figure 125 — Transmission and Differential, Left Side

Differential and Final Drives

A—FLYWHEEL HOUSING TIMING COVER
B—FLYWHEEL AND PLANETARY REDUCTION DRIVE GEAR HOUSING
C—PRESSURE REGULATOR TO VALVE BODY TUBE
D—TORQUE CONVERTER
E—LIFTING EYE
F—THIRD AND REVERSE SERVICE COVER
G—SECOND SPEED VALVE BODY AND SERVO COVER
H—BREATHER
I—CLUTCH PRESSURE GAGE CONNECTIONS
J—RIGHT AND LEFT REVERSE ANCHORS
K—RIGHT AND LEFT BRAKE ADJUSTING NUTS
L—DIFFERENTIAL COVER
M—DIFFERENTIAL
N—MOUNTING
O—UNIVERSAL JOINT FLANGE
P—ANCHOR ROD BRACKET
Q—BRAKE SHAFT OPERATING LEVER
R—SPRING
S—SCREEN COVER
T—STOP
U—BRAKE REAR ROD BELL CRANK
V—BRAKE OPERATING LEVER ROD
W—SPEED RANGE SELECTOR
X—TRANSMISSION

RA PD 344389B

Legend for Figure 125

TM 9-735

Part Three—Maintenance Instructions

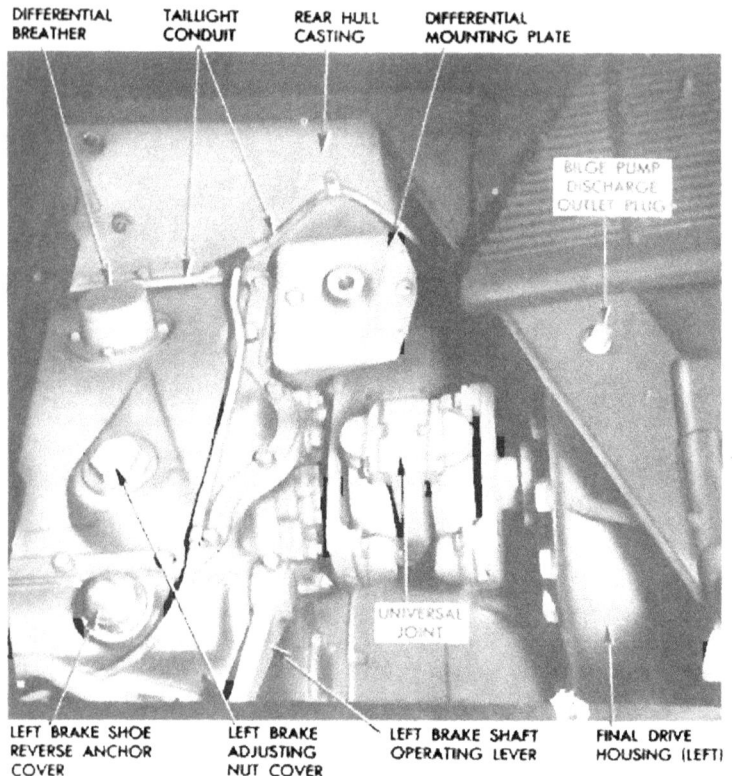

Figure 126 — Final Drive Universal Joint

(figs. 18 and 19). Long slots in the steering lever pull rod clevises permit each set of steering levers to operate independently of the other set (fig. 128). The long slots permit the cross shaft lever clevis pins to move freely in slot when opposite steering lever is in use. The pull rods from each steering lever connect to levers on the front cross shafts located on the hull floor directly behind the drivers' seats. The right steering lever pull rods connect to the levers on the ends of the right steering brake front cross shaft. The left steering lever pull rods connect to the levers on the ends of the left steering brake front cross shaft (fig. 128). Short rods from the parking brake mechanism (fig. 134) connect to the tops of levers mounted on each front cross shaft directly behind the front turret ring support. The right and left brake front rods are attached to the centers of these same levers. The right and left brake front rods extend rearward between the ammunition stowage boxes to levers on the inner ends of the right and left steering brake rear cross shafts beneath the battery box. From the outside ends of the rear cross shafts, jointed right and left brake rear rods extend through the engine compartment bulkhead, and along each side of the engine oil pan close to the hull floor to bell cranks on each side of the differential housing (fig. 132). From each bell crank, a brake operating lever rod extends upward and connects with the brake shaft operating levers.

b. **Adjustment of Steering Levers.** Disconnect steering lever pull rods from front cross shaft levers. Loosen jam nut on steering lever stop screws. Position all steering levers so that they are parallel and the bottom surface of the levers is ¾ inch above the stop screw stop. Adjust steering lever stop screws, and tighten jam nuts. Allow steering levers to rest in full-forward position against the stops, and recheck measurement. Adjust clevises on pull rods so that clevis pins will contact rear edge of clevis slot with steering levers and parking brake released. Position front cross shaft levers vertically and connect clevises. From the fully released to the fully applied position the distance is 12 inches, measured from the top-center of the lever.

c. **Removal of Steering Levers.** Remove clevis pins attaching pull rods to steering levers. Remove lever mounting bolt. Disconnect horn button switch from driver's right steering lever. Lift steering lever from bracket.

d. **Installation of Steering Levers.** Position lever in the housing on the seat frame with stop screw toward front of vehicle. Install lever mounting bolt and flat washer. Connect steering lever pull rod to lever. Install clevis pin. Adjust steering lever as outlined in subparagraph b preceding. Connect horn button switch to driver's right steering lever.

e. **Adjustment of Brake Rods.** The procedure for adjusting steering brake rods is given in the following steps: NOTE: *All adjustments of steering levers and control rods are to be made with the steering levers and control rods in fully released position.*

(1) STEERING LEVER PULL RODS. Disconnect both left steering lever pull rods from levers on left front cross shaft (fig. 128). Place left steering levers forward in fully released position against stops. Release parking brake. Disconnect brake front rods from front center cross shaft levers. Disconnect parking brake rod links from center cross shaft levers. Position left center cross shaft lever, so that the center of link pin hole in the top of cross shaft lever measures 3¼ inches from the front wall of the 90-mm ammunition stowage compartment. Loosen jam nuts, and adjust left steering lever pull rod

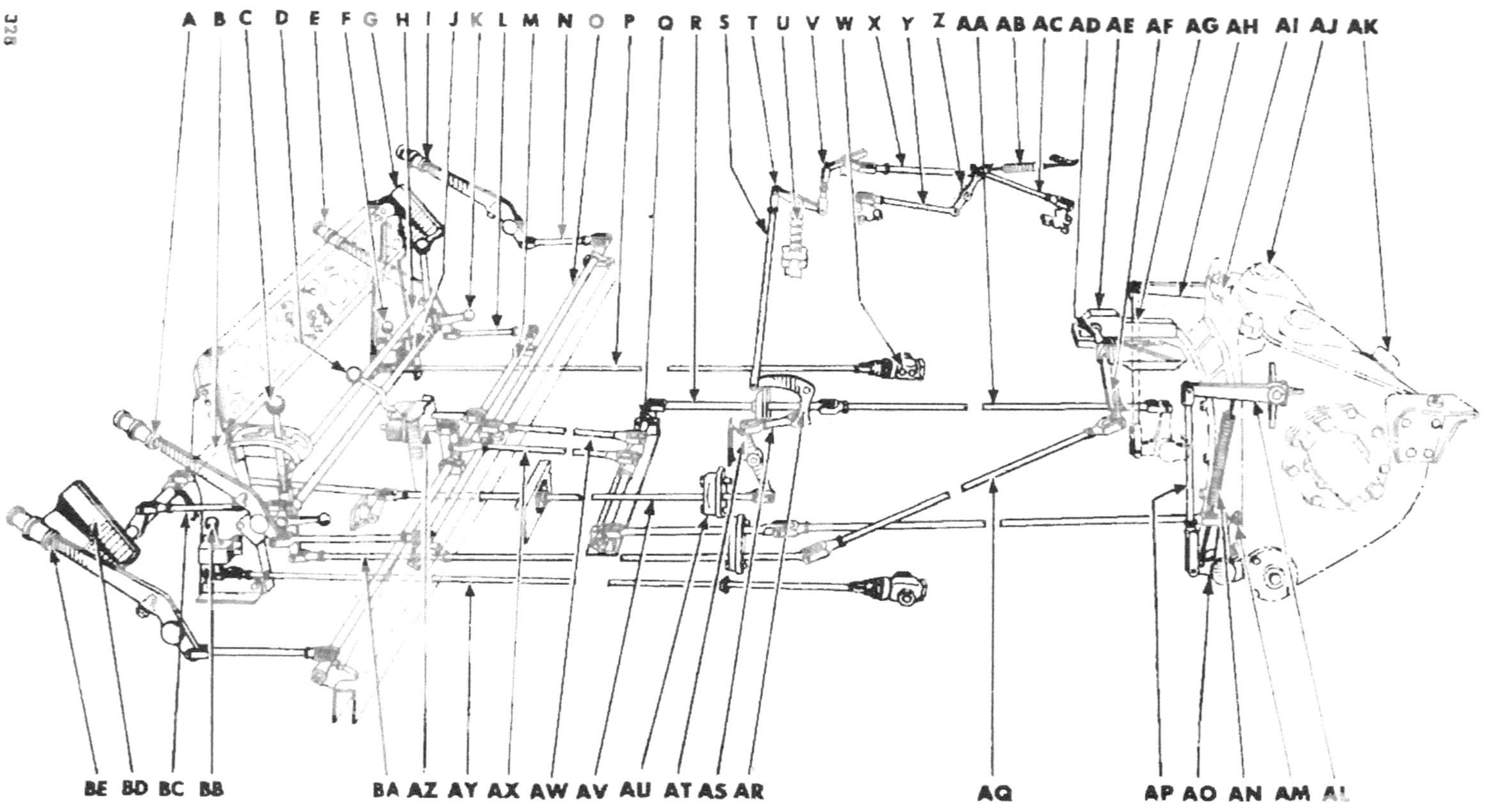

Figure 127 — Driving Controls and Linkage

RA PD 344654

Legend for Figure 127

A—DRIVER'S RIGHT STEERING LEVER
B—ACCELERATOR CROSS SHAFT
C—LEFT SPEED RANGE SELECTOR LEVER
D—PARKING BRAKE
E—ASSISTANT DRIVER'S LEFT STEERING LEVER
F—LEFT FUEL TANK SHUT-OFF VALVE CONTROL LEVER
G—ASSISTANT DRIVER'S ACCELERATOR
H—SPEED RANGE SELECTOR LEVER TIE LINK
I—ASSISTANT DRIVER'S RIGHT STEERING LEVER
J—THROTTLE AND SPEED RANGE SELECTOR LEVER CROSS SHAFTS
K—RIGHT THROTTLE
L—STEERING LEVER PULL ROD
M—FRONT CROSS SHAFT
N—STEERING LEVER PULL-ROD
O—REAR FRONT CROSS SHAFT
P—RIGHT FUEL TANK SHUT-OFF VALVE CONTROL ROD
Q—REAR CROSS SHAFTS
R—RIGHT REAR CROSS SHAFT RIGHT ROD
S—ENGINE THROTTLE BELL CRANK ROD
T—THROTTLE CONTROL ROD BELL CRANK
U—GOVERNOR BOOSTER
V—ROCKER ARM LINK
W—RIGHT FUEL TANK SHUT-OFF VALVE
X—THROTTLE CONTROL ROD
Y—THROTTLE VALVE LEVER ROD (FRONT)
Z—THROTTLE CONTROL ROD ARM
AA—RIGHT REAR BRAKE CONTROL ROD
AB—THROTTLE LINKAGE CONTROL SPRING
AC—THROTTLE VALVE LEVER ROD (REAR)
AD—SPEED RANGE SELECTOR SHAFT
AE—SECOND SPEED VALVE BODY AND SERVO COVER
AF—SPEED RANGE SELECTOR
AG—RIGHT BRAKE OPERATING LEVER ROD
AH—BRAKE SHAFT OPERATING LEVER
AI—RIGHT REVERSE ANCHOR COVER
AJ—RIGHT BRAKE ADJUSTING NUT COVER
AK—DIFFERENTIAL BREATHER
AL—LEFT BRAKE SHAFT OPERATING LEVER
AM—STOP
AN—RETURN SPRING
AO—BELL CRANK
AP—LEFT BRAKE OPERATING LEVER ROD
AQ—SPEED RANGE SELECTOR REAR CONTROL ROD
AR—ENGINE THROTTLE BELL CRANK
AS—THROTTLE CONTROL ROD LEVER LINK
AT—THROTTLE CONTROL ROD LEVER
AU—THROTTLE CONTROL ROD SUPPORT
AV—THROTTLE CONTROL ROD
AW—RIGHT CROSS SHAFT ROD
AX—LEFT CROSS SHAFT ROD
AY—LEFT FUEL TANK SHUT-OFF VALVE CONTROL ROD
AZ—PARKING BRAKE OPERATING ROD
BA—SPEED RANGE SELECTOR FRONT CONTROL ROD
BB—LEFT FUEL TANK SHUT-OFF VALVE CONTROL LEVER
BC—THROTTLE FRONT CONTROL ROD
BD—DRIVER'S ACCELERATOR
BE—DRIVER'S LEFT STEERING LEVER

RA PD 344654B

TM 9-735
164

Part Three—Maintenance Instructions

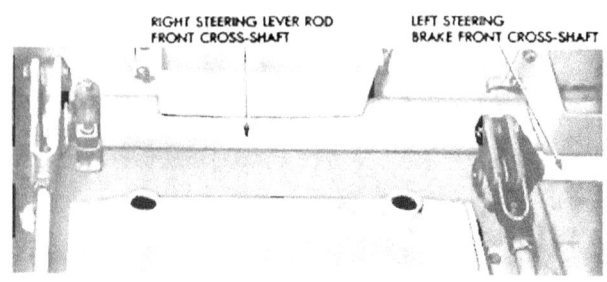

Figure 128 — Steering Lever Rods Front Cross Shafts

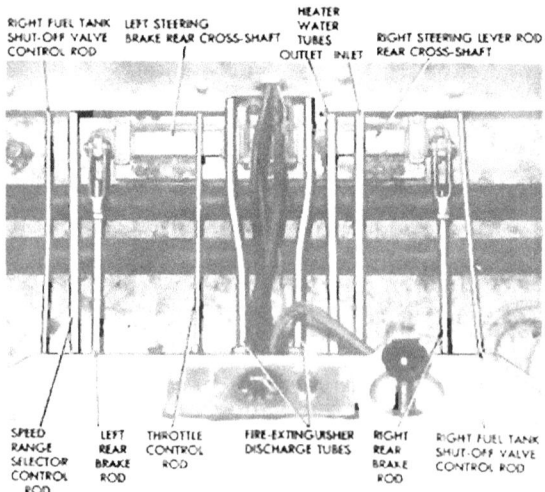

Figure 129 — Steering Lever Rods Rear Cross Shafts

330

Differential and Final Drives

clevises so that rear ends of slots in clevises will be in contact with rear sides of clevis pins. Install clevis pins. Tighten clevis jam nuts securely. Connect parking brake rod links. Repeat same operations for adjusting right steering lever pull rods with the following exception: Before connecting pull rods to cross shaft, center cross shaft lever must be positioned so that center of right front parking brake link pin measures 4 inches from rear edge of front turret ring support. Adjust brake front rods (step (2) following), and connect to front cross shaft levers by installing clevis pins.

(2) BRAKE FRONT RODS. Traverse turret to provide additional working space. Turn 24-volt master switch off. Remove battery box cover and batteries (par. 92 d). Remove battery box (par. 93 b). Disconnect brake front rods from front cross shaft. Place steering levers in maximum forward position against lever stops. Adjust steering lever pull rods as outlined in step (1) preceding. Rotate rear cross shaft toward rear of vehicle to bring the inside brake rod levers against their stops. Loosen jam nuts, and adjust clevises on front ends of brake front rods. Connect left brake front rod to lever on left front cross shaft, and right brake front rod to lever on right front cross shaft. Tighten clevis jam nuts securely. Install battery box (par. 93 c), batteries, and battery box cover. NOTE: *If brake rear rods are to be adjusted, do not install battery box or batteries.*

(3) BRAKE REAR RODS. Traverse turret to provide additional working space. Turn 24-volt master switch off. Remove battery box cover, batteries, and battery box (pars. 92 d and 93 b). Disconnect brake rear rods from levers on rear cross shaft and from bell cranks on differential. Adjust brake front rods as outlined in step (2) preceding. Rotate rear cross shaft toward rear of vehicle until inside cross shaft levers are back against their stops. Loosen jam nuts and adjust clevises so that clevis pin holes line up in cross shaft levers and bell cranks. Tighten jam nuts on brake rods securely. NOTE: *As brake rear rods are in two sections jointed by a clevis, brake rod clevises must be adjusted so that brake rod connections at rear of bulkhead do not strike against bulkhead and limit travel of rods.* Install battery box, batteries, and cover.

(4) BRAKE OPERATING LEVER RODS. Remove brake rod return spring. Disconnect brake operating lever rods from bell cranks (fig. 132). Disconnect brake rear rods from bell crank levers. Note position of brake shaft operating levers and adjust as required (step (5) below). Loosen jam nut on bell crank stop screw. With brake shaft operating levers in correct position, adjust operating lever rod clevis until rod measures 20¼ inches from center-to-center of clevis holes. Connect operating lever rod to vertical bell crank lever, and tighten rod jam nut securely. Loosen jam nut and adjust bell crank stop screw. Tighten stop screw jam nut. Connect brake rod return

331

spring. Adjust clevises on brake rear rods and connect to lower bell crank levers. Tighten clevis jam nuts securely.

(5) BRAKE SHAFT OPERATING LEVERS. Disconnect brake operating lever rods from bell cranks. Remove screw in split hub of operating lever, and remove lever from shaft. Position lever on serrated end of brake shaft so that index marks on lever hubs and index marks on brake shaft are alined. Install and tighten screw in split hub of operating lever. Adjust brake operating lever rod clevises as outlined in step (4) preceding. Connect operating lever rod to bell crank.

f. **Removal of Steering Lever Pull Rods.** Remove steering lever return spring. Disconnect pull rod at steering lever. Remove drivers' seat backs. Disconnect pull rod from front cross shaft lever.

g. **Installation of Steering Lever Pull Rods.** Place pull rod in position so slotted clevis will connect to front cross shaft lever (fig. 128). Insert clevis pin and install one end of return spring on clevis pin. Connect other end of steering lever return spring to seat bracket. Adjust pull rod clevis (subpar. e (1) preceding), and connect to front cross shaft lever. Install driver's seat back.

h. **Removal of Brake Front Rods.** Disconnect brake front rods from front cross shaft levers. Traverse turret to provide additional working space. Turn 24-volt master switch off. Remove outer and inner covers from tunnel between right and left 90-mm ammunition stowage compartments. Remove battery box cover, batteries, and battery box (par. 92 d and 93 b). Disconnect brake front rods from rear cross shaft levers. Move rods forward or back as far as possible to clear bottom edge of ammunition stowage compartment front wall. Lift front ends of rods out of tunnel, at the same time pulling rods forward to clear bottom edge of ammunition compartment rear wall. Remove rods through either driver's door.

i. **Installation of Brake Front Rods.** Place brake front rods in tunnel by reversing the removal procedure outlined in subparagraph h preceding. Rotate inside levers on rear cross shaft back against their stops. Adjust rod clevises, tighten jam nuts, and connect clevises to rear cross shaft levers. Adjust front clevises on brake front rods (subpar. e (2) preceding). Connect clevises to front cross shaft levers. Tighten clevis jam nuts securely. Install battery box, batteries, and battery box cover (pars. 92 e and 93 c). Install outer and inner covers in tunnel between 90-mm ammunition stowage compartments.

j. **Removal of Brake Rear Rods.** Traverse turret to provide additional working space. Turn 24-volt master switch off. Remove battery box cover, batteries and battery box (par. 92 d and 93 b). Disconnect brake rear rods from rear cross shaft levers. Remove

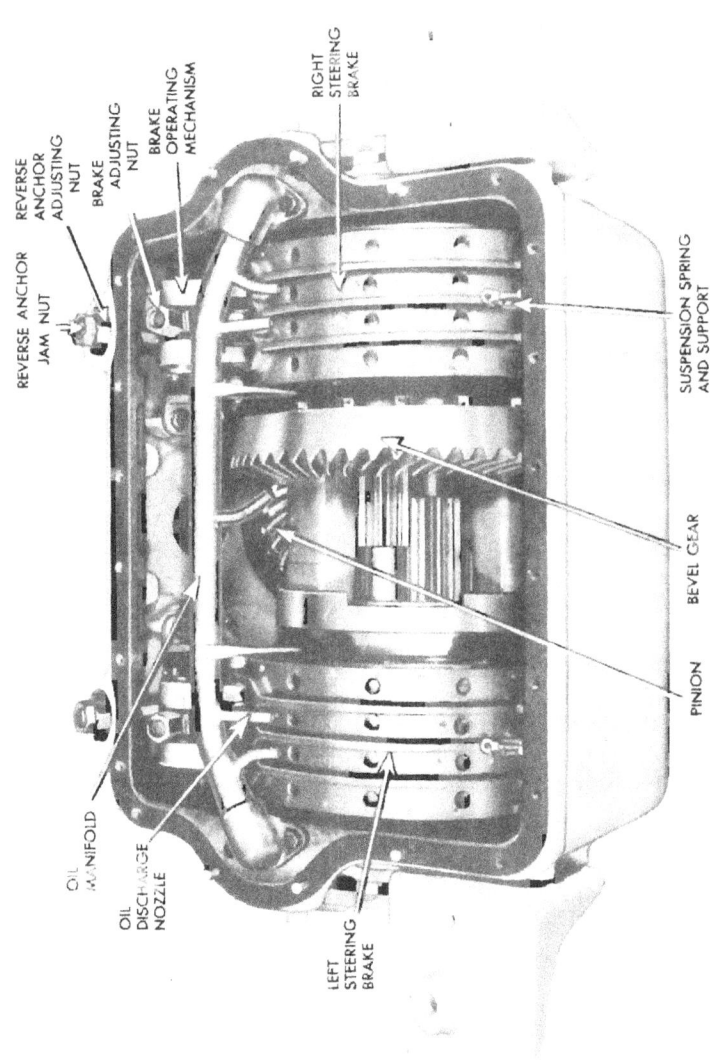

Figure 130 — Differential (Cover Removed)

clevises and jam nuts from rods. Disconnect rods from bell cranks on differential housing. Pull rods through bulkhead and lift out through engine compartment.

k. Installation of Brake Rear Rods. Remove bulkhead cover. Disconnect brake rear rods at joints. From fighting compartment carefully insert short length of rod through seal in bulkhead. Working through opening in bulkhead, extend long section of rod into engine compartment, and connect long and short sections with clevis pin. Adjust brake rear rods as outlined in subparagraph *e* (3) preceding. Connect brake rear rods to bell cranks on differential housing. Install bulkhead cover. Install battery box, batteries and battery box cover (par. 92 e and 93 c).

l. Removal of Brake Operating Lever Rod. Remove brake rod return spring. Remove clevis pin and disconnect operating lever rod from brake shaft operating lever and bell crank. Remove brake operating lever rod.

m. Installation of Brake Operating Lever Rod. Adjust brake operating lever rod and bell crank stop screw as outlined in subparagraph *e* (4) preceding. Place operating lever rod in position, and install clevis pin. Connect brake rod return spring.

n. Removal of Brake Shaft Operating Lever. Remove clevis pin from brake operating lever rod, and remove rod. Remove screw from split hub on brake shaft operating lever. Use a copper drift or soft hammer to drive lever off brake shaft serrations.

o. Installation of Brake Shaft Operating Lever. Position brake shaft operating lever (fig. 132) on brake shaft with index marks on lever and brake shaft alined. Install screw in split hub of lever and tighten securely. Adjust brake operating lever rod as outlined in subparagraph *e* (4) preceding. Install brake operating lever rod and clevis pin.

165. STEERING BRAKE.

a. Description. The steering brakes consist of the steering levers, brake rods and linkage, cross shafts, and brake shoe assemblies (fig. 130). The brake shoe assemblies are three-shoe, external-contracting type, and are mounted on brake drums, one on each side of the differential carrier within the differential housing. The brake shoes may be adjusted for lining wear or equalization of braking pressure by turning adjusting nuts as outlined in subparagraph *c*. Pulling back on one steering brake lever contracts the corresponding brake shoe assembly. This contracting action slows the respective brake drum which, in turn, slows the track on that side, while the speed of the opposite track is increased. This slowing action causes the vehicle to turn toward the side of the slower-moving track. Pulling back on

Differential and Final Drives

both levers simultaneously, and with equal pressure, applies the brakes to both drums which slows or stops the vehicle according to the pressure applied. Either or both steering brakes may be locked in applied position by operating the parking brake mechanism (par. 166).

b. **Inspection.** Inspection of steering brakes, according to procedure given in the following steps, will determine if braking efficiency may be restored with a minor brake adjustment (subpar. c following), or if a major brake adjustment is necessary (subpar. d following). If a minor brake adjustment does not restore brake efficiency, a major brake adjustment must be made (subpar. d following).

(1) INSPECT STEERING BRAKE LEVER FREE TRAVEL. Free travel is the distance the steering lever travels when pulled back, with light pressure only, from fully released position to a position where the brake shoes are felt just contacting the brake drums. The recommended free travel is 7 inches, and is measured at the tops of the levers. If free travel has increased to any great extent it indicates brake lining is worn or shoes are out of adjustment, and a minor brake adjustment (subpar. c following) is necessary.

(2) INSPECT BRAKE RODS, LINKAGE, AND BELL CRANK. Examine brake rods and linkage to be sure they are not loose or binding at any point. Note position of brake shaft operating lever, and measure length of brake operating lever rods (par. 164 e (4)). Measure settings of bell crank stop screws. Operate steering levers and observe return of bell cranks against stops. If bell cranks do not seat firmly against stops, it indicates brakes are not fully released. Note action of bell crank return springs. If springs are weak, replace with new springs, and operate brake levers again. If brake shoes still fail to release properly adjust brake shoes as outlined in subparagraph c following.

(3) INSPECT REVERSE ANCHORS. Inspect reverse anchor pins (fig. 130). If reverse anchor pin nut is raised from its seat, with brake levers in forward released position, it indicates brake shoe assembly is not fully released from drum. Adjust reverse anchor as outlined in subparagraph c (3) following.

(4) INSPECT BRAKE COOLING SYSTEM. Whenever brakes are adjusted always check operation of differential oil pump. NOTE: *It is of the utmost importance that the differential oil pump operate at full capacity at all times to provide sufficient cooling to the brake assemblies.* Remove both right and left brake adjusting nut covers. Start engine and run it at 1,000 to 1,200 revolutions per minute. Observe oil flow from oil discharge nozzles through the brake adjusting nut ports in the differential cover. If oil is flowing solidly and without air bubbles or foam, it indicates differential oil pump is

operating correctly. If oil from discharge nozzles contains foam or air bubbles it indicates an air leak in the suction tubing. The air bubbles or foam are diluting the oil and decreasing the amount of oil being circulated. Check for leaks and tighten all connections. Oil leaking from the differential breather is also a sure indication of air leaks in the suction tubing. Keep radiators and oil cooler cores clean. Plugged cores will cause high oil temperature which will seriously affect brake linings.

c. **Minor Brake Adjustment.** A minor brake adjustment consists of tightening adjusting nuts to compensate for brake lining wear, and to bring steering levers to correct (7 inches) and equal free travel.

(1) ADJUST BRAKE ADJUSTING NUTS. From fully released position pull both steering levers back until both brake shoes are felt to just contact the brake drums with a light pressure. Measure free travel of each steering lever. Remove the two brake adjusting nut covers (fig. 126). Adjust each steering brake nut using socket wrench (41-W-3329) until each steering lever has 7 inches of free travel. Turn adjusting nut so that after each turn a flat of the nut is in contact with the retainer clip to lock the nut in position. If brakes are badly out of adjustment, tighten adjusting nuts to 8 to 10 foot-pounds with a torque wrench. Back off each nut 6 flats or one full turn. Test operation of steering levers, brakes, and brake linkage. Measure steering lever free travel for 7 inches and adjust as described in step (2) following. Install brake adjusting nut covers.

(2) ADJUST STEERING BRAKE LEVER EQUAL TRAVEL. Pull back both steering brake levers with equal force and check for equal lever travel. If travel is unequal, loosen adjusting nut on brake controlled by lever having the least travel. Continue backing off nut until travel of both levers is equal. Install adjusting nut hole cover gaskets and covers.

(3) ADJUST BRAKE REVERSE ANCHORS. Remove covers from both right and left reverse anchors. Loosen reverse anchor adjusting nuts jam nuts. Tighten reverse anchor adjusting nuts to 8 to 10 foot-pounds with a torque wrench. Back off adjusting nuts 6 flats or one full turn. Install brake adjusting nut covers. Test operation of linkage. NOTE: *Reverse anchors must be adjusted whenever brake adjusting nuts are adjusted.*

d. **Major Brake Adjustment.** The procedure for a major brake adjustment includes removal of brake shoe assemblies and inspection of brake linings and drums. Drain oil from differential housing (par. 153 c). Remove brake shoe assembly (subpar. e following), and inspect brake shoe lining. If lining is excessively worn, or glazed and shiny, replace all brake shoes (subpar. f below). Examine brake

TM 9-735
165

Differential and Final Drives

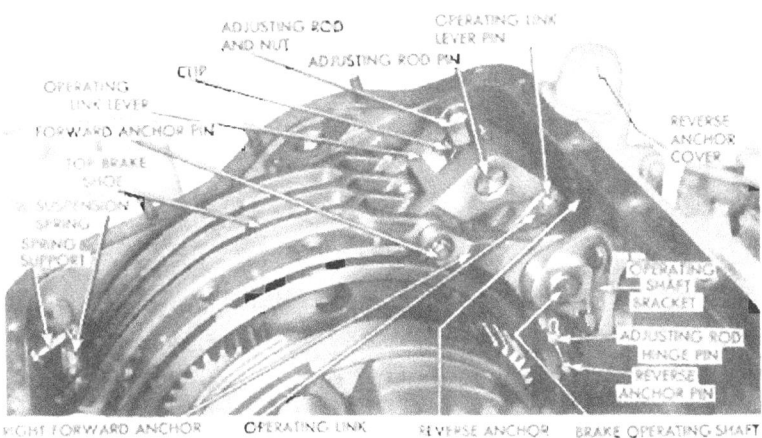

Figure 131 — Brake Shoe Linkage and Operating Mechanism

drums for scores, cracks, or any other condition which would warrant their replacement. If replacement of drums is indicated, notify higher authority. Install brake shoe assembly as outlined in subparagraph f following. Adjust steering levers to correct free and equal travel (subpar. e (1) and (2) preceding). Adjust reverse anchor bolt (subpar. e (3) preceding). Adjust brake operating lever rod (par. 164 e (4)). Fill differential with oil (par. 153 d). Adjust steering levers and control rods if necessary (par. 164 h and 164 e).

e. **Removal of Brake Shoe Assembly.**

(1) PREPARATORY. Remove rear center deck plate. Disconnect exhaust tubes from engine exhaust manifolds. Pull exhaust tubes through openings in baffle, and remove from engine compartment. Remove taillight conduit clips from differential housing. Leave clips attached to conduit, and move conduit out of way. Remove remaining differential cover nuts and bolts. Disconnect brake operating lever rod from brake shaft operating lever (fig. 132). Remove split-hub brake shaft operating lever from shaft. To remove the four tapered split dowels from dowel holes, strike cover next to dowel with a punch and hammer which will loosen dowel. Lift off differential cover and gasket.

(2) REMOVAL. Remove oil manifold and oil manifold gasket. Drain oil from differential (par. 153 e). Wipe out differential case. Remove suspension spring from rear of top brake shoe link. Remove brake shoe adjusting rod nut and retainer clip with socket wrench

TM 9-735

Part Three—Maintenance Instructions

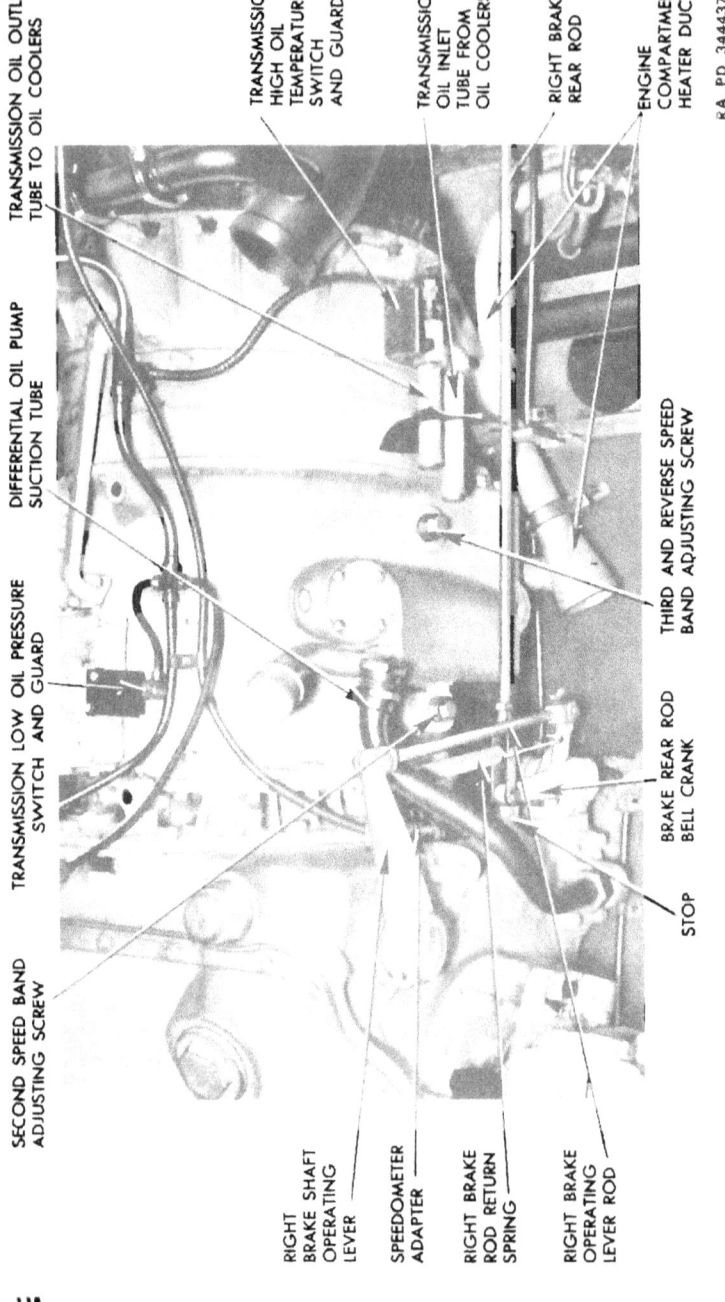

Figure 132 — Steering Brake Control Rods, Right Side of Differential

TM 9-735
165—166

Differential and Final Drives

(41-W-2573-400). Remove cotter pin from outside of top brake shoe link pin (fig. 131) and remove pin and both washers. Remove outside cotter pin from operating link lever pin. Remove pin and operating link lever. Remove spring from brake adjusting rod (fig. 131). Remove locking wire and two bolts from operating shaft bracket. Pull operating shaft bracket from end of brake lever operating shaft. Slide inner forward anchor from operating shaft. Pull brake lever operating shaft with outer forward anchor and operating link attached from differential housing (fig. 131). Pull up reverse anchor bolt and remove inside locking clip from reverse anchor pin and adjusting rod hinge pin (fig. 131). Remove pins and adjusting rod. Grasp top brake shoe and pull brake shoe assembly from differential housing. Separate brake shoes by removing link pins.

f. **Installation of Brake Shoe Assembly.**

(1) INSTALL BRAKE SHOE ASSEMBLY. Assemble the three brake shoes by installing the two hinge pins. Attach a strong wire to rear brake shoe. Pass wire down around brake drum. Attach end of wire to rear brake shoe. Start rear brake shoe down and around front of drum, and with wire pull brake shoe assembly around drum and up into position. Install adjusting rod, spring and hinge pin. Install reverse anchor pin and adjusting rod hinge pin. Install locking clip in pins. Install brake lever operating shaft, with outer forward anchor and operating link attached. Install inner forward anchor on brake operating shaft. Install operating shaft bracket, screws and locking wire. Install suspension spring to rear brake shoe. Remove pull wire used to position brake shoe assembly. Place operating link lever in position (fig. 131). Install operating link lever pin and top brake shoe link pin. Install retainer clip and adjusting rod nut with tapered end of nut down. Be sure adjusting rod pin is installed with flat portion uppermost so adjusting nut retainer clip will seat on flats on nut (fig. 131). Install brake operating shaft.

(2) ADJUSTMENTS AND COMPLETION OF INSTALLATION. Install brake operating lever rod and adjust (subpar. e (2) preceding). Install oil manifold with a new gasket. Adjust brakes as outlined in subparagraphs c and d preceding. Install a new differential cover gasket and install differential cover. Install oil outlet line bracket and taillight conduit clips. Install four tapered split dowels, two on each side of differential cover, and tighten all screws and nuts alternately to compress gasket evenly. Install exhaust tubes and connect to engine manifold. Fill differential system with oil (par. 153 d). Install rear center deck plate.

166. PARKING BRAKE MECHANISM.

a. **Description.** The parking brake mechanism locks the steering brakes in applied position. It is mounted on the turret front sup-

Figure 133 — Steering Brake Control Rods, Left Side of Differential

Differential and Final Drives

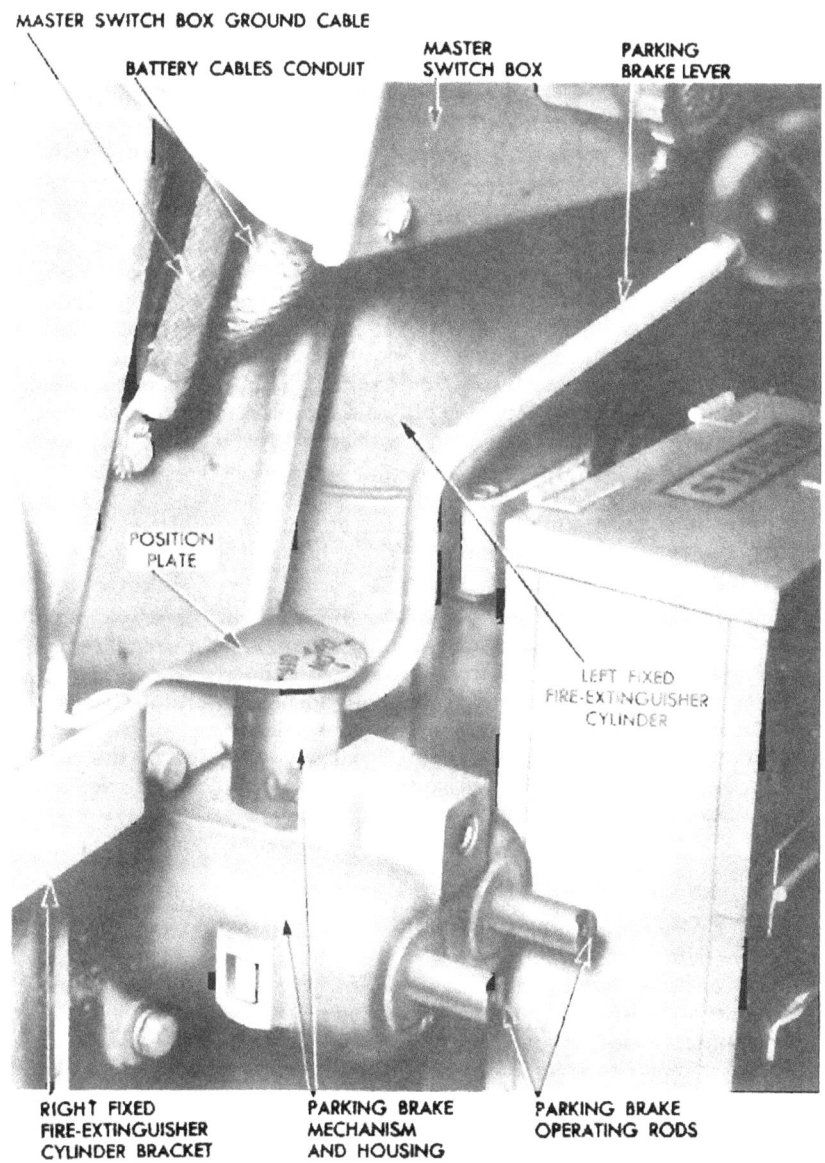

Figure 134 — Parking Brake Mechanism

port between the driver and assistant driver (figs. 12 and 134). Parking brake rods are connected to the brake front cross shaft levers at the rear of the turret front support. When the parking brake lever is in center position, and the steering brake levers have been pulled all the way back and released, the parking brake is on. It is released by swinging the lever all the way to right or left. For instructions on operating parking brake, refer to paragraph 14.

  b. **Removal of Parking Brake Mechanism.** Remove ammunition and stowage boxes in drivers' compartment located between instrument panel and turret front support. Disconnect both right and left parking brake rod links from front center steering brake cross shaft lever links at rear of the front turret ring support. Place brake lever in released position. Remove four screws attaching parking brake mechanism to turret front support. Pull mechanism away from support. As mechanism is pulled away both parking brake rod clevises will slide forward and stop against the opening in turret front turret ring support. Continue pulling parking brake mechanism and housing off rods. After mechanism has been removed, work rods out of front turret ring support toward rear of vehicle. When rods have been removed, reinstall them in parking brake mechanism for safe keeping.

  c. **Installation of Parking Brake Mechanism.** Remove parking brake rods from brake mechanism. Position mechanism and position plate on front side of front turret ring support. From rear of turret ring support work rods through openings. Install parking brake mechanism on both rods, and seat mechanism and housing into position against front of turret ring front support. Install the four attaching screws. Release parking brake lever. Connect clevises on parking brake rod ends to connecting links on steering brake front cross shaft levers. Test parking brake by operating.

167. FINAL DRIVES.

  a. **Description.** The final drives transmit the power from the differential to the track driving hubs and sprockets. Each final drive consists of a pinion shaft, pinion and gear, final drive shaft, and final drive housing. The final drive shaft is splined near the outer end and threaded at the extreme end to receive a large lock nut. Sprocket hubs are splined and are mounted on the final drive axles, hub spline matching axle spline. Sprocket assemblies are held in position by a lock washer and the large nut. An edge of the washer is turned over against a flat on the nut to form a lock washer. The gears and bearings in the final drive housing operate in a continuous bath of oil. Final drives are bolted to the outside of the hull side plates at the rear of the vehicle. A washer-shimmed anchor cap screw in the lower front face of each final drive housing, screwed into

Differential and Final Drives

the final drive housing, keeps final drives in correct position and prevents the housing from working on the mounting bolts.

b. Removal. When the final drive is to be removed with the hub and drive sprockets attached, follow procedure in step (1) following. When it is necessary to remove a final drive and install a new one, or to remove hub and sprockets before removing final drive from vehicle, follow procedure in step (2) following.

(1) REMOVE FINAL DRIVE WITH HUB AND SPROCKETS ASSEMBLED. Drain final drive (fig. 162). Remove rear fender (par. 186 d). Disconnect track and move it forward on track support rollers until clear of sprockets (par. 170 c). Attach final drive lifting plate eye (41-E-649-875) and final drive sling (41-S-3830-30) as shown in figure 135. Disconnect final drive pinion shaft from universal joint (par. 163 b). Remove final drive breather. Cover hole in differential to prevent entrance of dirt. Flatten washer and remove final drive shaft nut. Use hoist to support final drive. Remove final drive housing mounting bolts. From inside engine compartment remove anchor cap screw. Lift off final drive assembly and lower it on to blocks. Remove sling and lifting plate eye, and install cap screws previously removed to attach eye to final drive.

(2) REMOVE FINAL DRIVE WITH HUB AND SPROCKETS DISASSEMBLED. Remove hub and sprockets as an assembly by following procedure in paragraph d. Drain final drive (fig. 162). Attach lifting plate eye (41-E-649-375) as shown in figure 135. Remove final drive by following balance of procedure in step (1).

c. Installation. The final drive is installed with the hub and sprockets attached (step (1) following) or with the hub and sprockets removed, as in step (2) following.

(1) INSTALL FINAL DRIVE WITH HUB AND SPROCKETS ASSEMBLED. Attach lifting plate eye and sling (subpar. b (1) preceding). Coat mating surfaces on final drive and on hull with joint sealing compound. Raise final drive assembly into position and install attaching cap screws, but do not tighten screws. Pry final drive toward rear of vehicle and partially install anchor cap screw. Install required number of washers on anchor cap screw so there is a snug fit between final drive and hull. Tighten anchor cap screw securely. Tighten final drive attaching screws securely. Bolt final drive pinion shaft flange to universal joint flange and secure with locking wire. Install final drive breather. Remove lifting plate eye and sling, and install cap screws previously removed to attach eye. Install final drive shaft washer and lock nut. Bend edge of washer over one flat on nut. Fill final drive (par. 46). Connect and adjust track (par. 170 b and d). Install rear fender (par. 186 e).

(2) INSTALL FINAL DRIVE WITH HUB AND SPROCKETS DISASSEM-

TM 9-735

Part Three—Maintenance Instructions

Figure 135 — Removing Final Drive, Sprockets and Hub

BLED. Attach lifting plate eye and install final drive on vehicle according to procedure given in step (1) preceding. Install hub and sprockets (par. 186 c and e).

(3) RECORD INSTALLATION. Make proper entry of either installation, step (1) or (2) preceding, on W.D., A.G.O. Form No. 478, "MWO and Major Unit Assembly Replacement Record."

Differential and Final Drives

RA PD 335585

Figure 136 — Removing Track Drive Sprockets and Hub from Final Drive, Using Sling (41-S-3830-30)

168. DRIVING SPROCKETS AND HUBS.

a. **Description.** A driving sprocket wheel is splined to each final drive shaft. Teeth on the inner and outer sprockets engage the sprocket holes in the ends of the track links to drive and support the track. Sprockets can be interchanged on the hubs; or sprockets and wheel assemblies can be interchanged from one side of the vehicle to the other, to equalize wear on the driving faces of the sprocket teeth. Both right and left sets of sprockets on the vehicle should be reversed or replaced at the same time to ensure even distribution of driving wear. The driving sprocket hub is splined with a double-flange track link guide bolted to the middle section. The spline holds the driving sprocket assembly in alinement on the final drive hub. The double-flange provides a slot through which the track link guides pass to keep the track links in correct alinement on the sprocket.

b. **Removal of Driving Sprocket.** Disconnect track (par. 170 c) and pull upper end back out of the way. Attach lifting tackle to sprocket and take up slack. Flatten bent edges of locking washers away from screws and remove screws and lock washers. Lift sprocket and lower it to the ground. To remove inner sprocket, remove hub and inner sprocket as an assembly (subpar. d following). Stand assembly on end with inner sprocket uppermost. Flatten bent tabs of lock washers away from screws, and remove screws and lock washers. Attach lifting tackle to inner sprocket and remove it from hub.

c. **Installation of Driving Sprocket.** To install outer sprocket to hub with hub attached to vehicle, support sprocket in position against hub and track. Locking washers must be installed so that tabs may be bent *down* over edge of bosses on inside diameter of sprocket and one tab bent *up* against screw flat to properly secure the screw from turning. Install one screw and lock washer, in upper diameter of sprocket and then using screw as a pivot line up remaining bolt holes by moving sprocket. Install remaining screws and lock washers. Use torque wrench (41-W-3630) and tighten screws alternately to 80 foot-pounds torque. Bend one tab of each lock washer up against flat on screw head and one tab *down* over edge of sprocket boss on inside diameter of sprocket. Remove sprocket support. Connect track, and adjust tension (par. 170 b and d). The inner sprocket is installed in the same manner, except that sprocket must be installed with hub removed from vehicle (subpar. d following).

d. **Removal of Driving Sprocket Hubs.** Remove rear fender. Disconnect track and pull upper end out of the way (par. 170 d). Install cable sling (41-S-3830-30) around hub, and hook lifting tackle to eyes in sling (fig. 135). Take up slack in sling until there is a slight strain on the cable. Flatten large diameter washer on end of final drive shaft. Remove lock nut and washer from end of final drive shaft. Pull hub and sprocket assembly carefully off final drive shaft. Be careful not to damage splines. Lower hub and sprocket assembly on to blocking and unhook sling. Remove drive sprockets as outlined in subparagraph b preceding.

e. **Installation of Driving Sprocket Hubs.** Install inner and outer drive sprockets (subpar. c preceding). Attach cable sling to hub and lift assembly into position so that splines in hub mate with splines on end of final drive shaft. Push hub carefully on shaft so as not to damage splines. Install flat washer and lock nut on end of shaft. Bend one end of washer over flat on nut Remove lifting sling. Apply parking brake. Connect and adjust track (par. 170 h and d). Install rear fender.

Section XXXII

TRACKS AND SUSPENSION

169. DESCRIPTION OF SUSPENSION SYSTEM.

a. **Description.** The T26E3 is supported by large-diameter, rubber-tired wheels which roll on all-metal tracks (fig. 137). The wheels are bolted in pairs to the road wheel hubs, which are mounted on inner and outer tapered roller bearings on each road wheel arm spindle. Each road wheel arm is supported in an inner and outer needle bearing in a road wheel arm support (fig. 148). The road wheel arm supports are bolted at intervals along the bottom of the hull on each side of the vehicle. The road wheel arm supports also provide seats for the torsion bar spring anchor plugs which contain the ends of the torsion bar springs extending from the opposite side of the vehicle. The torsion bar springs, made of spring steel and serrated at each end, run crosswise of the vehicle from the road wheel arm to the torsion bar spring anchor plug in the opposite road wheel arm support. The serrations on one end of the torsion bar spring match with serrations in the hollow road wheel arm spindle and at the other end with serrations in the torsion bar spring anchor plug. One serration has been eliminated to facilitate removal and installation. Each spring is marked on the wheel arm end with an arrow showing direction of rotation and a letter symbol for correct installation (fig. 155). The torsion bar springs support the weight of the vehicle and also, through torsional resistance, or "twisting" action of the spring caused by the up-and-down movement of the road wheel arm and spindle, act as springs to cushion the up-and-down movement of the vehicle. Volute-type bumper springs are provided for all road wheel arms (figs. 137 and 139). These bumper springs stop the travel of the road wheel arms, should the track and wheels strike an obstacle of sufficient size to overcome the torsional resistance of the torsion bar springs and the control of the shock absorbers. Hydraulic shock absorbers (figs. 137 and 139) on each side of the vehicle control the movement of the two front and two rear road wheel arms. Shock absorbers are not provided for the two center, or intermediate wheel arms, as their location at the center of the vehicle makes shock control unnecessary. The track idler wheel is supported on an eccentric spindle in the upper end of the front road wheel arm.

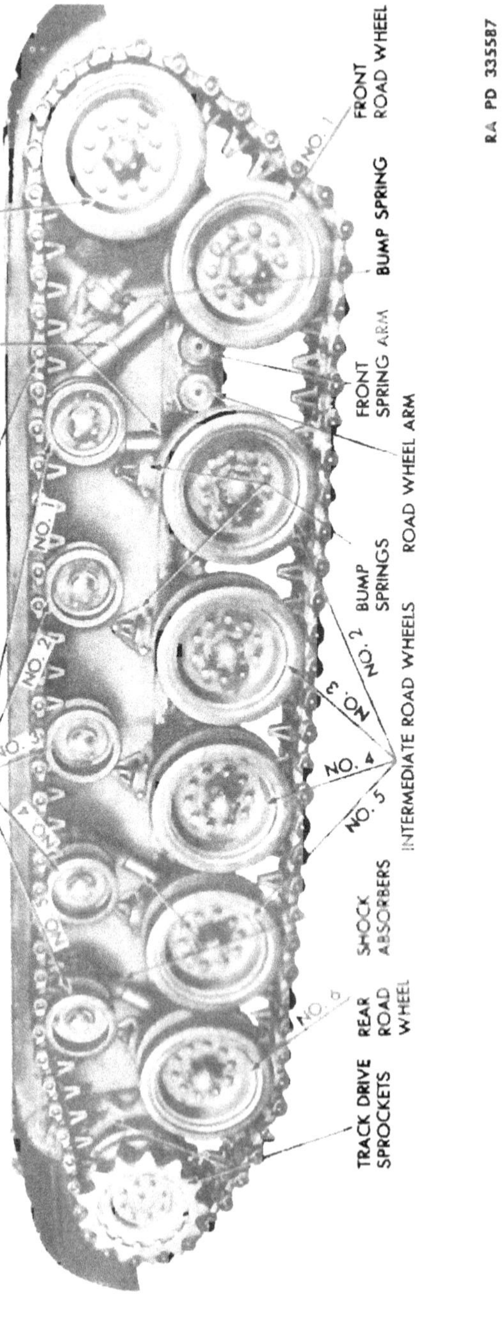

Figure 137 — Suspension System

Tracks and Suspension

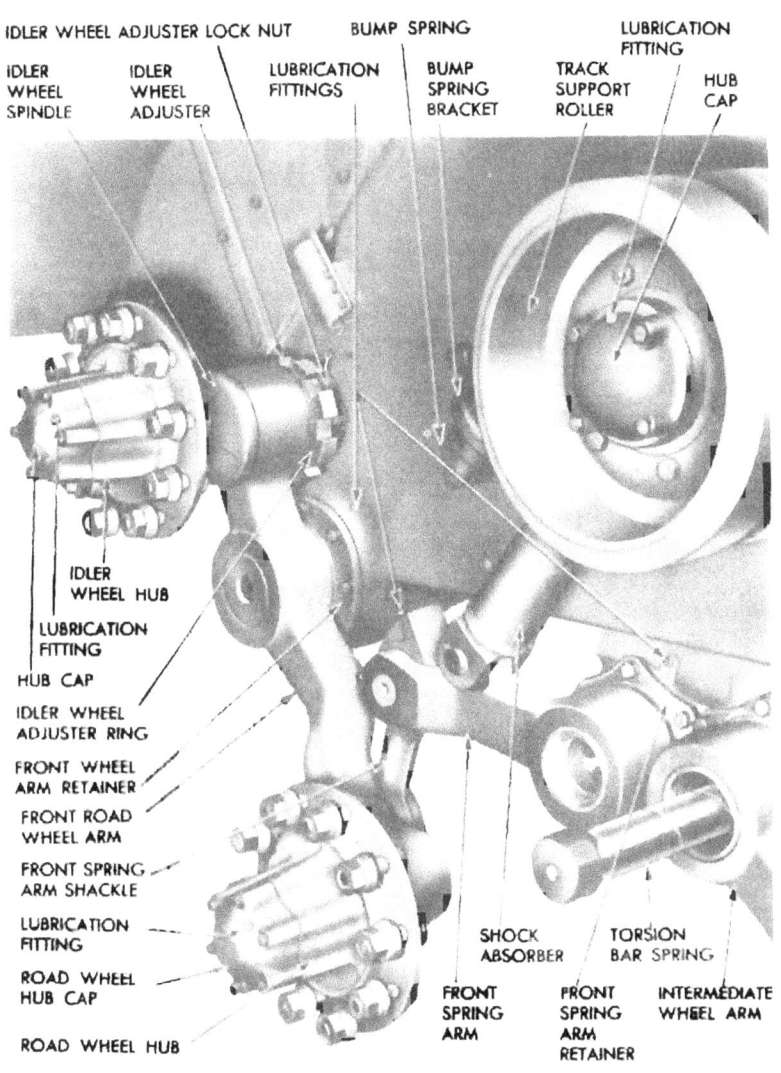

Figure 138 — Left Front Road Wheel Arm (Wheels Removed)

TM 9-735
169

Part Three—Maintenance Instructions

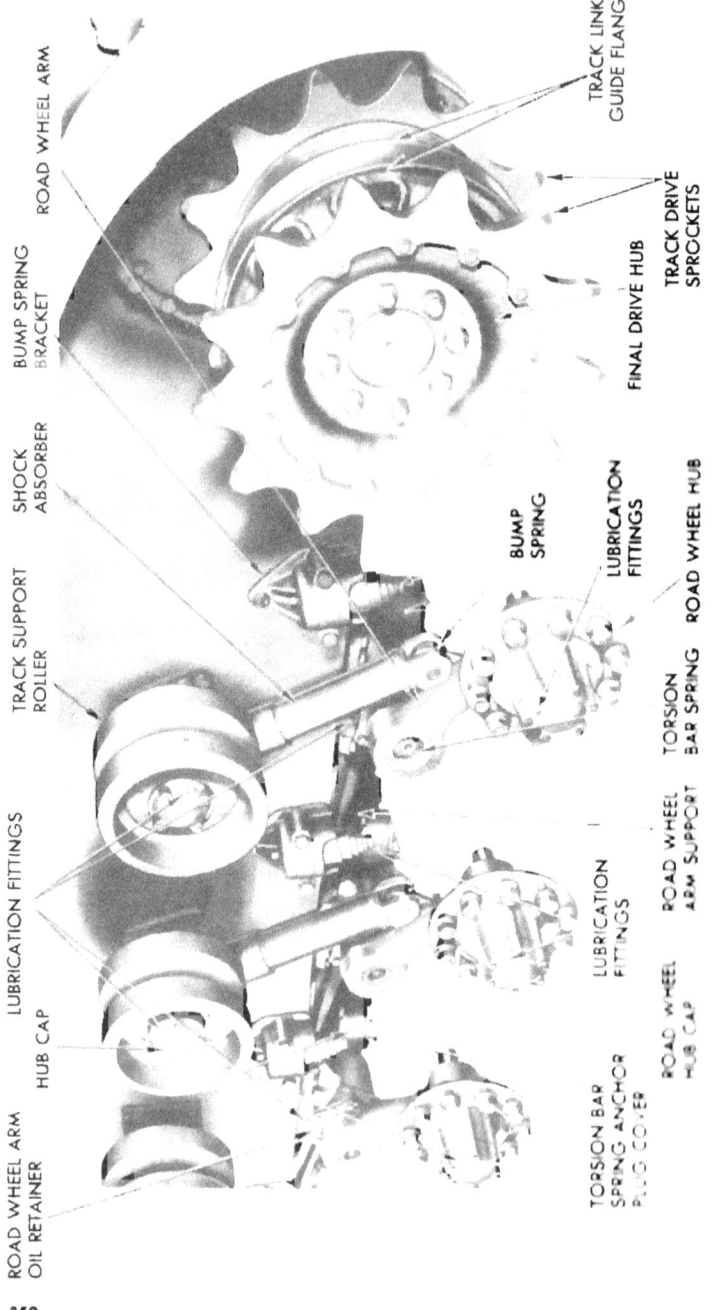

Figure 139 — Road Wheel Arms and Sprockets

Tracks and Suspension

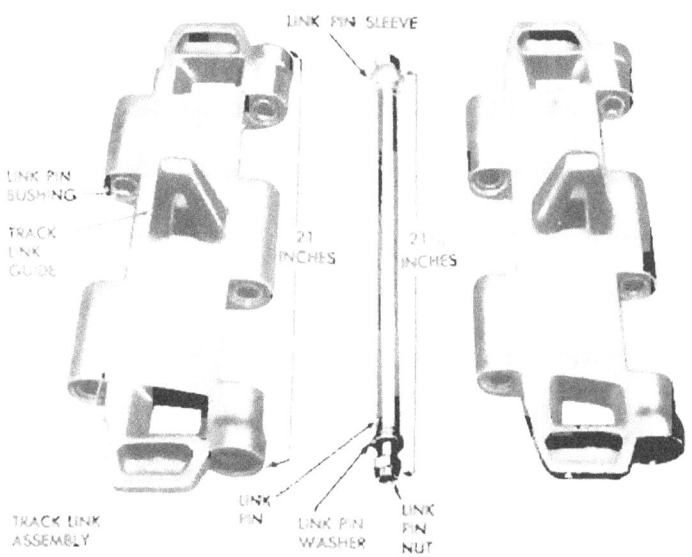

Figure 140 — Track Link Parts

The position of the spindle in the arm is adjustable to provide correct track tension. When the vehicle is in motion track tension is maintained by the forward movement of the idler wheel. As the lower end of the front road wheel arm raises when the track beneath the front road wheel passes over an obstacle, the arm pivots on the front road wheel arm spindle. This pivoting motion moves the idler wheel on the upper end of the front wheel arm forward and downward to take up the slack in the track caused by the obstacle raising the front road wheel. At the same time the front wheel arm shackle transmits the upward thrust to the front spring arm attached to the torsion bar spring in the front spring arm support (fig. 138). The front shock absorber also assists in controlling this upward movement if it gets beyond control of the torsion bar spring and the shock absorber. The front bump spring stops any movement beyond the control of these units.

170. **TRACKS.**

a. **Description.** The tracks consist of 82 all-metal links per track, with hardened surfaces at point of ground contact (fig. 137). Track links are male-and-female hinge type, fastened together by steel track link pins held in place by a hex-socket type link pin sleeve

TM 9-735
170

Part Three—Maintenance Instructions

Figure 141 — Checking Track Tension

serrated on one end, serrated link pin washer, and link pin nut (fig. 140). Track links are drilled through the mating shoulders of each link. Rubber-covered steel link pin bushings, with serrated ends, are installed in each hole. As the link pin nut is tightened, the serrations on the washers, sleeve, and bushings mate with each other and form one long bushing. Links are assembled so the flat inner surface of each link forms an angle of 14 degrees above the surface of the link to which it is connected. This permits the track to follow around the sprockets and idler wheel without abnormal twisting of the bushings. Tracks are installed so the two hinges of each track link face toward the front of the vehicle or forward. The T26E3 track links (fig. 140) extend beyond the track drive sprockets so the sprocket teeth enter holes at each end of the link instead of engaging lugs on the end of the track links as on other full-track laying vehicles. A track link guide in the center of each link guides the track as it passes between the dual road wheels, dual track support rollers, and the track guide flanges on the final drive hubs.

b. *Adjust Track Tension.* Track tension is adjusted by movement of the eccentric idler wheel spindle (fig. 147). *This spindle is designed to work in the downward arc only.* Before making an adjustment, allow vehicle to coast to a stop on level ground without applying brakes. Check track tension by installing a 1-inch spacer between the tops of No. 2 and No. 4 support roller tires and the track (fig. 141). If track has correct tension, No. 3 support roller will have ¼-inch clearance between top of roller tire and the track. If the track has incorrect tension, adjust as follows: Remove front fender and dust shield. Remove idler wheel adjuster lock nut cotter pin (fig. 147). Back off lock nut approximately ½-inch (fig. 142), using wrench (41-W-3250-875). Pry idler wheel adjuster toward lock nut until serrations are separated sufficiently to permit movement of the adjuster (fig. 147). Place track tension adjuster spanner wrench on adjuster and adjust track (fig. 143). Pull up on spanner to tighten track. Push down on wrench to loosen track. Recheck track tension as described in the first part of this paragraph. When correct track tension has been obtained, pry adjuster toward idler wheel until serrations on adjuster and on idler wheel spindle housing match. NOTE: *It may be necessary to turn adjuster slightly in either direction to match serrations.* Tighten adjuster lock nut. Install and spread new cotter pin in lock nut. Remove 1-inch spacers from between No. 2 and No. 4 support roller tires and track. Install front fender and dust shields.

c. *Disconnect Track.* Remove front dust shield and fender. Install the two track connecting fixtures (41-F-2995-155) on track links between idler and front load wheel (A, fig. 144). Remove track link pin nut and serrated washer (fig. 140). Screw up on track

TM 9-735
170

Part Three—Maintenance Instructions

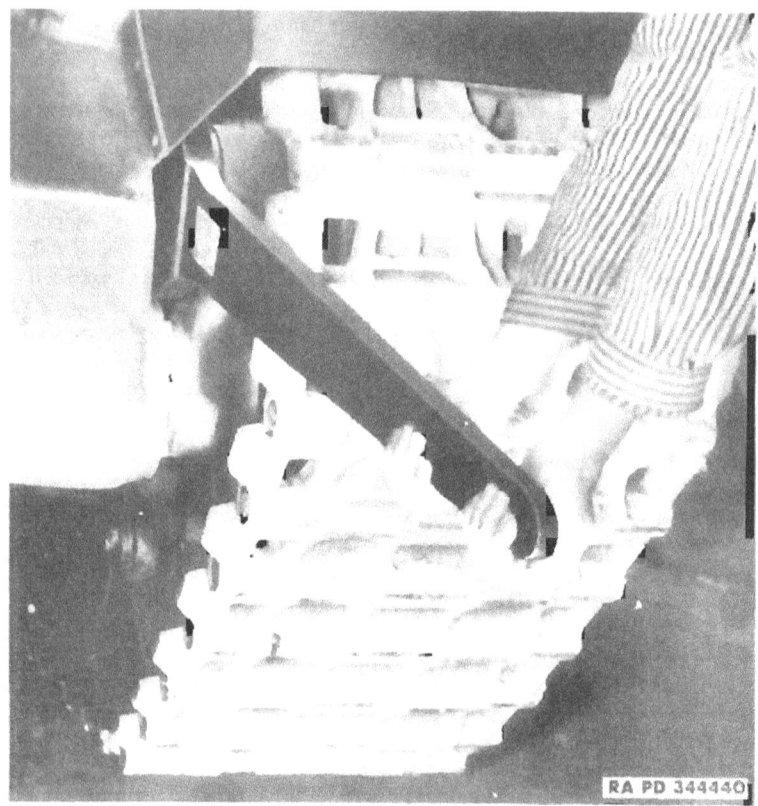

Figure 142 — Loosening Track Adjuster Lock Nut Using Wrench (41-W-1436-25)

connecting fixture until load on link pin is relieved. Install the track link pin removing fixture in holes in outer ends of the two adjacent links (A, fig. 145). Tighten the adjusting screw to fully compress the rubber bushings in the track links. Drive out the link, using the long drift which is part of the fixture (A. fig. 145). Remove the pin removing fixture. CAUTION: *Be sure no personnel is in line with link pin as it is driven from bushings, as it will be expelled with considerable force.* Lift the link pin sleeve out of inner end of track link. Install the two track link bushing pullers in the end bushings of adjacent links (B, fig. 145). Tighten the screws to expand collet securely in bushings. Turn the puller nuts to separate the serrations at inner ends of both outside bushings. CAUTION: *Do not operate*

TM 9-735
170

Tracks and Suspension

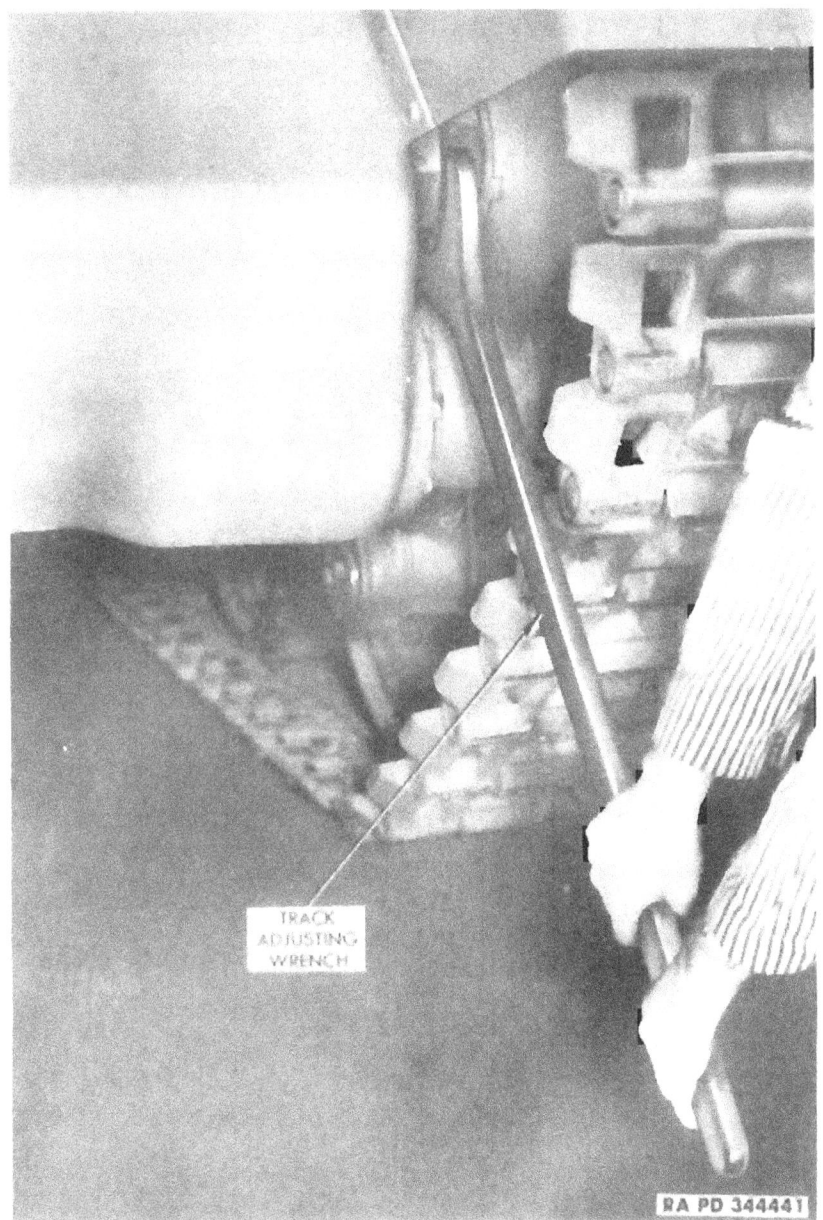

Figure 143 — Adjusting Track Tension

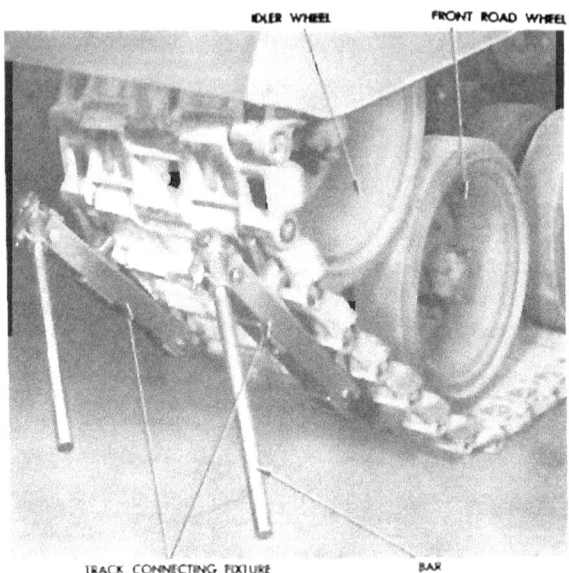

A—TRACK CONNECTING FIXTURE INSTALLED

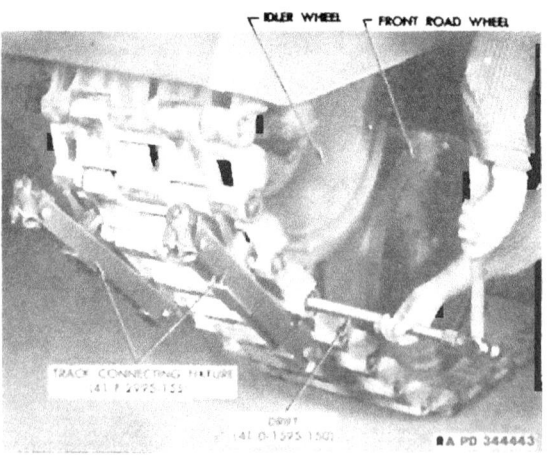

B—DRIVING OUT DRIVE LINK PIN

Figure 144 — Early Method of Disconnecting Tracks

puller more than necessary, as to do so may distort the bushings or loosen the rubber bond in the link. With pullers still in place, back off the track connecting fixtures until track is separated and then remove pullers and fixtures. B, figure 144, illustrates a method of removing the track link pin if special tools illustrated in figure 145 are not available. However, the links must be separated without damaging the serrations on the ends of the track link bushings. If serrations are damaged, the track pin and interior of bushings will rapidly wear, resulting in early failure of these parts.

d. **Connect Tracks.** Set parking brake on track not being connected. Install the bushing pullers and pull bushings away from guide as described in subparagraph c preceding. Mate male and female ends of track links and install track connecting fixtures (subpar. c preceding). Screw up on fixtures evenly until bushings are in line. Remove the bushing pullers. Install serrated link pin sleeve. Install track link pin pilot (41-P-402-400) on end of link pin, and install pin from inside of track. After sleeve and link pin have been installed, tighten track connecting fixtures as tight as possible. This will automatically position track links at an angle of 14 degrees. Remove pilot from end of track link pin and install serrated link pin washer and link pin nut, with round end of nut outward. Tighten nut securely so that serrations on sleeve, bushings, and washer match. Use a torque wrench and tighten nut from 280 to 300 foot-pounds. Loosen and remove track connecting fixtures and stow in brackets on left side of turret. Check track tension (subpar. b preceding). Install front dust shields and fender.

e. **Inspect Track.** See paragraph 53.

f. **Remove Track.** Remove front and rear fenders and dust shields. Disconnect track (subpar. d preceding). Block road wheels to prevent vehicle from moving. Set parking brake on track not being removed. Start engine and place speed range selector lever in reverse (R) speed range. Very slowly reverse the drive sprockets on the disconnected track. As sprockets revolve, sprocket teeth will pull disconnected track over support rollers toward rear of vehicle. Guide track links over rollers with crowbar. As track comes off sprockets, use crowbar and carry track away from vehicle. This prevents track from "piling up" behind final drive and jamming sprockets. As old track length and weight increase use additional crowbars as necessary, or stop vehicle and disconnect track (subpar. c preceding) in sections, and then continue removing track as before. As end of track drops clear of sprockets, place speed range selector lever in neutral (N) and stop engine. Carry end of track away from vehicle until track lies flat on ground.

g. **Install Track.** These instructions cover the situation when the vehicle is standing on part of the track, and the other part is

Part Three—Maintenance Instructions

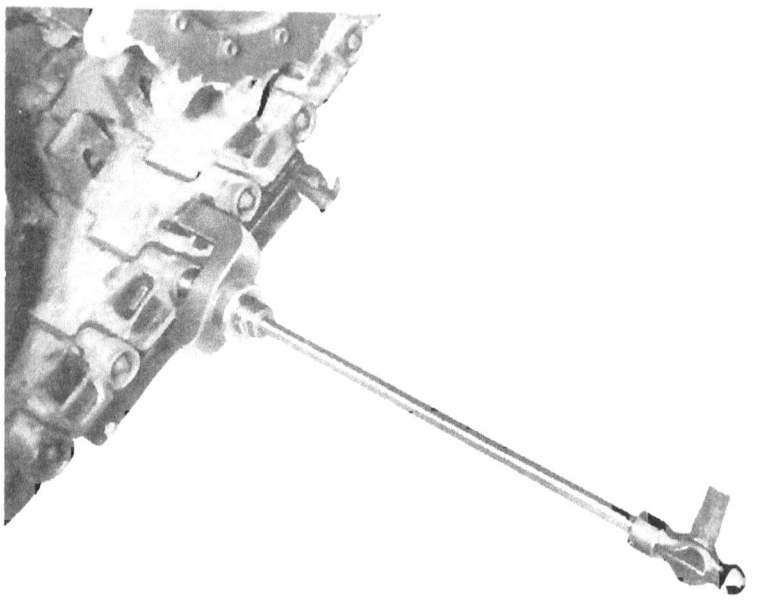

A—DRIVING OUT TRACK LINK PIN

B—SEPARATING SERRATIONS OF INNER ENDS OF OUTER BUSHINGS

RA PD 344442

Figure 145 — Late Method of Disconnecting Tracks

stretched out flat on the ground behind it. Place a crowbar under the seventh or eighth link from free end of track and carry end up to drive sprockets. Lift free end and place first three links on and engage with top drive sprocket teeth. Set parking brake and block opposite road wheels. With crowbar, hold track being installed in position on sprockets. Insert second crowbar between sprocket teeth and wedge end of bar in opening in track link guide flange on drive sprocket hub. Revolve final drive hub with bar and force free end of track over No. 5 support roller. Use another crowbar to guide track links over roller. Be sure speed range selector lever is in neutral (N). When three or four links have passed over roller, remove crowbar and start engine. Place speed range selector lever in first (1) speed range and very slowly revolve drive sprockets. As track moves forward, guide free end over remaining support rollers and idler wheel with crowbar. When track links are tight against sprocket teeth, move speed range selector lever into neutral (N) and stop engine. Connect track (subpar. d preceding). Adjust track tension (subpar. b preceding). Install front and rear fenders and dust shields. Set parking brake on track installed and remove road wheel blocks. NOTE: *Be sure arrows on links point to front of vehicle.*

h. **Install New Track with Old Track on Vehicle.** Remove front and rear fenders and dust shields. Remove old track (subpar. f preceding). Place new track on ground in a straight line extending from end of old track. Be sure track links or new tracks are pointing in the same direction as links on track. Connect new track to old track (subpar. d preceding). Place a wood block between the seventh or eighth track link guides from outer end of new track. Start engine and place speed range selector lever in reverse (R) speed range. Back vehicle slowly. When rear road wheel contacts wood block near end of new track, stop vehicle and engine and move speed range selector lever into neutral (N). Remove wood block. Disconnect old track from new track (subpar. e preceding). Install new track (subpar. g preceding). Connect track and adjust tension (subpar. d and h preceding). Install front and rear fenders and dust shields.

i. **Install New Track with Old Track Removed.** Place new track on the ground ahead of the vehicle and directly in line with the front road wheel, sides of the links with two hinges pointing toward rear of vehicle. Place a metal plate or plank on first track link nearest vehicle to act as a ramp; or dig a trench under first track link so link can drop down into depression and road wheels can roll directly onto track. Place a wood block between the third and fourth track link guides from outer end of new track. Drive two stakes into the ground at front end of track to prevent track from being pushed ahead when vehicle is moving up onto track. Remove front and rear fenders and dust shields. Tow or push vehicle ahead onto track until

TM 9-735
170

Part Three—Maintenance Instructions

Figure 146 — Front Road Wheel Lifting Tool in Position

front road wheel contacts wood block. Lift rear end of track and start it on drive sprockets as described in subparagraph g preceding. Install track (subpar. g preceding). Remove wood block. Connect track (subpar. d preceding). Adjust track tension (subpar. h preceding). Install front and rear fenders and dust shields.

j. **Install Thrown Track.** Follow procedure given in subparagraph i preceding. Replace any damaged track links as described in subparagraph k following.

k. **Install New Track Links.** Follow procedure given in subparagraphs c and d above for disconnecting and connecting track. Repeat operations in subparagraph c preceding to remove second track link pin. Install new track link (subpar. d preceding). Connect and adjust track tension. (subpars. d and h preceding). NOTE: *Be sure arrows on new link correspond in direction to arrows on other links.*

171. ROAD WHEELS, IDLER WHEELS, AND HUBS.

a. **Description.** Twenty-four interchangeable large-diameter, disk-type wheels with hard rubber tires support the vehicle and roll on the all-metal tracks. There are six pairs (dual) of road wheels on each side of the vehicle (fig. 4). An idler wheel of the same size, type, and construction as the road wheels is mounted at the upper end of the front road wheel arm on each side of the vehicle. Road wheels are referred to numerically, numbering from front of vehicle (fig. 155). The No. 1 road wheel is also called the front road wheel. The No. 2, 3, 4, and 5 road wheels are also known as the intermediate road wheels. The No. 6 wheel is also known as the rear road wheel. Each road wheel and idler wheel is individually mounted on a hub which, in turn, is mounted on roller bearings on the wheel arm spindles. Wheels are bolted together in pairs and their construction forms a groove in which the track link guides travel. Wheels and hubs are removed separately. Wheel hubs are only removed when it is necessary to inspect or replace bearings or oil retainers.

b. **Remove Intermediate and/or Rear Road Wheels.**

(1) LOOSEN HUB NUTS. With road wheel still on the track, loosen all hub nuts, but do not remove.

(2) RAISE ROAD WHEEL. Slide the upper end of intermediate road wheel lifting fixture onto inner end of wheel spindle. Place lower end of fixture on the track ahead of the wheel to be lifted. Drive the vehicle slowly forward making sure end of fixture engages in sprocket tooth hole in track link. Stop vehicle when fixture is vertical and wheel is fully raised.

(3) REMOVE ROAD WHEEL. Remove the hub nuts and washers and lift wheels off hubs.

c. **Install Intermediate and/or Rear Road Wheels.**

(1) ATTACH WHEELS TO HUB. Position the two wheels on the hubs. Install the washers and hub nuts. Tighten wheel nuts sufficiently to hold wheels securely.

(2) LOWER WHEEL ONTO TRACK. Drive the vehicle slowly forward until wheel is lowered onto track and fixture is disengaged from track link. Remove the fixture.

(3) TIGHTEN WHEEL HUB NUTS. Use torque wrench and tighten all hub nuts using 280 to 300 foot-pounds of torque.

d. **Remove Front Road Wheel.**

(1) LOOSEN HUB NUTS. Before raising wheel, loosen but do not remove the hub nuts.

(2) RAISE FRONT ROAD WHEEL. Release track tension (par. 170 b) so as not to stretch track or subject it to excessive strain when raising front road wheel. At inner side of track, position the

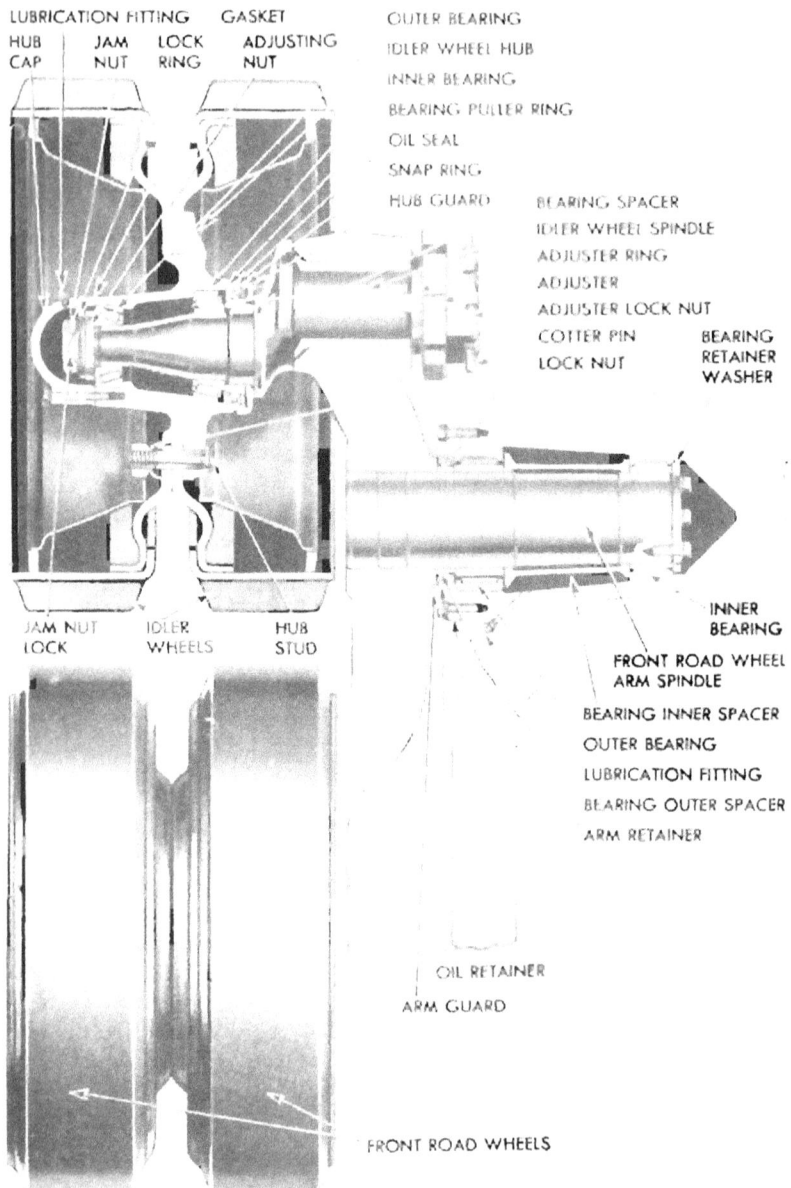

Figure 147 — Idler Wheel and Front Road Wheel Arm — Sectional View

Tracks and Suspension

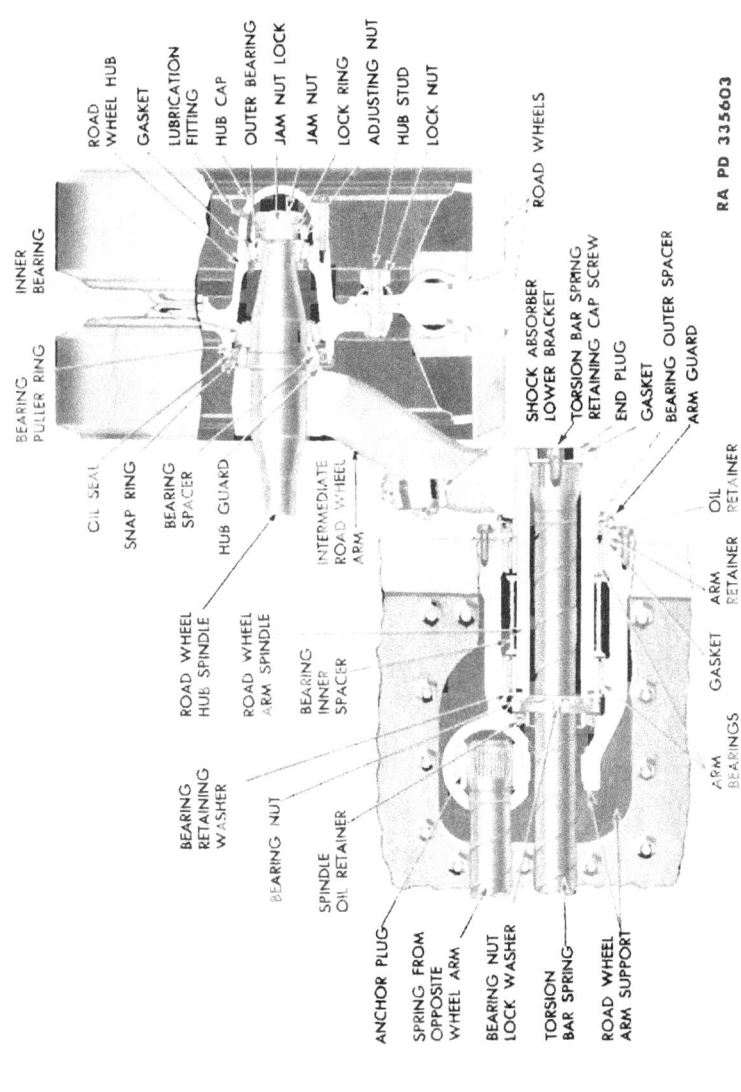

Figure 148 — Road Wheel Arm Support — Sectional View

TM 9-735
171

Part Three—Maintenance Instructions

Figure 149 — Installing Inner Bearing Cup in Road Wheel Hub

front road wheel lifting fixture (fig. 161) in a nearly horizontal position with the short tip of the rounded end on top of the eye of the front spring arm shackle which connects to the front road wheel arm spindle (fig. 160). Place the lower or forked end of the fixture on the track, with the lower fork resting in a sprocket tooth hole in the track link. Drive the vehicle slowly forward until fixture is in a vertical position and front road wheel is fully raised (fig. 146).

(3) REMOVE FRONT ROAD WHEEL. Remove the hub nuts and washers, and lift both wheels off hub studs.

e. Install Front Road Wheel.

(1) ATTACH WHEELS TO HUB. Position the two wheels on the hubs. Install the washers and hub nuts. Tighten wheel nuts sufficiently to hold the wheels securely.

TM 9-735

Tracks and Suspension

Figure 150 — Installing Oil Seal in Road Wheel Hub

(2) LOWER WHEEL ONTO TRACK. Drive the vehicle slowly backward until wheel is lowered onto track and fixture is disengaged from track link. Remove the fixture.

(3) TIGHTEN WHEEL HUB NUTS. Use torque wrench and tighten all hub nuts using 280 to 300 foot-pounds of torque. Adjust track tension (par. 170 b) before driving vehicle.

f. **Remove Idler Wheel.** Remove front fender and dust shield, loosen hub stud nuts, but do not remove. Disconnect track (par. 170 c) and move upper section back off idler wheel. Remove hub nuts, work idler wheel off hub studs and lower wheel to ground.

g. **Install Idler Wheel.** Lift idler wheel into position. Install wheel on hub studs. Install hub stud nuts finger-tight. Move track forward over idler wheel and connect and adjust track (pars. 170 d

TM 9-735
171

Part Three—Maintenance Instructions

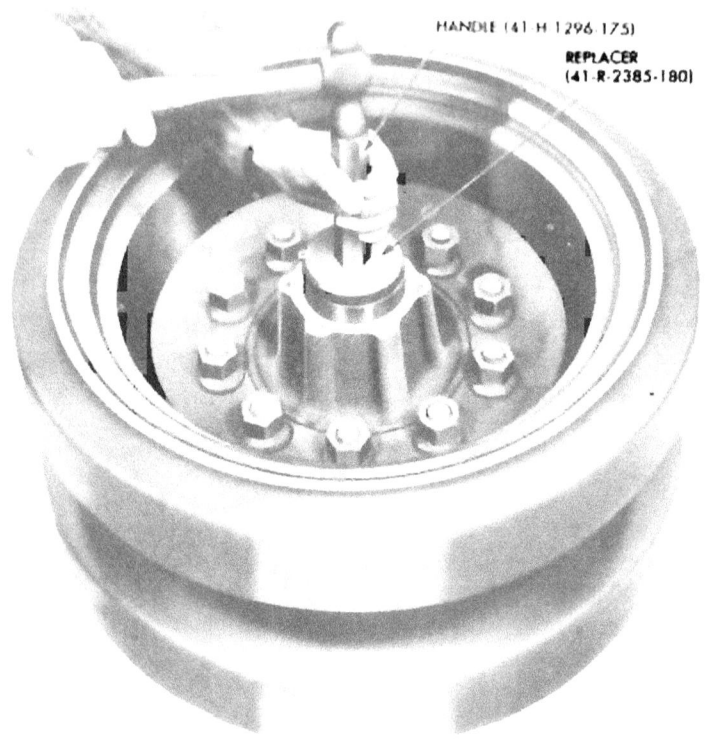

RA PD 335606

Figure 151 — Installing Outer Bearing Cup in Road Wheel Hub

and 170 b). Tighten hub nuts securely. Use a torque wrench and tighten each nut to from 280 to 300 foot-pounds. Install front fender and dust shield.

h. Remove Road or Idler Wheel Hub and Outer Bearing. Remove road and/or idler wheel as described in subparagraphs b and/or f preceding. Remove screws from hub cap and lift off cap. Remove gasket. Wipe out surplus grease. Flatten end of jam nut. Remove jam nut, using socket wrench (41-W-3058-470) on socket wrench sliding bar (41-B-312-200) with male head (41-H-1779-50). Remove lock ring. Unscrew outer bearing adjusting nut (fig. 152), using spanner wrench (41-W-3242-300). Install hub cap temporarily to prevent outer bearing dropping onto ground when hub is removed. Pull hub and bearing assemblies off spindle. Remove inner and outer

366

Tracks and Suspension

RA PD 335607

Figure 152 — Adjusting Road Wheel Bearings, Using Wrench (41-W-3242-300)

bearing assemblies (subpar. i following) and inspect parts (subpar. j following).

i. **Remove Hub Bearing Cups and Oil Retainer.** Remove wheel hub (subpar. h preceding). Remove temporarily installed hub cap and lift out outer bearing. Place hub on end and remove inner bearing snap ring (fig. 147). Drive out inner bearing oil retainer. NOTE: *Oil retainers, once removed, cannot be replaced.* Lift out inner bearing puller ring, and then lift out bearing. Wipe out hub and examine bearing cups. If cups are scored or damaged, remove them by driving out of hub with drift. If cups are in good condition, do not remove them. Clean and inspect bearings (subpar. j following).

TM 9-735
171

Part Three—Maintenance Instructions

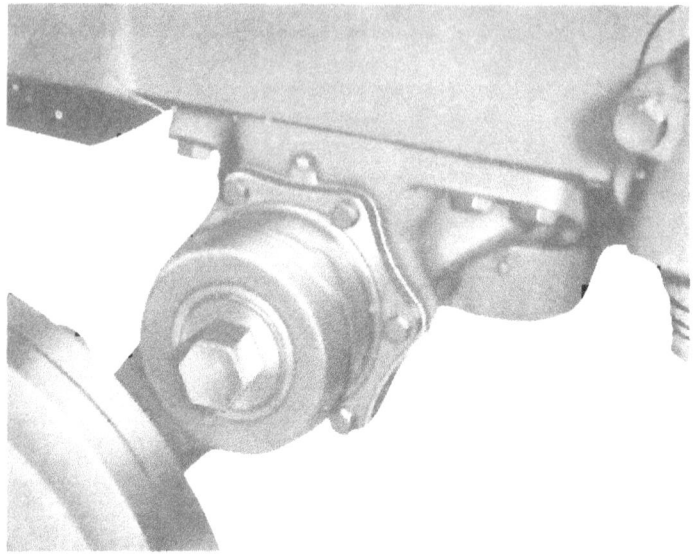

Figure 153 — Wrench (41-W-1961-125) Installed in End Plug

j. **Clean and Inspect Bearings.** Remove bearings and clean in dry-cleaning solvent. Clean old lubricant out of hub and from parts, using dry-cleaning solvent. Examine rollers, cage, cup, and cone for wear and scoring. Inspect condition of guard and bearing spacer on arm spindle. Examine bearing cones for wear, scoring, or loose fit. Inspect spindles and bearing cups for scoring or other damage. If bearings are damaged, replace with new bearings, cups, and cones. If spindles are out of line, scored, or otherwise damaged, replace wheel arm (par. 173 b and e).

k. **Install Hub Bearing Cups and Oil Retainer.** Install bearing cups with narrow lip outward. For inner bearing cup (fig. 149), use replacer (41-R-2384-967) and handle (41-H-1296-175). For outer bearing cup (fig. 151), use replacer (41-R-2385-180) and handle (41-H-1296-175). Pack bearing with grease as instructed in paragraph 46. Install bearing in cup. Install bearing puller ring (fig. 147). Install oil retainer with spring (or lip) side up (fig. 150), using replacer (41-R-2395-465) and handle (41-H-1296-175). CAUTION: *Use great care in installing oil retainer so as not to bend edge of lip down over coil spring. Be sure leather seal is in correct position.* Install snap ring. Be sure ring is seated securely in snap ring groove. Position hub on spindle and install outer bearing as described in subparagraph l following.

l. **Install Road or Idler Wheel Hub and Outer Bearing.** Be sure inner bearing is completely assembled, packed with grease, and correctly installed (fig. 147). Install hub on spindle, seating it firmly against bearing spacer and guard. Pack outer bearing with grease as instructed in paragraph 46. Seat bearing securely in hub (fig. 147). Install outer bearing adjusting nut. Tighten nut securely (fig. 152), using spanner wrench (41-W-3242-300). Place adjusting nut lock ring on spindle, and determine position of dowel pin on adjusting nut and nearest dowel pin hole in lock ring. Remove lock ring from spindle, and back off adjusting nut far enough so that when lock ring is reinstalled, dowel pin and dowel pin hole will mate. Install adjusting nut lock ring with dowel pin seated in dowel pin hole. Install jam nut lock. Install jam nut and tighten securely with socket wrench (41-W-3058-470). Turn up edge of jam nut lock against a flat on jam nut. Install new gasket. Coat edge of hub cap with joint sealing compound. Install hub cap. Install road or idler wheel (subpars. c or g preceding). Lubricate hub until grease can be seen at oil retainer as directed in paragraph 46.

172. TORSION BAR SPRINGS AND ANCHOR PLUGS.

a. **Description.** Torsion bar springs are made of high carbon steel, and have a protective covering which must not be removed (fig. 138). Serrations on the ends of the bars match with serrations in the road wheel arms, and in the torsion bar spring anchor plugs. NOTE: *To facilitate installation and load adjustment a serration has been eliminated on each end of each torsion bar spring (fig. 155).* The torsion bar springs, through a twisting motion or torsional resistance, act as a spring in controlling the upward thrust of the road wheel arms and thus cushion the vertical movement of the vehicle. Torsion bar springs require no attention except to be replaced when broken. Springs are identified for correct installation by a rotational arrow and a part number followed by a letter symbol stamped on the outside end of each bar (fig. 155). The letter following each part number indicates in which support housing the spring is installed. See table in figure 155.

b. **Testing for Broken Torsion Bar Spring.** Test for a broken torsion bar spring by placing end of a crowbar under road wheel connected to torsion bar spring suspected of being broken, and attempt to lift wheel. All road wheels have an equal amount of load. If wheel being tested can be lifted with crowbar it indicates torsion bar spring is broken. All pieces of a broken spring must be removed before installing a new spring. If end of spring is jammed in torsion bar spring anchor plug, road wheel arm support must be removed (par. 174 h), and jammed pieces of spring cleared from plug.

TM 9-735
172

Part Three — Maintenance Instructions

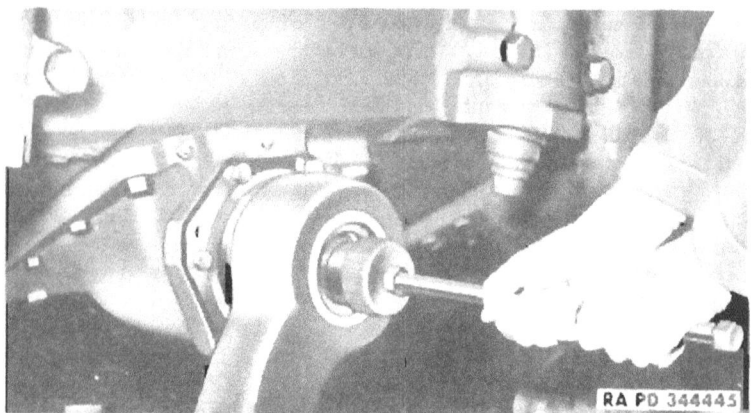

Figure 154 — Installing or Removing Torsion Bar Spring Using Bar (41-B-2378-950)

c. **Remove Torsion Bar Spring.** Remove road wheel (par. 171 b or d). Lower jack until road wheel arm is down in fully released, no-load position. Back off screws and loosen cover plate on anchor plug in opposite road wheel arm support. Remove torsion bar spring retaining cap screw. Using wrench (41-W-1961-125), remove torsion bar spring end plug (fig. 153). Install torsion bar spring removing bar (41-B-2378-950), and pull spring out of support (fig. 154). NOTE: *With removing bar still installed it may be necessary to work spring up and down with tool before it can be broken loose from road wheel arm serrations.* While spring is out of supports, remove anchor plug cover plate from opposite support, lift out plug, inspect, and reinstall. Do not tighten anchor plug cover cap screws.

d. **Install Torsion Bar Spring.** Jack up wheel arm with jack placed under stub axle and adjust arm to correct dimensions (subpars. e and f following). Install torsion bar spring, using bar (41-B-2378-950). NOTE: *See figure 155 for correct bar to install.* When installing spring line up blank serration in road wheel arm and on end of spring. Index mark on wheel arm shows location of blank serration in arm. Install spring with identifying marks facing out. Install torsion bar spring end plug, using wrench (41-W-1961-125). Install torsion bar spring retaining cap screw and lock washer. After torsion bar spring has been installed, hold arm in position with jack, and install road wheel (par. 171 c or e). Tighten anchor plug cover plate cap screws. Lower and remove jack.

e. **Adjust Road Wheel Arm and Torsion Bar Spring Tension.** There are four separate adjustments required to correctly adjust torsion bar spring tension on the T26E3. Instructions covering each adjustment are given in the following subparagraphs. NOTE: *All dimensions are plus or minus ¼ inch (fig. 155).*

(1) FRONT SPRING ARM (LEFT). Disconnect track (par. 170 c). Remove left idler wheel (par. 171 f). Remove left front road wheel (par. 171 d). Using suitable lifting equipment to support front spring arm, adjust idler and front road wheel linkage so front spring arm is in no-load or fully released position. Remove old torsion bar spring (subpar. e preceding). Adjust left front spring arm so inside center of front road wheel spindle measures 15 1/16 inches from bottom outside edge of hull side plate (fig. 155). Hold linkage in position and install torsion bar spring (subpar. d preceding) stamped "G" (fig. 155). Install road wheel (par. 171 e). Connect and adjust track (par. 170 b and d).

(2) FRONT SPRING ARM (RIGHT). Repeat procedure described in step (1) preceding. Adjust right front spring arm so inside center of front road wheel spindle measures 15 1/16 inches from bottom outside edge of hull side plate (fig. 155). Hold linkage in position and install torsion bar spring stamped "H" (subpar. d preceding) (fig. 155).

(3) INTERMEDIATE ROAD WHEEL ARMS (BOTH RIGHT AND LEFT). Remove road wheel (par. 171 b). Lower road wheel arm to no-load position and remove old torsion bar spring (subpar. e preceding). Adjust road wheel arm so inside center of road wheel spindle No. 2, 3, and 4 measures 14 7/16 inches from bottom outside edge of hull side plate and No. 5 measures 14 1/8 inches (fig. 155). Install torsion bar spring (subpar. d preceding). NOTE: *Refer to table in figure 155 for correct torsion bar springs to install.* Install road wheel (par. 171 e).

(4) ROAD WHEEL ARM NO. 6 (REAR—BOTH RIGHT AND LEFT). Disconnect track (par. 170 c), and move track back of drive sprocket to release pressure on rear road wheel. Remove road wheel (par. 171 b). Lower road wheel arm to no-load position, and remove old torsion bar spring (subpar. e preceding). Adjust road wheel arm so inside center of road wheel spindle measures 14 1/8 inches from bottom outside edge of hull side plate (fig. 155). Install torsion bar spring stamped "E" (subpar. d preceding) on right side, and torsion bar spring stamped "F" on left side (fig. 155).

f. **Torsion Bar Spring Anchor Plugs.** Anchor plugs are solid steel cylinders with a splined hole through the center. The splined end of the torsion bar spring is seated or anchored, in the splined hole of the plug. An anchor plug is located in each wheel arm support and is held in place by a cover plate.

Part Three—Maintenance Instructions

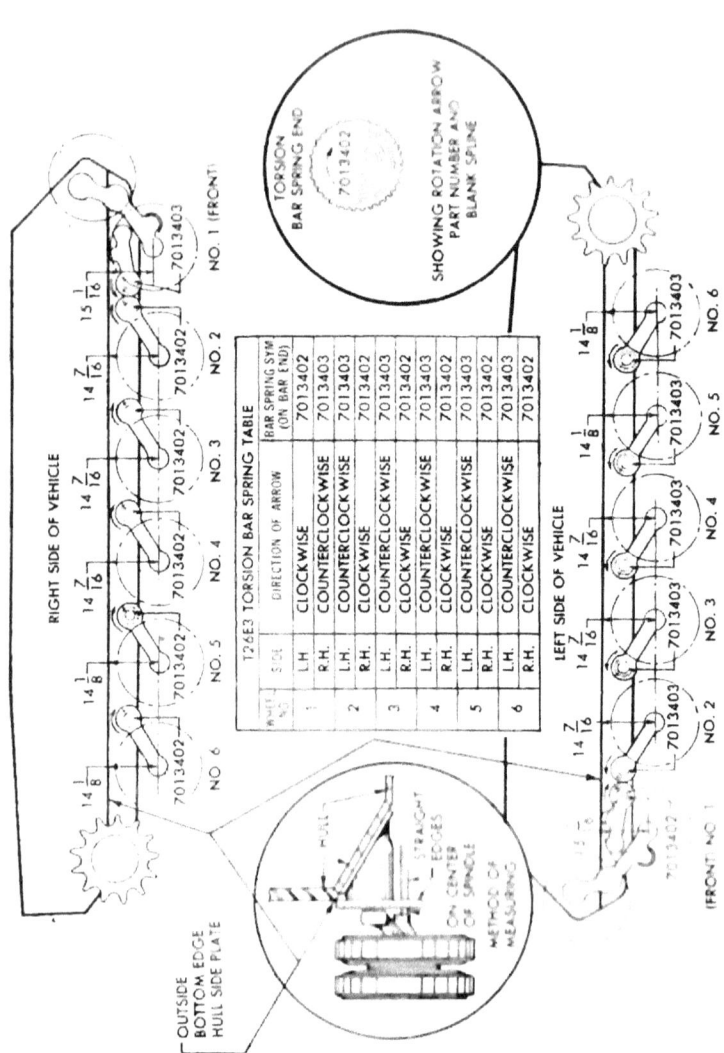

Figure 155 — Torsion Bar Spring Identification and Adjustment Chart

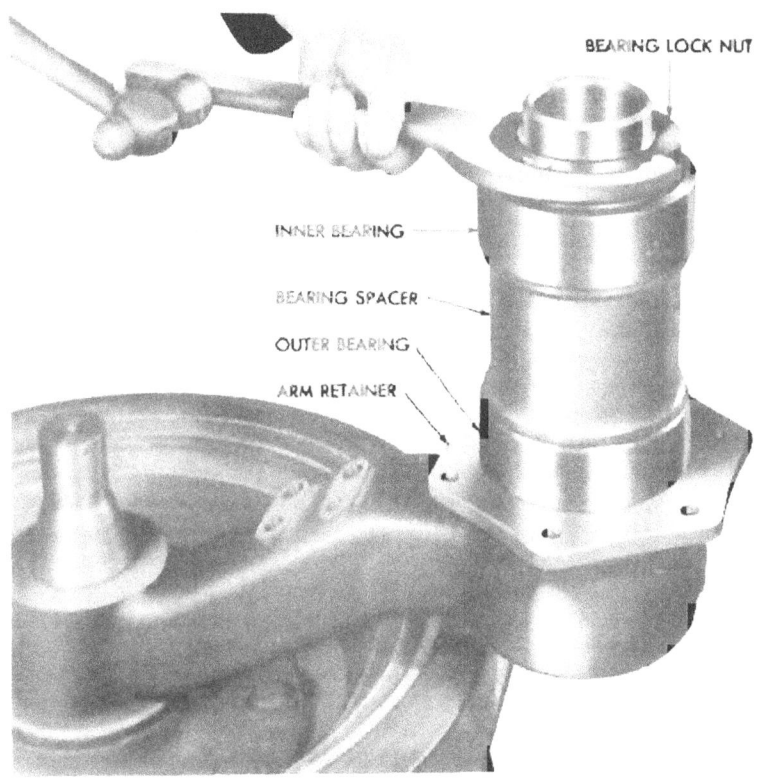

Figure 156 — Tightening Road Wheel Arm Bearing Nut, Using Spanner Wrench (41-W-3252-375)

g. **Remove Torsion Bar Spring Anchor Plug.** Remove torsion bar spring (subpar. e preceding). Remove anchor plug cover from bottom of wheel arm support. As cover is removed, let plug drop out of support.

h. **Install Torsion Bar Spring Anchor Plug.** Position anchor plug in wheel arm support. Install torsion bar spring (subpar. d preceding). Be sure splines on torsion bar spring end mate with splines in anchor plug. Seat torsion bar spring and anchor plug securely. Install anchor plug cover and gasket. Adjust road wheel arm and torsion bar spring tension (par. 172 e). Tighten cover screws securely.

173. ROAD WHEEL ARMS.

a. **Description.** The road wheel arm (fig. 148) is a heavy forging with a solid road wheel hub spindle in one end and a hollow road wheel arm spindle in the opposite end. An inner and outer needle bearing supports the hollow road wheel arm spindle in the road wheel arm support. The front road wheel is mounted on a solid road wheel hub spindle at the lower end of the front road wheel arm. This spindle is connected to the front spring arm by the front wheel arm shackle (fig. 160). The arm and both spindles are serviced as a complete assembly.

b. **Remove Road Wheel Arm.** Disconnect shock absorber from road wheel arm (par. 177 b). Remove road wheel (par. 171 b and d). Remove torsion bar spring (par. 172 c). Remove cap screws attaching road wheel arm retainer to road wheel arm support. Pull road wheel arm spindle and inner and outer bearings and inner bearing oil retainer out of support. Remove gasket. NOTE: *Inner and outer bearings and inner bearing oil retainer will remain on road wheel arm spindle as it is removed from support.* Inspect bearings (par. 171 j).

c. **Remove Road Wheel Arm Bearings and Oil Retainer.** Remove road wheel arm (subpar. b preceding). Lift off inner bearing oil retainer (fig. 148). NOTE: *When oil retainers have been once removed they cannot be reinstalled.* Flatten edge of lock washer. Use spanner wrench and remove bearing lock nut (fig. 156). Lift off lock washer. Remove bearing retainer ring. Lift off bearing. After inner bearing has been removed, lift bearing spacer off spindle. Lift off outer bearing assembly. Lift off road wheel arm retainer, and remove oil retainer by driving out with drift. Inspect bearings (par. 171 j).

d. **Install Road Wheel Arm Bearings and Oil Retainer.** Pack bearing with grease as directed in paragraph 46. Reverse removal procedure (subpar. c preceding). Install new oil retainer in wheel arm retainer (fig. 157), using replacer (41-R-2392-65) and handle (41-H-1296-175). NOTE: *Coat inside of inner race of bearing with micronized graphite (type 200-16) or light graphited grease (ORD Dept. AXS-683) before assembling.* Install inner bearing oil retainer using replacer (41-R-2395-90).

e. **Install Road Wheel Arm.** Install new gasket on face of road wheel arm retainer and cover with joint sealing compound. Be sure inner and outer bearings are completely assembled, packed with grease, and installed correctly (subpar. d preceding). Position road wheel arm assembly, and install spindle in road wheel arm support. Install road wheel arm retainer against support. Adjust arm to the correct dimension (par. 172 c). Install torsion bar spring and adjust

TM 9-735
173

Tracks and Suspension

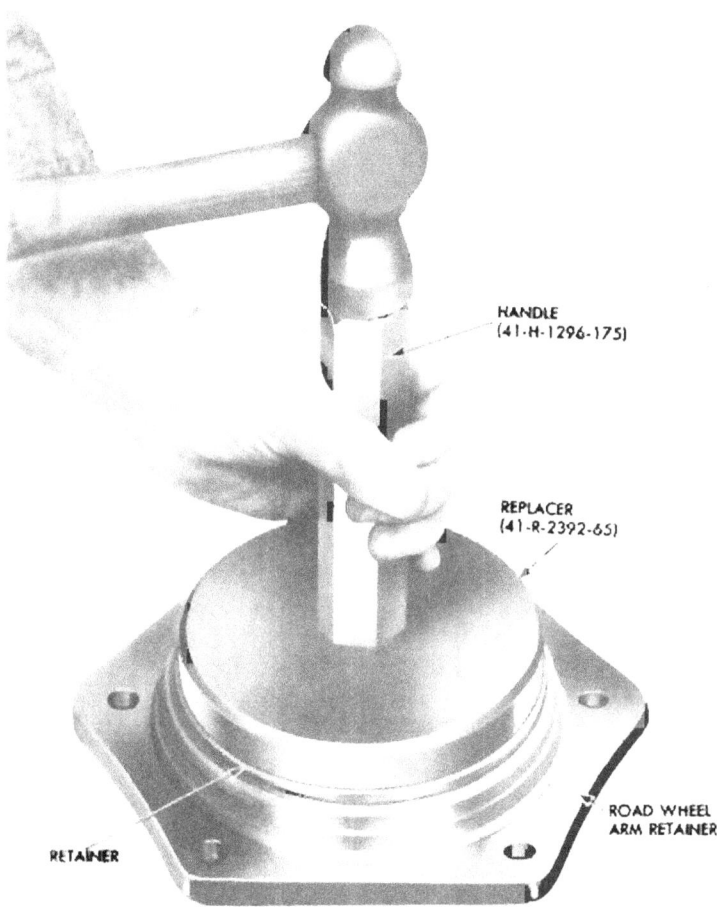

Figure 157 — Installing Oil Retainer in Road Wheel Arm Retainer

tension (par. 172 d and e). Install road wheel (par. 171 c or e). Connect shock absorber (par. 177 c). Lubricate bearings as directed in paragraph 46.

f. Remove Front Spring Arm. Remove front fender and dust shield. Disconnect track (par. 170 c). Place 20-ton jack under front corner of vehicle on side from which front spring arm is to be removed (fig. 158). Block up vehicle as a safety factor. Raise jack sufficiently to take vehicle weight off front road wheel. Remove

TM 9-735

Part Three—Maintenance Instructions

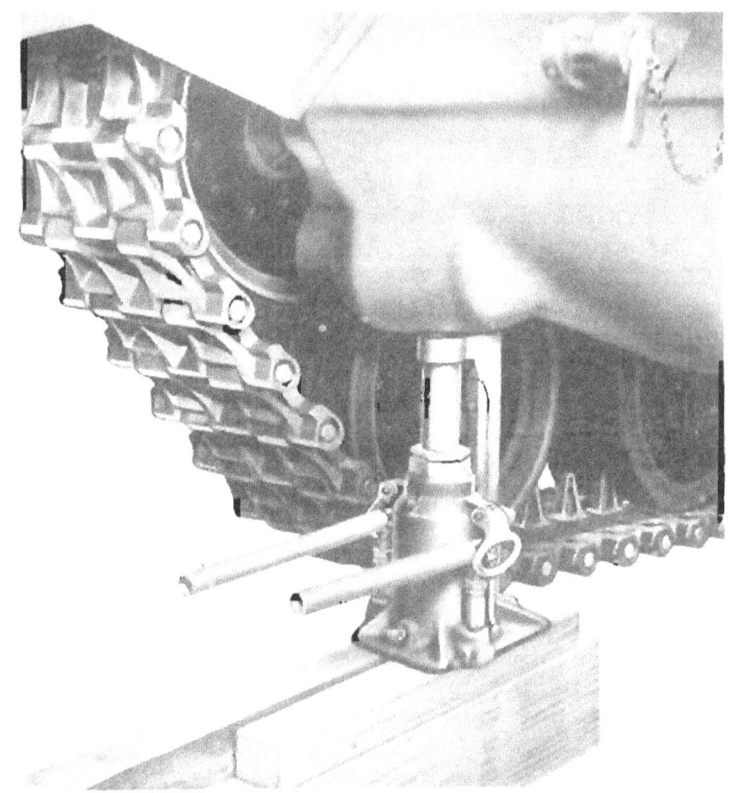

RA PD 335613

Figure 158 — Front of Vehicle Raised on Jack

(par. 171 f) or block up idler wheel. Remove front road wheel (par. 171 d). Disconnect shock absorber from front spring arm (par. 177 h). Place small hydraulic jack under front spring arm. Disconnect front spring arm shackle from front spring arm (par. 176 h). Lower jack until front spring arm is in no-load position. Remove jack. Remove torsion bar spring (par. 172 c). Remove front spring arm retainer from front spring arm support. Pull spindle out of support and remove front spring arm assembly (fig. 138) with suitable hoisting equipment. NOTE: *Inner and outer bearings will come away with spindle.* Remove gasket. Inspect bearings (par. 171 j).

Tracks and Suspension

g. **Remove Front Spring Arm Bearings and Oil Retainer.** Remove front spring arm (subpar. f preceding) (fig. 138). Follow removal procedure in subparagraph c preceding.

h. **Install Front Spring Arm Bearings and Oil Retainer.** Follow installation procedure described in subparagraph d preceding (fig. 148). Install front spring arm (subpar. i following).

i. **Install Front Spring Arm.** Install new gasket. Cover face of front spring arm retainer with joint sealing compound. Be sure bearings are completely assembled, packed with grease, and securely in place. Position front spring arm, and install spindle in front spring arm support (fig. 138). Install retainer against support. Adjust arm to correct dimension (par. 172 e). Install torsion bar spring (par. 172 d) and adjust tension (par. 172 e). Connect front spring arm shackle (par. 176 c). Install front road wheel (par. 171 e). Install idler wheel (par. 171 g) and/or remove blocking. Remove 20-ton jack and safety blocking from beneath vehicle. Connect and adjust track (pars. 170 d and 170 b). Install front fender and dust shield. Lubricate bearings as directed in paragraph 46.

174. ROAD WHEEL ARM SUPPORTS.

a. **Description.** Each road wheel arm spindle is supported in a road wheel arm support which is bolted to the hull slope (figs. 148 and 162). There are 10 supports spaced at intervals along both sides of the vehicle, 5 on each side (fig. 162). One end of each torsion bar spring is held in the anchor plug in each road wheel arm support (fig. 148). Each road wheel arm spindle is supported in the support on two needle bearings.

b. **Remove Road Wheel Arm Support.** Remove road wheel arm from road wheel arm support to be removed (par. 173 b). Remove corresponding road wheel from hub on opposite side of vehicle (par. 171). Remove both torsion bar springs from support to be removed. Remove cap screws attaching support to hull, and lift off support. Remove oil retainer and torsion bar spring anchor plug. NOTE: *If rear road wheel arm support is being removed, disconnect track (par. 170 c), and move track back off sprocket.*

c. **Install Road Wheel Arm Support.** Cover face of road wheel arm support with joint sealing compound. Install oil retainer, using replacer (41-R-2392-65). Install road wheel arm support against hull. Using torque wrench, tighten all cap screws to 150 foot-pounds. Install road wheel arms (par. 173 e). Adjust road wheel arms to correct dimensions (par. 172 e). Install and adjust torsion bar springs (par. 172 d). Install road wheels (par. 171). Connect and adjust track is disconnected (pars. 170 d and b).

TM 9-735
174

Part Three—Maintenance Instructions

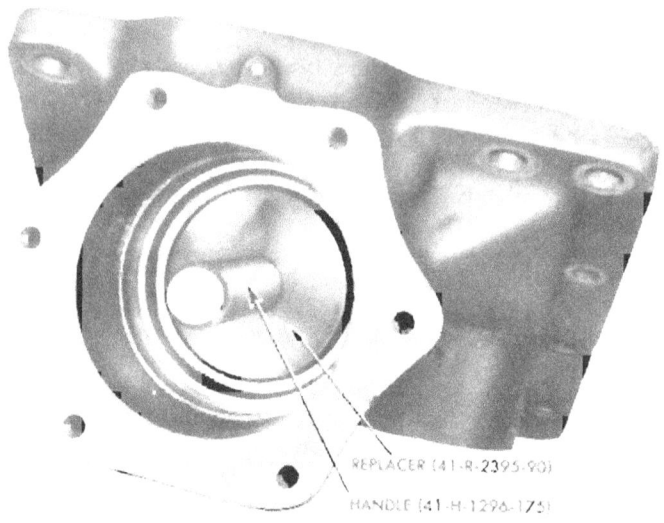

Figure 159 — Installing Inner Oil Seal in Road Wheel Arm Support

d. Remove Front Spring Arm Support. Remove right and left front fenders and dust shields. Place a 20-ton jack and wood blocking under each front corner of the vehicle (fig. 158). Raise jacks evenly to take weight of vehicle off front road wheels. Disconnect both tracks (par. 170 c). Remove both idler wheels (par. 171 f). Remove both front road wheels (par. 171 d). Remove both right and left No. 2 road wheels. Remove cap screws attaching drivers' seat assembly to support to be removed. NOTE: *It is not necessary to disconnect steering lever rods and remove seat assembly.* Remove the four torsion bar springs (par. 172 e). Remove cap screws attaching front spring arm support to hull and remove support. NOTE: *Oil retainers once removed cannot be reinstalled.*

e. Install Front Spring Arm Support. Cover face of support with joint sealing compound. Using suitable hoisting tackle, position front spring arm support against hull. Install cap screws and lock washers. Using torque wrench tighten all cap screws to 150 foot-pounds. Adjust front spring arms and No. 2 road wheel arms to correct dimensions (par. 172 e). Install and adjust torsion bar springs (pars. 172 d and 172 e). Install No. 2 road wheels and front

TM 9-735
174–175

Tracks and Suspension

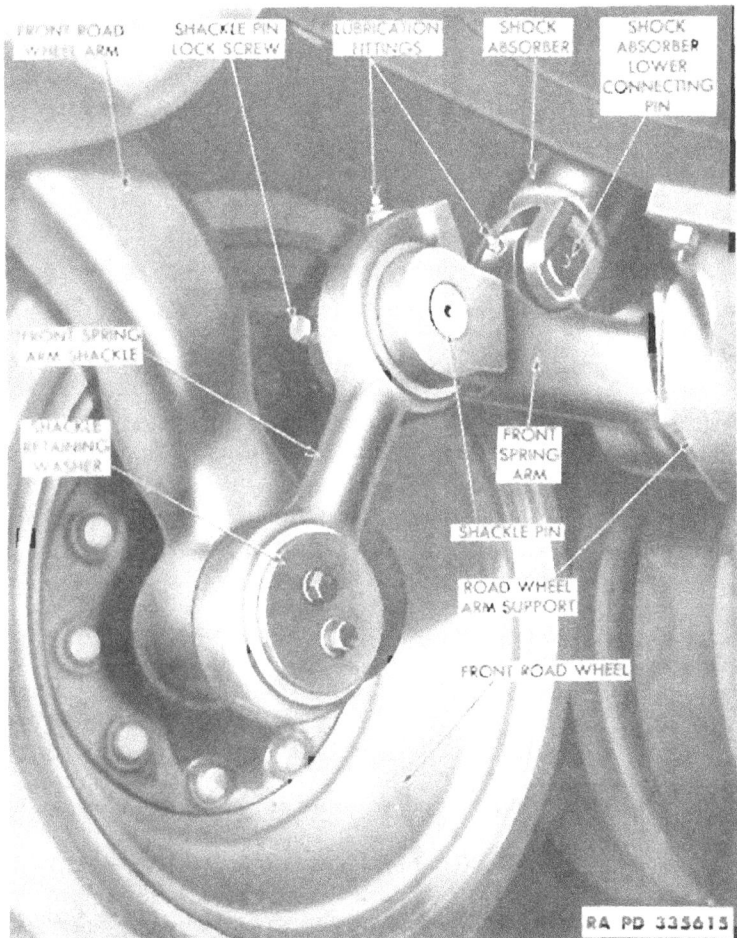

Figure 160 — Front Spring Arm Shackle

road wheels (par. 171). Install idler wheels (par. 171 g). Remove wood blocking and lower jacks. Connect and adjust both tracks (par. 170 b, c). Install front fenders and dust shields.

175. FRONT ROAD WHEEL ARM.

a. **Description.** A front road wheel arm is mounted in a support on each side of the vehicle, in the point of the "V" of the hull front slope (fig. 138). The idler wheel is mounted on a spindle at the upper end of the arm, and the front road wheel and front spring arm shackle are mounted at the lower end. The front road wheel arm pivots on a spindle mounted in bearings (fig. 147) in the front road wheel arm support, and thus maintains track tension as the track and front road wheel pass over obstructions or uneven terrain.

  b. Remove Front Road Wheel Arm. Remove front fender and dust shield. Disconnect track (par. 170 d). Jack up front corner of vehicle with 20-ton jack so weight of vehicle is off front road wheel (fig. 158). Install wood blocking or shoring as a safety factor. Remove idler wheel (par. 171 f). Remove front road wheel (par. 171 d). Remove front spring arm shackle (par. 176 b). Remove cap screws from front road wheel arm retainer. Support arm with suitable hoisting tackle, and pull arm spindle out of hull. Lower assembly to ground. NOTE: *Inner and outer bearings are attached to front road wheel arm spindle and will come out as part of the assembly when arm is removed (fig. 147).* Remove gasket.

  c. Remove Front Road Wheel Arm Spindle Inner Bearing. Follow procedure given in subparagraph b preceding for removing front road wheel arm (fig. 138). When arm has been removed, loosen screws and remove inner bearing retaining washer from end of spindle. Lift off bearing (fig. 147).

  d. Remove Front Road Wheel Arm Outer Bearing and Oil Retainer. Remove front road wheel arm (subpar. b preceding). Remove front road wheel arm spindle inner bearing (subpar. c preceding). Remove bearing spacer. Lift bearing off spindle. Lift off road wheel arm retainer. Inspect oil retainer, and if damaged or leaking, drive out of road wheel arm retainer with drift. NOTE: *Oil retainers once removed cannot be reinstalled.*

  e. Inspect Bearings. Follow inspection procedure given in paragraph 171 j.

  f. Install Front Road Wheel Arm Outer Bearing and Oil Retainer. Install oil retainer in front road wheel arm retainer, with spring (or lip) side of oil retainer up. Use replacer (41-R-2392-50) and handle (41-H-1296-175). Install road wheel arm retainer on spindle, being sure leather seal is in correct position. Use great care in installing oil retainer so as not to damage or bend lip or coil spring. Coat inside of bearing with micronized graphite (type 200-16) and pack bearing with general purpose grease as directed in paragraph 46. Install bearing on spindle. Install bearing spacer. Install inner bearing (subpar. g below). Install front road wheel arm shackle (par. 176 c). Install front road wheel (par. 171 e). Install idler wheel (par. 171 g). Connect and adjust track (pars. 170 d and 170 b). Lubricate assembly until grease can be seen at oil retainer as directed in paragraph 46.

Figure 161 — Front Road Wheel Lifting Tool

g. **Install Front Road Wheel Arm Inner Bearing.** Pack bearing with general purpose grease, as directed in paragraph 46. Install on front road wheel arm spindle. Coat inside of bearing race with micronized graphite (type 200-16) or light graphited grease (Ord. Dept. AXS 683) before installing. Install inner bearing retainer washer (fig. 147).

h. **Install Front Road Wheel Arm.** Be sure bearings are properly assembled, packed with grease, and correctly installed on the front road wheel arm spindle. Install new gasket. Coat face of retainer with joint sealing compound. With hoist position front road wheel arm and install spindle in front road wheel arm support. Install retainer against hull. Install front spring arm shackle (par. 176 c). Install front road wheel (par. 171 e). Install idler wheel (par. 171 g). Remove 20-ton jack and wood blocking. Connect and adjust track (par. 170 d, b). Install front fender and dust shield. Lubricate bearings as directed in paragraph 46.

176. FRONT SPRING ARM SHACKLE.

a. **Description.** The front spring arm shackle is a short arm connecting the lower end of the front road wheel arm to the front end of the front spring arm (fig. 160). Roller bearings are installed in an eye in each end of the shackle. A shackle pin connects the top eye (the upper part of which is forged to form the front bump spring bumper or striker plate) to the front spring arm yoke.

b. **Remove Front Spring Arm Shackle.** Remove front fender and dust shield. Disconnect track (par. 170 d). Place a 20-ton jack under front corner of vehicle, and raise it sufficiently to take weight of vehicle off front road wheel. Install wood blocking as a safety factor. Block up idler wheel. Remove front road wheel (par. 171 d). Remove screws and shackle retaining washer from lower end of shackle (fig. 160). At upper end of shackle loosen jam nut and back off shackle pin lock screw in end of front spring arm yoke. Drive out shackle pin connecting front spring arm shackle to front spring arm. Remove front spring arm shackle. NOTE: *Second echelon will not attempt to replace or service shackle bearings. Install new shackle as a complete assembly.*

c. **Install Front Spring Arm Shackle.** Position front spring arm shackle, striker plate up. On lower end of shackle install shackle retaining washer and cap screws. Position eye in upper end of shackle in front spring arm yoke, and install shackle pin. Tighten shackle pin lock screw and jam nut. Install front road wheel (par. 171 e). Remove blocking from idler wheel. Remove shoring from under vehicle and lower jack. Connect and adjust track (par. 170 d and h). Install front fender and dust shield.

177. SHOCK ABSORBERS.

a. **Description.** Shock absorbers are hydraulic plunger, double-acting type. They control the upward and downward movement of the road wheel arms, with the exception of the two center intermediate arms which have no shock control (fig. 137). The location of these arms at the center of the vehicle makes shock control unnecessary. The outer end of the shock absorber plunger forms an eye. This eye is attached to a fitting on the road wheel arm with a connecting pin. The top of the shock absorber cylinder shield forms a bracket which is bolted to the hull side plate. Shock absorbers are installed and removed as an assembly. Always replace leaking or "frozen" shock absorbers immediately.

b. **Remove Shock Absorbers.** Remove cotter pin from connecting pin which attaches shock absorber plunger eye to fitting on road wheel arm (fig. 139). Use drift to drive pin out. Remove screws attaching shock absorber bracket to hull side plate, and lift off shock absorber.

c. **Install Shock Absorbers.** Cover face of bracket with joint sealing compound, and install shock absorber. Using torque wrench tighten each screw to 190 to 200 foot-pounds. Place a short bar in plunger eye, and extend plunger until connecting pin can be inserted. Insert pin and install and spread new cotter pin.

178. BUMPER SPRINGS.

a. **Description.** The upward travel of each road wheel arm is limited by a volute-type bumper spring attached to the hull side plate (fig. 139). The volute spring is held in position in its mounting bracket by a cap screw, lock washer, and retaining washer at the top. A striker cap is provided at the bottom of apex of the spring.

b. Removal. Cut screw locking wire. Remove the short and two long screws which attach bumper spring bracket to hull side plate. Lift off bumper spring and bracket as an assembly. NOTE: *Volute spring can be removed from bracket with assembly either on or removed from vehicle by removing cap screw.*

c. Installation. Cover face of bracket with joint sealing compound and reverse removal procedure described in subparagraph b preceding. Using torque wrench, tighten each bumper spring bracket screw from 350 to 360 foot-pounds. Install locking wire.

179. SUPPORT ROLLERS.

a. Description. The upper section of each track is supported on five dual-type track rollers equipped with hard rubber tires. They are held in position by mounting brackets bolted to the upper hull side plates on each side of the vehicle (fig. 137). Each support roller consists of two interchangeable disk-type wheels bolted to a support roller hub. The hub is mounted on a spindle which is supported in an inner and outer roller bearing in the mounting bracket (fig. 139). Wheels, hub, and support are removed and installed as an assembly.

b. Remove Track Support Roller and Bracket Assembly. Remove dust shield. Place board across tops of road wheels below support roller to be removed. Install jack on board and raise section of track at support roller sufficiently to relieve track weight on roller. Remove locking wire. Remove lower bracket screws. Support roller assembly, and remove upper screws. Lower assembly to ground.

c. Install Track Support Roller and Bracket Assembly. Cover face of bracket with joint sealing compound. Reverse removal procedure as described in subparagraph b preceding. Lower jack and remove jack and board. Tighten screws alternately with torque wrench, using 280 to 300 foot-pound pull. Secure screws with locking wire.

Section XXXIII

HULL

180. HULL.

a. Description. The hull is a welded unit made up of armor steel castings, plates and/or sections welded together. The cast V-shaped front section (fig. 1) is welded to the hull floor at the bottom of the "V." The top of the "V" extends up and back to form the front of the vehicle and the roof of the drivers' compartment.

The rear hull casting forms the rear of the engine compartment. Both front and rear castings are welded to the center upper hull plates and hull side plates. A transverse casting extends across the engine compartment at the center, and is bolted to the side plates. Its purpose is to strengthen the hull structure at the rear of the vehicle and to support the fans, fan shrouds, oil coolers, radiators, and water expansion tank. The rear hull side plates, which have a large hole machined in them to receive the final drives, are welded at the front edge to the center hull side plates and at the rear edges to the rear upper hull casting. The floor side plates, or extensions, are welded to the hull floor at an angle, and slope upward to meet the hull side plates. The wheel arm supports extend through openings in the floor side plates and are bolted in position. Crossmembers or ribs welded to the hull floor plates on the inside reinforce the floor structure (fig. 61). They are also used to support units and equipment above the torsion bar springs. The fighting compartment is separated from the engine compartment by a bulkhead (fig. 103) welded in place and provided with a bulkhead opening cover. The bulkhead helps support the hull rear top plate and the turret ring at the rear of the turret opening and also serves as a firewall between engine and fighting compartments. Two escape doors in the front floor section, one in front of each driver's seat, provide emergency exits (fig. 14). Covers attached to the bottom of the hull floor on the outside protect the various drain plugs. Hull drain valves and bilge pumps installed in fighting and engine compartments (fig. 61) permit the hull to be quickly drained of any accumulated water, oil, or fuel. Center deck plates over the engine compartment, extending lengthwise at front and rear of the transverse housing, support the inner ends of the intake and exhaust doors (fig. 9). Two towing shackles at the front (fig. 1) and two at the rear (fig. 3) provide means of attaching tow bar or cable. A demountable pintle hook at the rear (E, fig. 49) is used to tow light vehicles, such as armored supply trailers. The hull is equipped with two lifting eyes at each end for use when hoisting the vehicle. Outriggers welded to the hull side plates support the equipment stowage boxes, fenders, and dust shields (fig. 6).

181. BULKHEAD COVER.

a. Description. The opening in the engine compartment bulkhead is sealed by a removable bulkhead cover. The cover is attached to the front side of the bulkhead by means of quick-release fasteners. A seal around the edge of the cover makes the cover air- and watertight.

b. Removal of Bulkhead Cover. Press in and turn to release cover fasteners. Lift cover away from bulkhead.

TM 9-735

Hull

Figure 162 — Hull Floor, Viewed From Below

c. **Installation of Bulkhead Cover.** Position cover on bulkhead so all fasteners are inserted in holes in bulkhead. Press in and turn fasteners to locked position.

182. DRAIN VALVES AND CONTROLS.

a. **Description.** Four poppet-type, spring-loaded drain valves are provided at different locations in the hull floor to drain the vehicle of any accumulated water, oil, or fuel. The driver's compartment drain valve is to the right of the center-line of the vehicle in front of the ammunition stowage compartment. The fighting compartment drain valve is at the left end of the battery box. The two engine compartment drain valves (right front and left rear) are operated by remote control handles behind the battery box.

b. **Reseating Drain Valves.** If drain valves leak due to improper seating, reseat valve on valve seat. Remove valve (subpar. *c* following). Coat surface of seat with coarse valve grinding compound. Insert valve in cage and while exerting upward pressure on valve from below, turn valve with circling motion to grind valve and seat. Wipe valve and seat clean and test for seating, using prussian blue to show high spots. Continue valve grinding until satisfactory seat has been obtained. Install valve (subpar. *d* following).

c. **Removal of Drain Valves.**

(1) REMOVE DRIVERS' COMPARTMENT DRAIN VALVE. Block valve in closed position from below. Compress coil spring to clear taper pin attaching knob to valve stem. Tap lightly on small end of taper pin to drive it out of stem and knob. Lift off knob and spring. Release valve by removing blocks, and withdraw valve from below.

(2) REMOVE FIGHTING COMPARTMENT DRAIN VALVE. Unhook valve pull chain from hook, and remove battery box (par. 93 b). To remove valve from cage follow procedure in step (1) above.

(3) REMOVE ENGINE COMPARTMENT FRONT DRAIN VALVE. Remove bulkhead cover. Remove clevis pin and disconnect control rod. To remove valve from cage follow procedure in step (1) above.

(4) REMOVE ENGINE COMPARTMENT REAR DRAIN VALVE. Open left exhaust doors. Remove clevis pin and disconnect control rod. To remove valve from cage follow procedure in step (1) above.

d. **Installation of Drain Valves.**

(1) INSTALL DRIVERS' COMPARTMENT DRAIN VALVE. Wipe valve and seat clean. Insert valve in cage, and from below block up valve in closed position. From inside vehicle place spring and then knob on stem. Compress spring, aline holes in knob and stem, insert tapered pin, small end first, and tap into place. Release compression on spring. Remove blocking from below valve. Lubricate valve stem with engine oil, and operate valve several times.

TM 9-735
182

Hull

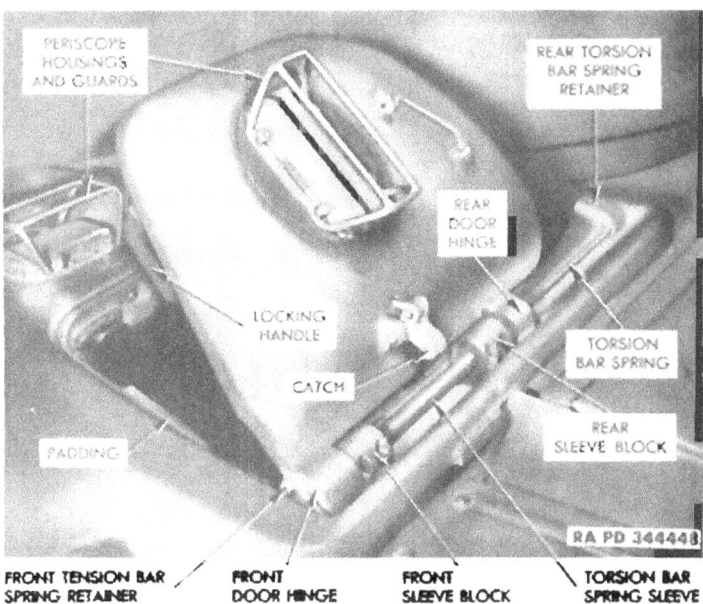

Figure 163 — Driver's Door

(2) INSTALL FIGHTING COMPARTMENT DRAIN VALVE. To install valve in cage follow procedure in step (1) above. Install battery box (par. 93 c). Hook chain ring on hook.

(3) INSTALL ENGINE COMPARTMENT FRONT VALVE. To install valve in cage follow procedure in step (1) above. Connect rod clevis to lever by inserting clevis pin. Install bulkhead cover.

(4) INSTALL ENGINE COMPARTMENT REAR VALVE. To install valve in cage follow procedure in step (1) above. Connect rod clevis to lever by inserting clevis pin. Lower and lock exhaust doors.

c. **Removal of Engine Compartment Drain Valve Control Rods.**

(1) REMOVE ENGINE COMPARTMENT FRONT DRAIN VALVE CONTROL ROD. Remove bulkhead cover. Loosen lock nut at clevis. Disconnect clevis from lever. Unscrew clevis and lock nut from end of control rod. At rear of bulkhead unscrew jam nut from control handle housing. Pull control rod and housing out through bulkhead into fighting compartment.

TM 9-735
182-183

Part Three—Maintenance Instructions

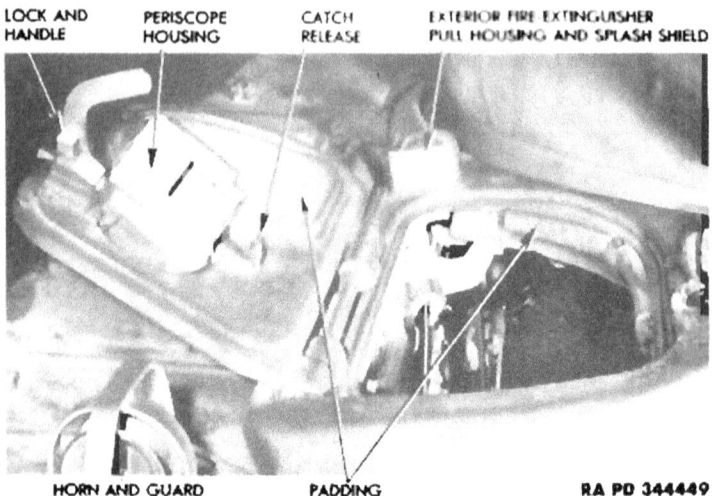

Figure 164 — Assistant Driver's Door

(2) REMOVE ENGINE COMPARTMENT REAR DRAIN VALVE CONTROL RODS. Open left exhaust door. To remove control rod follow procedure in step (1) above.

f. **Installation of Engine Compartment Drain Valve Control Rods.**

(1) INSTALL ENGINE COMPARTMENT FRONT DRAIN VALVE CONTROL ROD. From fighting compartment insert end of rod and housing through bushing in bulkhead. Install housing jam nut. Turn handle until it is horizontal and tighten jam nut at rear of bulkhead. Install clevis jam nut and clevis on end of rod. Position rod and aline holes in clevis with hole in lever. Insert clevis pin. Install bulkhead cover.

(2) INSTALL ENGINE COMPARTMENT REAR DRAIN VALVE CONTROL RODS. To install compartment rear drain valve control rods, follow procedure in step (1) above. Lower and lock exhaust doors.

183. DRIVERS' DOORS.

a. **Description.** Heavy cast armor-steel doors cover the openings in the front section roof above the driver's and assistant driver's stations (figs. 163 and 164). Each door is hinged at the outer edge and when in closed position seals against a gasket. Doors can be locked in closed position from the inside, or secured in open position by a catch on the outside of the door (fig. 163). A torsion bar spring counterbalances door weight as they are closed and opened. Each door is provided with a periscope mount.

b. **Adjustment of Driver's Door Spring.** Test action of torsion bar spring by opening and closing door several times. This test will determine whether more or less spring tension is needed to counterbalance the door. Use a pencil or scribe to make reference marks on top of bar and at retainer rear end. Loosen the two socket head screws which attach retainer to hull (fig. 163). Relieve torque in torsion bar spring by raising door until it is in a neutral position ahead of vertical. Securely block door in this position. Remove attaching screws and spring bar up to remove retainer. Reposition retainer on torsion bar spring one or two serrations away from reference marks, or further if required. This repositioning will provide more or less spring tension when retainer is reinstalled. Install retainer and tighten screws. Test action of door. Repeat adjustment procedure as required to provide proper spring loading.

c. **Removal of Drivers' Door.**

(1) REMOVE DRIVERS' DOOR FROM HULL. Raise door into neutral position ahead of vertical to relieve torque in torsion bar spring. Hold or block door securely in this position. Remove socket-head screws which attach torsion bar spring rear retainer to hull. Remove socket head screws which attach both male hinge blocks to hull (fig. 163). Lift door off hull.

(2) REMOVE TORSION SPRING BAR FROM DOOR. If same torsion bar spring is to be reinstalled, first reference mark both bar and retainer (subpar. b above). Slide rear retainer off bar. Remove bar from door, by driving it to the rear out of front retainer.

(3) REMOVE DRIVERS' DOOR HINGE PIN. Remove socket-head screws which attach torsion bar spring front retainer to door. Remove hinge pin by driving it out of door and through hinge blocks toward the front.

(4) REMOVE DRIVERS' DOOR LOCKING HANDLE AND CATCH. Remove cotter pin from locking handle hinge pin. Drive pin out of handle. Loosen lock nut at catch handle. Unscrew handle and lock nut from rod. Remove cotter pin from catch handle rod hinge pin. Drive out hinge pin. Remove cotter pin from catch hinge pin, and drive out hinge pin (fig. 163).

(5) REMOVE PERISCOPE HOUSING. Follow procedure in paragraph 206 e (5).

d. **Installation of Drivers' Doors.**

(1) INSTALL DRIVERS' DOOR LOCKING HANDLE AND CATCH. Position locking handle between bosses on inside of door. Install

hinge pin. Position catch spring between catch and door. Aline holes in catch and bosses on door. Install catch hinge pin (fig. 163). Insert catch handle rod through door from inside. Aline hole in catch with hole in end of rod, and insert hinge pin (fig. 163). Install lock nut on catch handle rod. Screw handle on rod until it is nearly tight and at right angle to hinge pin. Hold handle from turning, and tighten lock nut.

(2) INSTALL DRIVERS' DOOR HINGE PIN. Insert hinge pin through front hinge of door. Position hinge blocks on hinge pin as it is driven into position against snap ring in hinge at rear door. Place torsion bar spring front retainer in position, and install two socket-head screws (fig. 163).

(3) INSTALL TORSION BAR SPRING IN DOOR. From rear insert bar through hinge pin. Aline serrations and reference marks, if previously made. Drive bar into front retainer until it is flush with face of retainer. Aline serrations and reference marks (if any), and position rear retainer on bar.

(4) INSTALL DRIVERS' DOOR ON HULL. Place door on hull in closed position. Aline hinge blocks, and install socket-head attaching screws (fig. 163). Follow procedure in subparagraph b above to adjust drivers' door spring.

(5) INSTALL PERISCOPE HOUSING. Follow procedure in paragraph 206 e (6).

184. DRIVERS' SEATS.

a. Description. Form-fitting, tractor-type seats, with removable backs, are provided for driver and assistant driver. Seats can be adjusted in height to permit driver to operate vehicle with his head through hatch opening. Steering brake levers are mounted on brackets, which are a part of each seat frame. Each seat is also equipped with a safety-belt.

b. Removal of Drivers' Seats. Remove clevis pin at rear end of each brake lever to cross-shaft rod. Raise seat and remove screws which attach seat to support and front road wheel spring arm support housing. Lower seat and remove back. Lift seat out of vehicle through turret.

c. Installation of Drivers' Seats. With steering lever pull rods attached to steering brake levers, position seat on support and front road wheel spring arm support. Raise seat and install attaching cap screws, but do not tighten. Lower seat and shift position as required to clear brake cross shaft. Tighten cap screws. Connect steering lever pull rods to brake front cross-shaft levers. Install seat back.

185. ENGINE COMPARTMENT DOORS AND DECK PLATES.

a. **Description.** Eight grilled doors, hinged at the hull side plates, cover the engine compartment (fig. 6). The four doors forward of the transverse housing are known as the intake doors. The four doors at the rear of the transverse housing are known as the exhaust doors. All doors open outward. When closed intake doors lock against the front center deck plate and the exhaust doors against the rear center deck plate. Each door can be opened or closed separately. When fully opened the doors rest flat on the stowage boxes and serve as a platform on each side of the engine compartment. The exhaust tubes enter a flange at the back of the rear center deck plate. The exhaust outlet (fig. 181) is attached to the outside of this flange.

b. **To Open Engine Compartment Doors.** Raise latch handle and turn it until parallel with end of door. Pull up on door and lower it carefully against top of stowage box. WARNING: *Do not drop doors.*

c. **To Lower and Lock Engine Compartment Doors.** Raise door and carefully lower it into closed position with latch handle turned parallel with end of door. Turn handle until in locked position at right angles to end of door. Push handle down against door over retainer pin. WARNING: *Do not drop doors.*

d. **Removal of Engine Compartment Doors.** With door in closed position, remove cotter pins from ends of hinge pins. Drive hinge pins out of hinge. Unlock latch. Lift door off vehicle with latch handle and hinge pins in place in door part of hinges.

e. **Installation of Engine Compartment Doors.** Lower door into closed position on vehicle. Aline hinge pin holes, and drive hinge pins into place. Install and spread new cotter pins in ends of hinge pins. Lock door in closed position.

f. **Removal of Front Center Deck Plate.** Open intake doors. Remove deck plate screws. Raise front end to release bracket at rear underside of plate. Lift front end of plate until dowel pin clears and then pull plate forward off support on transverse housing.

g. **Installation of Front Center Deck Plate.** With front end raised, push rear end of deck plate onto support on transverse housing so bracket is under support and dowel pin is entered in slot in plate. As front end is lowered, slide plate to rear to aline bolt holes at front end. Install screws. Lower and lock intake doors.

h. **Removal of Rear Center Deck Plate.** Open exhaust doors. Remove nuts from studs at inner side of rear hull plate. Support front end of deck plate. Slide plate to rear while tapping exhaust tubes to ease them out of holes in flange. When pipes are clear of holes and front end of plate has cleared dowel pins, lift plate off vehicle.

i. **Installation of Rear Center Deck Plate.** Rest flange at rear of plate in opening in rear hull plate. Guide ends of exhaust tubes into holes in flange as plate is pushed forward over dowel pins at front end. Install brass nuts on attaching studs. Tighten nuts evenly to draw deck plate into contact with rear hull plate. Lower and lock exhaust doors.

186. FENDERS AND DUST SHIELDS.

a. **Description.** Heavy metal fenders cover the tracks at the front and rear with dust shields along the sides of the stowage boxes (figs. 3 and 4). The fenders are bolted to the hull along their inside edges and to the stowage box brackets or outriggers. The front and rear fenders can be removed in two sections or as assemblies. The dust shields are attached to the side stowage box brackets or outriggers and can be removed in three sections.

b. **Removal of Front Fenders.**

(1) REMOVAL OF FRONT FENDER BY SECTIONS. Remove front section of dust shield. Remove bolts which attach front section of fender to rear section and to the front slope. Lift off front section of fender. Remove front dust shield and fender attaching bolts and lift off rear section.

(2) REMOVAL OF FRONT FENDER ASSEMBLY. Remove headlight and stow in drivers' compartment if installed. Remove headlight guard from fender. Remove front dust shield. Remove rear edge of fender from stowage box support. Starting at rear, remove bolts attaching inner edge of fender assembly to hull. Support assembly at front as last bolt is removed. Pull assembly forward and slightly away from vehicle, and lift off. If necessary, fender assembly can be disassembled after removed.

c. **Installation of Front Fenders.**

(1) INSTALLATION OF FRONT FENDER BY SECTIONS. Raise rear section into position and install attaching bolts. Install front section of dust shield. Raise front section into position. Aline bolt holes and insert two bolts in rear edge of front section at top and in lower front corner to hull. Install remaining attaching bolts. Shift fender as required to aline it with hull. First tighten bolts attaching fender to hull, then tighten remaining bolts.

(2) INSTALLATION OF FRONT FENDER ASSEMBLY. Raise fender assembly into position. Aline bolt holes and insert bolts in bolt holes to attach rear edge to stowage box support and to lower front corner of hull. Install remaining bolts. Aline fender assembly with hull, and tighten all bolts. Install two bolts attaching headlight guard to

fender. Install both headlights, and operate light switch and siren switch to make sure units are working properly. If headlights are not to be used, stow in proper location after test. Install front dust shield.

d. **Removal of Rear Fenders.**

(1) REMOVAL OF REAR FENDER BY SECTIONS. Remove attaching bolts and lift off outside rear section. Remove dust shield. Remove bolts attaching rear inside section of fender to hull. Lift off section.

(2) REMOVAL OF REAR FENDER ASSEMBLY. Remove dust shield. Remove three bolts attaching front edge of fender to stowage box support. Starting at the front, remove the bolts which attach fender to hull, supporting fender at rear while removing last bolt. Pull fender to rear and slightly away from hull as it is being lowered, to clear lifting eye and taillight. After fender is removed it can be disassembled by removing the attaching bolts.

e. **Installation of Rear Fenders.**

(1) INSTALLATION OF REAR FENDER BY SECTIONS. Position inside rear section and install attaching bolts. Install dust shield. Position outside rear section and install three attaching bolts. Install two bolts attaching fender to hull. Tighten all bolts securely.

(2) INSTALLATION OF REAR FENDER ASSEMBLY. Raise fender assembly into position. Aline bolt holes, insert the three bolts attaching fender assembly to stowage box support, and install bolt in lower corner to hull. Install remaining nuts, bolts and tighten nuts finger-tight. Shift fender as required to aline it with hull. First tighten all fender to hull bolts, then tighten remaining bolts. Install dust shield.

f. **Removal of Dust Shield Front Section.** Remove all but the two extreme end bolts which attach dust shield to front fender section and to stowage box side rail. Remove side brace bolt. Support dust shield and remove two remaining bolts. Lift off shield.

g. **Installation of Dust Shield Front Section.** Reverse removal procedure described in subparagraph f above.

h. **Removal of Dust Shield Center Section.** Remove two bolts which attach each end of center section to front and rear sections. Remove all but the two end bolts which attach center section to stowage box side rail. Remove support brace bolt. Support center section, and remove two remaining bolts. Lift off center section.

i. **Installation of Dust Shield Center Section.** Raise center section into position. Install front upper corner bolt, in stowage box side rail, finger-tight. Aline bolt holes and install rear upper corner bolt, finger-tight. Install remaining nuts and bolts and tighten all bolts securely. Install side brace bolt.

j. **Removal of Dust Shield Rear Section.** Follow removal procedure described in subparagraph f above.

k. **Installation of Dust Shield Rear Section.** Follow the installation procedure described in subparagraph g above.

187. FIGHTING COMPARTMENT FLOOR PLATES.

a. **Description.** The 90-mm ammunition stowage box covers also serve as the fighting compartment floor (figs. 165 and 166). Covers are hinged for quick access to the 90-mm ammunition stowage compartments. Covers can be removed as required.

b. **Removal of Battery Box Cover Plate.** Remove stowage boxes from top of battery box cover. Unlatch battery box cover and lift off.

c. **Installation of Battery Box Cover Plate.** Reverse removal procedure described in subparagraph b preceding.

d. **Removal of Ammunition Compartment Covers.** Traverse turret so gunner's platform clears cover to be removed. Using handles, raise cover, and lift out or fold back out of the way.

e. **Installation of Ammunition Compartment Covers.** Lower or slide covers into position. Lock by swinging handle catch into locking position.

188. GUN TRAVELING LOCK.

a. **Description.** The gun traveling lock, hinged to the exhaust outlet (fig. 3), supports the gun when placed in traveling position. This method of support relieves the elevating mechanism of load-shock incidental to vehicle motion. A self-locking catch holds the traveling lock from swinging when in released or down position.

b. **Removal of Gun Traveling Lock.** Remove exhaust outlet (par. 132 b). Drive lock pin out of each attaching screw. Remove attaching screws.

c. **Installation of Gun Traveling Lock.** Insert two attaching cap screws through arms of gun traveling lock. Screw the cap screws into outlet until tight. Back off as required to aline retaining pin hole. Drive new retaining pin into outlet from attaching side. Install exhaust outlet (par. 132 c).

189. PROTECTIVE PADDING.

a. **Description.** Sponge-type protective padding is installed on the inner side of the drivers' doors and inside both door openings (fig. 164) to prevent accidental injury to personnel.

b. **Removal of Protective Padding.**

(1) REMOVE DRIVERS' DOOR PADDING. Scrape padding off door with scraper or putty knife. Thoroughly clean surface, using dry-cleaning solvent.

(2) REMOVE DRIVERS' DOOR OPENING PADDING. Remove screws which attach padding retainer to hull, and lift out sections of padding.

c. **Installation of Protective Padding.** If old padding is damaged install new material. If padding is undamaged, old padding may be reinstalled according to procedure following.

(1) INSTALL DRIVERS' DOOR PADDING. Coat one side of crash pad and similar area on inside of door with nonvulcanizing rubber cement. Press cemented side of crash pad firmly against door so edge is spaced evenly 1-inch from rear edge of door.

(2) INSTALL DRIVERS' DOOR OPENING PADDING. Place section of padding in position. Install attaching screws with lock washers.

190. STOWAGE BOXES.

a. **Description.** Three stowage boxes above each track provide protection and space for carrying material and equipment (fig. 6). Covers can be locked with padlocks to prevent tampering. The covers close against seals which prevent water and dust from entering boxes. Supports or channels welded to hull side plates support inner edge of the boxes. The outer edge of the boxes rests on a support connecting the outriggers which, in turn, support the ends of the boxes.

b. **Removal of Stowage Boxes.** Remove equipment and materiel as required to reach nuts and heads of attaching bolts at bottom of box. Remove dust shields as required to reach attaching bolts and nuts. Remove nuts and bolts. Lift box off vehicle.

c. **Installation of Stowage Boxes.** Position stowage box. Install attaching bolts and nuts. Install equipment and materiel previously removed. Install dust shields which were previously removed.

191. TRANSVERSE HOUSING.

a. **Description.** The transverse housing is a heavy casting made of armor steel. When in position it extends across the engine compartment at the center and is rigidly bolted to the hull side plates at each end (fig. 115). The housing strengthens the rear hull structure and supports the radiators, fans, fan shrouds, and oil coolers. The cooling system filler opening is in the center of the housing. Housing and all attached units are always removed as a complete assembly.

b. **Removal and Installation of Transverse Housing.** See paragraphs 140 b (1) thru (3) and 140 c (3) thru (5).

TM 9-735
191
Part Three—Maintenance Instructions

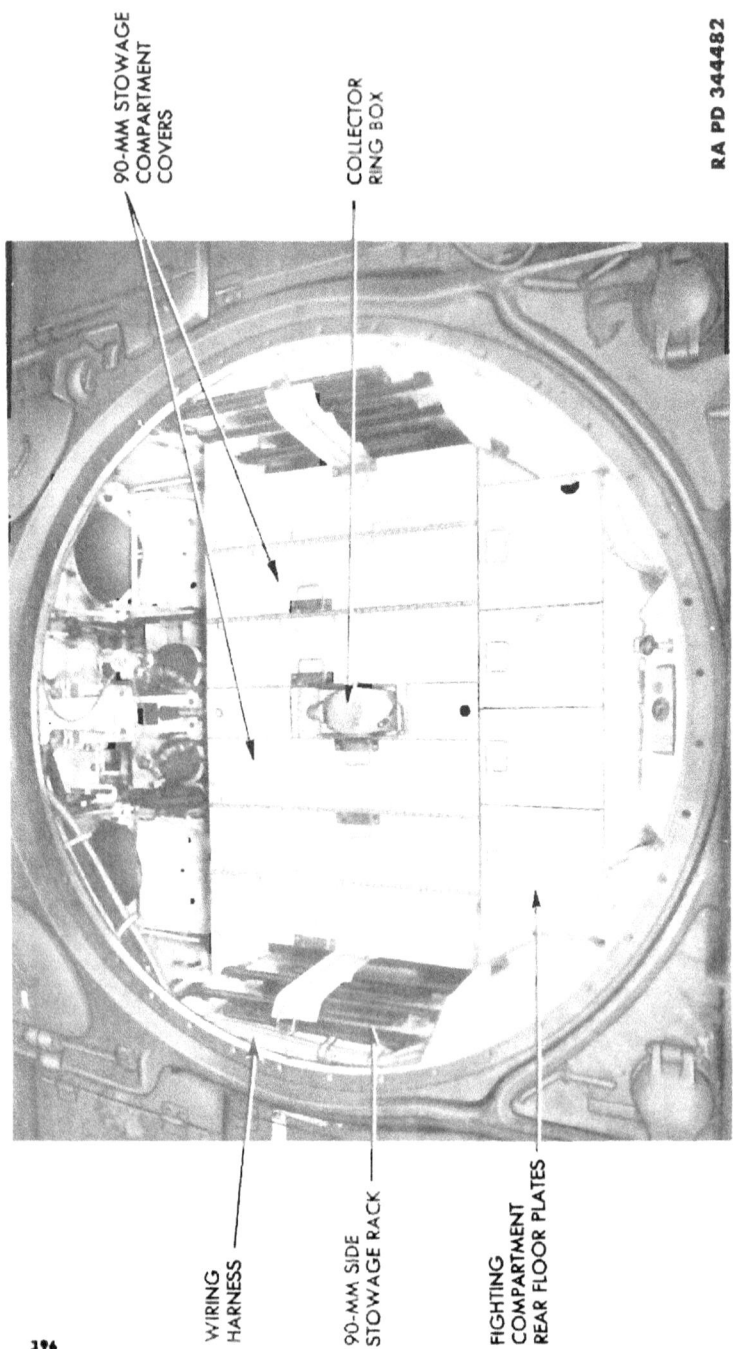

Figure 165 — 90-mm Stowage Compartment, Covers Closed

Figure 166 — 90-mm Stowage Compartment, Covers Open

192. ESCAPE DOORS.

a. **Description.** There is an escape door in the hull floor ahead of each driver's seat for use as an emergency exit. The door is held in place by a quick-release toggle-type bar, which allows door to drop to ground when release handle is pulled upward (fig. 14).

b. **Operation of Escape Door.** Pull upward on the release handle and let go (fig. 14). To install escape door reach down through door opening and lift door upward by lifting handle. Slide holding flange over edge of opening. Lift latch bar edge of door upward until door is seated, then swing release handle over and down against the door (fig. 14). CAUTION: *During inspection do not fully release escape door, as its weight makes it difficult to raise into position. Keep escape door release mechanism well oiled and mechanism and door edges free from dirt, grease, mud, rust, or any substance which will prevent quick release of the door. Do not stow gear or boxes on escape doors.*

Section XXXIV

TURRET

193. TURRET.

a. **Description.** The cast armor steel turret rotates on ball bearings in a turret race ring attached to the top of the hull. The weight of the 90-mm gun is counterbalanced by a large hollow extension at the rear of the turret (fig. 7). The radios and various other equipment are located in the interior of this extension (figs. 7 and 180). The commander and gunner enter the turret through the door in the commander's cupola (fig. 20). The loader's hatch is to the left of the commander's cupola (fig. 21). A collapsible cal. .50 machine gun mount and gun traveling clip are located directly behind the loader's hatch. When not in use gun mount and gun traveling clip can be folded down and strapped to turret roof (fig. 10). The gunner's seat, on which is mounted the gunner's foot rest or platform, is suspended from the turret rim, and rotates with the turret (fig. 23). The commander's and loader's seats are also attached to the turret rim (fig. 7). The commander's seat can be elevated and the loader's seat is demountable. The turret can be rotated a full 360 degrees by either manual or hydraulic traversing mechanisms (fig. 23). The turret can be securely locked in any position by operating the turret lock (fig. 7).

194. AZIMUTH INDICATOR.

a. **Description.** The azimuth indicator (fig. 23) is attached to the turret ring at right of the gunner and engages with turret ring gear. It is used to indicate the amount of turret rotation when aiming the gun. The azimuth indicator is illuminated by dry-cell batteries in a case mounted on the periscope rack (fig. 25).

b. Removal. Disconnect battery wire from indicator. Remove screws which attach indicator housing to turret ring. Remove screws which attach indicator bracket to base of turret. Remove screws which attach bracket to indicator.

c. Installation. Attach bracket to indicator by installing the two screws with lock washers. Position indicator on turret base, and install the bracket cap screws (B, fig. 49) with lock washers, but do not tighten. Install and tighten the two cap screws with lock washers in indicator housing. Tighten bracket screws. Attach battery wire number to indicator. Release turret lock. Rotate turret and operate indicator light switch to test operation of azimuth indicator.

195. LOADER'S HATCH.

a. Description. The loader's hatch, in the turret roof on the left side, is covered by an oval-shaped armor plate door (fig. 22). The front of the door is hinged to the turret roof and the door opens towards the front of the vehicle. The door is secured in fully opened position by a spring-loaded catch on the turret roof. When fully opened the door lies flat (fig. 2). Two large coil springs assist in controlling the opening and closing action of the door. A locking handle inside the door locks it when closed.

b. Removal. Disconnect door control springs. Raise door and support in vertical position. Drive out hinge pin dowel. Drive out hinge pin and remove door.

c. Installation. Position and support door in vertical position. Install hinge pin. Install hinge pin dowel pin. Connect door control springs. Test door for correct opening and closing action.

196. HYDRAULIC TRAVERSING MECHANISM.

a. Description. The hydraulic traversing mechanism includes the electric motor which drives the hydraulic pump, the control valve, hydraulic motor, oil reservoir, and tubes. Oil from the reservoir (fig. 167) is pumped under pressure to the control valve by the hydraulic pump driven by the electric motor. The valve controls the volume and direction of flow of the oil to the hydraulic motor on the manual traversing mechanism. The hydraulic motor is coupled to the manual traversing mechanism by the action of the shift lever, and oil pressure traverses the turret by driving the manual traversing mechanism. The oil from the hydraulic motor is returned to the reservoir. The power traversing control handle also contains firing switches for the 90-mm gun and the cal. .30 machine gun. A remote control turret traversing

TM 9-735
196

Part Three—Maintenance Instructions

Figure 167 — Turret Hydraulic Traversing Assembly

Turret

lever is connected by a flexible cable to the control valve. It is provided with a safety latch and can be used by the commander to traverse the turret hydraulically (fig. 24).

b. *Servicing Hydraulic Traversing Mechanism.* Inspect and service the hydraulic traversing mechanism at regular intervals as given in paragraph 46 and 53 according to the following instructions. When servicing the system every precaution must be taken to prevent dirt or foreign matter entering the system or contaminating the hydraulic oil.

(1) CHECK OIL LEVEL IN RESERVOIR. Every 250 miles, or more frequently as required (par. 46), check oil level and condition of oil in reservoir with oil level indicator. Oil level must be maintained at the "FULL" mark on the indicator.

(2) DRAIN HYDRAULIC OIL RESERVOIR. At overhaul, or more frequently as required, when oil becomes discolored, drain reservoir. Wipe off outside and bottom of reservoir. Place a clean container of 8-quart capacity under reservoir. Disconnect a tube at bottom of reservoir to drain reservoir. Allow reservoir to drain dry. Connect tube and tighten fitting.

(3) FILL HYDRAULIC OIL RESERVOIR. Fill hydraulic oil reservoir each time after it has been drained. Clean top of reservoir. Remove filler cap (D, fig. 47) and pour 8 quarts of hydraulic oil into reservoir as directed in paragraph 46. Inspect filler cap gasket, and install new gasket as required. Install filler cap. Operate mechanism for several 30-second intervals and recheck level (subpar. b (1) preceding). Add hydraulic oil as required to maintain level at "FULL" mark. CAUTION: *Do not use recoil oil.*

(4) CLEAN HIGH PRESSURE RELIEF VALVE. Slow or sluggish operation of the hydraulic traversing mechanism may be caused by a dirty or sticking high pressure relief valve. Lock the turret. Place oil container under reservoir and drain. Remove high pressure relief valve cap. Lift out gasket, shims (if used), retainer, relief valve spring, and plunger. Clean all parts thoroughly in dry-cleaning solvent. Examine valve plunger seat for scores and spring for weakness. Turn traversing control handle full to right and hold in this position. Turn traversing motor switch on only long enough to force sufficient oil from valve port to thoroughly flush out any foreign matter. Turn switch off and release control handle. Assemble valve parts by inserting spring guide into bushing retainer with flat side of guide down against bottom of retainer. Place spring in retainer on top of spring guide. NOTE: *Pressure relief valve spring has fewer coils of heavier wire than pump relief valve spring.* If shims were used, place them in cap. First insert plunger into valve seat bushing, and then install assembled parts, using new gasket on cap. Tighten cap securely. If

Part Three—Maintenance Instructions

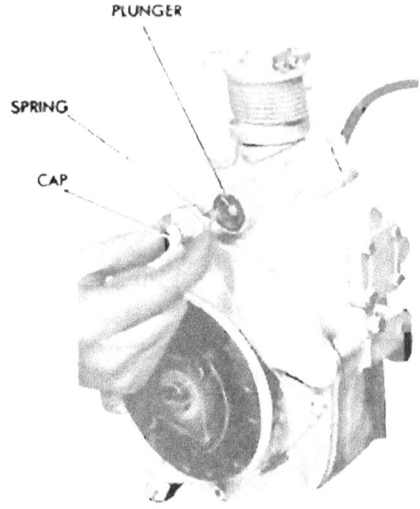

RA PD 344649

Figure 168 — Removing Hydraulic Pump Pilot Valve

pump relief valve is to be cleaned do not fill reservoir until cleaning operation has been performed. Fill reservoir and system release turret lock. Test operation of mechanism.

(5) CLEAN HYDRAULIC PUMP RELIEF VALVE. With correct amount of oil in the system, if turret fails to traverse when hydraulic mechanism is operated, or operation is sluggish or erratic, pump relief valve may be dirty or stuck open. Remove and clean valve marked No. 2 by following procedure in step (4) above. NOTE: *Pump relief valve spring has more coils of smaller wire than pressure relief valve spring.* Install parts in order. Fill hydraulic system. Release turret lock and test operation of mechanism.

(6) CLEAN HYDRAULIC PUMP PILOT VALVE. If turret fails to stop and continues to rotate to right with control handle in vertical position, pilot valve may be dirty or stuck open and should be cleaned (fig. 168). Lock turret and move shifter lever down. Remove locking wire, unscrew valve cap, and allow gasket, spring and plunger to drop out. If plunger does not drop out, push lightly on end with 1/8-inch diameter blunt instrument, at the same time turning control handle to left until plunger is free in housing. When plunger is free turn handle to right to move plunger down. Repeat operations as required to work plunger out of housing. Hold cloth or container under valve opening,

TM 9-735
196

Turret

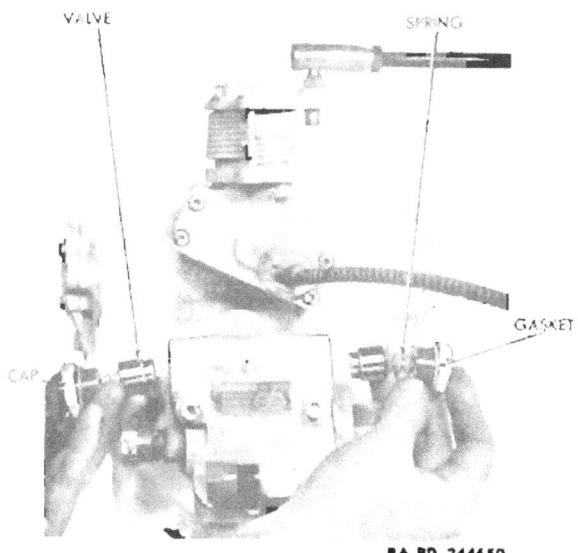

Figure 169 — Removing Hydraulic Pump Check Valve

and snap traversing motor switch on and off quickly several times to flush valve and valve seat. Thoroughly clean parts removed in dry-cleaning solvent. Place spring on plunger and install cap, using new gasket. Tighten cap and secure with locking wire. Release turret lock and check level of hydraulic oil (step (1) above). Test operation of hydraulic traversing mechanism.

(7) CLEAN HYDRAULIC PUMP CHECK VALVES. If turret traversing action becomes erratic or slow, or if the turret will traverse in one direction only, the hydraulic pump check valves may be dirty and one or both valves may be stuck open. Valves should be cleaned (fig. 169). Lock turret. Remove locking wire from check valve cap. Place cloth to catch oil and remove check valve cap and gasket. Lift out spring and disk check valve. Thoroughly clean all parts removed in dry-cleaning solvent. Examine all parts for scores. Clean valve seat in pump body. Insert valve and spring in body and install cap, using new gasket. Tighten cap and secure with locking wire. Repeat procedure for other valve. Check oil level (step (1) above). Release turret lock. Test operation of traversing mechanism.

(8) ADJUST CONTROL HANDLE POSITION. If turret has a tendency to "creep" or traverses slowly when hydraulic traversing mechanism is operating with the control handle in vertical or released

403

TM 9-735
196

Part Three—Maintenance Instructions

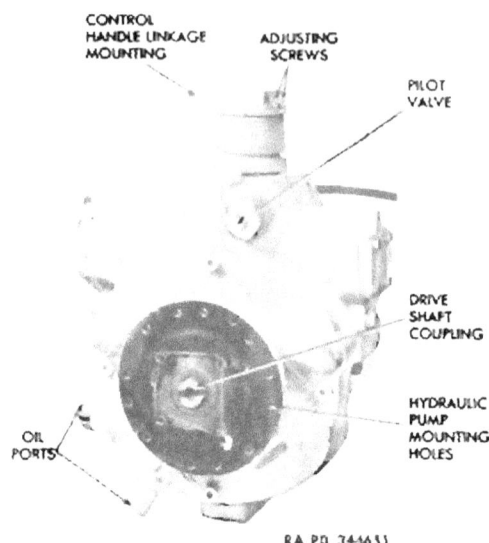

Figure 170 — Bottom of Hydraulic Traversing Pump

position, it may be necessary to adjust position of the hydraulic control (fig. 170). If turret "creeps" to the right, loosen right adjusting screw and tighten left adjusting screw an equal amount. Check adjustment with control handle in released position. Repeat adjustment as required to stop turret from creeping. Tighten both adjusting screws and secure with locking wire. Tighten handle clamp screw. Turn traversing mechanism electric motor on and place shift handle in "UP" position. If turret "creeps" to left, follow same procedure, except to loosen left screw and tighten right adjusting screw.

c. **Removal of Hydraulic Traversing Mechanism Pump.** The hydraulic traversing mechanism pump can be removed from the electric motor without removing the motor. Drain hydraulic system (subpar. b (2) preceding). Disconnect oil tubes from pump and control valve assembly (fig. 167) and loosen other ends of tubes as required. Disconnect control handle linkage rod from pump control. Remove locking wire and screws which attach pump to electric motor while supporting pump. Carefully pull pump off motor until drive shaft is clear.

d. **Installation of Hydraulic Traversing Mechanism Pump.** Install brass couplings on motor armature shaft. Aline groove in pump drive shaft with torque on brass coupling, and attach pump to motor

404

with cap screws (fig. 170). Tighten screws alternately, and secure with locking wire. Connect control handle linkage rod to pump control. Connect oil tubes to pump, and tighten all fittings moderately. Fill oil reservoir (subpar. h (3) above). Operate mechanism, and inspect all connections for leaks.

e. **Removal of Hydraulic Traversing Mechanism Hydraulic Motor.** Disconnect oil tubes from hydraulic motor (fig. 167). Tape ends of tubes. Remove nuts from studs which attach adapter to gear box. Lift off hydraulic motor and adapter as an assembly.

f. **Installation of Hydraulic Traversing Mechanism Hydraulic Motor.** Place new gasket on studs on gear box. Rotate adapter shaft to aline splines, and slide adapter and motor assembly down on gear box. Install castellated safety nuts, and tighten alternately to compress gasket uniformly. Connect oil tubes to motor and tighten fittings. Fill hydraulic system (subpar. h (3) above). Operate mechanism and inspect for leaks.

g. **Removal of Hydraulic Traversing Mechanism Reservoir.** Drain hydraulic traversing system (subpar. h (2) above). Disconnect all tubes from reservoir (fig. 167), and tape ends of tubes. Support reservoir and remove screws which attach it to bracket. Lift off reservoir.

h. **Installation of Hydraulic Traversing Mechanism Reservoir.** Position reservoir on bracket, and install the attaching screws. Connect all tubes to proper fittings (fig. 167), and tighten fittings moderately. Fill reservoir (subpar. h (3) above). Operate hydraulic traversing mechanism and check for leaks.

197. OIL TUBES.

a. **Description.** Oil tubes transfer the oil from the oil reservoir to the traverse pump, from traverse pump to the traverse motor, and from traverse motor back to traverse pump and drain tubes, into reservoir (fig. 167). Shear-type fittings are used, and if properly installed and tightened, will not require retightening. All tubes must be connected to the correct port; otherwise the system will not function. Each port has a number stamped near it to indicate tube to be connected to it. Tube fitting nuts should be drawn up securely with a 6- or 8-inch wrench when tightening, as abnormal tightening will damage fittings.

b. **Removal.** Turn turret lock handle to lock position, and traverse motor switch to "OFF" position. Drain reservoir into a clean container by disconnecting tube at port marked "13" on the bottom of oil reservoir. If oil is to be used again, cover container to keep out dirt and foreign matter. Disconnect the tubes, one at a time, and tag each tube with number of port from which it was disconnected. Clean the fitting and end of tube. Cover tube openings with tape.

c. **Installation.** Place all tubes in position, checking the installation diagram and tags to determine correct connections. Remove tape and clean off end of tubes and fittings to be sure no dirt or lint enters system. Examine all tubes and fittings carefully. If any are crimped, bent or damaged, replace with a new tube of equal size and length. Cut ends of new tube square, and remove all burs. Grease tube end, ring, and fitting. Place nut and ring over tube, with blunt end of ring facing nut. Press tube firmly into fitting, and tighten nut securely. Fill oil reservoir with hydraulic oil. Place traverse shift lever in manual control position, turn traverse motor switch to "ON" and "OFF" positions several times to permit pump to deliver oil throughout system before operating pump continuously. Check oil level, and refill with hydraulic oil. If level has dropped appreciably, clean all units and check for oil leaks.

198. MANUAL TRAVERSING MECHANISM.

a. **Description.** The manual traversing mechanism consists of the gear box, gear box extension, and manual drive with brake (fig. 23). The gear box contains the gear train which transmits the power from the hydraulic motor or manual drive, through the action of the shift lever, to the drive pinion which constantly is engaged with the turret ring gear. The manual drive is equipped with a two-shoe expanding brake which is operated by the lever on the side of the control handle.

b. **Adjustment of Manual Drive Brake Shoes.** Adjustment for brake shoe wear is accomplished by turning the square-headed screw, below the hole in top of drive mechanism, clockwise to expand the shoes closer to the drum. Adjust position of shoes so they will not contact drum until lever is depressed ½ inch.

c. **Removal of Manual Traversing Mechanism.** Remove hydraulic traversing mechanism hydraulic motor with adapter from manual traversing mechanism gear box (par. 196 e), if necessary. If motor does not need to be removed, disconnect oil tubes from motor and proceed as follows: Turn off 24-volt master switch and disconnect micro-switch wire. Remove locking wire and cap screws which attach gear box to side of turret flange. Support mechanism, and remove cap screws which attach gear box to top of turret flange. Lift mechanism off turret. Count shims, if used, and save for reinstallation.

d. **Installation of Manual Traversing Mechanism.** Position mechanism on base of turret. If shims were used, install the same number in the same locations as previously installed. Install the two vertical cap screws. Place cone-shaped split washers on the two horizontal cap screws, and tighten cap screws. Connect wire and conduit to micro-switch and turn on 24-volt master switch. Test rotation of turret by operating hand mechanism before securing lower cap screws with locking wire, as shim adjustment may be required if pinion binds on turret ring gear. Connect oil tubes to hydraulic motor, and tighten fittings moderately. Fill oil reservoir (par. 196 b). Test operation of hydraulic mechanism and inspect all connections for leaks.

e. *Removal of Brake Shoes.* Remove brake assembly by removing nut from splined shaft in the center of the cover. Press release lever against brake handle to release brake, while moving cover back and forth until cover assembly can be removed from the housing. Disconnect springs from the shoes, and remove shoes. Remove shoe-spreading plungers and adjusting screws.

f. *Installation of Brake Shoes.* Coat anchor plungers lightly with general purpose grease before installing. Assemble shoe-spreading anchor plungers in anchor assemblies, place shoes in position, and install the two brake shoe springs. Press the brake lever against the brake handle, to release the brake. Assemble cover and shoe assembly over splined shaft. Install nut and washer and adjust brake shoes (subpar. e preceding).

199. PROTECTIVE PADDING.

a. *Description.* Protective padding is installed on the base of the turret at the rear, to prevent injury to the loader's hand when serving the gun.

b. *Removal of Turret Protective Padding.* Remove cap screws which attach padding mounting plate to turret base and lift off padding.

c. *Installation of Turret Padding.* Position padding on turret base. Install and tighten attaching cap screws.

200. TURRET LOCK.

a. *Description.* A sliding-bolt type turret lock engages the turret ring gear to lock the turret securely in any position (fig. 7).

b. *Removal.* Place gun in traveling lock. Release turret lock. Remove locking wire and screws which attach lock to turret ring. Lift off lock. NOTE: *If same lock is to be installed and shims were used, the same number and size shims must be reinstalled.*

c. *Installation.* Place lock in position on turret ring. Install attaching cap screw. Test operation of lock. If it binds, either loosen attaching cap screws and shift position of lock, or remove lock and adjust number of shims to allow lock bolt to fully enter ring gear without binding. After lock is properly installed, tighten cap screws and install locking wire (C. fig. 49).

201. TURRET SEATS.

a. Description. The gunner's seat, mounted on the gunner's platform, is adjustable horizontally and to height (fig. 23). It is provided with a removable back rest. The commander's seat, mounted on the turret ring, is adjustable in height only (fig. 7). The loader's seat on the left side is installed on pins on the turret ring. When not in use it is stowed behind the "ready-rounds" rack.

b. Removal of Turret Seats. Remove bolts or cap screws which attach seat to turret platform or turret ring. Lift seat off.

c. Installation of Turret Seats. Position seat and install attaching screws or bolts. When installing commander's seat, tighten vertical cap screws first.

Section XXXV
AUXILIARY ENGINE AND CONTROLS

202. AUXILIARY ENGINE.

a. Description. When the main engine is not running, the auxiliary engine (fig. 32) is used. It drives the generator to maintain the charged condition of the batteries, so as to provide current to operate the hydraulic traversing mechanism and other electrical units. The belt, at the front end of the generator which is mounted on top of the auxiliary engine, permits the auxiliary engine to drive the generator by means of a pulley with an over-running clutch when the main engine is stopped. This clutch also permits the pulley to "free-wheel" when the main engine is driving the generator through the pulley at the rear end of the generator when the auxiliary engine is not running. The rear pulley also "free-wheels" when the auxiliary engine is operating and the main engine is stopped. The auxiliary engine is a 4-cylinder, 4-cycle, liquid-cooled, 13.6-horsepower, gasoline engine. It is provided with an electric choke, magneto ignition, and an electric starter operated from the control box at the rear of the fighting compartment (fig. 29). The auxiliary engine is supplied with fuel from the right fuel tank, and its cooling system is connected to the main engine cooling system. In addition to the starter, a hand cranking mechanism is provided for emergency use (fig. 30). Oil for the auxiliary engine is carried in its own wet-sump type engine oil pan. Oil pressure is indicated by an electric gage in the control box.

TM 9-735

Auxiliary Engine and Controls

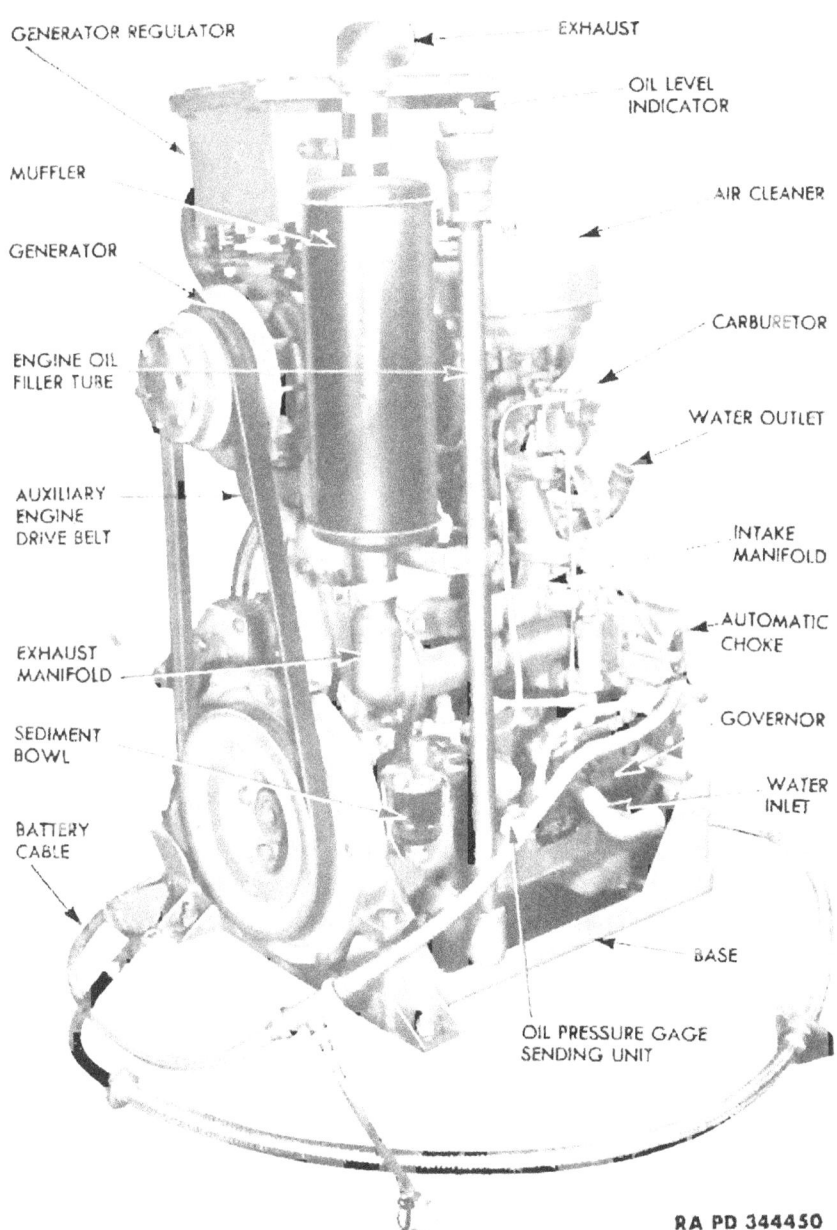

Figure 171 — Auxiliary Engine and Generator — Left Side View

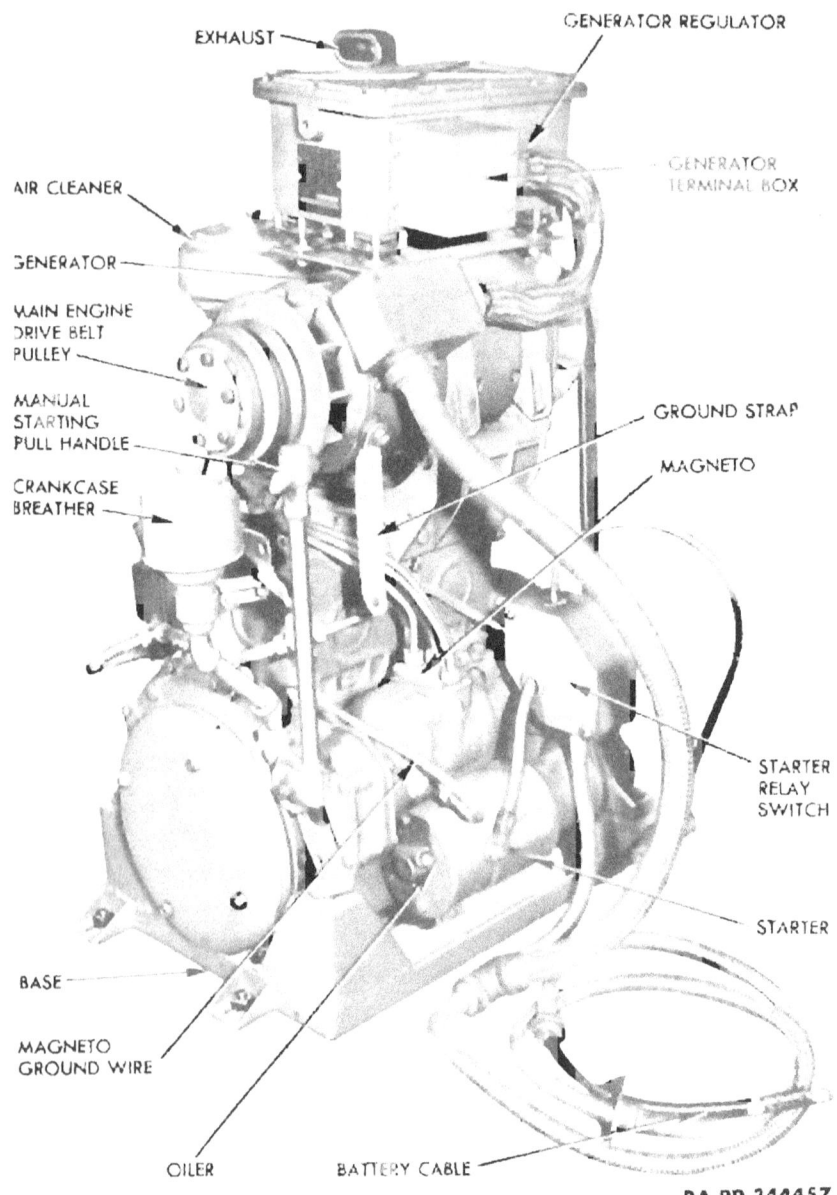

Figure 172 — Auxiliary Engine and Generator

Auxiliary Engine and Controls

b. **Servicing Auxiliary Engine.** The auxiliary engine must be serviced periodically according to instructions given in the following steps:

(1) CHECK AUXILIARY ENGINE OIL LEVEL. Check oil level daily. Raise rear right intake door and lift breather cap with oil level indicator attached out of oil filler tube (fig. 31). Read oil level on indicator. Add engine oil to "FULL" mark if required (par. 46). Install oil level indicator. Lower and lock intake door.

(2) DRAIN AUXILIARY ENGINE OIL. Drain oil only when hot. From below vehicle remove drain plug cover from right side of hull (fig. 162). Wipe off oil pan around drain plug. Place suitable container with 1-gallon capacity under drain plug. Remove drain plug and allow oil pan to drain dry. Clean drain plug, inspect gasket to make sure it is serviceable, and install drain plug. Make sure drain plug cover gasket is serviceable, and install cover.

(3) FILL AUXILIARY ENGINE WITH OIL. Raise right rear intake door. Remove crankcase breather cap with oil indicator attached (fig. 31). Pour 3 quarts of specified oil (par. 46) down oil filler tube. Install oil level indicator and breather tube cap. Start and run engine 3 minutes. Stop engine, and check oil level (step (1) above), adding oil to "FULL" mark, if necessary.

(4) CLEAN AUXILIARY ENGINE CRANKCASE VENTILATOR VALVE. This valve is attached to the center of the intake manifold (fig. 171). Open right intake doors. Disconnect fan drive propeller shaft from fan drive pulley (par. 158 h). Lower shaft into out-of-way position. At bottom of valve housing unscrew fitting at end of tube. At top of housing unscrew hexagon cap from elbow nipple. Remove housing with valve. Lift valve from housing. Thoroughly clean valve and housing in dry-cleaning solvent. Wipe parts dry and assemble. Screw valve housing on nipple at top, and connect tube at bottom. Connect fan drive propeller shaft (par. 158 c).

c. **Removal of Auxiliary Engine Drive Belt.** At left side of generator loosen lock nut and turn adjusting screw to lower generator. Work belt off pulley at front end of generator (fig. 171). Then work belt off engine pulley and out between pulley and fuel tank. NOTE: *It may be necessary to loosen or remove engine mounting bolts so engine can be pried to rear to provide clearance at front to remove belt.*

d. **Installation of Auxiliary Engine Drive Belt.** Work belt into position between engine pulley and fuel tank, and place it in pulley groove. Carefully work belt over generator pulley and into groove. Tighten mounting bolts if previously loosened. Turn adjusting screw to raise generator until a finger force of 8 to 10 pounds applied at the middle of the belt with one finger will deflect belt $\frac{3}{8}$-inch.

Tighten lock nut. Belt arrows must point in direction of travel. Install belts with arrow pointing in direction of rotation.

e. **Removal of Auxiliary Engine with Generator.** The auxiliary engine and generator can be removed from the vehicle as an assembly. To remove generator separately follow procedure in paragraph 83 d. The procedure for removing the auxiliary engine generator assembly follows:

(1) REMOVE COOLING UNIT FROM VEHICLE. Follow procedure in paragraph 139 b.

(2) DISCONNECT ADJUSTER MECHANISM. Remove bolts which attach generator regulator guard to hull side plate, and lift off guard. Remove clevis pin and disconnect adjuster control link from auxiliary engine. To remove adjuster remove clevis pin which attaches adjuster assembly to hull side plate bracket.

(3) DISCONNECT ELECTRICAL CONDUITS AND WIRES. Remove cover from generator terminal box (fig. 74). Disconnect wires from generator to cranking motor relay and to battery. Disconnect conduit elbow from generator terminal box, and pull wires out of box. Disconnect conduit shielding cap, and remove wire from oil pressure gage sending unit (fig. 171). Disconnect wire from automatic choke. Tilt engine away from hull side plate, and support the engine. Disconnect generator to starter relay switch cable. Disconnect starter button wire from starter relay. Remove conduit shielding cap and disconnect wire from magneto ground terminal. Move disconnected conduits into out-of-way position.

(4) LIFT AUXILIARY ENGINE FROM VEHICLE. Disconnect fuel tube from fuel pump (fig. 59). Remove the four engine mounting bracket attaching bolts or remove the four bolts which attach mounting bracket to hull floor. Pass rope sling around engine under each end of oil pan and attach to hoist. Take up slack in sling. Make certain all conduits, tubes and attaching parts are disconnected. Lift engine slowly, guiding it so it does not strike any parts or units. Carefully lower engine to blocks placed under oil pan to avoid damage to oil pan.

(5) REMOVE AUXILIARY ENGINE BASE. Remove four bolts which attach auxiliary engine base to hull floor.

(6) ATTACH AUXILIARY ENGINE TO BASE. Place engine on base. Install four mounting bolts (using lock washers) and tighten nuts. Remove sling.

f. **Installation of Auxiliary Engine with Generator.** The procedure for installing the auxiliary engine generator assembly follows:

(1) INSTALL AUXILIARY ENGINE BASE IN VEHICLE. Place base in position on engine compartment floor. Install four attaching bolts and tighten securely.

Auxiliary Engine and Controls

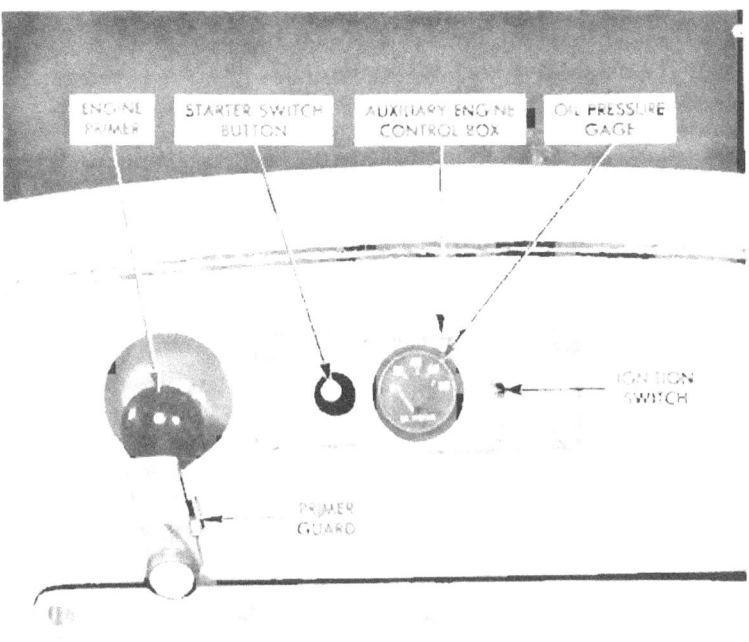

Figure 173 — Auxiliary Engine Control Box

(2) ATTACH AUXILIARY ENGINE TO BASE. Guide engine so it does not strike adjuster mechanism or other units as it is lowered into position on base. Make all electrical connections. Install mounting bolts. Connect fuel tube to fuel pump. Remove rope sling.

(3) CONNECT CONDUITS AND WIRES. At right side of engine connect starter button wire to starter relay (fig. 172). Connect wires from generator to starter relay and to battery. Connect ground wire to magneto and attach conduit. At left side connect wire to oil pressure gage sending unit, and attach conduit shielding cap. Connect wire to automatic choke. Feed cables into generator terminal box. Attach cables to terminals. Install terminal box cover. Attach conduit to nipple on box. Install auxiliary engine mounting bolts with lock washers and tighten all bolts evenly.

(4) INSTALL DRIVE BELTS. Install all drive belts. Be sure arrows point in direction of travel.

(5) CONNECT ADJUSTER AND CHECK AUXILIARY ENGINE DRIVE BELT TENSION. Connect adjuster and test auxiliary engine drive belt for proper tension and adjust if required (subpar. d above).

(6) INSTALL COOLING UNIT IN VEHICLE. Follow procedure in paragraph 139 c. After "left" propeller shaft has been connected to right fan drive pulley, adjust generator belt tension (par. 86 h). Install generator regulator guard, and attach ground strap. Connect inlet and outlet water hoses.

(7) RECORD INSTALLATION OF AUXILIARY ENGINE. Make proper entry on W.D., A.G.O. Form No. 478, "MWO and Major Unit Assembly Replacement Record."

203. AUXILIARY ENGINE CONTROLS.

a. Description. The auxiliary engine controls which include magneto ignition switch, starter button, and oil pressure gage are enclosed in a control box at the top of the bulkhead at the rear of the fighting compartment (fig. 57). The ignition switch, which opens the magneto ground circuit, also operates the electric fuel shut-off valve (par. 118 a). The starter button operates the starter relay switch to close the starter circuit. The oil pressure gage is connected to a sending unit on the left side of the auxiliary engine crankcase (fig. 171). An automatic choke aids quick starting in cold weather.

b. Removal of Auxiliary Engine Control Box and Units. The procedure to remove the auxiliary engine control box and the units from the box is given in the following steps:

(1) REMOVE AUXILIARY ENGINE CONTROL BOX FACE PLATE FROM BULKHEAD. Remove bulkhead cover. Turn 24-volt master switch off. Remove screws which attach face plate to bulkhead (fig. 174). Pull face plate out and down.

(2) REMOVE AUXILIARY ENGINE STARTER BUTTON. Note wire numbers and remove wires from terminals (fig. 174). Remove lock nut from face of plate. Lift out starter button.

(3) REMOVE AUXILIARY ENGINE OIL PRESSURE GAGE. Note wire numbers and remove wires from terminals (fig. 174). Remove nuts which attach mounting bracket, and lift off bracket. Withdraw gage and gasket from face of plate.

(4) REMOVE AUXILIARY ENGINE IGNITION SWITCH. Note wire numbers and remove wires from terminals (fig. 174). Mark top of switch. Remove screws which attach switch to face plate.

(5) REMOVE AUXILIARY ENGINE CONTROL BOX FACE PLATE. Lift gasket off face plate, and remove face plate.

(6) REMOVE WIRING CONDUITS FROM AUXILIARY ENGINE CONTROL BOX. Remove lock nut from end of conduit inside control box, and slide lock nut off wires. Reaching through bulkhead open-

Auxiliary Engine and Controls

ing, withdraw conduit with nipple and wires from each end of control box.

(7) REMOVE AUXILIARY ENGINE CONTROL BOX FROM BULKHEAD. Remove the countersunk screws which attach box to bulkhead (fig. 174). Push box to the rear, and remove it through bulkhead opening.

c. **Installation of Auxiliary Engine Control Box and Units.** The procedure for installing the auxiliary engine control box and units follows:

(1) INSTALL AUXILIARY ENGINE CONTROL BOX IN BULKHEAD. Pass box through bulkhead opening and raise it into position. Install countersunk attaching screws (fig. 174).

(2) ATTACH WIRING CONDUITS TO AUXILIARY ENGINE CONTROL BOX. Insert wire through hole in each end of box. Slide nipple and lock nut over wires, and screw nut on nipple until tight.

(3) INSTALL GASKET ON AUXILIARY ENGINE CONTROL BOX FACE PLATE. Cement gasket (fig. 174) to back of face plate with nonvulcanizing rubber cement.

(4) INSTALL AUXILIARY ENGINE IGNITION SWITCH. Insert switch through face plate from rear with end marked "TOP" at the top. Install attaching screws. Connect wires to proper terminals.

(5) INSTALL AUXILIARY ENGINE OIL PRESSURE GAGE. Place gasket on gage. Insert gage through front of face plate. Install mounting bracket with nuts finger-tight. Position gasket evenly under edge of gage and tighten bracket nuts. Connect wires to proper terminals.

(6) INSTALL AUXILIARY ENGINE STARTER BUTTON. Insert starter button through face plate from rear. Install lock nut on switch. Attach wires to proper terminals.

(7) INSTALL AUXILIARY ENGINE CONTROL BOX FACE PLATE ON BULKHEAD. Position face plate on bulkhead. Install attaching screws. Install bulkhead cover.

(8) TEST OPERATION OF AUXILIARY ENGINE CONTROLS. Operate auxiliary engine controls to start engine. Observe action of oil pressure gage. Stop engine.

d. **Removal of Auxiliary Engine Oil Pressure Gage Sending Unit.** Disconnect conduit shielding cap from sending unit, and remove wire (fig. 171). Unscrew sending unit from auxiliary engine crankcase.

e. **Installation of Auxiliary Engine Oil Pressure Gage Sending Unit.** Screw sending unit firmly into place on auxiliary engine crankcase (fig. 171). Connect wire to sending unit and attach conduit

TM 9-735
203-204

Part Three—Maintenance Instructions

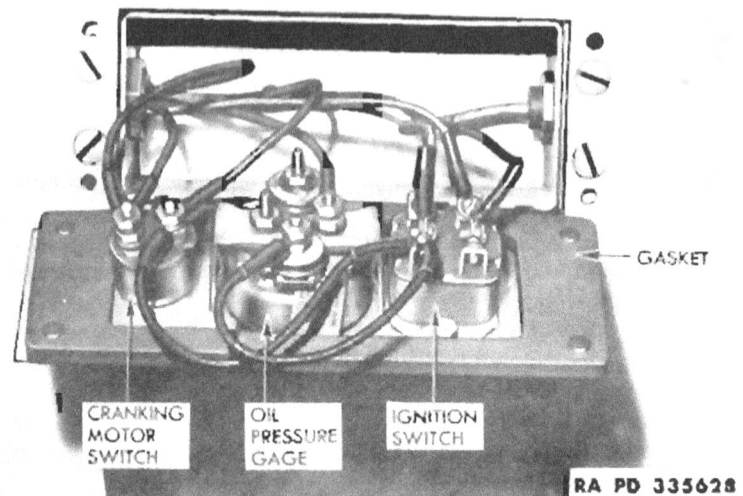

Figure 174 — Auxiliary Engine Control Box — Face Plate Removed

shielding cap. Start and run auxiliary engine to test operation of sending unit. Stop engine.

f. Removal of Auxiliary Engine Starter Relay Switch. Follow procedure in paragraph 202 e. Tilt engine toward center of vehicle and block in position. Note wire numbers and terminals and remove cables and wires from starter relay switch. Remove screws which attach switch to mounting bracket and slide switch out of bracket.

g. Installation of Auxiliary Engine Starter Relay Switch. Insert switch through hole in mounting bracket with name plate at top. Install attaching screws. Connect starter and battery cables. Connect switch button lead wire and ground wire. Remove blocks and lower engine into position on mounts. Tighten mounting nuts. Complete installation by following procedure in paragraph 202 e.

Section XXXVI

FIRE EXTINGUISHER SYSTEM

204. DESCRIPTION AND INSPECTION.

a. **Description.** The vehicle is equipped with a carbon-dioxide (CO_2) fixed fire-extinguisher system for smothering fires in the en-

TM 9-735
Fire Extinguisher System

gine compartment. The system consists of two 10-pound cylinders, control head, copper tubing, discharge nozzles and an exterior pull handle and cable. A 10-pound carbon-dioxide cylinder is mounted in a bracket on each side of the turret ring support (fig. 175). Both cylinders are discharged simultaneously by operating either the lever on the control head on the right cylinder, or by pulling the exterior remote control red pull handle in the housing behind the assistant drivers' door opening on top of the hull (fig. 164). Cylinders are connected by tubes to discharge nozzles in the engine compartment. Each cylinder is equipped with a safety valve (protected by a red seal) which will automatically discharge the gas when pressure becomes excessive due to heat. WARNING: *Handle all charged carbon dioxide cylinders with great care. Do not bump, jar, or drop.*

b. **Inspection of Fire-extinguisher System.** Inspect the fire-extinguisher system periodically in accordance with instructions in paragraphs 49, 52 and 53. Procedure for inspecting fire-extinguisher system follows:

(1) INSPECT FIRE-EXTINGUISHER CONTROLS. Examine control head on right cylinder (fig. 175). If safety pin seal wire is broken, it indicates cylinders have been discharged. If stop pin in control head shaft is not in vertical position, in line with arrow on housing, the cylinders have been discharged. Inspect remote control pull handle. If it has been pulled out or seal wire broken, the cylinders have been discharged. Remove cylinders and install fully charged ones (subpar. e following). While cylinders are removed, lubricate and test operation of exterior pull handle control.

(2) INSPECT FIRE-EXTINGUISHER CYLINDERS. Examine each cylinder valve assembly to make sure red safety cap seal is in place (fig. 175). If seal is missing, the cylinder has been discharged through the safety valve. Remove cylinder and install a fully charged one (par. 205 e). Remove both cylinders and carefully weigh each cylinder. Subtract weight stamped on valve body from weight shown on scales. If difference is less than 9 pounds, have cylinder recharged, or install a fully charged cylinder (par. 205 e) containing 10 pounds of carbon dioxide. Make certain cylinders are securely mounted. WARNING: *Handle all carbon dioxide cylinders, with great care. They are as dangerous as a loaded shell. Do not drop, jar, or bump.*

(3) INSPECT FIRE-EXTINGUISHER TUBES AND DISCHARGE NOZZLES. Examine tubes for loose connections, dents, or fractures. Tighten all loose connections. Remove damaged tubes and install new ones (par. 205 e). Inspect all discharge nozzles to make sure they are properly directed, not damaged, or clogged. Clean discharge nozzles and adjust their position, or remove damaged nozzles and install new ones. CAUTION: *Be sure nozzles are not plugged.*

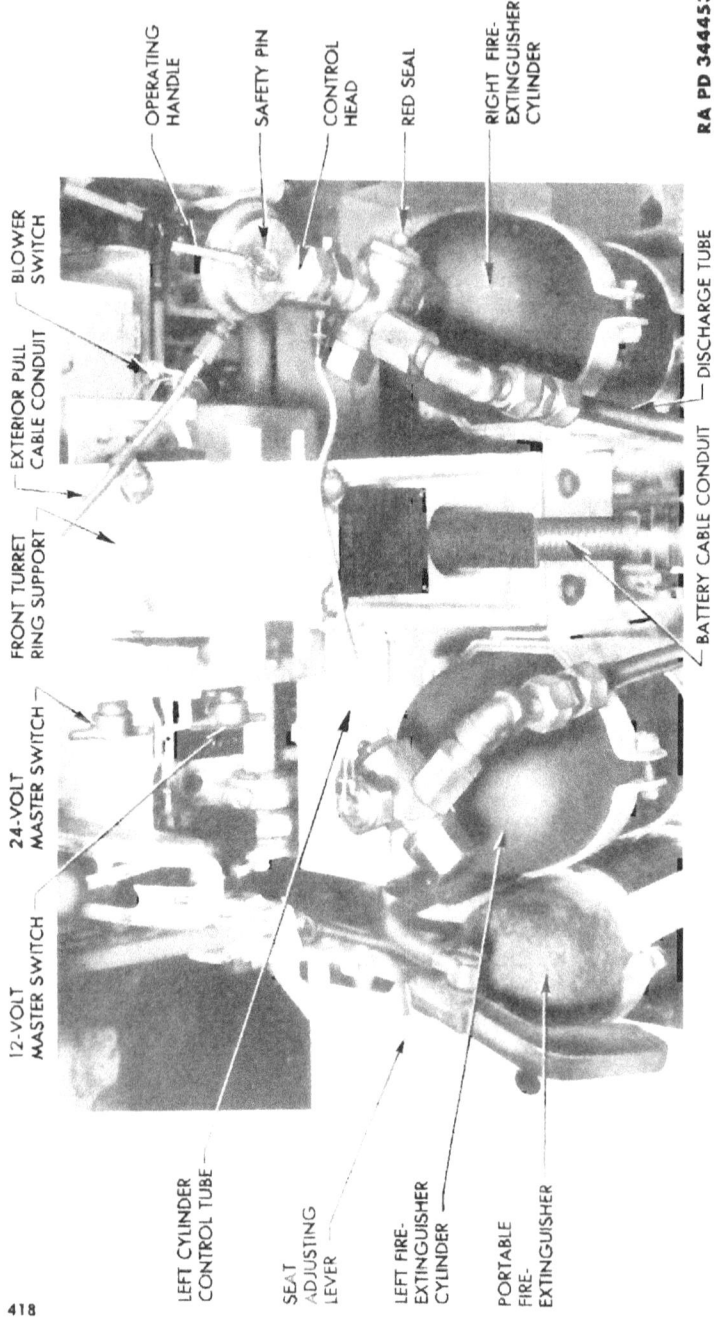

Figure 175 — Fire Extinguisher Cylinders Installed

TM 9-735
205

Fire Extinguisher System

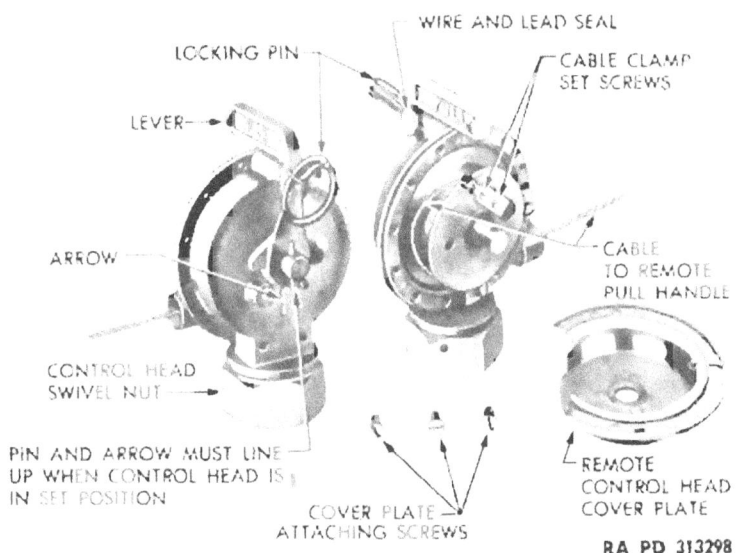

Figure 176 — Fire Extinguisher Cylinders Control Head

205. REMOVAL AND INSTALLATION.

a. **Removal of Fire-extinguisher Controls.** Remove control head from right cylinder to prevent accidental discharge. Cap opening in valve body. Remove screws which attach control head cover (fig. 176), but do not remove cable housing. Lift out cable clamp, and remove cable clamp set screws. Unscrew connector which attaches conduit to control head. Free cable from clamp. Remove conduit and cable from control head. Pull out handle to remove cable from conduit.

b. **Installation of Fire-extinguisher Controls.** Feed pull handle cable into conduit. Pull end of cable out of upper section of conduit. Feed end of cable into lower section of conduit. Pull cable tight at lower end of conduit, making sure handle is seated in clip in guard at top of hull, and cable is in groove in pulley at top of turret ring support. Attach handle with sealing wire. Attach conduit to control head. Take up slack in cable, and insert and lock it firmly in clamp (fig. 176). NOTE: *Use narrow-blade screwdriver to tighten set screws.* Wrap cable around drum, and position clamp in control head. Install cover on control head. Install control head on right cylinder.

419

c. **Removal of Fire-extinguisher Cylinders.** Unscrew lower coupling which attaches control head to right cylinder. Unscrew valve body from left cylinder. Lift off control head connected to the valve body by the pressure discharge tube. Remove clamp bolts and lift out cylinders. WARNING: *Handle all carbon dioxide cylinders with great care. They are as dangerous as a loaded shell. Do not drop, jar, or bump.*

d. **Installation of Fire-extinguisher Cylinders.** Slide cylinders into place in bracket. Install and tighten clamp bolts. Place valve body on left cylinder, and control head on right cylinder (fig. 175). Hold valve or control head, and tighten connector nuts. NOTE: *If cylinder has been discharged, the control head must be reset before it is installed.* Insert pin or nail in hole in control head shaft. Turn shaft counterclockwise until stop pin in lever end of shaft is in vertical position in line with arrow on housing. CAUTION: *Cylinder must not be dropped nor struck, nor subjected to temperatures above 140° F.*

e. **Removal of Fire-extinguisher Tubes and Discharge Nozzles.** The following general instructions apply to the removal of the fire-extinguisher tubes and discharge nozzles. When removing tubes or nozzles use wrenches with snug fit. Hold elbow or fitting from which the connector is being removed. Take care not to bend, crimp, or dent tubes as they are being removed from the vehicle. Protect them from damage after removal.

f. **Installation of Fire-extinguisher Tubes and Nozzles.** The following general instructions apply to the installation of the fire-extinguisher tubes and discharge nozzles. Preformed flared copper tubes with fittings in place will be used to make replacements. Take care not to bend, crimp, or dent tubes as they are worked into position. Make sure the flare on tube and the seat in connection are perfectly clean. Use snug-fitting wrenches. Connect both ends of tube by screwing connectors on finger-tight. Install attaching clamps. Then tighten compression fittings without excessive force while holding elbow or coupling from turning. When installing discharge nozzles, make sure they are clean and discharge port is not clogged.

Section XXXVII

ACCESSORY EQUIPMENT

206. **ACCESSORY EQUIPMENT**

a. **Heater.**

(1) DESCRIPTION. A hot water heater with electric circulating fan is installed in the drivers' compartment (fig. 35). It is equipped

with a water shut-off valve and is controlled by the heater switch on the instrument panel.

(2) REMOVAL. Drain engine cooling system (par. 134 b). At heater motor terminal disconnect wire. Disconnect water tubes from shut-off valve and heater core elbow (fig. 35). Support heater and remove bolts which attach heater to mounting brackets. Lift heater out.

(3) INSTALLATION. Position heater on mounting brackets, and install attaching bolts. Connect wire to heater motor terminal. Connect water tubes to shut-off valve and heater core elbow. Fill engine cooling system (par. 134 d).

b. **Horn and Horn Switch.**

(1) DESCRIPTION. An electric, motor-driven horn is mounted on the right front fender (fig. 1). The motor requires no lubrication, and the horn is replaced as an assembly. The horn button and switch are mounted on the driver's right steering lever below the rubber grip (fig. 18).

(2) REMOVAL OF HORN. Remove clip from conduit, and disconnect conduit and wire at connection at horn. Remove bolts which attach horn to fender. Lift horn out from under guard.

(3) INSTALLATION OF HORN. Place horn in position under guard. Install attaching bolts. Attach conduit and wire to horn terminal, and install conduit clip. If tactical situation permits test operation of horn.

(4) REMOVAL OF HORN SWITCH. Turn 24-volt master switch off. Remove bolts which attach switch to steering lever. Turn switch over, remove cover and note wire numbers. Remove wires from terminals.

(5) INSTALLATION OF HORN SWITCH. Attach wires to switch terminals (fig. 91). Install switch cover and position switch on right steering lever. Install attaching bolts. Turn 24-volt master switch on and test operation of switch.

c. **Spotlight.**

(1) DESCRIPTION. The spotlight, when mounted on the remote control handle in the roof of the turret (fig. 26), can be used to throw a powerful beam of light in any direction. It may also be used as a recognition or signal light when attached to the pistol-type handle connected to the reel inside the turret (fig. 36).

(2) REPLACEMENT OF SPOTLIGHT LAMP-UNIT. Remove spotlight from remote control head (subpar. c (1) below) or from handle (subpar. e below). Raise spring clips which attach shield to light, and remove shield. Remove screw which attaches rim to light body at bottom. Pull out on bottom of rim and lift off at top. Loosen

Part Three—Maintenance Instructions

terminal screws, and remove wires from terminals. Remove retainer ring and lamp-unit from rim. Position new lamp-unit in unit and install retainer ring. Connect wires to terminals on lamp-unit, and tighten screws. Insert tang at top of rim into slot in light body, and swing bottom of rim into position against light body. Install and tighten rim attaching screw. Raise spring clips and place shield on light in locked position with sight at top. Install spotlight on control head (subpar. c (4) following) or on handle (subpar. e following), and test operation of light.

(3) REMOVAL OF SPOTLIGHT AND/OR CONTROL HANDLE.

(a) Remove Spotlight from Control Head. At base of spotlight turn wing nut to release lock, and pull out on wing nut. Slide spotlight off control head shaft.

(b) Remove Spotlight Control Handle. Inside turret disconnect spotlight extension socket from outlet at dome light (fig. 26). Pull out on lock pin above handle knob shaft, and slide handle off shaft.

(c) Remove Spotlight Shaft. From outside vehicle lift shaft out of housing and mounting boss. Remove protector cap from retainer, and install cap on mounting boss (fig. 9).

(d) Remove Spotlight Shaft Housing. Inside turret remove the two screws in housing base (fig. 26). Tighten wing nut on housing base, and unscrew housing from mounting base.

(4) INSTALLATION OF SPOTLIGHT AND/OR CONTROL HANDLE.

(a) Install Spotlight Shaft Housing. Screw spotlight shaft housing securely into mounting boss from inside turret. Loosen wing nut on housing base (fig. 26). Turn base to aline screw holes, and install the two set screws in housing base.

(b) Install Spotlight Shaft. From outside turret remove protector cap from mounting boss, and install cap on retainer (fig. 9). Insert shaft into mounting boss.

(c) Install Spotlight Control Handle. Inside turret slide control handle on to end of shaft while holding out on lock pin. Turn handle on shaft to aline hole, and release lock pin so it seats in locked position (fig. 26). Install spotlight wire connector in outlet socket.

(d) Install Spotlight on Control Head. Slide spotlight on to control head shaft with lock pin in released position. Rotate light on shaft and release lock pin, by turning wing nut, so it seats in locked position. Turn 24-volt master switch on and test operation of spotlight.

1. Removal of Spotlight from Pistol Grip. Turn wing nut at base of light and pull out on lock pin. Remove handle from light, and stow it in bracket by securing clip (fig. 36).

TM 9-735
206

Accessory Equipment

2. *Installation of Spotlight on Pistol Grip.* Release stowage bracket clip and remove handle from bracket (fig. 36). Turn wing nut at base of spotlight, and pull out on lock pin. Insert handle in spotlight base. Aline lock pin hole, and release lock pin so it seats in locked position. Turn 24-volt master switch on and test operation of spotlight.

d. **Ventilator Blower.**

(1) DESCRIPTION. An electric motor-driven blower is mounted on the roof the drivers' compartment (figs. 18 and 19) ahead of the turret front ring support. When in operation fresh air is drawn from outside of the vehicle and forced into the drivers' and fighting compartment. During operation the blower maintains a slight air pressure within the vehicle. Dust is separated from the air by the centrifugal action of the blower, and discharged through a hose to the outside of the vehicle. The direction of air flow is controlled by a movable outlet which is equipped with a damper-type valve.

(2) REMOVAL OF VENTILATOR BLOWER. Turn 24-volt master switch off. Disconnect compartment light (fig. 18). Loosen hose clamp and pull hose off blower housing. Unscrew conduit coupling nut, and disconnect conduit from terminal box. Disconnect wire from terminal. Loosen all mounting cap screws. Remove the two front mounting caps, and pull blower forward until free of rear mounting bracket.

(3) INSTALLATION OF VENTILATOR BLOWER. Inspect seal ring on blower inlet, and install new ring if required. Slide blower rear rubber bushing into mounting bracket. Raise blower into position, and attach two mounting bracket caps. Do not tighten cap screws. Install plug in terminal box receptacle and connect wire and conduit. Tighten conduit coupling nut. Install terminal box cover. Position discharge hose on housing and tighten hose clamp. Install compartment light. Turn 24-volt master switch on and test operation of blower ventilator.

e. **Vision Equipment.**

(1) DESCRIPTION. The vehicle is equipped with seven periscopes and six commander's cupola viewing prisms to provide protected vision for all members of the crew when doors and hatches are closed. The four drivers' periscopes and the loader's periscope are mounted in holders and can be rotated and tilted as desired. The commander's periscope in the cupola door (fig. 20) is rotated by turning the door race plate. The telescopic gunner's periscope (par. 218) is linked to and moves vertically with the gun. All periscope heads are designed to shatter when struck, to prevent damage to the periscope or holder. The heads are readily replaced from a supply of spare heads carried in the vehicle. The six heavy, laminated, bulletproof

glass viewing prisms in the commander's cupola (fig. 20) may also be quickly replaced with spares stowed in the turret.

(2) REPLACEMENT OF PERISCOPE HEAD. Remove periscope from holder (subpar. c following). Turn the half-ring latches (fig. 37), on both sides of periscope nearest head, counterclockwise as far as possible and pull head out. Press new head into periscope until firmly seated, and turn both latches clockwise as far as possible to lock head in place. Install periscope in holder (subpar. d following).

(3) REMOVAL OF PERISCOPE FROM HOLDER. Slide safety latch on back of holder to unlocked position. Support periscope and loosen mounting nut. Pull periscope out of holder. Stow periscope in rack and tighten mounting nut. Return safety latch to locked position.

(4) INSTALLATION OF PERISCOPE IN HOLDER. Loosen mounting nut and remove periscope from rack. Move safety latch to unlocked position. Insert periscope in holder and seat securely. Tighten mounting nut and move safety latch to locked position.

(5) REMOVAL OF PERISCOPE HOLDER. The four drivers' and the loader's periscope holders are removed from the housings by the procedure in step (1) following. The commander's periscope holder is removed from the cupola door by the procedure in step (2) following.

(a) *Remove Driver's and Loader's Periscope Holder.* While supporting holder, remove eight screws which attach the two bottom segments to top segment. Remove the segments. Lower holder and remove trunnion bushings (fig. 177). Lift off top segment with seal.

(b) *Remove Commander's Periscope Holder.* Support holder and move both trunnion levers outward to release holder (fig. 20). Stow holder in rack. Insert protector plate in holder opening, and move both trunnion levers inward to lock plate in place.

(6) INSTALLATION OF PERISCOPE HOLDER. The four drivers' and the loader's periscope holders are installed in the housings by procedure in step (1) following. The commander's periscope holder is installed in the cupola door race plate by procedure in step (2) following.

(a) *Install Driver's and Loader's Periscope Holder.* Thoroughly clean surfaces of top and bottom segments, and surface of door around and in mounting hole. Clean and inspect holder trunnion bushings (fig. 177). Use new bushings if required. Install bushings on trunnions. Lubricate bushings with SAE 10 engine oil. Saturate bushing oil wicks in lower segments with SAE 10 engine oil. Inspect felt seal; if required, use new seal. Install seal around machined edge of top segment. Place top segment in door, and hold it in place by resting door in opened position against block in contact with top

TM 9-735

Accessory Equipment

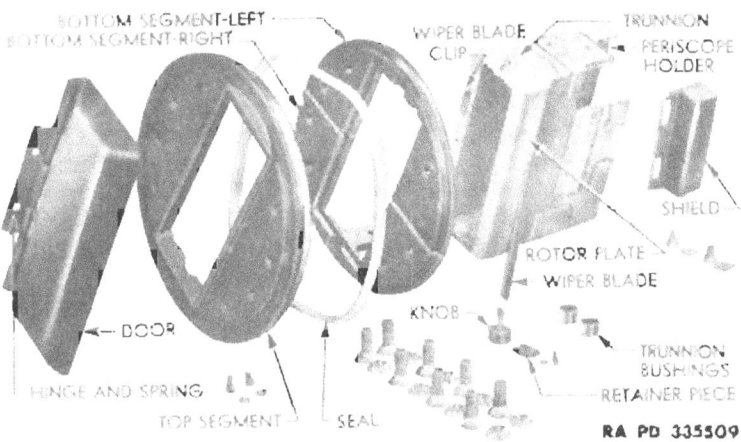

Figure 177 — Periscope Housing Disassembled

segment. Place holder in inner side of top segment so that trunnion bushings enter seats in segment. Inspect rotor plate gaskets in bottom segments, and install new gaskets if required. Position bottom segments around holder on over bushing. Install eight screws and tighten evenly and securely. Test periscope housing for binding by using scale at free end of holder to measure force required to pivot housing on trunnions. If force exceeds 5 pounds, remove bottom segments and add shims between segments until force is five pounds or less, but sufficient friction remains to hold the holder in desired position. Close door and remove block. Test housing to see that it rotates freely without excessive play.

(*b*) *Install Commander's Periscope Holder.* Move both trunnion levers outward while supporting protector plate. Remove and stow plate. Insert holder in opening, and move both trunnion levers inward to lock holder in place (fig. 20).

(7) REMOVAL OF VIEWING PRISM. Remove the two end screws. Unscrew center screw to loosen prism wedge. Remove screw and wedge. Slide prism out of opening.

(8) INSTALLATION OF VIEWING PRISM. Clean prism opening. Insert new prism in opening. Install wedge under prism with lip of wedge between shoulder and head of center screw. Turn screw in to hold wedge in place. Install the two end cap screws, and tighten all screws evenly.

PART FOUR — AUXILIARY EQUIPMENT

Section XXXVIII
GENERAL

207. SCOPE.

a. Part four contains information for guidance of personnel responsible for operation of this equipment. It contains only the information necessary for using personnel to properly identify, connect, and protect such auxiliary equipment while being used or transported with the main equipment. Detailed instructions on this equipment are contained in separate technical manuals.

Section XXXIX
ARMAMENT DESCRIPTION

208. SCOPE.

a. This section contains instructions for the operation of the 90-mm Gun M3 in the Combination Gun Mount T99E2 (fig. 5). It lists the authorized ammunition for the gun, and contains informa-

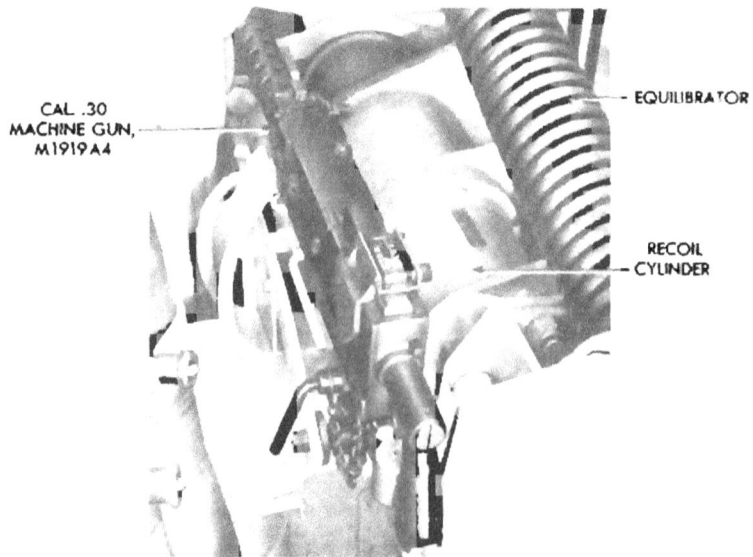

Figure 178 — Cal. .30 Machine Gun M1919A4 Installed in Turret

TM 9-735
208

ARMAMENT DESCRIPTION

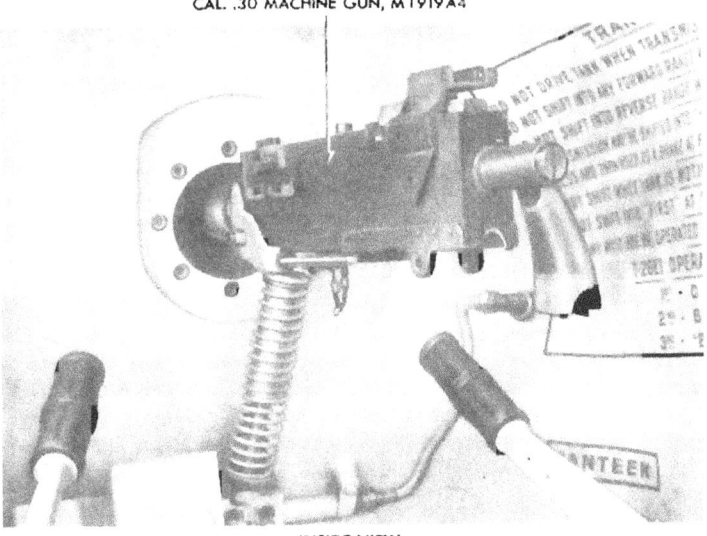

Figure 179 — Cal. .30 M1919A4 Machine Gun Installed in Bow Mount

Part Four—Auxiliary Equipment

Figure 180 — Cal. .50 HB Antiaircraft Gun M2 Installed

tion on the sighting equipment. Refer to paragraphs 45 and 46 for lubrication instructions on the armament.

b. The other armament of the heavy tank T26E3 is as follows: cal. .30 M1919A4 machine gun mounted in the Combination Gun Mount T99E2 (fig. 178); cal. .30 M1919A4 machine gun mounted in a bow mount in the right front side of the vehicle (fig. 179); and a cal. .50 HB machine gun M2 mounted in a bracket-type mount on top of the turret (fig. 180).

209. CHARACTERISTICS.

a. The armament on the heavy tank T26E3 is employed chiefly against enemy tanks and other ground objectives. The turret can be traversed 360 degrees manually or by hydraulic traversing mechanism. The 90-mm gun can be elevated 20 degrees and depressed 10 degrees. The cal. .30 machine gun in the combination gun mount is elevated and traversed with the 90-mm gun.

b. The cal. .30 machine gun in the bow mount is elevated and traversed manually and fired by the conventional trigger. Refer to FM 23-50 for complete information on its operation.

c. The cal. .50 machine gun is employed chiefly against enemy aircraft, is elevated and traversed manually, and is fired by the conventional trigger. The machine gun can be traversed independently of the turret. Refer to FM 23-65 for complete information on its operation.

Section XL
ARMAMENT OPERATING INSTRUCTIONS

210. PLACING THE 90-MM GUN IN FIRING POSITION.

a. Remove the muzzle cover.

b. Remove traveling lock from gun tube by unscrewing traveling lock handle and swinging the top half of the lock away from the gun tube (fig. 181). Unlatch the cradle traveling lock in turret (fig. 183) and elevate the gun so that gun traveling lock can be lowered. Latch turret gun traveling lock in position against turret roof (fig. 184). Screw the gun traveling lock handle back in place.

c. Disengage the turret lock by turning the handle, pulling it out, and rotating ¼ turn clockwise (fig. 187).

211. INSPECTION BEFORE FIRING.

a. Check Oil Level in Recoil Cylinder. To check oil level, depress gun 5 degrees and remove the filling plug from each recoil cylinder (fig. 184). The oil level in each cylinder should reach the bottom of the hole. If the oil does not reach this level, fill until level is reached. Replace plugs. The recoil mechanism uses special recoil oil for all temperatures.

b. Check path of recoil to make sure that it is free from all obstructions.

c. Open breech and inspect bore of gun for cleanliness.

d. Bore Sighting. Refer to paragraph 223.

212. TRAVERSING TURRET.

a. Power Operation. Before traversing turret by means of the hydraulic mechanism, be sure personnel are clear of rotating parts. Move gearshift lever, in front of gunner and under traversing gear mechanism, to the "UP" position or power position (fig. 185). It may be necessary to turn slightly, the manual traversing handle on top of traversing mechanism to engage gears. Turn master turret electrical switch to the "ON" position to start electric drive motor (fig. 186). Disengage the turret lock by turning the handle, pulling out handle, and rotating it ¼ turn clockwise (fig. 186). CAUTION: *It is important to have turret lock completely disengaged before traversing turret, and to lock turret in position when the need for*

Part Four—Auxiliary Equipment

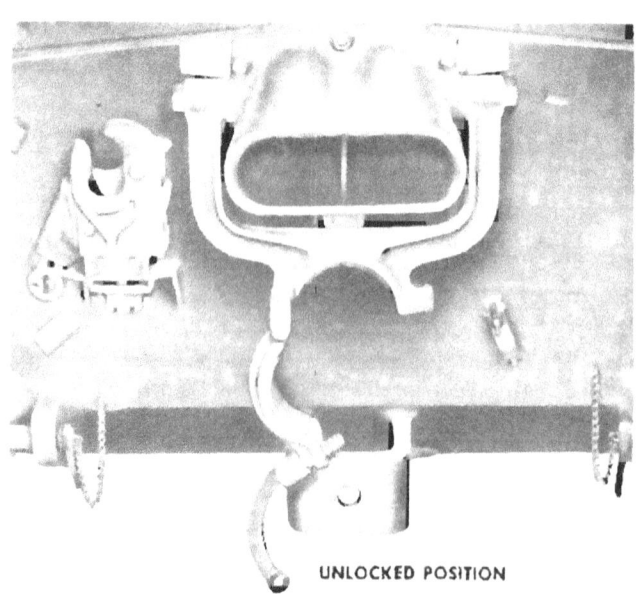

Figure 181 — 90-mm Gun Traveling Lock

TM 9-735
212

Armament Operating Instructions

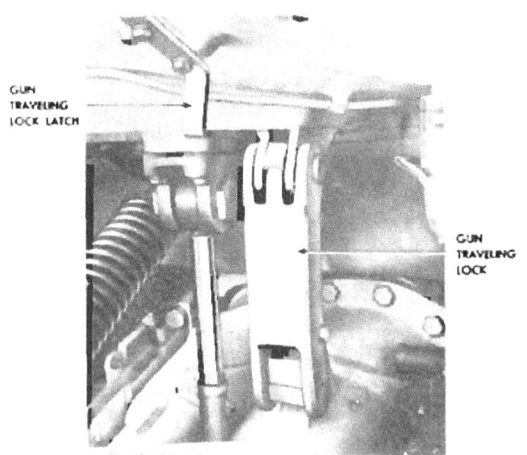

Figure 182 — Turret Gun Traveling Lock in Traveling Position

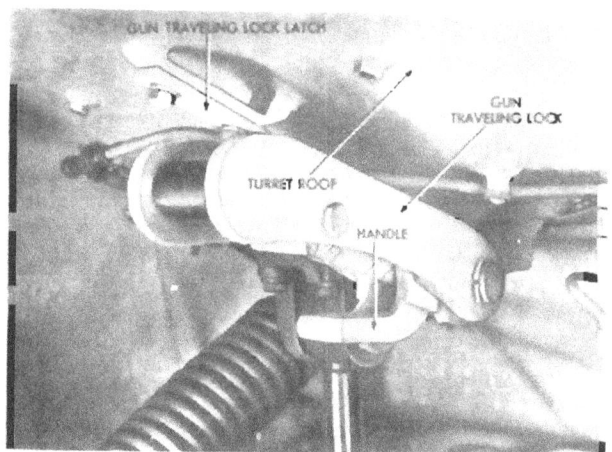

Figure 183 — Turret Gun Traveling Lock in Unlocked Position

Part Four—Auxiliary Equipment

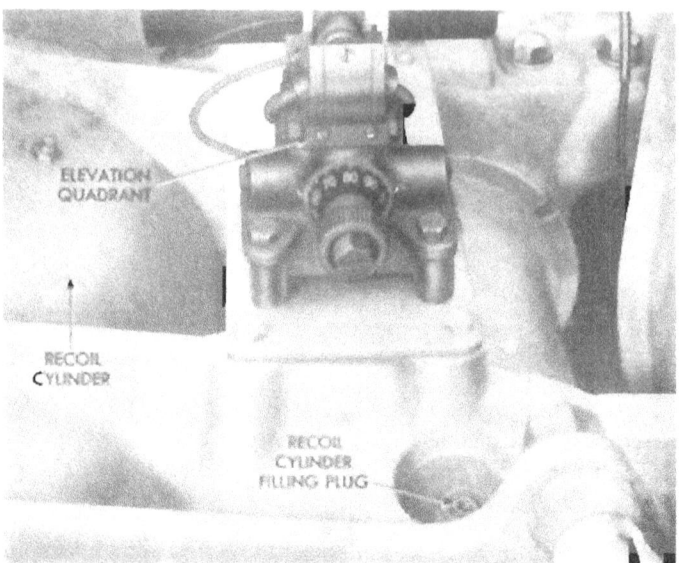

Figure 184 — Recoil Cylinder Filling Plug

traversing has been completed. The power traverse control handle on the traverse pump to the right of the gun elevating handwheel controls the speed and direction of turret traverse (rotation) (fig. 187). To traverse turret to the right, rotate (turn) the power traverse control handle so that the top is to the right. To traverse turret to the left, turn the power traverse control handle so that the top is to the left. The further the handle is rotated from its normal position, the faster the speed of traverse. To stop turret traverse, release power traverse control handle or bring it to the upright or neutral position. Turret rotation can be reversed instantly without damage to the power mechanism.

b. **Commander's Hydraulic Traverse Control Handle.** Mounted on a bracket on the right underside of the turret roof directly over the azimuth indicator and to rear of the compass is the commander's hydraulic traverse control handle (fig. 188). If the commander so desires, he may take traverse control of the turret away from the gunner, since operation of the handle automatically cuts out the gunner's hydraulic traverse control handle. To operate the commander's hydraulic traverse control handle, depress button in end of handle and move handle forward to traverse turret to the left, and toward rear to traverse turret to the right.

TM 9-735
212

Armament Operating Instructions

A—ELEVATING HANDWHEEL
B—TRAVERSING MECHANISM
C—TURRET TRAVERSE GEARSHIFT LEVER

RA PD 86812

Figure 185 — Elevating and Traversing Mechanisms

TM 9-735

Part Four—Auxiliary Equipment

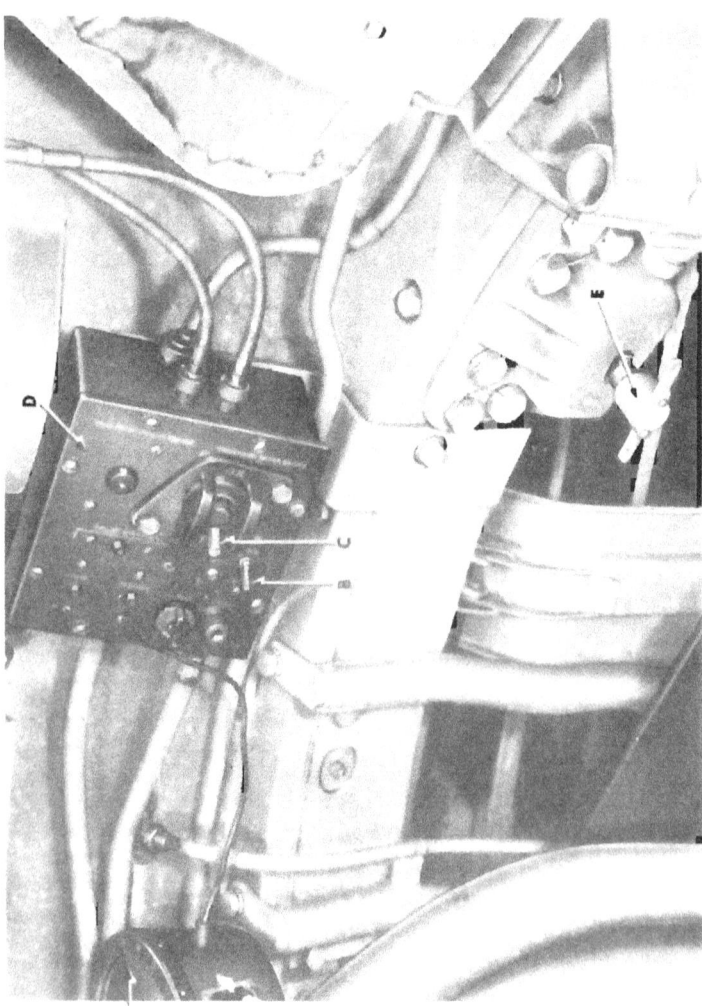

A—AZIMUTH INDICATOR
B—GUN ELECTRICAL FIRING SWITCH
C—MASTER TURRET ELECTRICAL SWITCH
D—TURRET CONTROL BOX
E—TURRET LOCK HANDLE

RA PD 86809

Figure 186 — Turret Control Box

TM 9-735
212

Armament Operating Instructions

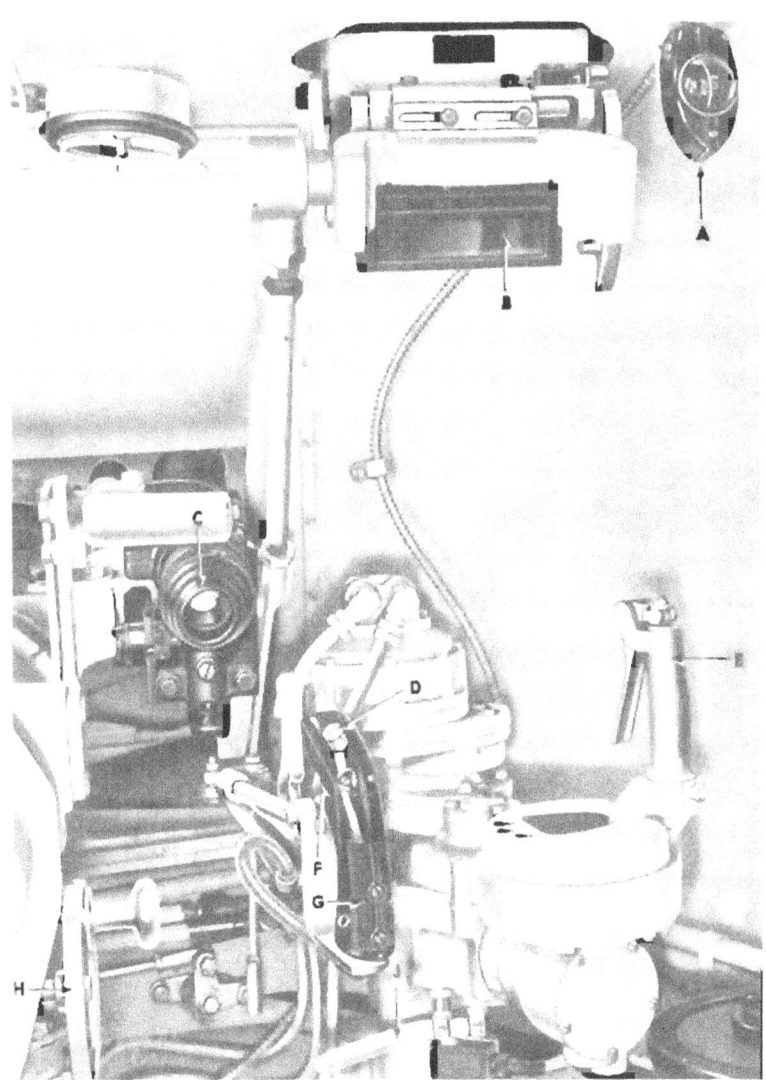

A—COMPASS
B—PERISCOPE
C—DIRECT SIGHT TELESCOPE
D—CAL. .30 MACHINE GUN ELECTRICAL FIRING TRIGGER
E—MANUAL TURRET TRAVERSE CONTROL HANDLE
F—90-MM GUN ELECTRICAL FIRING TRIGGER
G—HYDRAULIC TURRET TRAVERSE CONTROL HANDLE
H—ELEVATING HANDWHEEL
J—TURRET TRAVERSE GEARSHIFT LEVER

RA PD 86811

Figure 187 — Gunner's Control

Part Four—Auxiliary Equipment

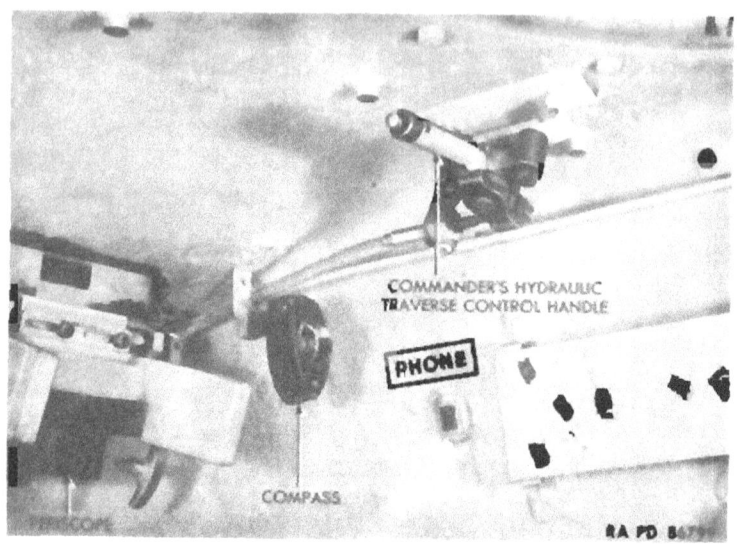

Figure 188 — Commander's Hydraulic Traverse Control Handle

c. **Manual Operation.** If the electric power fails, or if the hydraulic traversing system will not operate due to some other cause, the turret can be rotated by the manual traverse control handle located on top of the traversing gear mechanism to the right and forward of the gunner's seat (fig. 187). Disengage the turret lock (fig. 186). Move gearshift lever under traversing gear mechanism to the "DOWN" or manual position (fig. 187). It may be necessary to turn, slightly, the manual traverse control handle on top of traversing mechanism to engage gears. The manually operated gears are then engaged with the turret ring gear and pinion. Grip the manual traverse control handle and its associate lever on top of gear mechanism to release brake in gear box. Turn manual traverse control handle to the right, or clockwise, to rotate turret clockwise, and to the left, or counterclockwise, to rotate the turret to the left or counterclockwise (fig. 187). The rate of speed of turret rotation is dependent upon the speed at which the handle is operated. If the manual traverse control handle turns freely and fails to move the turret, the gearshift lever is in the "UP" or power position, and should be moved to the "DOWN" or manual traverse position. NOTE: *If the tank is on an incline when the turret lock is disengaged and the shift lever is not in the "UP" or power traverse position and the*

Armament Operating Instructions

Figure 189 — Azimuth Indicator

traversing pump is not in operation, the weight of the gun and mount will cause the turret to turn rapidly until the gun reaches the lowest position. Always traverse the turret 360 degrees manually with the manual traverse control handle to be sure the turret is free.

d. **Azimuth Indicator** (fig. 189). An azimuth indicator is mounted to the right of the gunner's seat, and gives the position of the gun in traverse.

213. ELEVATING OR DEPRESSING GUN.

a. The gun is elevated or depressed by turning the elevating handwheel located on the right side of the gun mount, slightly forward and to the left of the gunner's seat (figs. 185 and 190). A clockwise rotation of the handwheel will elevate the gun. Maximum elevation is 20 degrees and maximum depression is 10 degrees.

b. An elevation quadrant is located on the right top side of the gun mount and gives position of the gun in elevation. Refer to paragraph 221 for information on its operation.

214. OPERATING BREECH MECHANISM.

a. **Opening Breech.** To open breech, grasp the grip portion of the breech operating handle, unlatch, and press the handle down

TM 9-735
214

Part Four—Auxiliary Equipment

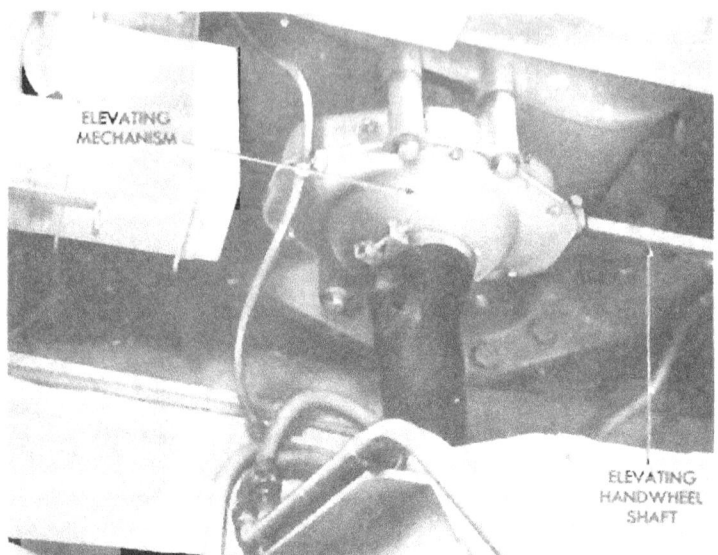

Figure 190 — Elevating Mechanism

until the extractors lock the breechblock in the open position (fig. 191). Keep hands out of breech ring when breech is open. Return breech operating handle to the closed (latched) position immediately after opening breech in order to avoid injury to personnel and mechanism.

b. **Closing Breech.** Close the breech by unlatching the breech operating handle and pressing it down. Bear sufficient weight on the handle to overcome the tension of the closing spring and release the extractors from their locking position by pressing them forward with the base of an empty cartridge case or a block of wood. The breechblock is then free to be eased up into its closed position by means of the operating handle which should finally be latched in place (fig. 192).

c. **Semiautomatic Operation.** Under ordinary conditions it will be necessary to open the breech only at the start of firing operations. A cam attached to the crank arm within the breech mechanism is operated by the recoil of the gun after firing, and serves to open the breech and extract the empty cartridge case during the recoil and return of gun to battery.

TM 9-735
215—216

Armament Operating Instructions

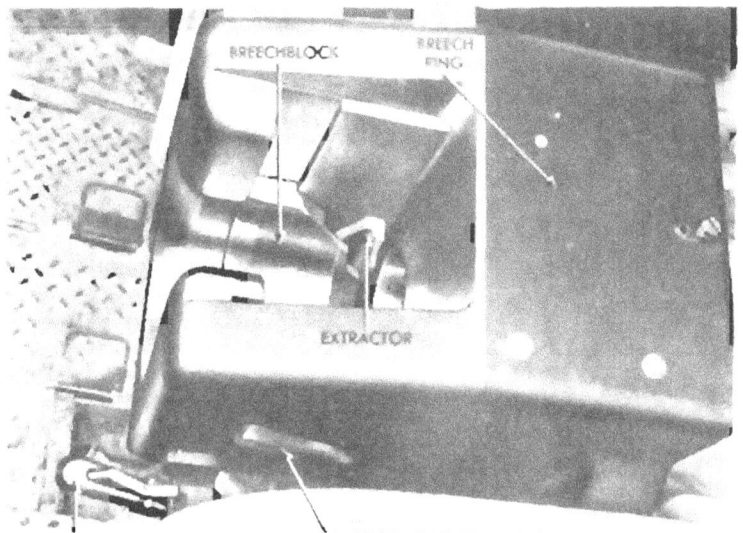

Figure 191 — Breech Open

215. LOADING THE GUN.

a. Before loading, open breech and examine the bore of the gun to see that it is clear and free from foreign material. Be sure to return the breech operating handle to its closed position. Refer to paragraphs 225 and 226 for information on authorized ammunition and preparation of ammunition for firing.

b. To load the gun, place a round in the breech with the nose protruding in the bore, and impel the round into the chamber with sufficient force to cause the flange of the cartridge case to drive the extractors forward and automatically close the breech. NOTE: *The loader's hand should be moving upward as he shoves the projectile home, in order to clear breechblock as it is automatically closed.*

216. FIRING THE GUN.

a. The 90-mm gun and the cal. .30 machine gun can be fired electrically, or manually in case of failure of the electrical system. A firing switch is located in the turret control box to the right and rear of the gunner's seat (fig. 186). This switch controls the circuit which energizes the solenoids which fire the 90-mm gun and cal. .30 machine gun. There is also a foot firing switch mounted in a switch box on the turret floor convenient to the gunner. The switch

TM 9-735

Part Four—Auxiliary Equipment

BREECH OPERATING HANDLE   RA PD 86806

Figure 192 — Breech Closed

operates the 90-mm Gun solenoid by pressing the foot up against the switch (fig. 27). Each of these two guns can also be fired manually.

b. To fire the 90-mm gun, pull the safety lock lever, mounted in the firing mechanism located on the right side of the breech ring to the "UP" or firing position. Throw the firing switch to the "ON" position which will cause the indicating light in the control box to glow. To fire the 90-mm gun electrically, depress the firing trigger in the front face of the hydraulic traverse control handle (fig. 187). If the solenoid does not function, the 90-mm gun may be fired manually by pulling the hand firing lever to the rear. This lever is located on the right side of the gun mount just in front of the solenoid (fig. 193).

c. The caliber .30 machine gun can be fired electrically by means of the cal. .30 machine gun electrical firing trigger mounted on top of the hydraulic traverse control handle (fig. 187). If the gun fails to fire electrically, it can be fired manually by means of the conventional trigger on the gun.

d. If the 90-mm gun fails to fire using either the electrical or manual means, the position of the safety lock lever should be checked. Failure to fire may be due to the gun staying out of battery, failure

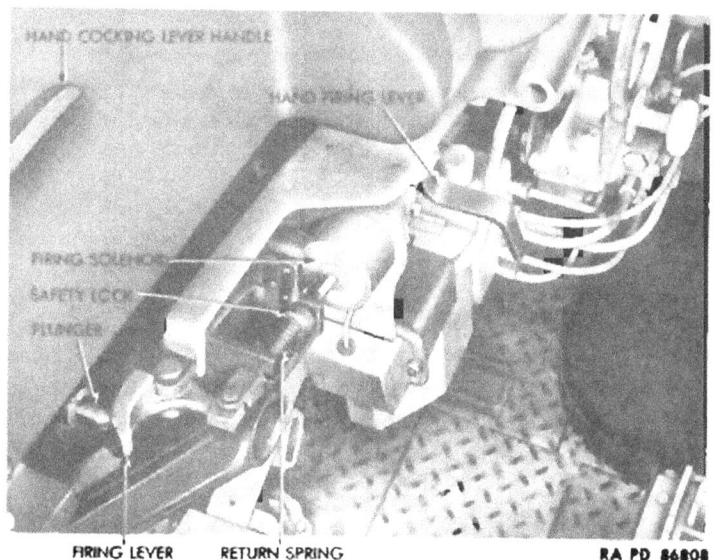

Figure 193 — 90-mm Gun Firing Mechanism (Recoil Guard Removed)

of the firing mechanism, failure of the breech to close, or defective ammunition. If the gun is in battery, recock by means of the hand cocking lever handle located on the right side of the breech ring (figs. 191, 192, and 193), and attempt to fire. CAUTION: *In case of a misfire, open the firing switch immediately before recocking.* If the gun still fails to fire after three attempts, wait 30 seconds before opening the breech; then remove round, reload, and attempt to fire again. After need for firing is completed, throw the firing switch to the "OFF" position.

217. PLACING GUN IN TRAVELING POSITION.

a. Push safety lock lever down to its "SAFE" position. This locks the firing mechanism so that the gun cannot be fired.

b. Open the breech and check to be sure that a round has not been left in gun and that the bore is clear. Close breech.

c. Separate the halves of the gun traveling lock by unscrewing the traveling lock handle. Swing the traveling lock up and position tube in lower half. Then swing top half in place over tube and lock by screwing the handle in place (fig. 181).

d. Place the gun in traveling position and secure with gun traveling lock (figs. 182 and 183).

e. Clean and lubricate gun and install muzzle cover.

f. Lock the turret in traveling position by pulling out and turning turret lock handle ¼ turn counterclockwise (fig. 186).

Section XLI

SIGHTING AND FIRE CONTROL EQUIPMENT

218. CHARACTERISTICS.

a. The sighting equipment includes a telescope and telescope mount for direct fire, a periscope for observation and direct fire, and an elevation quadrant for laying the gun in elevation for indirect fire.

b. For direct fire, telescope M71C with telescope mount T90 and instrument light M33 is used. Telescope M70P with instrument light M32 or M39C is supplied as a spare.

c. The periscope and related equipment depend on whether or not the tank turret has a bulge.

(1) Tank turrets constructed with a bulge are equipped with periscope mount T113. Periscope M10N is the standard periscope. Periscope M4A1 with telescope M77F is furnished as a spare for periscope M10N. When so used, it requires an adapter.

(2) Tank turrets, without the bulge, use the periscope mount T117. Periscope M10F is the standard periscope. Periscope M8A1 with telescope M80D is substitute standard issue for the mount.

(3) Instrument light M30 is furnished with telescope M80D and telescope M77F.

d. Elevation quadrant M9 is used for indirect laying of the weapon in elevation.

e. Binocular M13, fuze setter M14, and gunner's quadrant M1 are treated in TM 9-575.

219. TELESCOPE MOUNT T90 WITH TELESCOPE M71C.

a. Telescope mount T90 (fig. 194) is mounted at the right side of the gun and is equipped with range and deflection adjusting mechanisms for bore sighting the telescope, an adjustable head rest, and a clamp for instrument light M33 on the left side of the mount.

b. Telescope M71C (fig. 195) used for direct fire is mounted in the telescope mount T90. The reticle pattern (fig. 196) is graduated for use when firing armor-piercing-capped projectile M82; gradua-

TM 9-735
219

SIGHTING AND FIRE CONTROL EQUIPMENT

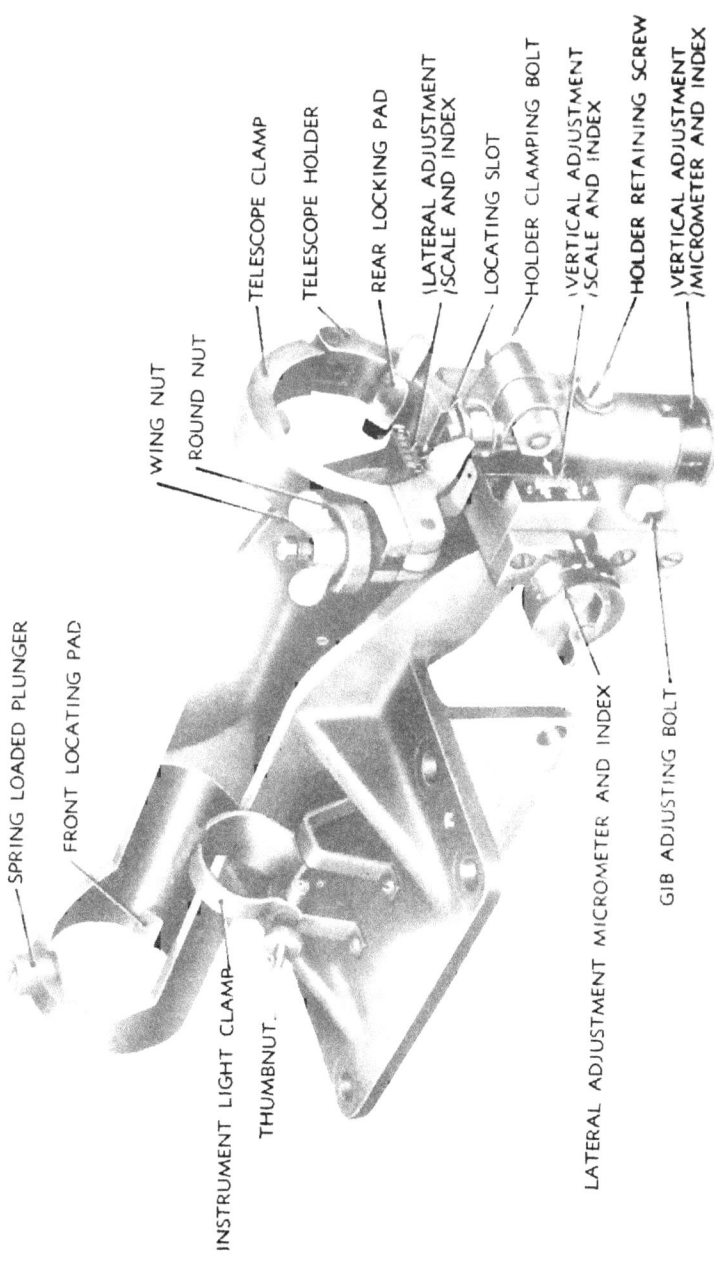

Figure 194 — Telescope Mount T90 — Left Rear View

TM 9-735
219

Part Four—Auxiliary Equipment

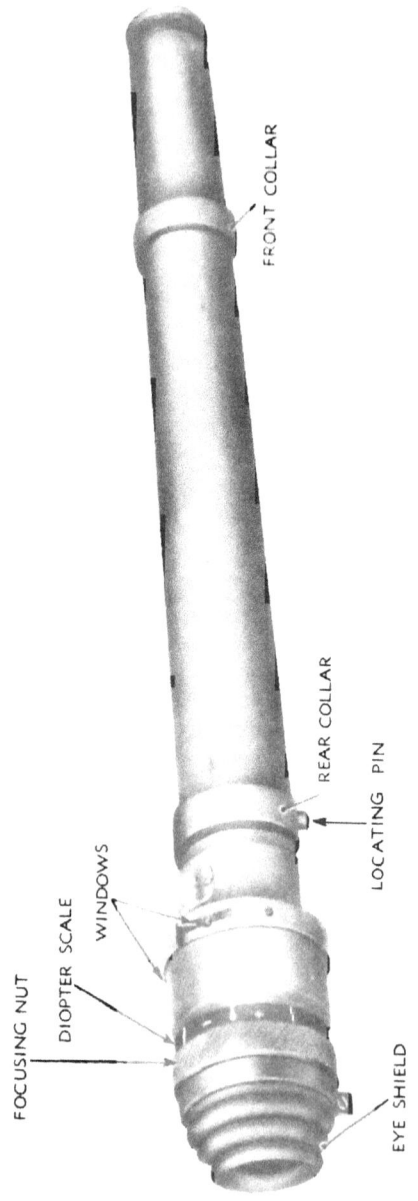

Figure 195 — Telescope M71C

444

tions are based on data from firing table 90-C-2, part 2A. The cross of the pattern represents zero range and zero deflection; range markings are in hundreds of yards.

c. To aim the gun, rotate the elevating handwheel and traverse the turret until the image of the target is at the point on the reticle representing the required range and deflection. The gunner's eye should be approximately 1¼ inches from the telescope eyelens during observation.

220. PERISCOPE M10F WITH PERISCOPE MOUNT T117.

a. Periscope M10F (fig. 197), used for direct fire against moving targets, contains two built-in optical systems. One is for observation of general terrain, the one-power system; the other a six-power system, for laying on distant targets. The one-power system may be used in an emergency to supplant the six-power system.

b. The reticle pattern (fig. 198) for the one-power system appears in a horizontal view of 42 degrees, 10 minutes and in a vertical field of view of 8 degrees, 10 minutes. The reticle is graduated for armor-piercing-capped projectile M82 based on data contained in Firing Table 90-C-2, Part 2A. The cross represents zero range and zero deflection. The broken line extending downward below the cross is at a slight angle to and offset from the vertical center to compensate for side jump and drift. The inscription "90-M82" indicates the ammunition. The reticle pattern for the six-power system (fig. 198) appears in a horizontal and vertical field of view of 10 degrees, 20 minutes. The pattern is identical to that of telescope M71C (fig. 198), with the exception that the pattern of the periscope reticle has the part number 7672807 etched on its lower edge.

c. Periscope mount T117 is mounted in the turret to hold periscope M10F. Mounting studs on the periscope position it in the mount being held in slots by two locking levers. The levers work together and are supported in either one of two positions, one retracted, the other in viewing position. Only the periscope head projects through the tank armor when ready for operation. Linkage connects the mount to the gun so the periscope moves as the gun is elevated or depressed to maintain periscope relation to the gun as positioned when bore sighted.

d. For laying the gun on targets close to the tank, observe through the window for the one-power system, and bring the image of the target to the point of the reticle representing required range and deflection by rotating the elevating handwheel and traversing the turret. Be sure the tank power source is plugged into a 24-volt light socket. For laying on distant targets, bring the image of the target to the required reticle point by means of the elevating handwheel and traversing turret.

Part Four—Auxiliary Equipment

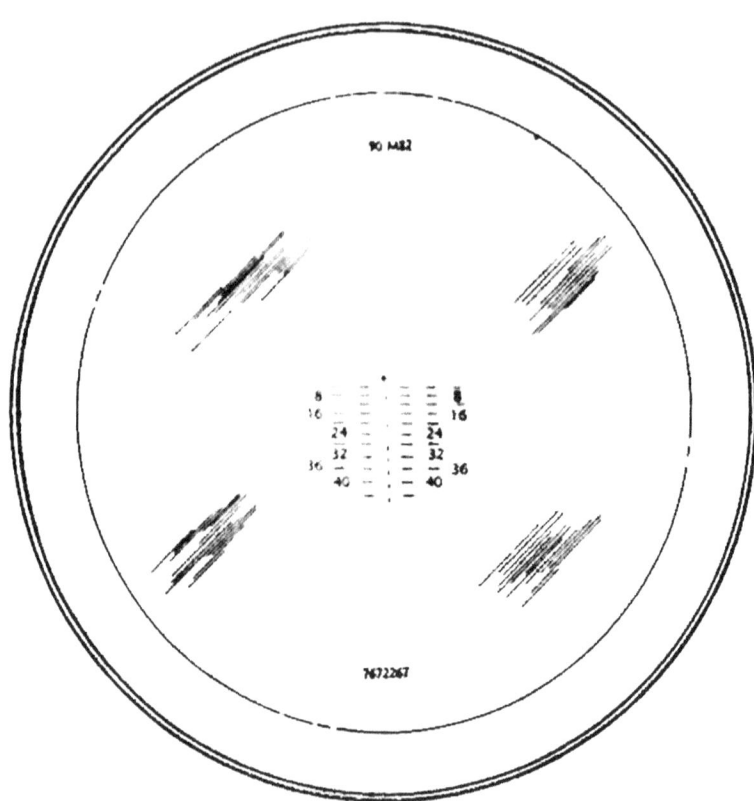

Figure 196 — Reticle Pattern for Telescope M71C

221. ELEVATION QUADRANT M9.

a. Elevation quadrant M9 (fig. 199) is used to lay the 90-mm gun M3 in elevation for indirect fire.

b. To operate the quadrant M9, set off the elevation angle on the coarse (100-mil interval) scale and the micrometer (1-mil interval). As the quadrant has two scales and two micrometer indexes, use the micrometer index on the side corresponding to the scale in use.

222. INSTRUMENT LIGHTS M33 AND M30.

a. Instrument light M33 consists principally of a body tube and rheostat connected to a lamp bracket by a lead wire and lead wire body. A rheostat knob on the end of the body turns the light on

TM 9-735

Sighting and Fire Control Equipment

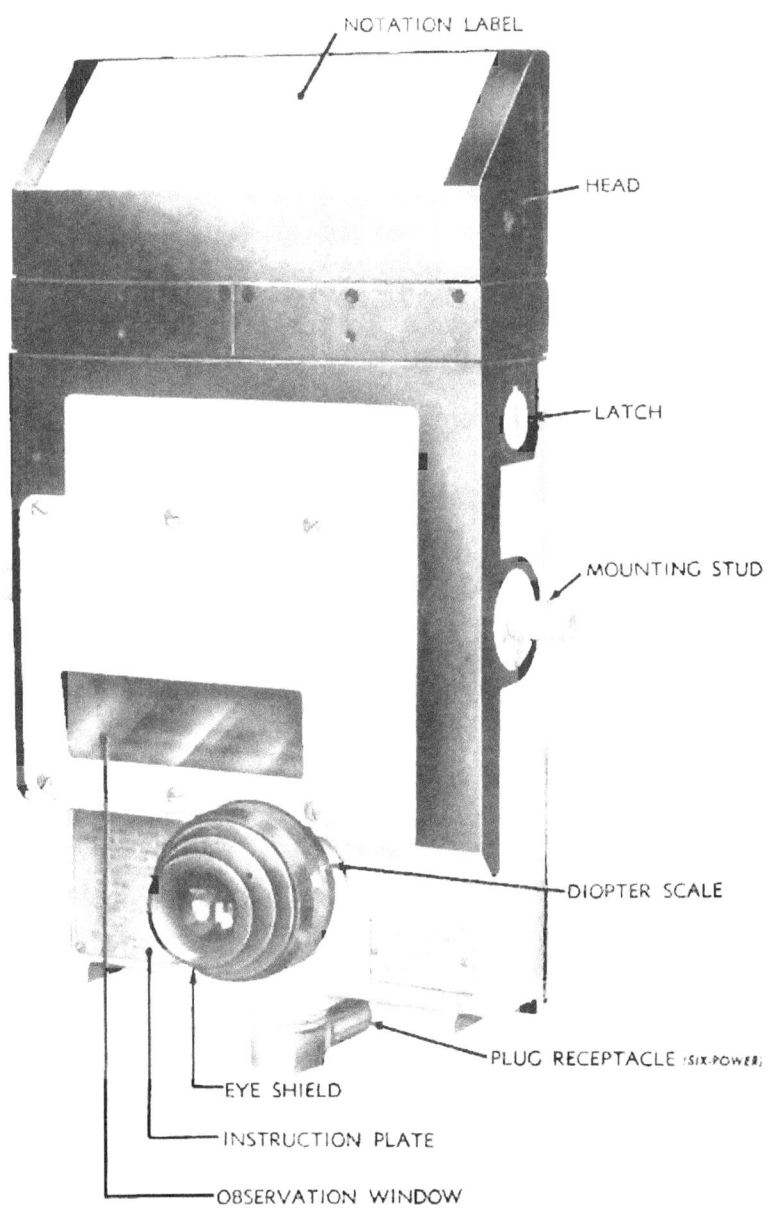

RA PD 31451

Figure 197 — Periscope M10F — Right Rear View

TM 9-735
222

Part Four—Auxiliary Equipment

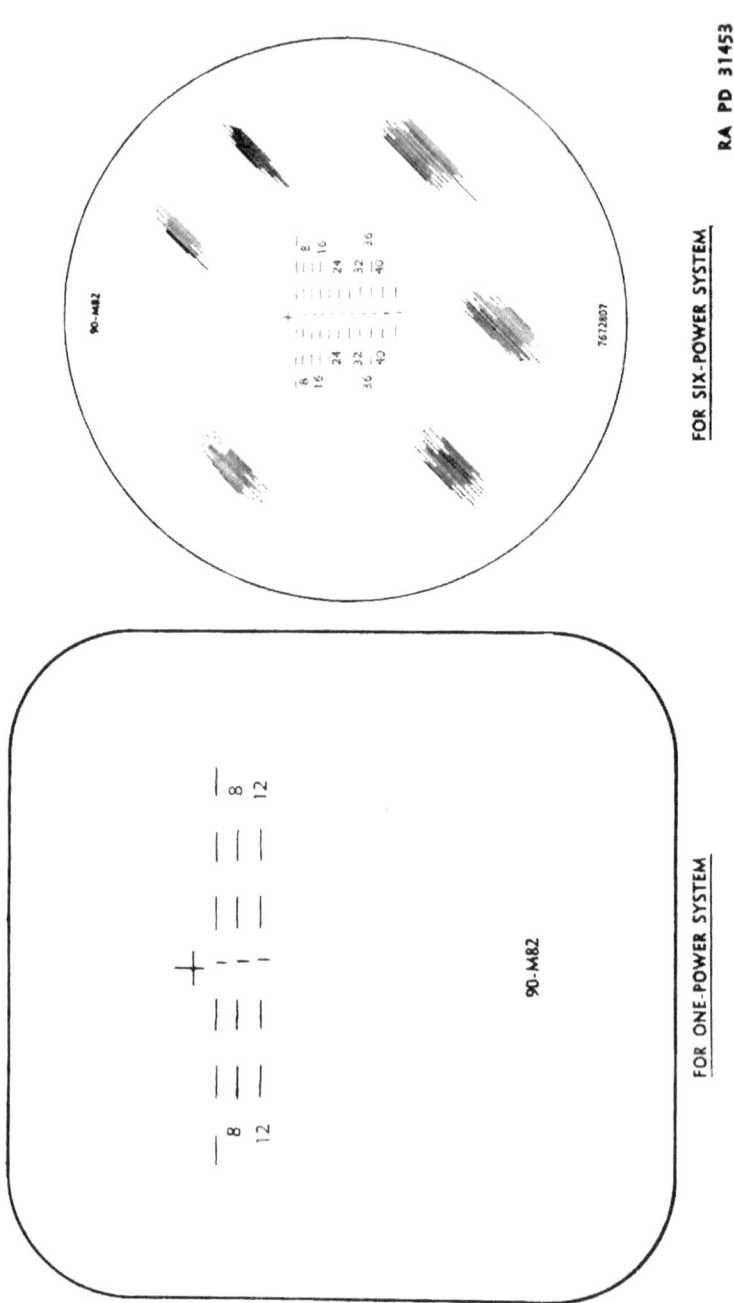

Figure 198 — Reticle Patterns For Periscope M10F

TM 9-735

Sighting and Fire Control Equipment

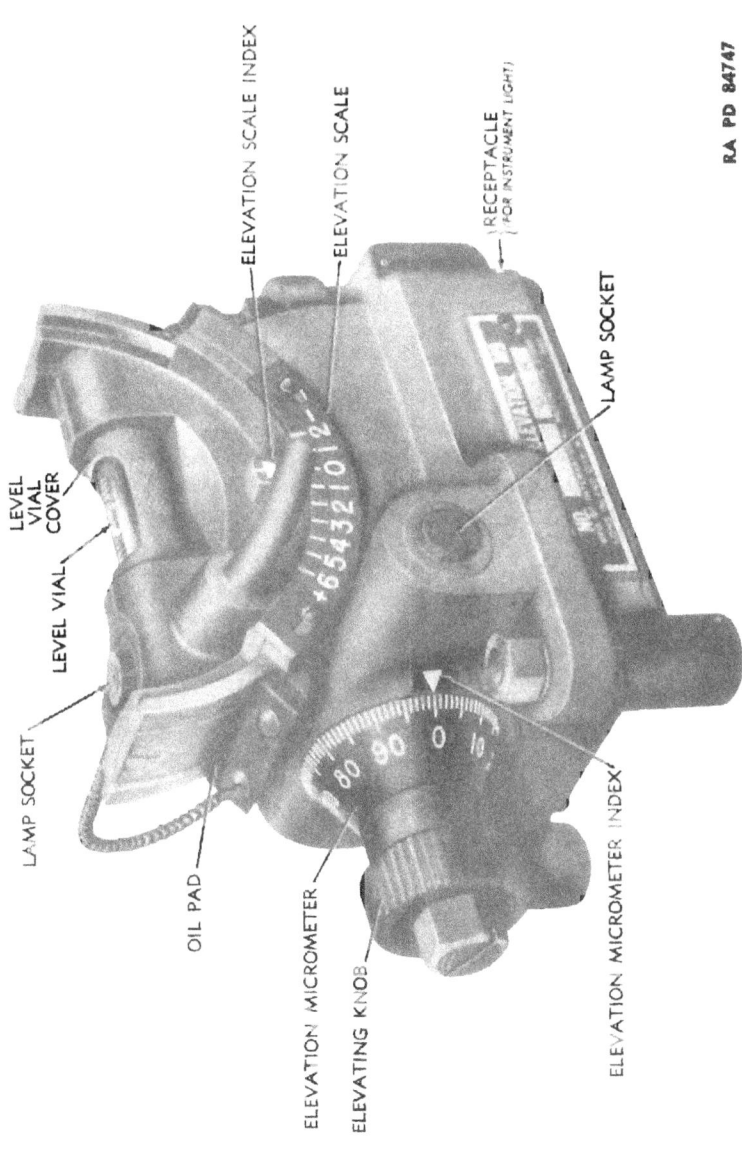

Figure 199 — Elevation Quadrant M9 — Right Side View

and off and regulates the intensity of the illumination. The lamp bracket engages a slot in the top of the telescope.

b. Instrument light M30 is identical with instrument light M33 except that the light M30 is provided with a plug assembly instead of the lamp bracket assembly and lead wire body.

c. The lights are used to illuminate the reticle patterns during night operation or under poor visibility conditions.

223. BORE SIGHTING.

a. Bore sighting is done to test the alinement of sighting equipment for parallelism with the gun bore. For expediency, choose a well-defined, fixed object at least 1,000 yards distant. Bore sights should be inserted and used in this operation.

b. Open the breech of the gun and while looking through the barrel, aline the gun on the distant object chosen. With the telescope M71C in position on its mount, observe and note the reticle point for required point-blank range and zero deflection with respect to the aiming point. If they do not coincide, move the line of sighting of the telescope by turning the micrometers on the mount (fig. 194).

224. PERISCOPE M6.

a. The periscope M6 (fig. 200) is used by the driver and assistant driver of the vehicle for observation from the interior of the tank. The periscope is secured in a holder by a latch mechanism and a locking knob. The heads of these periscopes are constructed of plastic materials so they will shatter into small pieces if struck by a projectile. The heads are secured to the periscope body with an eccentric mechanism which is so arranged that the heads are readily replaced with the spare heads which are provided.

b. To remove the periscope from the holder, open the latch on the front of the holder and loosen the locking knob on the front of the periscope. Pull the periscope straight out of the holder.

c. To remove the head from the periscope body, turn the two eccentric assemblies located on the sides and upper part of the periscope body until the head clamp is completely disengaged from the latches of the eccentric mechanism. Lift the head from the periscope body. When replacing the head, position it on the top of the periscope body with the window facing the front side of the periscope. Turn the eccentric assemblies until the reference arrow on each eccentric matches the corresponding arrow on the periscope body. See that the handles of the eccentric mechanism lie flat on the sides of the periscope body.

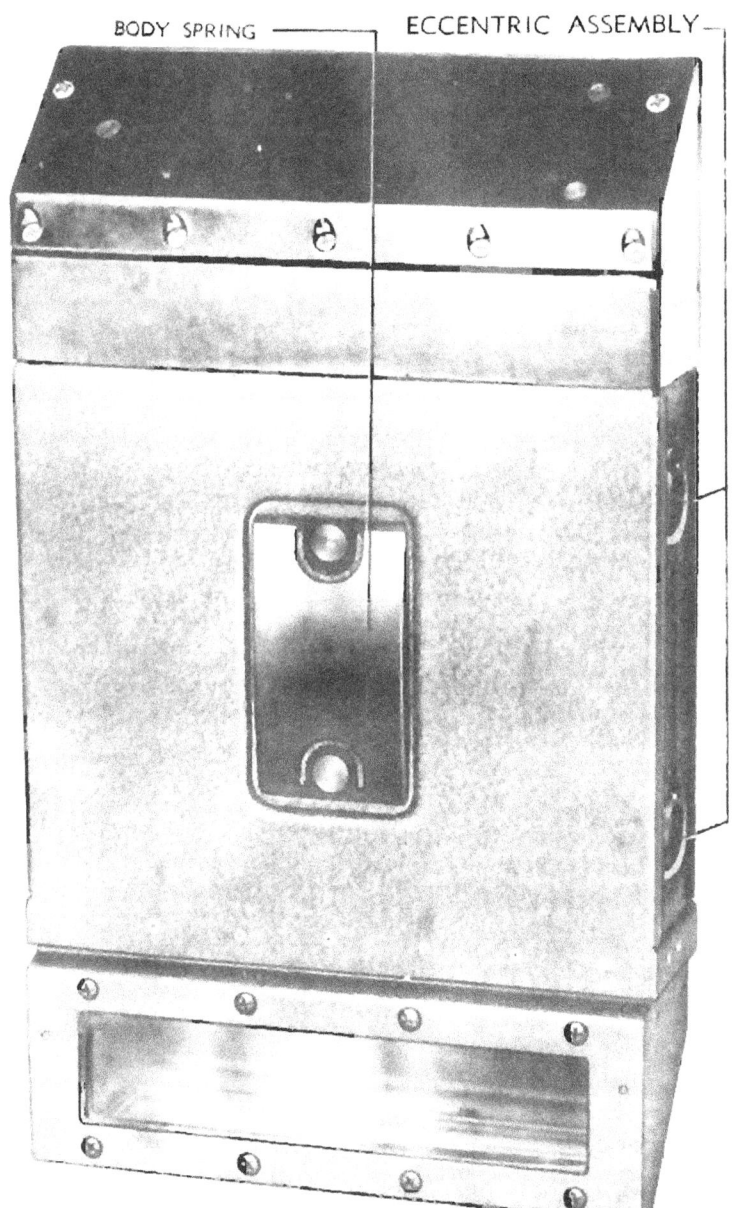

Figure 200 — Periscope M6 — Rear View

Part Four — Auxiliary Equipment

Section XLII

AMMUNITION

225. AUTHORIZED AMMUNITION.

a. Authorized ammunition for the 90-mm gun M3 is listed in Table I. It will be noted that standard nomenclature which completely identifies the ammunition, is used in the listing. The use of standard nomenclature is mandatory for purposes of record except where use of Ammunition Identification Code (A.I.C.) symbols is prescribed to expedite requisitioning and reporting. Identification is provided for by painting and marking on the round itself and on packing in accordance with the basic scheme described in TM 9-1900.

TABLE I — AUTHORIZED AMMUNITION

Standard Nomenclature[1]	Action of Fuze	Nominal Weight of Projectile as Fired (lb)
Service Ammunition		
PROJECTILE, fixed, A.P.C., M82, NH, w/FUZE, B.D., M68, and TRACER, 90-mm gun[2]	Delay	24.11
PROJECTILE, fixed, A.P.C., M82, NH, w/TRACER, 90-mm gun[2]	None	23.40
SHELL, fixed, H.E., M71, NH, w/FUZE, P.D., M48, 90-mm gun	SQ and delay[3]	23.29
SHELL, fixed, H.E., M71, NH, w/FUZE, P.D., M48A2, SQ and 0.05-sec. delay, 90-mm gun	SQ and delay[3]	23.29
SHELL, fixed, H.E., M71, NH, w/supplementary bursting charge and FUZE, P.D., M48A2, SQ and 0.05-sec. delay, 90-mm gun[4]	SQ and delay[3]	23.20
SHOT, fixed, A.P., M77, NH, w/TRACER, 90-mm gun	None	23.40

Ammunition

TABLE I — AUTHORIZED AMMUNITION — Contd.

Standard Nomenclature[1]	Action of Fuze	Nominal Weight of Projectile as Fired (lb)
Practice Ammunition		
SHELL, fixed, practice, inert loaded, M71, NH, w/FUZE, dummy or inert, M—, 90-mm gun	Inert	23.40
Blank Ammunition		
AMMUNITION, blank, 90-mm gun	Inert	....
Drill (Dummy) Ammunition		
CARTRIDGE, drill, M12, w/FUZE, M44A2, 90-mm gun	..	....

A.P.—armor-piercing
A.P.C.—armor-piercing capped
B.D.—base detonating
H.E.—high-explosive
NH—nonhygroscopic (flashing)
P.D.—point detonating
SQ—superquick

1—Nomenclature listed in this table refers to current standard rounds for use against ground targets. Earlier rounds were loaded with FNH propellent powder. Such rounds are indicated in nomenclature by the omission of the letters "NH" from the nomenclature. Identification between the types of rounds is provided for by marking "NH" on all packings holding NH-loaded rounds.

2—For future manufacture, armor-piercing-capped rounds will be loaded for a muzzle velocity of 2,800-2,850 feet per second. Earlier rounds have 2,650 feet per second muzzle velocity. Some rounds with 2,800-2,850 feet per second muzzle velocity may have "SC" marked on the cartridge case, signifying "supercharge." Markings on the later lots will show the muzzle velocity.

3—The M48 fuze has 0.05 second delay. The M48A2 may have either 0.05 or 0.15 second delay, depending on the lot. Identification is provided for by marking the time of delay on the body of the M48A2 fuze following the nomenclature of the fuze. The short delay is the authorized delay for use in the M3 90-mm gun.

4—These shells have a deep fuze cavity which has been adapted for standard fuzes and boosters by inserting a supplemental bursting charge, reducing the fuze cavity to the dimensions of the standard boosters. The shells may be made suitable for use with special fuzes in the field by removing the standard fuze and booster assembly and the supplemental charge.

5—FUZE, dummy, M73, or FUZE, dummy, M44A2, or an inert service fuze of the M48 series may be fitted to the shell as shipped.

226. PREPARATION FOR FIRING.

a. General. The rounds are ready for firing upon removal of packing material, except for adjustment of fuzes as described below. Rounds prepared for firing but not fired should be restored to the original condition and packing, and appropriately marked. Such rounds should be used first in subsequent firings in order to keep stocks of opened packings at a minimum.

b. P.D. Fuzes M48, M48A1, and M48A2. These fuzes are fitted with a slotted "setting sleeve" and two registration lines—one marked "SQ" (superquick) and the other "DELAY"—for fuze setting. As shipped, the fuzes are set "SQ" and need no adjustment for this action. To set for delay, it is only necessary to turn the setting sleeve so that the slot is alined with "DELAY." (A delay pellet incorporated in the delay action train provides for the delay action.) Delay action is always operative and will function if the superquick action fails when the fuze is set "SQ." The setting may be made or changed at will with fuze wrench M7A1 or similar instrument at any time before firing. This can be done in the dark by noting the position of the slot—parallel to the fuze axis for "SQ," at right angles thereto for "DELAY."

c. B.D. Fuze M68. This fuze requires no adjustment.

Section XLIII

COMMUNICATION SYSTEMS

227. RADIO.

a. General. Heavy Tank, T26E3 is equipped with two transmitting and receiving radios. Radio SCR-528 (fig. 201) which consists of two separate units, a transmitter and a receiver is for intertank communication. Radio AN/VRC-3 (fig. 202), which consists of a transmitter and a receiver in one case, is for two-way communication in combat areas between vehicle and ground troops. Both sets are shock-mounted on individual mountings, or bases, and are located on shelves in the turret overhang. Radio SCR-528 is installed in the right side behind the commander's station. Radio AN/VRC-3 is installed in the left side behind the loader's station. Power for both radios is supplied by the 24-volt circuit. Operating voltage of each set is stamped on a plate attached to each set.

b. Description of Radio SCR-528. This set consists of two separate units, transmitter BC-603 and receiver BC-604. Both units are shock-mounted on a common base or mounting (Mounting, FT-

TM 9-735
227

Communication Systems

Figure 201 — Radio SCR-528 — Installed

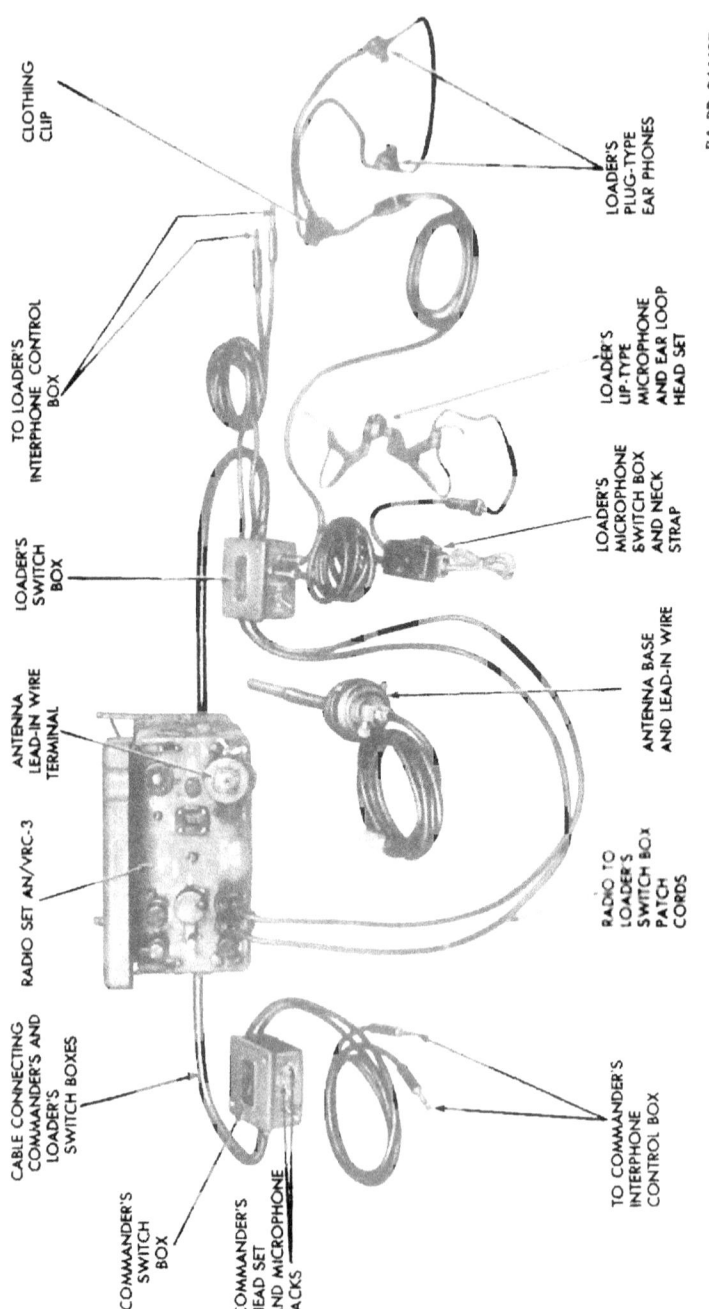

Figure 202 — Radio AN/VRC-3 — Installed

237-A) which is bolted to a shelf in the right side of the turret overhang behind the commander's station. The control panel of each unit faces to the right or toward the outside of the vehicle (fig. 201). So the radio operator can see and operate the frequency selection buttons and tuning controls, a stainless-steel mirror is provided (fig. 201). It is attached to a bracket on the right wall of the turret overhang. A combination red and white compartment light attached to the roof of the overhang provides illumination. Radio messages can be distributed over the vehicle inter-communication system.

c. **Description of Mounting, FT-237-A.** The mounting is a hollow rectangle with a low wall along one side and at one end (fig. 204). The metal is formed so as to provide a space between the walls and bed plates for the distribution of wiring. Multiple pin female-type connectors are installed along the side wall for connecting the multiple pin male-type connectors on the transmitter and receiver. A central terminal strip is located between the bed plates at the end of the mounting opposite the end wall. It is accessible by unlocking and lifting the hinged door in the upper bed plate. Two terminal posts are mounted on the end wall. The one marked "ANT" is for connecting the antenna wire lead and the marked "GRND" is for a ground wire connection. Guide pins on the side wall are for the installation of the receiver.

d. **Connecting Power Feed and Interphone Wires.** Connect power feed and interphone wires to turret radio terminal box (fig. 205) and central terminal strip in mounting. Lead interphone wire out of top outlet in turret terminal box and power feed wire out of bottom outlet.

e. **Installing Radio SCR-528.** Install mounting FT. 237-A in turret overhang with antenna and ground terminal posts toward rear of overhang. Connect wire from antenna base to antenna terminal post on mounting. NOTE: *The bolts attaching the mounting to the turret overhang shelf are usually sufficient ground for the set, so it may not be necessary to install a ground wire. However, this can only be determined after the set is installed and operated. If mounting bolts do not provide sufficient ground, remove set and install a ground wire between mounting ground terminal post and a convenient bolt or screw in the turret overhang.* Position transmitter BC-603 at rear of mounting so that control panel faces right side of turret and multiple pin connectors on unit mate with multiple pin connectors on mounting wall. Tighten transmitter mounting screws securely. Install receiver BC-604 in front of and in the same manner as the transmitter. In addition to mating the connectors securely be sure guide pins on side wall seat fully in guide pin holes in receiver. Turn on 24-volt master switch and tune and test radio.

TM 9-735
227

Part Four—Auxiliary Equipment

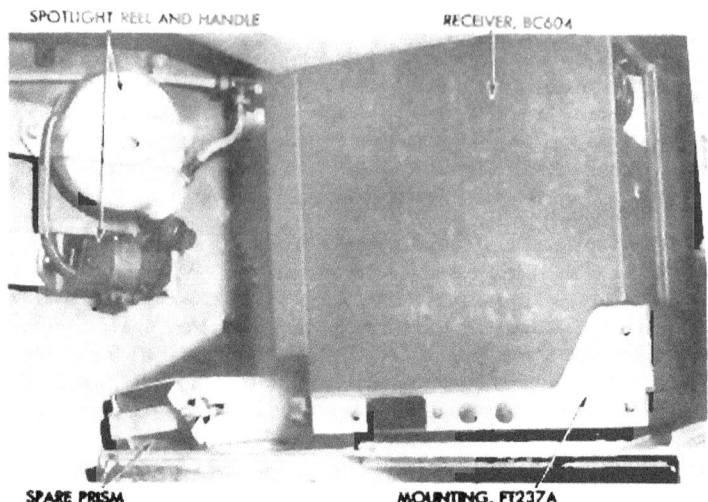

Figure 203 — Radio SCR-528 — Installed, End View

f. **Removal of Radio SCR-528.** Reverse installation procedure described in subpar. e above.

g. **Description of Radio AN/VRC-3.** This radio set is installed in the left side of the turret overhang. It is the standard infantry pack type radio communication set known as the "walkie-talkie (fig. 202). It is for tank-infantry communication and is a two-way set, receiving and transmitting. It can be operated on either a battery which is contained within the set, or by a vibrator power pack which connects to the radio terminal box. It operates on 24-volts. The tank commander and the loader are the only two in the vehicle crew who operate this set. The AN/VRC-3 is not channelled through the interphone system. The loader "listens-in" and when the vehicle is called switches over to interphone and notifies the commander who then switches over his control box to receive on this set. When "listening-in" the loader is off the vehicle interphone circuit. The vehicle commander is the only one who transmits on the AN/VRC-3 radio set. Loader's and commander's control boxes are mounted next to the standard interphone control boxes and patch cords connect the two boxes. The tank-infantry set control boxes operate in the same manner as the vehicle interphone control boxes, and are connected to the commander's and loader's interphone control boxes by patch cords.

TM 9-735

Communication Systems

Figure 204 — Mounting, FT-237-A, For Radio SCR-528

TM 9-735
228

Part Four—Auxiliary Equipment

Figure 205 — Turret Radio Terminal Box

228. ANTENNA.

a. **General.** The vehicle is equipped with two, 3-section, whip-type antenna which are mounted on top of the turret overhang (fig. 210). The antenna for radio AN VRC-3 is located at the rear of the loader's hatch to the left of the cal. .50 machine gun mount. The antenna for radio SCR-528 is located at the extreme center rear of the turret overhang. Each antenna is provided with a mast base which consists of a rubber-covered helical spring for flexibility, porcelain insulator, wire lead terminal, mounting plate, a retaining nut and a jam nut.

b. **Mast Sections.** Each mast or antenna is in three sections and tapers from bottom to top (fig. 210). When the three sections are

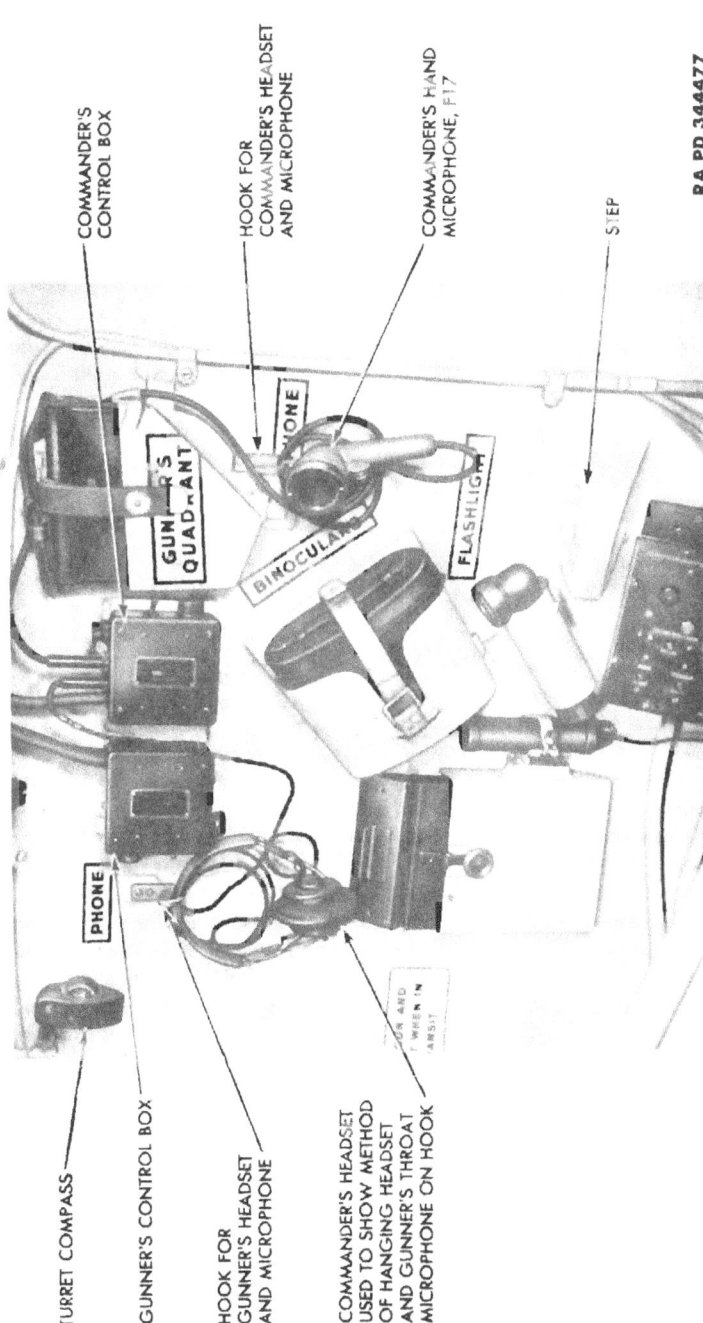

Figure 206 — Commander's and Gunner's Interphone

screwed together to form the mast the assembly looks not unlike a buggy whip, tapering from a comparatively wide diameter at the butt to almost a point at the opposite end. The Mast Base, AB-15/GR, is secured to the turret roof. Mast sections are numbered MS 116, MS 117 and MS 118 and are made of high tensile strength steel with the ends which join together marked in enamel of the same color. The mast body bears the type number. When not in use store mast or antenna sections in roll BG 56.

c. Use of **Phantom Antenna**. A phantom antenna contained in a cannister and marked Phantom Antenna, A-62 (fig. 211), is mounted in a bracket behind the SCR-528 radio mirror (fig. 204). WARNING: *The phantom antenna is to be used when tuning radio SCR-528 under ALL combat conditions. Under combat conditions do not tune set any other way.* Its use eliminates the high frequency noises broadcast when tuning the set in the normal manner. These noises can be picked up by enemy listening posts and the position of the vehicle quickly determined. To use phantom antenna follow instructions on plate on cannister. Tune transmitter as per instructions for 3-section aerial. NOTE: *Coaxial cable connection on phantom antenna is used when vehicle is equipped with 7-foot coaxial antenna cable.*

229. INTERPHONE SYSTEM.

a. General. The vehicle is equipped with a 5-station interphone system (figs. 206 through 209). The system not only permits conversation between members of the crew, but also enables all crew members to individually receive all radio orders and messages direct. The vehicle commander is the only crew member who can transmit. It consists of the necessary wiring, head sets, interphone control boxes, throat-type microphones, and a hand-type microphone. Each crew member is provided with a control box mounted on a bracket at his station, a throat-type microphone, a headset and a hook for hanging up headset and microphone. Both headset and microphone extension wires or cords have a plug at the end for plugging into the jacks in the control boxes. Plugs are of different shape and size to permit quick correct plug-ins and to eliminate the possibility of making wrong connections. Each control box is provided with a toggle-type switch for switching to "INT" for interphone conversation or to "RADIO" for radio reception. The sound amplifier or volume control knobs are provided on each control box. One knob is for radio volume control and the second knob is for interphone. The vehicle commander only is provided with a hand-type F17 microphone. Radio broadcasting, in addition to interphone conversation, is only possible through the hand-type microphone as the commander's control box is the only box wired for radio transmission. To talk on the interphone, or to transmit, it is necessary for the commander to shift his micro-

Communication Systems

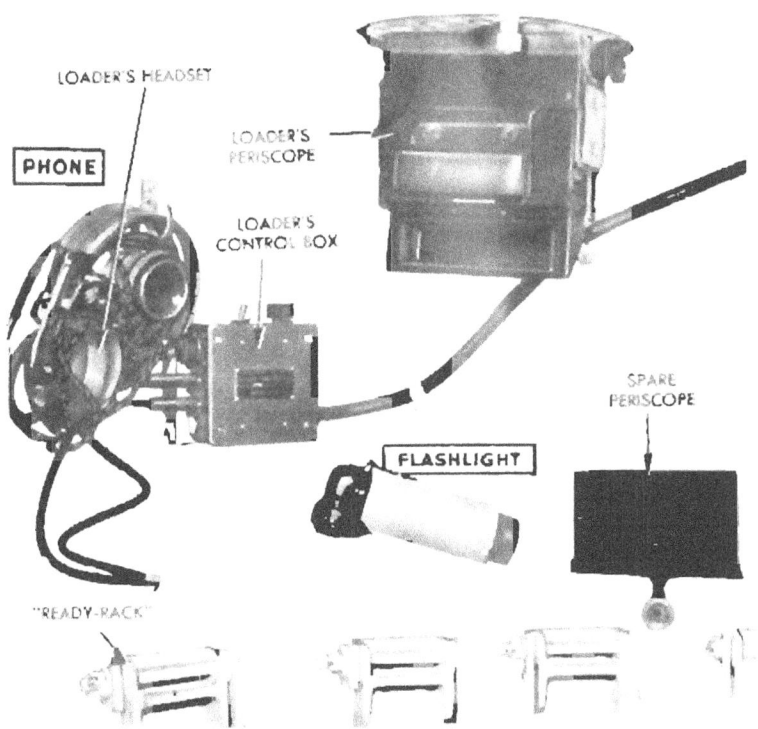

Figure 207 — Loader's Interphone Control Box

phone plug to the proper connection and to move his control box toggle-switch to either "INT" or "RADIO." The interphone system operates on the 24-volt circuit controlled by the 24-volt master battery switch. The interphone system is paramount in the vehicle. Any crew member can cut off all radio reception or transmitting by switching the toggle switch on his control box to interphone. Throwing the toggle switch back to RADIO reopens the radio circuit.

230. INSPECTIONS.

a. General. To ensure the continued and efficient operation of the radios and interphone system, frequent and careful inspections must be made of the various units and wiring.

b. Radio.

(1) MOUNTINGS. Inspect radio mounting screws to see that they are tight, and that shock mountings are in good condition. Rock

TM 9-735
230

Part Four—Auxiliary Equipment

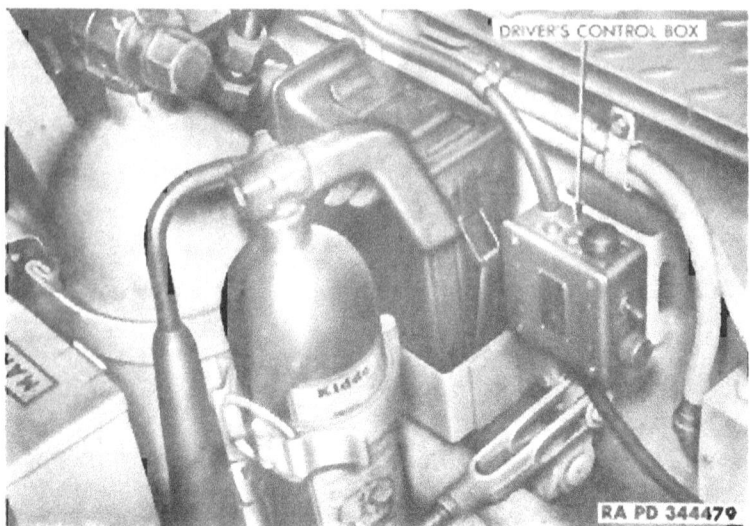

Figure 208 — Driver's Interphone Control Box

units to determine if they bump against anything. Be sure all connections are clean, dry and tight and that wiring is in good condition. Make sure units are securely fastened to mounting.

c. **Antenna.**

(1) MAST SECTIONS. Inspect antenna mast sections. Be sure they are screwed, clamped, or taped securely together and are not damaged or excessively bent.

(2) HELICAL SPRING. Inspect base helical spring. Be sure it maintains a vertical position and is not damaged to prevent flexibility.

(3) MAST. See that mast retaining nut and jam nut secure mast base to top of turret. Be sure porcelain insulator is present and in good condition.

(4) LEADS TO SETS. Inspect leads to radios. Be sure nothing interferes with them or that there is nothing present which could damage the wires. Be sure connections are clean and tight.

d. **Wires and Connections.**

(1) Inspect all wires which connect radios and interphones. Make sure they are not damaged and that they are properly secured in clips. Inspect all connections for tightness. Make sure headset

Communication Systems

Figure 209 — Assistant Driver's Interphone Control Box

and microphone cords are not twisted or knotted to prevent free movement. Inspect jack to see that they are not damaged or bent.

e. **Interphone Control Boxes, Microphones and Headsets.**

(1) Inspect control boxes for damage, looseness, and correct operation. Handle microphones and headsets with care. Hang them on hooks provided when not in use.

f. **Covers.**

(1) Be sure all covers for the protection of the radios are present in the vehicle. Install covers when equipment is not in use. See that zippers and fasteners are in good condition and operate properly.

231. **PRECAUTIONS.**

a. **Antenna.** Tie antenna back securely when vehicle is in motion and radio is not in use to prevent damage to antenna. Make sure antenna is vertical and not touching anything when radio is in use. When operating in woods or thick bush dismount and stow antenna and base. Cover base opening with cover plate. In woods or thick brush stop vehicle, if tactical situation permits, install antenna and base and operate radio(s). Dismount and stow antenna and base when through with radio(s). When going under bridges or other obstructions disjoint antenna as necessary to clear obstacle,

TM 9-735

Part Four—Auxiliary Equipment

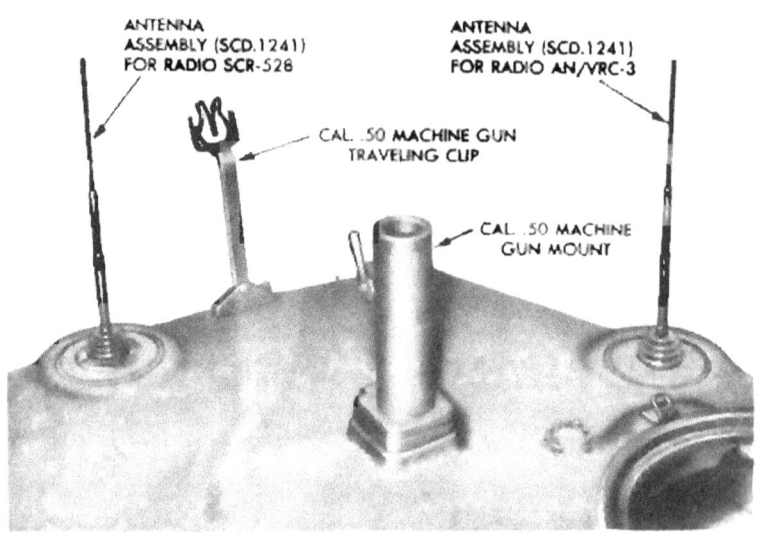

Figure 210 — Radio Antenna Assemblies

if tactical situation permits. Disjoint antenna as necessary when aboard decked landing craft.

b. Radio.

(1) Keep radios covered when not in use to prevent entrance of dust and moisture. Keep all hatch doors closed and securely fastened.

(2) Turn off all radio switches when sets are not in use. Do not turn off 24-volt battery switch with radio switch on.

(3) Do not store equipment on or behind radios where it can prevent action of shock mountings or damage connections.

(4) Do not start engine with radio switch on.

c. Batteries and Generating System.

(1) Make sure batteries are properly charged at all times to ensure satisfactory operation of the radios. Low batteries will cause sets to be noisy which will result in poor reception.

(2) See that all battery cables and terminals are in good condition, clean, and tight.

(3) Test operation of generator and regulator (par. 62). Excessive charge will damage radio and make it noisy.

TM 9-735

Communication Systems

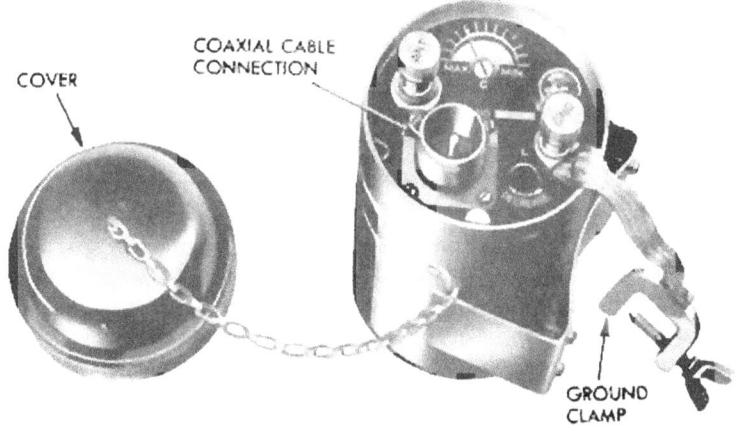

Figure 211 — Phantom Antenna, Showing Instruction Plate and Antenna with Cover Removed

(4) Do not race engine while radio is in operation.

d. **Interphone Equipment.** This equipment is sensitive and delicate. CAUTION: *Do not drop or misuse it.* Keep microphones and headsets on respective hooks when not in use. Keep cords clear of moving equipment, parts, and doors.

APPENDIX

Section XLIV

SHIPMENT AND LIMITED STORAGE

232. GENERAL INSTRUCTIONS.

a. Preparation for domestic shipment of the vehicle is the same as preparation for limited storage. Preparation for shipment by rail includes instructions for loading and unloading the vehicle, blocking necessary to secure the vehicle on freight cars, clearance, weight, and other information necessary to properly prepare the vehicle for rail shipment. For more detailed information and for preparation for indefinite storage refer to AR 850-18.

233. PREPARATIONS FOR LIMITED STORAGE OR DOMESTIC SHIPMENT.

a. Vehicles to be prepared for limited storage or domestic shipment are those ready for immediate service but maintained ready for immediate service, or those to be out of service for less than thirty days. If vehicles are to be indefinitely stored after shipment by rail, they will be prepared for such storage at their destination.

b. If the vehicles are to be placed in limited storage, take the following precautions.

(1) LUBRICATION. Lubricate the vehicle completely (par. 46).

(2) COOLING SYSTEM. If freezing temperature may normally be expected during the limited storage or shipment period, test the coolant with a hydrometer and add the proper quantity of antifreeze compound to afford protection from freezing at the lowest temperature anticipated during the storage or shipping period. Completely inspect the cooling system for leaks.

(3) BATTERY. Check battery and terminals for corrosion and if necessary, clean and thoroughly service battery (par. 92).

(4) ROAD TEST. The preparation for limited storage will include a road test of at least 5 miles, after the battery, cooling system, and lubrication service, to check on general condition of the vehicle. Correct any defects noted in the vehicle operation before the vehicle is stored, or note on a tag attached to the steering levers, stating the repairs needed or describing the condition present. A written report of these items will then be made to the officer in charge.

(5) FUEL IN TANKS. It is not necessary to remove the fuel from the tanks for shipment within the United States, nor to label the tanks under Interstate Commerce Commission Regulations. Leave fuel in the tanks except when storing in locations where fire ordnances or other local regulations require removal of all gasoline before storage. If fuel is to be left in fuel tanks, the following precautions against gum formation must be taken:

(a) The fuel system must be free from accumulated gum. Unless the vehicle is entering its first storage and has never been issued for use, inspect and clean the fuel pump valve; carburetor accelerator pump plunger, venturi tube choke and throttle valves, float mechanism; fuel lines; fuel tanks; fuel filters; fuel shut-off valves; and screens.

(b) If gum is present in the above parts, it can best be removed by benzol, acetone, alcohol, or a mixture of these solvents. Deposited gum is not readily soluble in fresh gasoline. When gum has dried, it may be necessary to resort to mechanical means to remove it.

(c) Parts which cannot be thoroughly cleaned and freed from the gum deposit without damage should be replaced.

(d) After cleaning and reassembling, fill fuel tank half full of fresh gasoline which has not been long in storage.

(e) Add three containers (12 oz) of gum-preventive compound to the right fuel tank, and four containers (16 oz) of gum-preventive compound to the left fuel tank.

(f) Fill the fuel tank to capacity and operate the vehicle for at least 5 minutes.

(6) BREECH MECHANISM. When possible, partially disassemble the breech mechanism and dip, spray, or brush the parts with light rust-preventive compound. Assemble the breech mechanism.

(7) GUN TUBE. Clean the bore of the 90-mm gun with dry-cleaning solvent and thoroughly dry. Swab the bore with light rust-preventive compound. Seal the muzzle with nonhygroscopic adhesive tape. Install the muzzle cover, if available, and seal with nonhygroscopic adhesive tape. If a muzzle cover is not available, wrap waterproof barrier wrapping paper over the tape and seal with nonhygroscopic adhesive tape.

(8) EXTERIOR OF VEHICLES. Remove rust appearing on the vehicle exterior with flint paper. Repaint painted surfaces whenever necessary to protect wood or metal. Coat exposed polished metal surfaces susceptible to rust with light rust-preventive compound. Close firmly all doors, hatches, and vision slots. Make sure paulins are in place and firmly secured. Leave rubber mats, such as floor mats, where provided, in an unrolled position on the floor, and not rolled or curled up. Equipment such as pioneer tools and fire ex-

tinguishers will remain in place in the vehicle. For treatment of small arms carried on or within vehicles, refer to the pertinent technical manuals.

(9) INSPECTION. Make a systematic inspection, just before shipment or temporary storage, to ensure all above steps have been covered and that the vehicle is ready for operation on call. Make a list of all missing or damaged items and attach it to the steering wheel. Refer to "Before-operation Service" (par. 49).

(10) BRAKES. Release brakes and check the wheels.

c. Inspections in Limited Storage. Vehicles in limited storage will be inspected weekly for condition of battery. If water is added to battery when freezing weather is anticipated, recharge the battery with a portable charger, or remove the battery for charging. Do not attempt to charge the battery by running the engine. If freezing temperature is expected, add the proper quantity of antifreeze compound to cooling system to afford protection from freezing. Remove any rust from vehicle with flint paper.

234. LOADING AND BLOCKING FOR RAIL SHIPMENT.

a. Preparation. In addition to the preparation described in paragraph 228, when Ordnance vehicles are prepared for domestic shipment, the following preparation and precautions will be taken.

(1) EXTERIOR. Cover the body of the vehicle with a canvas cover supplied as an accessory.

(2) BATTERY. Disconnect the battery to prevent its discharge by vandalism or accident. This may be accomplished by disconnecting the positive lead, taping the end of the lead, and tying it back away from the battery.

(3) BRAKES. The brakes must be applied and the transmission placed in low gear after the vehicle has been placed in position with a brake wheel clearance of at least 6 inches (A, fig. 212). Locate the vehicles on the car in such a manner as to prevent the car from carrying an unbalanced load.

(4) LABEL CARS. All cars containing Ordnance vehicles must be placarded "DO NOT HUMP."

(5) FREIGHT CARS. Ordnance vehicles may be shipped on flat cars, and door box cars, side door box cars, or drop end gondola cars, whichever type car is the most convenient.

b. Facilities for Loading. Whenever possible, load and unload vehicles from open cars under their own power, using permanent end ramps and spanning platforms. Movement from one flat car to another along the length of the train is made possible by cross-over plates or spanning platforms. If no permanent end ramp is available, an improvised ramp can be made from railroad ties. Vehicles may

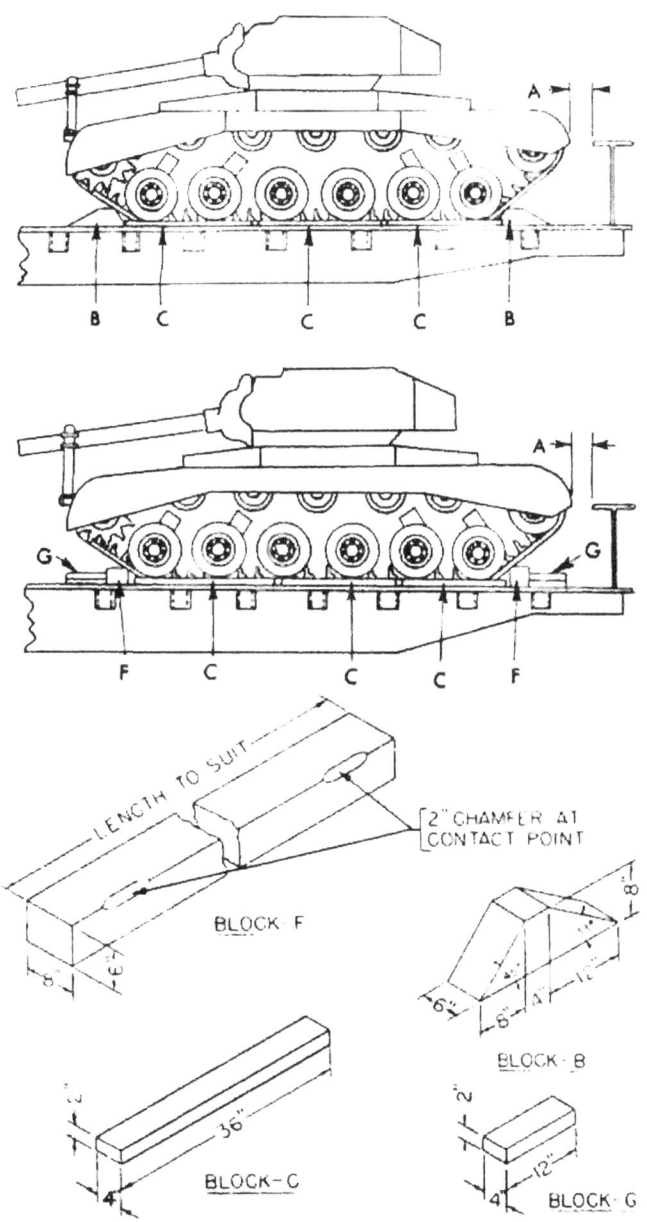

Figure 212 — Blocking Requirements for Rail Shipment

be loaded in gondola cars without drop ends by using a crane. In case of shipment in side-door cars, use a dolly type jack to warp the vehicles into position within the car.

c. **Securing Vehicles.** In securing or blocking a vehicle, three motions, lengthwise, sidewise, and bouncing must be prevented. There are two approved methods of blocking the vehicles on freight cars, as described below.

(1) METHOD ONE. Place four blocks (B, fig. 212), one to the front and one to the rear of each track. Nail the heel of each block to the car floor with five 40-penny nails. Nail the portion of each block which is under the track to the car floor with two 40-penny nails. Locate three blocks (C) on each side of the vehicle on the outside of each track. Nail each block to the car floor with three 40-penny nails. These blocks may be located on the inside of the tracks if conditions warrant.

(2) METHOD TWO. Place two blocks (F, fig. 212), one to the front and one to the rear of the tracks. These blocks are to be at least as long as the over-all width of the vehicle at the car floor. Locate eight blocks (G) against the blocks (F) to the front and to the rear of each track. Nail the lower block to the floor with three 40-penny nails and the top block to the lower block with three 40-penny nails.

d. **Shipping Data.**

Length, overall (gun in traveling position)	22 ft 4 in.
Width, over-all	11 ft 5 in.
Height, over-all	9 ft 1 in.
Area of car floor occupied per vehicle	258.68 sq ft
Volume occupied per vehicle	2,328 cu ft
Shipping weight per vehicle	84,010 lb

Section XLV
REFERENCES

235. PUBLICATIONS INDEXES.

The following publications indexes should be consulted frequently for latest changes to or revisions of the publications given in this list of references and for new publications relating to materiel covered in this manual:

a. Introduction to Ordnance Catalog (explaining SNL system) .... ASF Cat. ORD 1 IOC

b. Index (index to SNL's) .... ASF Cat. ORD 2 OPSI

c. Index to Ordnance Publications (listing FM's, TM's, TC's, and TB's of interest to ordnance personnel, FSMWO's, OPSR, BSD, S of SR's, OSSC's, and OFSB's and including alphabetical listing of ordnance major items with publications pertaining thereto) .... OSFB 1-1

d. List of Publications for Training (listing MTP's, MR's, TR's, FM's, TM's, MWO's, SB's, WDTB's, and FT's .... FM 21-6

e. List of administrative and supply publications (listing MP's, MWO's, SB's, RR's, and War Department Pamphlets) .... W.D. Pamphlet 12-6

f. List of Training Films, Film Strips, and Film Bulletins (listing TF's, FS's, and FB's by serial number and subject) .... FM 21-7

g. Military Training Aids (listing graphic training aids, models, devices, and displays) .... FM 21-8

236. STANDARD NOMENCLATURE LISTS.

a. **Ammunition.**

Ammunition, blank, for pack, light and medium field, tank, and antitank artillery .... ORD 11 SNL R-5

Ammunition, fixed and semifixed, including subcaliber, for pack, light and medium field, aircraft, tank, and antitank artillery, including complete round data .... ORD 11 SNL R-1

TM 9-735
Appendix

	Ammunition instruction material for pack, light and medium field, aircraft, tank, and antitank artillery	ORD 11 SNL R-6
b.	**Armament.**	
	Gun, 90-mm, M3	SNL C-69
	Gun, machine, cal. .30, Browning, M1919A4, fixed and flexible, M1919A5, fixed and M1916A6, flexible	SNL A-6
	Gun, machine, cal. .50, Browning, M2, heavy barrel, fixed and flexible; and ground mounts	SNL A-39
	Mount, gun, 90-mm, M4; and mount, combination, gun, T99, T99E1, and T99E2	SNL D-39
c.	**Maintenance.**	
	Cleaning, preserving and lubricating materials; recoil fluids, special oils, and miscellaneous related items	ORD 5 SNL K-1
	Soldering, brazing and welding material, gases and related items	SNL K-2
	Tools, maintenance, for repair of automatic guns, automatic gun aircraft materiel, automatic and semiautomatic cannon, and mortars	ORD 6 SNL A-35
	Tool-sets, for ordnance service command automotive shops	SNL N-30
	Tool-sets (common), specialists and organizational	SNL G-27 (Section 2)
	Tool-sets (special), automotive and semi-automotive	SNL G-27 (Section 1)
d.	**Sighting Equipment.**	
	Periscopes, telescopes for periscopes, and direct sighting telescopes for use in tanks	SNL F-235
	Quadrant, gunner's, M1 (mils)	SNL F-140
e.	Tank, heavy T26E3	SNL G-226

References

237. EXPLANATORY PUBLICATIONS.

a. **Firing Tables.**

Gun, 90-mm, AA, M1, firing shell H.E., M71 projectile, A.P.C., M82, and shot, A.P., M77 Shell H.E., M71, w/FUZE, P.D., M48 Shot, A.P., M77	FT 90-C-3
Shell, H.E., M71, w/FUZE, time, mechanical, M43 (all modifications)	FT 90AA-B-3

b. **Fundamental Principles.**

Ammunition, general	TM 9-1900
Automotive electricity	TM 10-580
Auxiliary fire control instruments (field glasses, eyeglasses, telescopes, and watches)	TM 9-575
Basic maintenance manual	TM 38-250
Browning machine gun, cal. .30, HB, M1919A4 (mounted in combat vehicles)	FM 23-50
Browning machine gun, cal. .50, HB, M2 (mounted in combat vehicles)	FM 23-65
Driver selection and training	TM 21-300
Driver's manual	TM 10-460
Electrical fundamentals	TM 1-455
Field artillery and field mortar ammunition	OFSB 3-3
Fuels and carburetion	TM 10-550
Military motor vehicles	AR 850-15
Motor vehicle inspections and preventive maintenance services	TM 9-2810
90-mm gun M3 mounted in combat vehicles	TM 9-374
Ordnance service in the field	FM 9-5
Precautions in handling gasoline	AR 850-20
Qualifications in arms and ammunition training allowances	AR 775-10
Radio fundamentals	TM 11-455
Radio set SCR 610	TM 11-615
Range regulations for firing ammunition for training and target practice	AR 750-10
Small arms ammunition	OFSB 3-5
Small arms ammunition	TM 9-1990
Small arms, light field mortars and 20-mm aircraft guns	TM 9-2200
Standard military motor vehicles	TM 9-2800

Appendix

	Targets, target materials, and rifle range construction	TM 9-855
	The radio operator	TM 11-454
	U. S. Rifle, cal. .30, M1903	FM 23-10

c. **Maintenance and Repair.**

Cleaning, preserving, sealing, lubricating and related materials issued for ordnance materiel	TM 9-850
Maintenance and care of pneumatic tires and rubber treads	TM 31-200
Ordnance maintenance: Accessories for tank engine, Model GAA V-8 (Ford)	TM 9-1731C
Ordnance maintenance: Ford tank engine, Model GAA V-8	TM 9-1731B
Ordnance maintenance: Hydraulic traversing mechanism (Oilgear) for medium tanks M4 and Modifications	TM 9-1731G
Ordnance maintenance: Power train for medium tanks T25E1 and T26E1	TM 9-1735A
Ordnance maintenance: Speedometers, tachometers, and recorders	TM 9-1829A
Ordnance maintenance: Tracks, suspension, hull and turret for medium tanks T25E1 and T26E1	TM 9-1735B

d. **Protection of Materiel.**

Camouflage	FM 5-20
Decontamination	TM 3-220
Decontamination of armored force vehicles	FM 17-59
Defense against chemical attack	FM 21-40
Explosives and demolitions	FM 5-25

e. **Storage and Shipment.**

Ordnance company, depot	FM 9-25
Ordnance storage and shipment chart—Group G—major items	OSSC-G
Registration of motor vehicles	AR 850-10
Rules governing the loading of mechanized and motorized army equipment, also major caliber guns, for the United States Army and Navy, on open top equipment published by Operations and Maintenance Department of Association of American Railroads.	
Storage of motor vehicle equipment	AR 850-18

INDEX

A

Accelerator
  adjustment of controls 262
  description
    accelerator 39, 241
    controls 241
    road test 136

Accessories
  equipment 420
  preliminary service 31

Accessory drives (See Drives)

Adapters, carburetor (See Carburetors)

After-operation and weekly service 118

Air cleaners
  description 265
  lubrication 93
  removal and installation 268
  road test 133
  servicing
    after operation 120
    at halt 118
    care (dusty conditions) 86
    general servicing 265
    preliminary servicing 31

Air intake system (See Intake systems)

Ammeter
  description 48, 207
  removal and installation 207
  road test 125
  servicing
    before operation 114
    preliminary 33
    trouble shooting 159

Ammunition
  authorized rounds 452
  on-vehicle equipment 22
  preparation for firing 454

Angle drives (See Drives)

Antenna
  description 19, 460
  inspection 464
  precautions in use 465
  use of phantom antenna 462

Antifreeze compound 83, 84

Armament
  description 427
  operating instructions 429

At-halt service 117

Auxiliary engine (See Engine)

Auxiliary equipment 69, 426

Azimuth indicator 398, 437

B

Batteries
  description
    batteries 208
    box 210
    wiring system 214
  maintenance and inspection
    after operation 119, 208
    care (low temperatures) 85
    general instructions 208
    preliminary service 31
    wiring system 214
  precautions in use 466
  removal and installation
    batteries 210
    box 212
    wiring system 214
  road test 136
  trouble shooting
    batteries 151, 160, 163
    wiring system 161

Before-operation service 113

Belts, preliminary service 31

Blackout lights (See Lights)

Blower, radio interference suppression 244

Bore sighting 450

Bores, lubrication 94

Brakes
  adjustment
    rods 327
    shoes 406
    steering brake 336
  description
    parking and service brakes 36
    parking brake mechanism 339
    steering brake 334

TM 9-735

Index

B — Contd.	Page
Brakes — Contd.	
inspection	335
installation	
front and rear rods	332, 333
operating lever	333
operating lever rods	334
parking brake mechanism	343
shoe assembly	339
shoes	402
removal	
front and rear rods	332
operating lever	334
operating lever rods	334
parking brake mechanism	342
shoe assembly	337
shoes	407
road test	126, 138
run-in test	35
shipment and storage	470
Breather caps, preliminary service	31
Breathers	
description	303
installation	305
lubrication	93
removal	304
road test	132, 138
servicing	
after operation	121
care (dusty conditions)	86
general instructions	304
Breech mechanism	
closing breech	438
lubrication	93
opening breech	437
shipment and storage	469
Bulkhead cover	384
Bumper springs	
description	382
removal and installation	383

C

	Page
Capacitors, removal and installation	246
Carburetors	
adjustment	
carburetor	258
control linkage	262
description	
adapters	260
carburetors	257
removal and installation	
adapters	260
carburetors	259
road test	133, 141
trouble shooting	147, 153, 154
Circuit breakers	
description	51
installation	226, 241
removal	226, 240
trouble shooting	162
(See also Generator)	
Collector ring, road test	140
Commander's cupola	59
Communication system	
description	454
on-vehicle equipment	18
tabulated data	12
Compartment, fighting	
description of floor plates	394
installation	
drain valves	387
floor plates	394
removal	
drain valves	386
floor plates	394
road test	137
Compartment covers, ammunition	394
Compartment lights (See Lights)	
Compass	73
Compression testing (engine)	174
Conduits	
description	234
removal and installation	236
road test	132
Connections, towing	
before-operation service	115
preliminary service	32
Control rods, description	321
Controls	
description (vehicle)	36
service (engine controls)	34

Index

C — Contd.

	Page
Cooling system	
antifreeze compound	288
care (dusty conditions)	87
cleaning	288
cold-weather operation	83
description	283
draining	284
filling and flushing	286
inspection	284, 335
shipment and storage	468
trouble shooting	155
Cooling unit	293
Crankcase	
cleaning	411
road test	128
(See also Breathers)	
Cylinder heads	
description	270
installation of gaskets and seals	277
removal of gaskets and seals	270
Cylinders	
checking oil level	429
road test	133

D

	Page
Data	4
Deck plates	391
Decontaminator, road test	144
Demolition	87
Differential	
description	
differential	320
oil pump	311
lubrication	
differential	310
oil screen	95
removal and installation of oil pump	311
road test	129, 138
Distributor points, adjustment	195
Doors	
description	
drivers'	388
engine compartment	391
escape	42, 398

	Page
operation (escape doors)	398
removal and installation	
drivers'	390
engine compartment	391
Drain valves	
description	386
installation	
control rods	388
valves	386
removal	
control rods	387
valves	386
Drive belt tension adjustment	205
Drives	
description and removal	
fan and angle drives	314
final drives	343
installation	
angle drives	314
fan drives	315
final drives	343
removal	
angle drives	314
final drives	343
road test	
accessory drives	134
final drive	129, 139
Driving	
instructions	55
precautions	56
Driving light switch (See Switches)	
Driving sprockets and hubs	345
During-operation service	116
Dust shields	393

E

	Page
Electric traversing motor	
description and removal	242
installation	243
Electrical system	164
Elevating mechanism	
road test	127
run-in test	35

TM 9-735

Index

E — Contd.

	Page
Engine	
data	5, 171
description	
auxiliary engine	
controls	414
engine	69, 408
main engine	166
lubrication	
auxiliary engine starter	93
crankcases	93
engine	303
in cold weather	82
observation	53
operation	
auxiliary engine	69
engine (low temperatures)	81, 86
radio interference suppression	244
removal and installation	
auxiliary engine	
controls	414, 415
drive belt	411
engine with generator	412
starter relay switch	416
compartment drain valves	386, 387
temperature gage and warning signals	291
road test	
auxiliary engine	140
compartment	134
compression	132
engine	127, 131
engine idle	127, 137
power unit	135
temperature gage	126
run-in test	35
servicing	
after operation	119, 122
auxiliary engine	410
preliminary service	33
starting	
above freezing	52
below freezing	53, 86
by towing vehicle	57
trouble shooting	146
tune-up	172
(See also Breathers; Oil filter; Oil pan; and Oil pump)	

	Page
Equipment	
on vehicles	19
road test	143
(See also Auxiliary equipment)	
Exhaust system	
description	
outlet	270
system	247, 268
tubes	269
removal and installation	
outlet	270
tubes	261
road test	135
trouble shooting	152
(See also Manifolds)	
Expansion tank	290
Extinguishers, fire (See Fire extinguisher system)	

F

	Page
Fan Drives (See Drives)	
Fans and shrouds	
adjustment of fan belt	292
description	
belts	292
fans and shrouds	297
removal and installation	
fan belts	293
fans	299
shrouds	298
Fenders	32, 392
Final drives (See Drives)	
Fire control equipment	442
Fire extinguisher system	
description	
extinguishers	74
system	416
inspection	417
on-vehicle equipment	20
operation	74
removal and installation (controls)	419
road test	135, 144
servicing	
after operation	119
before operation	113
preliminary	31

Index

F — Contd.

	Page
Firing mechanism, lubrication	93
Flag set	19
Flares	20
Flywheel pilot bearing	94
Fording or landing	87

Fuel
  data ... 8
  maintenance
    after operation ... 119
    at-halt ... 117
    preliminary service ... 31
    road test ... 134, 141
    trouble shooting ... 152
Fuel cut-off button, trouble shooting ... 148, 154
Fuel cut-off switch (See Switches)
Fuel filters ... 31, 252
Fuel gage
  description ... 44, 254
  removal and installation
    gage ... 254
    sending unit ... 255
  road test ... 126
  servicing before operation ... 34, 115
  trouble shooting ... 162
  (See also Switches)
Fuel pump
  description ... 256
  removal and installation ... 256
  road test ... 134
  testing ... 175
Fuel shut-off valves
  description ... 39, 253
  installation ... 254
  removal ... 253
Fuel system
  description ... 247
  operation below 0° F ... 80
Fuel tanks
  description
    shut-off valve levers ... 37
    tanks ... 247
  filling and draining ... 249
  installation ... 251
  removal ... 250
Fuel tubes ... 255

G

	Page
Gage, oil pressure (See Oil pressure gage)	
Gear case	94

Generating system
  description, removal, and installation ... 207
  precautions in use ... 466
  trouble shooting ... 159, 164
Generator
  care (low temperatures) ... 85
  description
    circuit breaker ... 206
    generator ... 201
  inspection and lubrication ... 201
  installation
    belt tension adjuster ... 206
    belts ... 205
    brushes ... 204
    circuit breaker ... 206
    generator ... 203
  polarizing ... 203
  radio interference suppression ... 244
  removal
    belt tension adjuster ... 206
    belts ... 205
    brushes ... 204
    generator ... 201
  road test ... 141
Generator regulator
  description and removal ... 204
  installation ... 205
  radio interference suppression ... 244
  road test ... 137
  trouble shooting ... 156
Governor ... 260
Guns
  after-operation service ... 122
  demolition ... 88
  elevating or depressing ... 437
  equipment ... 20
  inspection before firing ... 429
  loading and firing ... 439
  lubrication ... 94
  placing in
    firing position ... 429
    traveling position ... 441

TM 9-735

Index

G — Contd.	Page
Guns — Contd.	
road test	127, 142
run-in test	35
shipment and storage	469
spare parts	26
tools	19
(See also Switches)	

H

Hatch, loader's	60, 399
Hatch hoods, driver's	73
Headlights (See Lights)	
Heater	
description	75, 420
operation	75
removal and installation	421
(See also Switches)	
Horn	
description	
horn	421
horn button	43
preliminary service	34
removal and installation	421
road test	126
(See also Switches)	
Hose connections	302
Housing, transverse	395
Hubs	
installation	369
removal	366
Hull	
description	
hull	383
radio terminal box	244
removal and installation (capacitors)	246
servicing	
after operation	121
at halt	117
before operations	115

I

Idler wheels	
removal and installation	365
road test	130

	Page
Ignition system	
description	188, 196
removal and installation	196
road test	132
trouble shooting	149, 164
Inspection	463
Instrument panel	
description	43, 51, 220
installation	
capacitor	246
face plate	222
panel	220
radio interference suppression	244
removal	
capacitor	245
face plate	222
panel	220
(See also Lights)	
Instruments	
description	36
maintenance	223
run-in test	34
trouble shooting	161
Intake systems	
description	
air intake system	265
fuel system	247
trouble shooting	152
Interphone system	462, 467

L

Lamps (See Lights)	
Leaks	
preliminary service	34
road tests	128, 136
run-in test	35
Lights	
adjustments (headlights)	216
description	
blackout driving lights	42, 218
compartment lights	219, 241
headlights	41, 215
inspection lights and spotlights	76
instrument lights	446
marker lights	218
spotlights	421
taillights	218
wiring system	214

Index

L — Contd.

Lights — Contd.
installation and removal
  blackout lights ... 218
  compartment lights ... 219, 241
  indicator lamp lights ... 241
  spotlight ... 422
  taillight ... 218
  wiring system ... 214
maintenance and inspection
  lights ... 34, 115
  wiring system ... 214
operation (spotlights and signal lights) ... 77
road test ... 139
trouble shooting ... 160, 162
(See also Warning signals and Switches)

Linkage, steering
after-operation service ... 121
at-halt service ... 117
preliminary service ... 33

Lock, hydrostatic, trouble shooting ... 146

Lock, turret (See Turret)

Lubrication order ... 92

Lubrication system
check of oil level
  differential ... 310
  transmission ... 309
cleaning (differential) ... 311
description
  differential ... 310
  engine ... 303
  transmission ... 308
draining and filling
  differential ... 310, 311
  engine ... 303
  transmission ... 309
lubrication instructions
  detailed instructions ... 92
  for ordnance personnel ... 95
lubrication servicing
  preliminary service ... 32
  temperatures below 0° F ... 81, 82
road test ... 144
trouble shooting ... 156

M

Magnetos
description ... 188
installation ... 190
removal ... 189
road test ... 141, 132
trouble shooting ... 150, 156
(See also Switches)

Maintenance
first echelon ... 113
general instructions ... 89, 112
second echelon ... 123

Manifolds
description
  exhaust manifolds ... 269
  water manifolds ... 301
removal and installation
  exhaust manifolds ... 269
  water manifolds ... 301
road test ... 133

Master switch box ... 212

Master switches (See Switches)

Mount T90, telescope ... 442

Mount T117, periscope ... 445

O

Octane rating ... 8

Oil
checking oil level
  engine ... 303
  recoil cylinders ... 429
  transmission ... 309
data ... 8
servicing
  after operation ... 119
  at halt ... 117
  preliminary service ... 31
trouble shooting
  oil consumption ... 158
  oil level ... 148

Oil coolers ... 134, 312

Oil filter
cleaning and lubrication ... 95, 306
description and testing ... 305
removal and installation ... 306
road test ... 134

TM 9-735

TM 9-735

Index

O — Contd.

	Page
Oil pan	306
Oil pressure	
testing	316
trouble shooting	156
Oil pressure gage	
description	45, 307
removal and installation	307
road test	125
servicing	
before operation	114
preliminary	33
Oil pressure warning signals (See Warning signals)	
Oil pump	
description	
differential	311
engine	306
removal and installation	
differential	311
engine	307
Oil screen (transmission), cleaning	309
Oil supply, spare	144
Oil temperature warning signals (See Warning signals)	
Oil tubes	
description and removal	312, 319, 405
installation	313, 319, 406
Oilcan points	95
Operation	
at temperatures below 0° F.	80
road test	142
under dusty conditions	86
under ordinary conditions	52

P

	Page
Padding, protective	
hull	394
turret	407
Panel light switch (See Switches)	
Periscopes	
description	445, 450
removal and installation	79, 424
road test	138
use	79
Pistol port (See Turret)	
Power train, trouble shooting	156

	Page
Power unit	
installation	184
removal	175
Pressure gage (See Oil pressure gage)	
Pressure relief valve	401
Primer	39, 257
Prism, viewing	425
Pump valves, hydraulic	402

Q

Quadrant, elevation	446

R

	Page
Radiators	
description and removal	296
installation	297
road test	135
Radio interference suppression	
description and data	244
maintenance	245
replacement of bond straps	246
road test	140
trouble shooting	163, 165
Radio terminal box	
description	236
radio interference suppression	
hull	244
turret	245
removal and installation	
box	236
capacitors	246
Radios	
description	
mounting	457
radios	454, 458
inspection of mountings	463
installation	457
on-vehicle equipment	20
precautions in use	466
removal	458
Rations	20
Recoil cylinders, check of oil level	429
Reports and records	1
Reverse anchors	335, 336
Road test	125
Rollers, support	383
Run-in test	34

Index

S

	Page
Sand shields	32

Seats
- data ... 12
- description
  - driver's seats ... 42, 390
  - turret seats ... 60, 408
- removal and installation
  - driver's seats ... 390
  - turret seats ... 408

Service upon receipt of equipment ... 30

Shipment and storage ... 468

Shock absorbers ... 382

Shunt, ammeter, removal and installation ... 207

Sighting equipment ... 23, 442

Solenoids, gun firing, removal and installation ... 241

Spare parts ... 26

Spark plugs
- description ... 195
- removal and installation ... 196
- road test ... 132, 140
- testing ... 196
- trouble shooting ... 150

Specific gravity tests ... 208

Speed range selector
- controls ... 37, 317, 318

Speed ranges ... 54

Speedometer system
- description
  - speedometer ... 46
  - speedometer system ... 226
- lubrication ... 95
- radio interference suppression ... 245
- removal and installation
  - adapter ... 227
  - drive shaft ... 227, 228
  - receiving unit ... 229
  - sending unit ... 228, 229
  - wiring ... 229

	Page
road test	126
trouble shooting	162

Spotlight (See Lights)

Spring arm, front ... 375, 377, 378

Starting system
- care (low temperatures) ... 85
- description
  - starter ... 199
  - starting system ... 197
- removal and installation (starter and wiring) ... 200
- road test ... 132, 136
- trouble shooting ... 150
  (See also Switches)

Steering levers
- adjustment ... 327
- description
  - controls ... 36
  - levers ... 321
  - pull rods ... 332
- removal and installation
  - levers ... 327
  - pull rods ... 332

Steering linkage (See Linkage, steering)

Stowage boxes ... 395

Support rollers ... 383

Suspension
- after-operation service ... 121
- at-halt service ... 117
- before-operation service ... 115
- description ... 347
- preliminary service ... 32
- road test ... 130
- run-in test ... 35

Switch box, master ... 212

Switches
- description
  - booster ... 43

TM 9-735

TM 9-735

Index

S — Contd.

Switches — Contd.
	Page
driving light and heater	48
fuel cut-off	45
fuel gage	44
gun (firing)	243
horn	423
instrument panel	50
magneto	45, 188
master switch	39, 212
speedometer-tachometer	46
starting system	45, 197

removal and installation
driving light	219
fuel gage	255
gun (firing)	240, 241, 243
heater	226
horn	423
instrument panel	226
master switch	212, 213
starting system	197, 198
traversing motor	240

road test	139
trouble shooting	160, 162

(See also Warning signals)

T

Tachometers
description
tachometer	46
tachometer system	229
lubrication	95
radio interference suppression	245

removal and installation
adapter	229
drive shaft	229, 232
sending unit	232
tachometer	223
wiring	234
road test	126
servicing	34, 115
trouble shooting	161

Taillights (See Lights)

	Page
Tarpaulin	32
Telescope M71C	442

Temperature gage (engine)
description	45, 291
removal and installation	291
road test	126
servicing	33
trouble shooting	156

Temperature warning signals (See Warning signals)

Temperatures
road test	127
run-in test	35
Terminal block	240, 241
Thermostats	292

Throttle
adjustment of linkage	263
description	37, 261
road test	137

Tires, servicing
after operation	120
at halt	118
before operation	115

Tools
on-vehicle tools	14
road test	143
special tools	89

Torsion bar spring
adjustment of tension	371

description
anchor plug	371
spring	369

removal and installation
anchor plug	373
spring	370, 390
testing	369

Towing connections (See connections, towing)

Towing the vehicle	57

T — Contd.

	Page
Track links, road test	129, 144
Track suspension, care of	82, 87
Tracks	
adjustment of tension	353
after operation service	120
at halt service	118
weekly operation service	118
data	8
description	351
disconnecting	353
installation	359, 360
preliminary service	32
road test	128, 131
run in test	35
Transmission	
adjustment of bands	316
description	
oil tubes	319
transmission	315
warning signals	309
lubrication	
screen	95
transmission	308
removal of oil tubes	319
road test	139
run in test	35
trouble shooting	148, 162
Traversing mechanism	
road test	127
run in test	35
(See also Turret)	
Traversing motor, electric (See Electric traversing motor)	
Traversing motor, turret (See Turret traversing motor)	
Traversing motor switch (See Switches)	
Traversing pump (turret), trouble shooting	163
Trouble shooting	148
Turret	
adjustment of brake shoes	406
data	12

	Page
description	
electrical system	234
gear box and drive	63
pistol port	68
switch box	240
traversing mechanism	61, 399, 406
turret	398
wiring	234
inspection and maintenance	
after operation service	66, 122
electrical system	234
traversing mechanism	82, 401
operation (traversing mechanism)	63, 64, 429, 432, 436
radio interference suppression	245
removal and installation	
brake shoes	407
capacitors	246
switch box	240
traversing mechanism	404, 405, 406
wiring	236
road test	138
trouble shooting	63, 165
Turret lock	66, 407

U

Universal joints	
description	321
installation	185, 321
lubrication	95
removal	181, 321
servicing	33

V

Vehicle	
care and servicing	
after operation	86, 122
cold weather	86
data	8
demolition	88
description	3, 36
driving	53, 86
operation (cold weather)	86
shipment and storage	472
Ventilator blower	77, 423
Vents, road test	132
Vision equipment	
description	423
servicing	32, 78, 121

Index

W

	Page
Warning signals	
description	45, 46
removal and installation	291, 292 308, 310
road test	126
service	
before operation	114
preliminary	33
trouble shooting	162
Water pump	301, 302
Water supply	
at-halt service	117
capacities data	8
preliminary service	31
Water tubes	302
Wheels	
adjustment of arm	371

	Page
description	
arm support	377
arms	374, 379
wheel	361
preliminary service	32
removal and installation	
arm support	377
arms	374, 380, 381
wheels, road	361
road test	130
Wheels, idler (See Idler wheels)	

IN HIGH DEFINITION
NOW AVAILABLE!

COMPLETE LINE OF WWII AIRCRAFT FLIGHT MANUALS

WWW.PERISCOPEFILM.COM

©2013 Periscope Film LLC
All Rights Reserved
ISBN#978-1-937684-43-3
www.PeriscopeFilm.com

www.ingramcontent.com/pod-product-compliance
Lightning Source LLC
Chambersburg PA
CBHW070159240426
43671CB00007B/492